BIOLOGY

Biology FIFTH EDITION

PART 1   Biology of Cells

HELENA CURTIS

N. SUE BARNES

WORTH PUBLISHERS, INC.

Male wood duck (Aix sponsa) in breeding plumage, drying its wings, photographed on the St. Croix River, Minnesota. Animals, such as the wood duck, in which the males and females differ in appearance are said to be sexually dimorphic ("having two forms"). In such species, the males are characteristically polygamous. (© Scott Nielsen)

BIOLOGY, FIFTH EDITION, PART 1

COPYRIGHT © 1968, 1975, 1979, 1983, 1989 BY WORTH PUBLISHERS, INC.

ALL RIGHTS RESERVED

LIBRARY OF CONGRESS CATALOG CARD NUMBER: 88-51041

ISBN: 0-87901-394-X (CLOTHBOUND)

ISBN: 0-87901-435-0 (PART 1, PAPERBACK)

ISBN: 0-87901-436-9 (PART 2, PAPERBACK)

ISBN: 0-87901-437-7 (PART 3, PAPERBACK)

PRINTED IN THE UNITED STATES OF AMERICA

PRINTING: 2 3 4 5 6 7   YEAR: 9 0 1 2 3 4

EDITOR: SALLY ANDERSON

PRODUCTION: SARAH SEGAL

ART DIRECTOR: GEORGE TOULOUMES

LAYOUT DESIGN: PATRICIA LAWSON, DAVID LOPEZ

DESIGN: MALCOLM GREAR DESIGNERS

ILLUSTRATOR: SHIRLEY BATY

PICTURE EDITORS: DAVID HINCHMAN, ANNE FELDMAN, ELAINE BERNSTEIN

TYPOGRAPHY: NEW ENGLAND TYPOGRAPHIC SERVICE, INC.

COLOR SEPARATION: CREATIVE GRAPHIC SERVICES CORPORATION

PRINTING AND BINDING: VON HOFFMANN PRESS, INC.

COVER PHOTOGRAPH: WOOD DUCK (© SCOTT NIELSEN)

FRONTISPIECE: TROPICAL FOREST, COSTA RICA (© GARY BRAASCH)

ILLUSTRATION ACKNOWLEDGMENTS BEGIN ON PAGE A–0, CONSTITUTING

AN EXTENSION OF THE COPYRIGHT PAGE.

WORTH PUBLISHERS, INC.

33 IRVING PLACE

NEW YORK, NEW YORK 10003

This book is dedicated to those whose creative and painstaking studies have contributed to our understanding of biology.

Contents in Brief

Giraffe with oxpeckers

Contents

Ladybug on flower of shepherd's needle

SECTION 5 Biology of Plants   611

Preface

In the twenty years since the first edition of *Biology* appeared, the science of biology has been characterized by ever accelerating change, including not only a flood of new information but also new ideas and unifying concepts. Some areas of biology have undergone metamorphoses before our very eyes, while others have attained a new maturity. This has presented us—now at our word processors—and you—in the classroom and laboratory—with new challenges and opportunities as we work together to provide students with a solid foundation in the principles of biology while simultaneously sharing with them the excitement of the contemporary science.

With this, the Fifth Edition, *Biology* becomes both one of the oldest and one of the newest introductory biology textbooks. One of our principal goals in preparing this edition has been to maintain a balance between the old and the new. This has required not only a willingness to discard material, but also considerable care that we not eliminate or slight material that, although not new, is essential if students are to be adequately prepared to understand current and future developments in biology. Simultaneously, we have, of course, wanted to be as up to the minute as possible, without becoming merely trendy. At a time when important discoveries are published almost continuously, there is a temptation to become so engrossed in the new that we lose sight of the fact that the majority of today's students are, like their predecessors, coming to the formal study of contemporary biology for the first time. The clear explication of the basic principles of biology, with pertinent and readily understood examples, has become increasingly important with each passing year. Thus, specific topics for detailed treatment have been selected on the basis of their centrality to modern biology, their utility in illuminating basic principles, their importance as part of the requisite store of knowledge of an educated adult as we approach a new century, and their inherent interest and appeal to students. Throughout, we have tried to provide the underlying framework and arouse student curiosity so that a foundation is laid for those areas—very diverse—in which you may wish to give more extensive coverage in the classroom or laboratory than is possible in any introductory textbook, regardless of its length.

The central, essential foundation of biology is, of course, evolution, the major organizing theme of this text as of all modern biology texts. As in previous editions, the stage is set in the Introduction, which focuses on the development of the Darwinian theory. New to the Introduction is a section previewing the other major unifying principles of modern biology that are also recurring themes throughout the text; also new is a brief overview of the diversity of life. Both are designed to provide students with a broad framework before they begin their study of the details on which modern biology is built. In this edition, we have also strengthened the introductory discussion of the nature of science, and, throughout the text, we have included more information about how biologists know what they know and how scientists in general go about their business.

After the Introduction, this edition, like previous editions, follows the levels-of-organization approach. Part 1 deals with life at the subcellular and cellular levels, Part 2 with organisms, and Part 3 with populations, ending with a survey of the distribution of life on earth. Each part is divided into two or three sections. A significant amount of restructuring has occurred in the sequence of chapters within certain sections and within the chapters themselves.

One of the most striking aspects of the enormous burst of new discoveries in molecular and cell biology since the Fourth Edition is the power of these discoveries to explain processes that previously could only be described—and in the most general terms. The immune response, olfaction and color vision, events at the synapse, the summing of information by individual neurons, and differentiation and morphogenesis in animal development are just a few of the many phenomena whose secrets are being revealed by studies at the molecular and cellular level. These revelations depend, in large part, on what is now a flood tide of reports identifying specific membrane proteins, their amino acid sequences, their three-dimensional structures, and, in many cases, the nucleotide sequences of the genes coding for the proteins and the location of these genes within the genome. Because these discoveries, although fascinating in and of themselves, are of such value in explaining organismal phenomena, we have generally chosen to defer their discussion to later sections of the text, where their significance will be most readily grasped by students. Molecular and cell biology have, in many ways, come of age, and it seems to us that the essential task in these early sections of the book has become the clear communication of the underlying principles on which so much is now being built—rather than a catalog of the latest new discoveries, which will soon be superseded by even more exciting ones.

The extraordinary pace of discovery in genetics, principally as a result of recombinant DNA technology, requires, with each new edition, a major rethinking of Section 3. Responses to our surveys indicate that, however tempting it might be to begin the section with molecular genetics and to reduce the coverage of classical genetics, doing so could make this most exciting area of modern biology less accessible to students. Thus, as in previous editions, we begin our consideration of genetics with Mendel and take an essentially historical approach to the development of the powerful science we know today. Within that overall framework, however, there has been the addition of a significant amount of new material, coupled with a number of internal reorganizations that we believe provide greater clarity and a smoother conceptual development in our coverage of molecular genetics.

Part 2, Biology of Organisms, has also undergone many changes, particularly in the early chapters of Section 4 and in Sections 5 and 6. In the previous edition, Section 4, The Diversity of Life, was significantly expanded. The enthusiasm with which the revised section was received—plus our own continuing awe at the incredible variety of living organisms—led to our decision to retain the expanded section intact. We have, however, made major revisions in the first chapter of the section, dealing with the classification of organisms, and minor revisions throughout the section.

The organization of Section 5, the Biology of Plants, has long been problematic. It has been difficult to find a sequence that would flow logically, coordinate well with laboratory programs, and—most important—captivate students with the beauty and biological accomplishments of plants without overwhelming them with the vocabulary necessary for an accurate description of the living plant. In this edition we have chosen to begin the section with the familiar—the flower—and with the dynamic process of plant reproduction, a sequence that flows directly from the discussion of plant evolution and diversity in Section 4. In the following chapter, the anatomy of the plant body is considered in conjunction with another dynamic process, the development of the embryo into the mature sporophyte. The two chapters on plant hormones and plant responses in the Fourth Edition have now been merged into one integrated chapter; new understandings of the physiological processes of plants have made such a separation increasingly artificial.

In Section 6, the Biology of Animals, we have retained the overall organizational scheme and problem-solving approach of the Fourth Edition, while significantly revising many chapters. As noted previously, animal physiology is one of the principal areas in which enormous and rapid progress is being made as a result of new discoveries at the molecular and cellular level, and we have tried to capture and share with students as much of the current excitement as possible. Although we have continued to use the human animal—inherently fascinating to most students—as our representative organism in these chapters, we have strengthened the comparative thread and made explicit much comparative material that was previously implicit.

Part 3, the Biology of Populations, covers what G. E. Hutchinson aptly described as "the ecological theater and the evolutionary play." Modern evolutionary theory and ecology are so intertwined that any separation of the two is arbitrary. We believe, however, that the student's understanding of modern ecology is deepened and enriched if it is preceded by a knowledge of the mechanisms of evolution.

In Section 7, Evolution, the five chapters of the previous edition have been reworked into four. As in the Fourth Edition, the section begins with a chapter that reviews the key points of Darwin's theory, examines the types of evidence that support evolution, and considers the changes that have occurred in evolutionary theory since Darwin's original formulation. This is followed by extensively revised chapters on the genetic basis of evolution, natural selection, and the origin of species. Then follow two chapters, also heavily revised, on the evolution of the hominids and on animal behavior and its evolution. Many of you have told us that you prefer to cover human evolution while the discussion of evolutionary mechanisms is still fresh in students' minds, and we have accordingly shifted that chapter to this section from the end of the book. Behavior is a topic for which little, if any, time is available in many courses, but it holds great interest for students, professors, and these authors alike. We have tried to provide students with a solid introduction to the contemporary study of behavior and then to focus on topics that we think are most likely to be of immediate interest and appeal to them.

Section 8, Ecology, has also been extensively revised, as we attempt to track the continual shifts, rethinkings, and controversies that characterize this most vibrant science. As in the Fourth Edition, the section moves from population dynamics, through the interactions of populations in communities and ecosystems, to the overall organization and distribution of life on earth. The text ends with a consideration of the tropical forests—the most complex and most seriously threatened of all ecological systems.

Each section ends with suggestions for further reading. Scientifically speaking, the selections are arbitrary. They were chosen not as documentation for statements in the book or as fuller presentations of difficult subjects, but rather because of their accessibility to students. Our hope is that at least some students will continue reading on their own, preferably reports not yet published about discoveries just now being dreamed of.

A number of new supplements accompany this edition of *Biology*. Of particular interest is *More Biology in the Laboratory,* by Doris R. Helms of Clemson University, an expanded version of *Biology in the Laboratory,* which accompanies the Fourth Edition of *Invitation to Biology*. A detailed Preparator's Guide accompanies the lab manual. Other supplements include *BioBytes,* a series of computer simulations by Robert Kosinski of Clemson University, a Study Guide and a Test

Bank by David J. Fox of the University of Tennessee, a new computerized test-generation system, a new and greatly expanded Instructor's Resource Manual by Debora Mann of Clemson University, and an extensive set of acetate transparencies, most of them in color.

As with previous editions, we have been deeply dependent on the advice of consultants and reviewers. In addition to her work on the new laboratory manual, Dori Helms played a major role in the revisions of the genetics section, the plant section, and the development chapter in animal physiology. She has generously shared with us her extensive knowledge, her wealth of experience in the classroom and laboratory, and her enthusiasm—all of which have been marvelous resources that we have greatly appreciated.

We are also deeply indebted to Rita Calvo of Cornell University, who reviewed a series of revisions of the genetics section; to Jacques Chiller of the Lilly Research Laboratories, who has been an invaluable source on contemporary immunology; to Mark W. Dubin of the University of Colorado, who guided us through our revision of the integration and control chapters of animal physiology; and to Manuel C. Molles, Jr., of the University of New Mexico, and Andrew Blaustein of Oregon State University, both of whom made major contributions to our revision of the evolution and ecology sections.

In addition, we have been greatly assisted by advice and counsel from the following reviewers:

BRUCE ALBERTS, University of California Medical School, San Francisco
WILLIAM E. BARSTOW, University of Georgia
CHARLES J. BIGGERS, Memphis State University
WILLIAM L. BISCHOFF, University of Toledo
ROBERT BLYSTONE, Trinity University
LEON BROWDER, University of Calgary
RALPH BUCHSBAUM, Pacific Grove, California
JAMES J. CHAMPOUX, University of Washington
JAMES COLLINS, Arizona State University
JOHN O. CORLISS, University of Maryland
MICHAEL CRAWLEY, Imperial College at Silwood Park, Ascot, England
CHARLES CURRY, University of Calgary
FRED DELCOMYN, University of Illinois, Urbana-Champaign
RUTH DOELL, San Francisco State University
RICHARD DUHRKOPF, Baylor University
DAVID DUVALL, University of Wyoming
JUDI ELLZEY, University of Texas, El Paso
ROBERT C. EVANS, Rutgers University, Camden
RAY F. EVERT, University of Wisconsin
KATHLEEN FISHER, University of California, Davis
ROBERT P. GEORGE, University of Wyoming
URSULA GOODENOUGH, Washington University, St. Louis
PATRICIA GOWATY, Clemson University
LINDA HANSFORD, Baltimore, Maryland
JEAN B. HARRISON, University of California, Los Angeles
STEVEN HEIDEMANN, Michigan State University
MERRILL HILLE, University of Washington
GERALD KARP, San Francisco, California
JOHN KIRSCH, University of Wisconsin
ROBERT M. KITCHIN, University of Wyoming
KAREL LIEM, Harvard University
JANE LUBCHENCO, Oregon State University
R. WILLIAM MARKS, Villanova University

LARRY R. McEDWARD, University of Washington
SUE ANN MILLER, Hamilton College
RANDY MOORE, Wright State University
BETTE NICOTRI, University of Washington
JAMES PLATT, University of Denver
FRANK E. PRICE, Hamilton College
EDWARD RUPPERT, Clemson University
TOM K. SCOTT, University of North Carolina, Chapel Hill
LARRY SELLERS, Louisiana Tech University
DAVID G. SHAPPIRIO, University of Michigan
JOHN SMARRELLI, Loyola University, Chicago
GILBERT D. STARKS, Central Michigan University
IAN TATTERSALL, American Museum of Natural History
ROBERT VAN BUSKIRK, State University of New York, Binghamton
ERIC WEINBERG, University of Pennsylvania
JOHN WEST, University of California, Berkeley
ARTHUR WINFREE, University of Arizona

As always, the preparation of a new edition is a staggering and complex task, and its successful completion has depended on the efforts of many highly talented individuals. In particular, we wish to thank Shirley Baty, who, in addition to preparing many new illustrations for this edition, has reworked virtually all of the Fourth Edition art as we converted the book to full color throughout; David Hinchman, Anne Feldman, and Elaine Bernstein, who have located an enormous number of marvelous new photographs and micrographs; John Timpane, who prepared the comprehensive index; George Touloumes and the members of his staff who are responsible for the design and layout of each page of the book; Sarah Segal, who has managed the production process and somehow kept us all on course; and Sally Anderson, our extraordinary editor, and her capable assistant, Lindsey Bowman. Sally's editorial expertise, her thorough knowledge of biology in general and of this text in particular, and her long experience in working with us both have played an incalculable role in the successful completion of this revision. And, a special thank you to Bob Worth, whose vision and constant support have made it all possible.

Finally, we want to thank all of the professors and students who have written to us, some with criticisms, some with suggestions, some with questions, and some simply because they enjoyed the book. These letters serve to remind us of how privileged we are to be writing for young people. We continue to appreciate their curiosity, their energy, their imaginativeness, and their dislike of the pompous and pedantic. We hope we serve them well.

New York   Helena Curtis

December, 1988   N. Sue Barnes

An added note: As you may have noticed, with this edition of *Biology,* N. Sue Barnes is listed as coauthor. This is a recognition long overdue. Sue has been a member of the team for eleven years now. Over this period of time, she has assumed increasing responsibility for the revisions of both *Biology* and *Invitation to Biology* (on which she has been listed as coauthor for the last two editions). The Fifth Edition of *Biology* would have been impossible without her. In addition, I wish to express my personal gratitude for her integrity, patience, fortitude, and good spirits—and for the fact that she always comes through.

H.C.

BIOLOGY

Introduction

In 1831, the young Charles Darwin set sail from England on what was to prove the most consequential voyage in the history of biology. Not yet 23, Darwin had already abandoned a proposed career in medicine—he described himself as fleeing a surgical theater in which an operation was being performed on an unanesthetized child—and was a reluctant candidate for the clergy, a profession deemed suitable for the younger son of an English gentleman. An indifferent student, Darwin was an ardent hunter and horseman, a collector of beetles, mollusks, and shells, and an amateur botanist and geologist. When the captain of the surveying ship H.M.S. *Beagle*, himself only a little older than Darwin, offered passage for a young gentleman who would volunteer to go without pay, Darwin eagerly seized this opportunity to pursue his interest in natural history. The voyage, which lasted five years, shaped the course of Darwin's future work. He returned to an inherited fortune, an estate in the English countryside, and a lifetime of independent work and study that radically changed our view of life and of our place in the living world.

THE ROAD TO EVOLUTIONARY THEORY

That Darwin was the founder of the modern theory of evolution is well known. Although he was not the first to propose that organisms **evolve**—or change—through time, he was the first to amass a large body of supporting evidence and the first to propose a valid mechanism by which evolution might occur. In order to understand the meaning and significance of Darwin's theory, it is useful to look at the intellectual climate in which it was formulated.

Aristotle (384–322 B.C.), the first great biologist, believed that all living things could be arranged in a hierarchy. This hierarchy became known as the *Scala Naturae*, or ladder of nature, in which the simplest creatures had a humble position on the bottommost rung, man occupied the top rung, and all other organisms had their proper places in between. Until the late nineteenth century, many biologists believed in such a natural hierarchy. But whereas to Aristotle living organisms had always existed, the later biologists (at least those of the Occidental world) believed, in harmony with the teachings of the Old Testament, that all living things were the products of a divine creation. They believed, moreover, that most were created for the service or pleasure of mankind.

That each type of living thing came into existence in its present form—specially and specifically created—was a compelling idea. How else could one explain the astonishing extent to which every living thing was adapted to its environment and to its role in nature? It was not only the authority of the Church but also, so it seemed, the evidence before one's own eyes that gave such strength to the concept of special creation.

I–1 *When Charles Darwin visited the Galapagos archipelago, he found that each major island had its own variety of tortoise, so distinct from the others that it was easily recognized by local sailors and fishermen. This was one of the clues that led him to the formulation of the theory of evolution.*

The Galapagos consists of 13 volcanic islands that pushed up from the sea more than a million years ago. The major vegetation is thornbush and cactus, and the original black basaltic lava is often visible, as it is beneath the lumbering feet of this tortoise on Hood Island—"what we might imagine the cultivated parts of the Infernal regions to be," young Darwin wrote in his diary.

Among those who believed in divine creation was Carolus Linnaeus (1707–1778), the great Swedish naturalist who devised our present system of nomenclature for species, or kinds, of organisms. In 1753, Linnaeus published *Species Plantarum,* which described, in two encyclopedic volumes, every species of plant known at the time. Even as Linnaeus was at work on this massive project, explorers were returning to Europe from Africa and the New World with previously undescribed plants and animals and even, apparently, new kinds of human beings. Linnaeus revised edition after edition to accommodate these findings, but he did not change his opinion that all species now in existence were created by the sixth day of God's labor and have remained fixed ever since. During Linnaeus's time, however, it became clear that the pattern of creation was far more complex than had been originally envisioned.

Evolution before Darwin

The idea that organisms might evolve through time, with one type of organism giving rise to another type of organism, is an ancient one, predating Aristotle. A school of Greek philosophy, founded by Anaximander (611–547 B.C.) and culminating in the writings of the Roman Lucretius (99–55 B.C.), developed not only an atomic theory but also an evolutionary theory, both of which are strikingly similar to modern conceptions. The work of this school, however, was largely unknown in Europe at the time that the science of biology, as we know it today, began to take form.

In the eighteenth century, the French scientist Georges-Louis Leclerc de Buffon (1707–1788) was among the first to propose that species might undergo changes in the course of time. He suggested that, in addition to the numerous creatures that were produced by divine creation at the beginning of the world, "there are lesser families conceived by Nature and produced by Time." Buffon believed that these changes took place by a process of degeneration. In fact, as he summed it up, ". . . improvement and degeneration are the same thing, for both imply an alteration of the original constitution." Buffon's hypothesis, although vague as to the way in which changes might occur, did attempt to explain the bewildering variety of creatures in the modern world.

Another doubter of fixed and unchanging species was Erasmus Darwin (1731–1802), Charles Darwin's grandfather. Erasmus Darwin was a physician, a gentleman naturalist, and a prolific writer, often in verse, on both botany and zoology. He suggested, largely in asides and footnotes, that species have historical connections with one another, that animals may change in response to their environment, and that their offspring may inherit these changes. He maintained, for instance, that a polar bear is an "ordinary" bear that, by living in the Arctic, became modified and passed the modifications along to its cubs. These ideas were never clearly formulated but are interesting because of their possible effects on Charles Darwin, although the latter, born after his grandfather died, did not profess to hold his grandfather's views in high esteem.

The Age of the Earth

It was geologists, more than biologists, who paved the way for modern evolutionary theory. One of the most influential of these was James Hutton (1726–1797). Hutton proposed that the earth had been molded not by sudden, violent events but by slow and gradual processes—wind, weather, and the flow of water—the same processes that can be seen at work in the world today. This theory of Hutton's, which was known as uniformitarianism, was important for three reasons. First, it implied that the earth has a long history, which was a new idea to eighteenth-century Europeans. Christian theologians, by counting the successive generations since Adam (as recorded in the Bible), had calculated the maximum

I–2 *Charles Darwin in 1840, four years after he returned from his five-year voyage on H.M.S.* Beagle. *In his later book,* The Voyage of the Beagle, *Darwin made the following comments about his selection for the voyage: "Afterwards, on becoming very intimate with Fitz Roy [the captain of the* Beagle], *I heard that I had run a very narrow risk of being rejected on account of the shape of my nose! He . . . was convinced that he could judge of a man's character by the outline of his features; and he doubted whether anyone with my nose could possess sufficient energy and determination for the voyage. But I think he was afterwards well satisfied that my nose had spoken falsely."*

I-3 *While the* Beagle *sailed up the west coast of South America, Darwin explored the Andes on foot and horseback. He saw geological strata such as these, discovered fossil sea shells at about 3,700 meters (12,000 feet), and was witness to the upheaval of the earth produced by a major earthquake that occurred while he was there. In 1846, he published a book on his geological observations in South America. Strata are now seen as pages in evolutionary history.*

age of the earth at about 6,000 years. As far as we know, no one since the followers of Anaximander (whose school maintained that the earth was infinitely old) had thought in terms of a longer period. Yet 6,000 years is far too short for major evolutionary changes to take place, by any theory. Second, the theory of uniformitarianism stated that change is itself the *normal* course of events, as opposed to a static system interrupted by an occasional unusual event, such as an earthquake. Third, although this was never explicit, uniformitarianism suggested that there might be alternatives to the literal interpretation of the Bible.

The Fossil Record

During the latter part of the eighteenth century, there was a revival of interest in fossils, which are the preserved remains of organisms long since deceased. In previous centuries, fossils had been collected as curiosities, but they had generally been regarded either as accidents of nature—stones that somehow looked like shells—or as evidence of great catastrophes, such as the Flood described in the Old Testament. The English surveyor William Smith (1769–1839) was among the first to study the distribution of fossils scientifically. Whenever his work took him down into a mine or along canals or cross-country, he carefully noted the order of the different layers of rock, known as geological strata, and collected the fossils from each layer. He eventually established that each stratum, no matter where he came across it in England, contained characteristic kinds of fossils and that these fossils were actually the best way to identify a particular stratum in a number of different geographic locations. (The use of fossils to identify strata is still widely practiced, for instance, by geologists looking for oil.) Smith did not interpret his findings, but the implication that the present surface of the earth had been formed layer by layer over the course of time was an unavoidable one.

Like Hutton's world, the world seen and described by William Smith was clearly a very ancient one. A revolution in geology was beginning; earth science was becoming a study of time and change rather than a mere cataloging of types of rocks. As a consequence, the history of the earth became inseparable from the history of living organisms, as revealed in the fossil record.

Catastrophism

Although the way was being prepared by the revolution in geology, the time was not yet ripe for a parallel revolution in biology. The dominating force in European science in the early nineteenth century was Georges Cuvier (1769–1832). Cuvier was the founder of vertebrate paleontology, the scientific study of the fossil record of vertebrates (animals with backbones). An expert in anatomy and zoology, he applied his knowledge of the way in which animals are constructed to the study of fossil animals, and he was able to make brilliant deductions about the

I-4 *Particular strata, even though widely separated geographically, have characteristic assemblages of fossils. These fossil trilobites from the Devonian period (360 to 408 million years ago) were found in strata in (a) Ohio, (b) Oklahoma, and (c) upstate New York.*

(a)

(b)

(c)

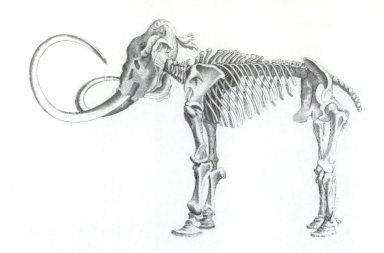

I–5 *A drawing by Georges Cuvier of a mastodon. Although Cuvier was one of the world's experts in reconstructing extinct animals from their fossil remains, he was a powerful opponent of evolutionary theories.*

I–6 *According to Lamarck's hypothesis —now known to be in error—as giraffes stretched to reach the high branches, their necks lengthened, and this acquired characteristic was transmitted to their offspring.*

form of an entire animal from a few fragments of bone. Today we think of paleontology and evolution as so closely connected that it is surprising to learn that Cuvier was a staunch and powerful opponent of evolutionary theories. He recognized the fact that many species that had once existed no longer did. (In fact, according to modern estimates, considerably less than one percent of all species that have ever lived are represented on the earth today.) Cuvier explained the extinction of species by postulating a series of catastrophes. After each catastrophe, the most recent of which was the Flood, new species filled the vacancies.

Cuvier hedged somewhat on the source of the new animals and plants that appeared after the extinction of older forms; he was inclined to believe they moved in from parts unknown. Another major opponent of evolution, Louis Agassiz (1807–1873), America's leading nineteenth-century biologist, was more straightforward. According to Agassiz, the fossil record revealed 50 to 80 total extinctions of life, followed by an equal number of new, separate creations.

The Concepts of Lamarck

The first modern scientist to work out a systematic concept of evolution was Jean Baptiste Lamarck (1744–1829). "This justly celebrated naturalist," as Darwin himself referred to him, boldly proposed in 1801 that all species, including *Homo sapiens,* are descended from other species. Lamarck, unlike most of the other zoologists of his time, was particularly interested in one-celled organisms and invertebrates (animals without backbones). Undoubtedly it was his long study of these forms of life that led him to think of living things in terms of constantly increasing complexity, each species derived from an earlier, less complex one.

Like Cuvier and others, Lamarck noted that older rocks generally contained fossils of simpler forms of life. Unlike Cuvier, however, Lamarck interpreted this as meaning that the more complex forms had arisen from the simpler forms by a kind of progression. According to his hypothesis, this progression, or evolution, to use the modern term, is dependent on two main forces. The first is the inheritance of acquired characteristics. Organs in animals become stronger or weaker, more or less important, through use or disuse, and these changes, according to Lamarck's proposal, are transmitted from the parents to the progeny. His most famous example was the evolution of the giraffe. According to Lamarck, the modern giraffe evolved from ancestors that stretched their necks to reach leaves on high branches. These ancestors transmitted the longer necks—acquired by stretching—to their offspring, which stretched their necks even longer, and so on.

The second, equally important force in Lamarck's concept of evolution was a universal creative principle, an unconscious striving upward on the *Scala Naturae* that moved every living creature toward greater complexity. Every amoeba was on its way to man. Some might get waylaid—the orangutan, for instance, had been diverted from its course by being caught in an unfavorable environment—but the will was always present. Life in its simplest forms was constantly emerging by spontaneous generation to fill the void left at the bottom of the ladder. In Lamarck's formulation, Aristotle's ladder of nature had been transformed into a steadily ascending escalator powered by a universal will.

Lamarck's contemporaries did not object to his ideas about the inheritance of acquired characteristics, which we, with our present knowledge of genetics, know to be false. Nor did they criticize his belief in a metaphysical force, which was actually a common element in many of the concepts of the time. But these vague, untestable postulates provided a very shaky foundation for the radical proposal that more complex forms evolved from simpler forms. Moreover, Lamarck personally was no match for the brilliant and witty Cuvier, who relentlessly attacked his ideas. As a result, Lamarck's career was ruined, and both scientists and the public became even less prepared to accept any evolutionary doctrine.

DEVELOPMENT OF DARWIN'S THEORY

The Earth Has a History

The person who most influenced Darwin, it is generally agreed, was Charles Lyell (1797–1875), a geologist who was Darwin's senior by 12 years. One of the books that Darwin took with him on his voyage was the first volume of Lyell's newly published *Principles of Geology*, and the second volume was sent to him while he was on the *Beagle*. On the basis of his own observations and those of his predecessors, Lyell opposed the theory of catastrophes. Instead, he produced new evidence in support of Hutton's earlier theory of uniformitarianism. According to Lyell, the slow, steady, and cumulative effect of natural forces had produced continuous change in the course of the earth's history. Since this process is demonstrably slow, its results being barely visible in a single lifetime, it must have been going on for a very long time. What Darwin's theory needed was time, and it was time that Lyell gave him. In the words of Ernst Mayr of Harvard University, the discovery that the earth was ancient "was the snowball that started the whole avalanche."

The Voyage of the *Beagle*

This, then, was the intellectual climate in which Charles Darwin set sail from England. As the *Beagle* moved down the Atlantic coast of South America, through the Strait of Magellan, and up the Pacific coast, Darwin traveled the interior. He explored the rich fossil beds of South America (with the theories of Lyell fresh in his mind) and collected specimens of the many new kinds of plant and animal life

I–7 **(a)** *A reproduction of the* Beagle, *sailing off the coast of South America.* **(b)** *Cutaway view of the ship. Only 28 meters in length, this "good little vessel" set sail on its five-year voyage with 74 people aboard. Darwin shared the poop cabin with a midshipman and 22 chronometers belonging to Captain Fitz Roy, who had a passion for exactness. Darwin's sleeping space was so confined that he had to remove a drawer from a locker to make room for his feet.*

(a)

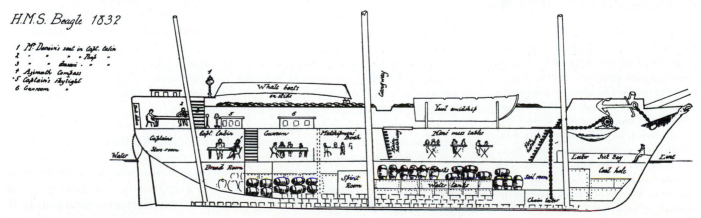

(b)

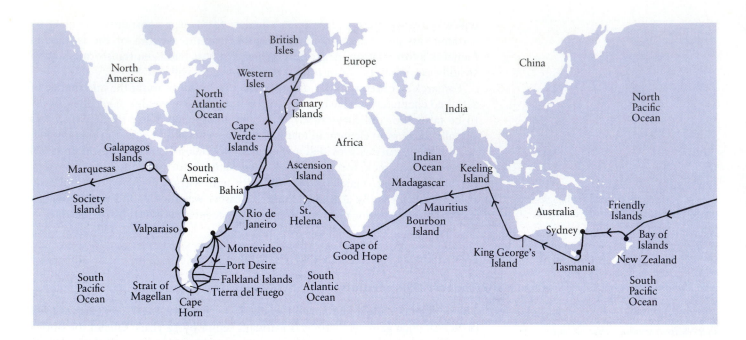

I–8 *The Beagle's voyage. The ship left England in December of 1831 and arrived at Bahia, Brazil, in late February of 1832. About 3¹/₂ years were spent along the coast of South America, surveying and making inland explorations. The stop at* the Galapagos Islands was for slightly more than a month, and, during that brief time, Darwin made the wealth of observations that were to change the course of the science of biology. The remainder of the voyage, across the Pacific to New Zealand and Australia, across the Indian Ocean to the Cape of Good Hope, back to Bahia once more, and at last home to England, occupied another year.

I–9 *A distinguishing feature of the Galapagos tortoise is the shape of its carapace, or shell, which varies according to its island of origin. The tortoises found on the islands with comparatively lush vegetation are characterized by a domed shell, shown here, which affords protection of the tortoise's soft parts as it makes its way through the thick undergrowth. The high arch at the front of the saddleback shell (see Figure I–1) enables the tortoise to reach upward in search of food; such shells are typical of tortoises living on arid islands where food may be scarce.*

he encountered. He was impressed most strongly during his long, slow trip down one coast and up the other by the constantly changing varieties of organisms he saw. The birds and other animals on the west coast, for example, were very different from those on the east coast, and even as he moved slowly up the western coast, one species would give way to another.

Most interesting to Darwin were the animals and plants that inhabited a small, barren group of islands, the Galapagos, which lie some 950 kilometers off the coast of Ecuador. The Galapagos were named after the islands' most striking inhabitants, the tortoises (*galápagos* in Spanish), some of which weigh 100 kilograms or more. Each island has its own type of tortoise; sailors who took these tortoises on board and kept them as convenient sources of fresh meat on their sea voyages could readily tell which island any particular tortoise had come from. Then there was a group of finchlike birds, 13 species in all, that differed from one another in the sizes and shapes of their bodies and beaks, and particularly in the type of food they ate. In fact, although clearly finches, they had many characteristics seen only in completely different types of birds on the mainland. One finch, for example, feeds by routing insects out of the bark of trees. It is not fully equipped for this, however, lacking the long tongue with which the true woodpecker flicks out insects from under the bark. Instead, the woodpecker finch uses a small stick or cactus spine to pry the insects loose.

From his knowledge of geology, Darwin knew that these islands, clearly of volcanic origin, were much younger than the mainland. Yet the plants and animals of the islands were different from those of the mainland, and in fact the inhabitants of different islands in the archipelago differed from one another. Were the living things on each island the product of a separate special creation? "One might really fancy," Darwin mused at a later date, "that from an original paucity of birds in this archipelago one species had been taken and modified for different ends." This problem continued, in his own word, to "haunt" him.

(a)

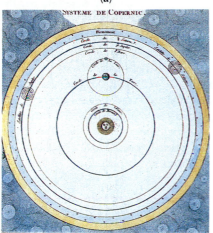

(b)

I–10 (a) *A view of the universe first proposed by the early Greeks and accepted throughout the Middle Ages. In this colored woodcut from Martin Luther's* Bible, *dated 1534, earth is in the center of the universe, surrounded by a layer of air containing clouds, stars, planets, the sun, and the moon. Beyond this is an outer layer of fire.* **(b)** *The solar system, as proposed by Nicholas Copernicus. In 1543, Copernicus set forth in* De Revolutionibus *the new concept that the sun, not the earth, is the center of the solar system. His theory was supported by the German astronomer Johannes Kepler (1571–1630), who discovered the laws of planetary motion, and by the Italian Galileo Galilei (1564–1642). The latter spent the last 10 years of his life confined to his home for heresy because of his advocacy of Copernican beliefs.*

The Darwinian Theory

Darwin was an assiduous and voracious reader. Not long after his return, he came across a short but much talked about sociological treatise by the Reverend Thomas Malthus that had first appeared in 1798. In this essay, Malthus warned, as economists have warned ever since, that the human population was increasing so rapidly that it would soon be impossible to feed all the earth's inhabitants. Darwin saw that Malthus's conclusion—that food supply and other factors hold populations in check—is true for all species, not just the human one. For example, Darwin calculated that a single breeding pair of elephants, which are among the slowest reproducers of all animals, would, if all their progeny lived and reproduced the normal number of offspring over a normal life span, produce a standing population of 19 million elephants in 750 years, yet the average number of elephants generally remains the same over the years. So, although a single breeding pair could have, in theory, produced 19 million descendants, it did, in fact, produce an average of only two. But why these particular two? The process by which the two survivors are "chosen" Darwin called **natural selection.**

Natural selection, according to Darwin, was a process analogous to the type of selection exercised by breeders of cattle, horses, or dogs. In artificial selection, we humans choose individual specimens of plants or animals for breeding on the basis of characteristics that seem to us desirable. In natural selection, the environment takes the place of human choice. As individuals with certain hereditary characteristics survive and reproduce and individuals with other hereditary characteristics are eliminated, the population will slowly change. If some horses were swifter than others, for example, these individuals would be more likely to escape predators and survive, and their progeny, in turn, might be swifter, and so on.

According to Darwin, inherited variations among individuals, which occur in every natural population, are a matter of chance. They are not produced by the environment, by a "creative force," or by the unconscious striving of the organism. In themselves, they have no goal or direction, but they often have positive or negative adaptive values; that is, they may be more or less useful to an organism as measured by its survival and reproduction. It is the operation of natural selection —the interaction of individual organisms with their environment—over a series of generations that gives direction to evolution. A variation that gives an organism even a slight advantage makes that organism more likely to leave surviving offspring. Thus, to return to Lamarck's giraffe, an animal with a slightly longer neck may have an advantage in feeding and thus be likely to leave more offspring than one with a shorter neck. If the longer neck is an inherited characteristic, some of these offspring will also have long necks, and if the long-necked animals in this generation have an advantage, the next generation will include more long-necked individuals. Finally, the population of short-necked giraffes will have become a population of longer-necked ones (although there will still be variations in neck length).

As you can see, the essential difference between Darwin's formulation and that of any of his predecessors is the central role he gave to variation. Others had thought of variations as mere disturbances in the overall design, whereas Darwin saw that variations among individuals are the real fabric of the evolutionary process. Species arise, he proposed, when differences among individuals within a group are gradually converted into differences between groups as the groups become separated in space and time.

The Origin of Species, which Darwin pondered for more than 20 years after his return to England, is, in his own words, "one long argument." Fact after fact, observation after observation, culled from the most remote Pacific island to a neighbor's pasture, is recorded, analyzed, and commented upon. Every objection is weighed, anticipated, and countered. *The Origin of Species* was published on November 24, 1859, and the Western world has not been the same since.

Darwin's Long Delay

Darwin returned to England with the *Beagle* in 1836. Two years later, he read the essay by Malthus, and in 1842 he wrote a preliminary sketch of his theory, which he revised in 1844. On completing the revision, he wrote a formal letter to his wife requesting her, in the event of his death, to publish the manuscript (which was some 230 pages long). Then, with the manuscript and letter in safekeeping, he turned to other work, including a four-volume treatise on barnacles. For more than 20 years following his return from the Galapagos, Darwin mentioned his ideas on evolution only in his private notebooks and in letters to his scientific colleagues.

In 1856, urged on by his friends Charles Lyell and botanist Joseph Hooker, Darwin set slowly to work preparing a manuscript for publication. In 1858, some 10 chapters later, Darwin received a letter from the Malay archipelago from another English naturalist, Alfred Russel Wallace, who had corresponded with Darwin on several previous occasions. Wallace presented a theory of evolution that exactly paralleled Darwin's own. Like Darwin, Wallace had traveled extensively and also had read Malthus's essay. Wallace, tossing in bed one night with a fever, had a sudden flash of insight. "Then I saw at once," Wallace recollected, "that the ever-present variability of all living things would furnish the material from which,

by the mere weeding out of those less adapted to the actual conditions, the fittest alone would continue the race." Within two days, Wallace's 20-page manuscript was completed and in the mail.

When Darwin received Wallace's letter, he turned to his friends for advice, and Lyell and Hooker, taking matters into their own hands, presented the theory of Darwin and Wallace at a scientific meeting just one month later. (Darwin described Wallace as "noble and generous," as indeed he was.) Lyell and Hooker read four papers from Darwin's notes of 1844, excerpts from two letters written by Darwin, and Wallace's manuscript. Their presentation received little attention, but for Darwin the floodgates were opened. He finished his long treatise in little more than a year, and the book was finally published. The first printing was a mere 1,250 copies, but they were sold out the same day.

Why Darwin's long delay? His own writings, voluminous though they are, shed little light on this question. But perhaps his background does. He came from a conventionally devout family, and he himself had been a divinity student. Perhaps most important, his wife, to whom he was deeply devoted, was extremely religious. It is difficult to avoid the speculation that Darwin, like so many others, found the implications of his theory difficult to confront.

Alfred Russel Wallace (1823–1913). As a young man, Wallace explored the Malay archipelago for eight years, covering about 22,500 kilometers (14,000 miles) by foot and native canoe. During his stay there, he collected 125,000 specimens of plants and animals, many of them previously unknown. His book about his Malay travels bears this inscription: "To Charles Darwin, Author of 'The Origin of Species,' I dedicate this book, not only as a token of personal esteem and friendship but also to express my deep admiration for his genius and his works."

(a)

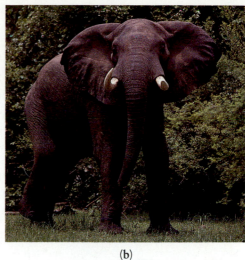

(b)

I–11 *According to biochemical tests, made possible by new genetic engineering techniques, there is a close evolutionary link between* (a) *the woolly mammoth, a creature that roamed North America, Asia, and Europe thousands of years ago, and* (b) *the modern elephant. Several years ago, a baby woolly mammoth that died about 40,000 years ago was found frozen in Siberia. Its tissues were so perfectly preserved that the exact structure of certain key molecules could be determined and compared with the structure of the same molecules in living elephants.*

Acceptance of Darwin's argument revolutionized the science of biology. "The theory of evolution," in the words of Ernst Mayr, "is quite rightly called the greatest unifying theory in biology." As we shall see throughout this text, it is the thread that links together all the diverse phenomena of the living world. It also deeply influenced our way of thinking about ourselves. With the possible exception of the new astronomy of Copernicus and Galileo in the sixteenth and seventeenth centuries, no revolution in scientific thought has had as much effect on human culture as this one. One reason is, of course, that evolution is in contradiction to the literal interpretation of the Bible. Another difficulty is that it seems to diminish human significance. The new astronomy had made it clear that the earth is not the center of the universe or even of our own solar system. Then the new biology asked us to accept the proposition that, as far as science can show, we are not fundamentally different from other organisms in either our origins or our place in the natural world.

Challenges to Evolutionary Theory

Today, with almost no exceptions, modern biologists are convinced by a vast body of accumulated evidence that the earth has a long history and that all living organisms, including ourselves, arose in the course of that history from earlier, more primitive forms. This accumulated evidence consists of an interlocking fabric of thousands upon thousands of pieces of data concerning past and present organisms, including not only anatomical structure but also physiological and biochemical processes, patterns of embryonic development, patterns of behavior, and most recently, the sequences of genetic information encoded in the DNA molecules of the chromosomes.

Yet, as everyone who reads a newspaper or watches television knows, evolutionary theory remains a matter of lively public controversy. Moreover, the advocates of special creation—who maintain that each species was created separately—seek to lend strength to their arguments from the fact that scientists ask many questions about evolution. They point out that even among scientists, evolution is "only a theory," and that even leading scientists do not agree on this "theory." Much of the confusion surrounding this controversy stems from the very definition of the word "theory," and from a misunderstanding of the nature and limitations of the scientific process, topics that we shall consider later in this Introduction. Among biologists, there is almost unanimous agreement that evolution has occurred in the past and continues to occur today. As we shall see in Section 7, however, the details and relative importance of the different processes involved in evolutionary change are currently the subject of intense research and discussion among biologists.

UNIFYING PRINCIPLES OF MODERN BIOLOGY

The foundations of modern biology include not only evolution but also three other principles that are so well established that biologists seldom discuss them among themselves. You could read widely in the current biological literature without seeing any of them mentioned, yet it is impossible to understand either the ideas or the data of contemporary biology without being aware of them. These principles, like evolution, will be discussed in greater detail in the course of this text and will recur as major themes, but you should have them in mind from the outset.

All Organisms Are Made Up of Cells

One of the fundamental principles of biology is that all living organisms are composed of one or more similar units, known as **cells.** This concept is of tremendous and central importance to biology because it emphasizes the basic sameness of all living systems. It therefore brings an underlying unity to widely varied studies involving many different kinds of organisms.

The word "cell" was first used in a biological sense some 300 years ago. In the seventeenth century, the English scientist Robert Hooke, using a microscope of his own construction, noticed that cork and other plant tissues are made up of small cavities separated by walls. He called these cavities "cells," meaning "little rooms." However, "cell" did not take on its present meaning—the basic unit of living matter—for more than 150 years.

In 1838, Matthias Schleiden, a German botanist, came to the conclusion that all plant tissues consist of organized masses of cells. In the following year, zoologist Theodor Schwann extended Schleiden's observations to animal tissues and proposed a cellular basis for all life. In 1858, the idea that all living organisms are composed of one or more cells took on an even broader significance when the great pathologist Rudolf Virchow generalized that cells can arise only from preexisting cells: "Where a cell exists, there must have been a preexisting cell, just as the animal arises only from an animal and the plant only from a plant. . . . Throughout the whole series of living forms, whether entire animal or plant organisms or their component parts, there rules an eternal law of continuous development."

From the perspective provided by Darwin's theory of evolution, published in the following year, Virchow's concept takes on an even larger significance. There is an unbroken continuity between modern cells—and the organisms they compose—and the primitive cells that first appeared on earth more than 3 billion years ago.

All Organisms Obey the Laws of Physics and Chemistry

Until fairly recently, many prominent biologists believed that living systems are qualitatively different from nonliving ones, containing within them a "vital spirit" that enables them to perform activities that cannot be carried on outside the living organism. This concept is known as vitalism and its proponents as vitalists.

In the seventeenth century, the vitalists were opposed by a group known as the mechanists. The French philosopher René Descartes (1596–1650) was a leading proponent of this point of view. The mechanists set about showing that the body worked essentially like a machine; the arms and legs move like levers, the heart like a pump, the lungs like a bellows, and the stomach like a mortar and pestle. Although these simple mechanical models provided much insight into the functioning of the animal body, by the nineteenth century the debate about the distinctiveness of living systems had moved beyond them. The argument became centered on whether or not the chemistry of living organisms was governed by the

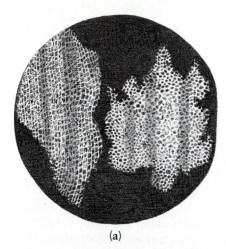

(a)

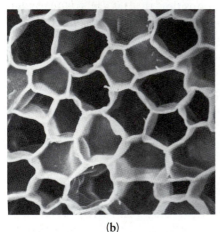

(b)

I-12 (a) *Robert Hooke's drawings of two slices of a piece of cork, reproduced from his* Micrographica, *published in 1665, and* (b) *a scanning electron micrograph of a slice of cork. Hooke was the first to use the word "cells" to describe the tiny compartments that together make up an organism. The cells in these pieces of cork have died—all that remain are the outer walls. As we shall see in subsequent chapters, the living cell is filled with a variety of substances, organized into distinct structures and carrying out a multitude of essential processes.*

I–13 *Wöhler was a student in this labor-atory at Giessen, in what is now West Germany. It was one of the first where practical work in chemistry could be done.*

same principles as the chemistry performed in the laboratory. The vitalists claimed that the chemical operations performed by living tissues could not be carried out experimentally in the laboratory, categorizing reactions as either "chemical" or "vital." Their new opponents, known as reductionists (since they believed that the complex operations of living systems could be reduced to simpler and more readily understandable ones), achieved a partial victory when the German chemist Friedrich Wöhler (1800–1882) converted an "inorganic" substance (ammonium cyanate) into a familiar organic substance (urea). On the other hand, the claims of the vitalists were supported by the fact that, as chemical knowledge improved, many new compounds were found in living tissues that were never seen in the nonliving, or inorganic, world.

In the late 1800s, the leading vitalist was Louis Pasteur, who claimed that the changes that took place when fruit juice was transformed to wine were "vital" and could be carried out only by living cells—the cells of yeast. In spite of many advances in chemistry, this phase of the controversy lasted until almost the turn of the century. However, in 1898, the German chemists Eduard and Hans Büchner showed that a substance extracted from yeast cells could produce fermentation outside the living cell. (This substance was given the name enzyme, from *zyme*, the Greek word meaning "yeast" or "ferment.") A "vital" reaction was demonstrated to be a chemical one, and the subject was eventually laid to rest. Today it is generally accepted that living systems "obey the rules" of chemistry and physics, and modern biologists no longer believe in a "vital principle."

The realization that living systems obey the laws of physics and chemistry opened a new era in the history of biology. Increasingly organisms were studied in terms of their chemical composition and the chemical reactions that take place within their bodies. These studies, which continue today at an extraordinary pace, have yielded a vast amount of information and provide an essential foundation for contemporary biology. Perhaps their greatest test came about 40 years ago. One of the most striking characteristics of living things is their capacity to reproduce, to generate faithful copies of themselves. By about 1950, this capacity had been shown to reside in a single type of chemical molecule, deoxyribonucleic acid (DNA). The race to discover the structure of this molecule began, and the question in everyone's mind was whether or not the structure of this one "simple" molecule could possibly explain the mysteries of heredity. As it turned out, and as we shall discuss in Section 3, it could.

I-14 *The acquisition of energy resources to power life processes is of fundamental importance for all living organisms. These cheetahs, photographed in Kenya, have caught and killed a springbok, on which they will feed until sated. The remainder of the carcass—and the stored chemical energy it contains—will then be abandoned by the cheetahs, which rarely return to a previous day's kill. Cheetahs typically hunt in small groups, stealthily stalking their prey and then rushing at the intended victim at high speed. Over short distances, the cheetah is the fastest land animal in the world. It can attain speeds of 110 kilometers per hour but can maintain such speeds for little more than 30 seconds (covering a distance of about half a kilometer).*

All Organisms Require Energy

Among the laws of physics that are pertinent to biology are the laws of thermodynamics. They state simply that (1) energy can be changed from one form to another but it cannot be created or destroyed, that is, the total energy of the universe remains constant; and (2) all natural events proceed in such a way that concentrations of energy tend to dissipate or become random. A heated object, which is one example of concentrated energy, loses its heat to its surroundings.

A living system, which is a concentration of energy of another kind, can maintain itself in the face of this tendency only by a constant intake of energy. Living organisms are experts at energy conversion. The energy they take in—whether in the form of sunlight or chemical energy stored in food—is transformed and used by each individual cell to do the work of the cell. This work includes powering not only the numerous processes that constitute the activities of the organism but also the synthesis of an enormous diversity of molecules and cellular structures. In the course of the cell's work, the energy may be further transformed to energy of motion, to heat energy, or even back to light energy again. It is ultimately dissipated, and the organism must take in more energy.

This flow of energy is the essence of life. A cell can be best understood as a complex of systems for transforming energy. At the other end of the biological scale, the structure of the biosphere—that is, the entire living world—is determined by the energy exchanges occurring among the groups of organisms within it. Similarly, evolution may be viewed as a competition among organisms for the most efficient use of energy resources.

THE FORMS OF LIFE

One of the principal consequences of the evolutionary competition is an incredible diversity in the living world. It is estimated that we share this planet with more than 5 million different species of organisms. These different organisms exhibit great variety in the organization of their bodies, in their patterns of reproduction, growth, and development, and in their behavior.

I–15 (a) *The smallest and most numerous living organisms known to biology are the prokaryotes, which include the bacteria and closely related forms. Despite their small size, bacteria, like all living things, are highly organized and contain a variety of structures. Neisseria gonorrhoeae (dark spheres), the causative agent of gonorrhea, is shown here being ingested by white blood cells.*

(b) *Although significantly more complex than prokaryotes, protists are generally quite small. This drop of water, from a freshwater lake, contains at least four different kinds of protists. The bright yellow-green organism is an alga, which manufactures its own food by photosynthesis. The larger, blue organisms are protozoans; they feed on bacteria and algae.*

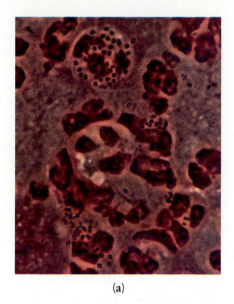

(a)

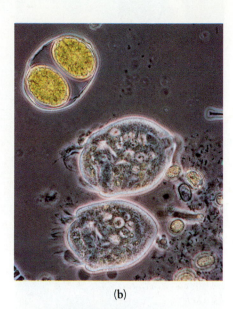

(b)

I–16 *This mushroom, growing on a forest floor, is the reproductive structure of a fungus. The bulk of the body of the fungus consists of underground filaments through which nutrients are absorbed.*

Despite the seemingly overwhelming diversity of living organisms, it is possible to group them in ways that reveal not only patterns of similarity and difference but also historical relationships among different groups. In Section 4, we shall consider these patterns and relationships in some detail. Before reaching Section 4, however, we shall encounter a marvelous variety of organisms. Thus you will find it helpful to be aware of the five major categories, or kingdoms, into which we group organisms in this text.

The first of these, which includes the earliest life forms known to have appeared on this planet, is the kingdom Monera. It comprises the smallest and simplest of all organisms, the bacteria and their relatives (Figure I–15a). Each individual organism consists of a single structural unit—a cell. This group of one-celled organisms makes up the prokaryotes. The term "prokaryote" means "before the nucleus" and refers to the internal organization of the cells, which have neither a clearly defined nucleus nor the other structures that can be found in all other kinds of cells. The first prokaryotes made their appearance at least 3.5 billion years ago, when the earth was very different from the green planet we know today, and prokaryotes were the earth's sole inhabitants for more than 2 billion years.

The second major category of life is known as the kingdom Protista. Protists are also mostly one-celled, but the cells are structurally very different from the prokaryotes. They are known as eukaryotes, meaning "truly nucleated." The cells of all the organisms in the other three kingdoms are also eukaryotes. Many biologists believe that the transition from the prokaryotic to the eukaryotic cell was the largest and most significant event in the history of life, second only in biological importance to the first appearance of living systems. The protists are an extremely varied collection of organisms (Figure I–15b), and this kingdom includes the most structurally complex and versatile of all cells. Examples of protists are amoebas, paramecia, and the many forms of algae.

The third kingdom is made up of the fungi, including such organisms as molds, yeasts, and mushrooms (Figure I–16). Their mode of existence is very different from that of all other living things. Fungi digest complex macromolecules, which they may find in soil, water, cotton, leather, or even on the surface of the human skin, into smaller molecules. They then absorb these smaller molecules into their bodies, typically composed of masses of fine filaments, the surfaces of which are in direct contact with the nutrient source.

I–17 *Among the largest and oldest of all living things are the giant sequoias of North America. Members of the plant kingdom, sequoias typically have trunks more than 10 meters in circumference and are sometimes more than 75 meters tall. These particular trees were already growing a thousand years ago, long before Europeans settled in North America.*

I–18 *Among the animals, the insects are noted for both their numbers and their variety. A major reason for their success, biologically speaking, is the diversity of their life styles. This is a cetonid beetle, a resident of the tropical forests of western Africa. Note its spiny legs, which are raised in defense. Adults of this species are thought to feed on the pollen and nectar of flowers. The eggs are laid in rotting wood, and the larvae (the immature insects) develop there.*

The other two kingdoms are, of course, the plants and the animals, Plantae and Animalia. Plants are most concisely defined as many-celled organisms that collect energy from light—sunshine. They then transform this energy into the complex molecules of which their bodies are composed (Figure I–17). These molecules, which include sugars, proteins, and oils, are the energy sources for animal life.

The fifth kingdom, the animals, includes those life forms that are many-celled and that depend on other forms—mostly plants or other animals—for their sustenance (Figure I–18). From our anthropocentric point of view, animal usually means mammal, but actually most animals are invertebrates. More than one and a half million different kinds of animals have been recorded, of which 95 percent are invertebrates and more than a million of these are insects.

In little more than a century, our knowledge of the diversity of organisms, past and present, of the processes occurring within their bodies, and of their interrelationships with one another has rapidly outstripped that gained in all the previous centuries of human inquiry. This explosion of knowledge, which continues at an ever-accelerating pace, is the direct consequence of that particular form of inquiry that we call science (from the Latin *scientia,* "having knowledge").

THE NATURE OF SCIENCE

Science, whether biological or other, is a way of seeking principles of order. Art is another way, as are religion and philosophy. Science differs from these others in that it limits its search to the natural world, the physical universe. Also, and perhaps even more significant, it differs from them in the central value it gives to observation (particularly that structured kind of observation called experimentation). Scientists begin their search by accumulating data—evidence—and trying to fit these data into systems of order, conceptual schemes that organize the data in some meaningful way. Accumulating and ordering data are not two steps; they go on simultaneously. Or to put it another way, the accumulation of data is undertaken by scientists as a way of answering a question, or of supporting or rejecting an idea. The data may be generated by systematic observation, including deliberate, planned experiments (of which we shall see many examples in the chapters that follow); they may also be gleaned retrospectively from reevaluation of one's earlier systematic observation or from reevaluation of verifiable information recorded by others. (Darwin, like many biologists before and since, made copious use of all of these methods.)

The great discoveries in science are not merely the addition of new data but the perception of new relationships among the available data—in other words, the development of new ideas. The ideas of science are categorized in ascending order of validity as hypotheses, theories, and principles or laws. Lower on the scale than the hypothesis is the hunch, or educated guess, which is how most hypotheses begin. A hunch becomes a hypothesis—and therefore an idea that can be investigated scientifically—when and only when it is stated in such a way that it is potentially testable, even if the test cannot be done immediately. The testing of a hypothesis can often be done quite promptly, but sometimes it is, of necessity, long delayed. For example, some current hypotheses about the interactions that determine the structure of tropical forests cannot be tested until many more data have been gathered by biologists working in the forests. Similarly, a number of hypotheses about the organization of the cell could not be tested until after the development of the electron microscope.

A hypothesis can sometimes be tested directly. For example, one can determine whether caterpillars are repelled by a particular substance, isolated from plant leaves, that is hypothesized to protect the leaves from the caterpillars' predation.

I-19 *In a swamp in Peru, two biologists, Terry Erwin and Linda Sims of the Smithsonian Institution, are collecting data on the population structure of the tropical forest. Erwin is shooting a line high into the treetops as the first step in insect collection. Many of the specimens they have found are wholly new to science.*

This type of test frequently involves a controlled experiment, in which two groups of organisms (or cells) are exposed to conditions that are identical in all respects except the one being tested. Often, however, the most important tests of a hypothesis are indirect. A basic assumption of science is that the universe is consistent and that its component parts interact with and affect one another in understandable and predictable ways. An essential part of the testing of a hypothesis is the logical deduction of what other things should follow if the hypothesis is correct. The deduction constitutes a prediction of what should be observed if the hypothesis is correct and indicates the type of data that need to be gathered in order to test it.

Although a key predictive test may demonstrate that a hypothesis is false or may indicate that it must be modified, such a test can never definitely prove, once and for all, that a hypothesis is true—simply because we can never be certain that we have examined all of the relevant evidence. However, repeated successful tests of a hypothesis—either directly or in terms of the consequences that would follow if the hypothesis were correct—provide strong evidence in favor of the hypothesis.

When a scientist has collected sufficient data to support a particular hypothesis, he or she then reports the results to other scientists; such a report usually takes place at a scientific meeting (such as the meeting at which Darwin's and Wallace's papers were read), or in a scientific publication, such as a journal or a book (for example, *The Origin of Species*). If the data are sufficiently interesting or the hypothesis important, the observations or experiments will be repeated in an attempt to confirm, deny, or extend them. Hence, scientists always report the methods that they used in gathering and analyzing their data as well as their conclusions.

When a hypothesis of broad, fundamental importance has survived a number of independent tests, involving a diversity of data, it is generally referred to as a theory. Thus, a theory in science has a somewhat different meaning from the word "theory" in common usage, in which "just a theory" carries with it the implication of a flight of fancy, a hunch, or an abstract notion, rather than a carefully formulated, well-tested proposition. A theory that has withstood repeated testing over a period of time becomes elevated to the status of a law or principle, although not always identified as such. The "theory" of evolution, which has been tested and retested, directly and indirectly, for the past 130 years, is an example. As far as scientists are concerned, it is a basic principle of biology, just as is the cell "theory." However, our knowledge of many of the details of cellular structure and function and of the details of the evolutionary process is in the stage of theory, or even hypothesis.

Because the subject matter of biology is enormously diverse, biologists utilize a wide variety of approaches in their studies. Careful and systematic observation, of the sort practiced by the nineteenth-century naturalists, Darwin among them, remains a cornerstone. It is now supplemented by an enormous array of technological innovations that began with the microscope. The experimental procedures of chemistry are essential for studying the physiological processes occurring within organisms and their constituent cells. The study of populations of organisms and their interactions depends on the same kind of statistical mathematics employed by economists and is enhanced by computers that can analyze large quantities of data very rapidly. Charting the past course of evolution depends not only on the work of paleontologists in the field and in the laboratory but also on the intellectual tools of the historian and the homicide detective. As we shall see in the course of this text, there is no single "scientific method" in biology; there are instead a multiplicity of methods, with the particular method used in any given instance being determined by the question to be answered.

Some Comments on Science and Scientists

Science and Theory

The power of theories is that they combine many generalizations and other theories into networks of interlocking ideas that point to the future.

J. Bronowski, *The Common Sense of Science,* Random House, Inc., New York, 1959.

On principle, it is quite wrong to try founding a theory on observable magnitudes alone. It is the theory which decides what we can observe.

A. Einstein, from J. Bernstein, "The Secrets of the Old Ones, II," *The New Yorker,* March 17, 1973.

The Scientific Method

Indeed, scientists are in the position of a primitive tribe which has undertaken to duplicate the Empire State Building, room for room, without ever seeing the original building or even a photograph. Their own working plans, of necessity, are only a crude approximation of the real thing, conceived on the basis of miscellaneous reports volunteered by interested travelers and often in apparent conflict on points of detail. In order to start the building at all, some information must be ignored as erroneous or impossible, and the first constructions are little more than large grass shacks. Increasing sophistication, combined with methodical accumulation of data, make it necessary to tear down the earlier replicas (each time after violent arguments), replacing them successively with more up-to-date versions. We may easily doubt that the version current after only 300 years of effort is a very adequate restoration of the Empire State Building; yet, in the absence of clear knowledge to the contrary, the tribe must regard it as such (and ignore odd travelers' tales that cannot be made to fit).

E. J. DuPraw, *Cell and Molecular Biology,* Academic Press, Inc., New York, 1968.

Scientists at Work

Scientists are like pickpockets. God has all the secrets in his pockets, and we try to pick them. You make an assumption in science—and it *is* an assumption—that there are fundamental laws you can find out. You have an idea you think can be proved and you try to prove it. Depending on how it goes, you make a step forward or you make a fool of yourself. Nature doesn't care whether you're right or wrong. Nature is the way it is, and you had better be smart enough to get a little glimpse.

Abraham Pais, Rockefeller University

Scientists at work have the look of creatures following genetic instructions; they seem to be under the influence of a deeply placed human instinct. They are, despite their efforts at dignity, rather like young animals engaged in savage play. When they are near to an answer their hair stands on end, they sweat, they are awash in their own adrenalin. To grab the answer, and grab it first, is for them a more powerful drive than feeding or breeding or protecting themselves against the elements.

It sometimes looks like a solitary activity, but it is as much the opposite of solitary as human behavior can be. There is nothing so social, so communal, so interdependent. An active field of science is like an immense intellectual anthill; the individual almost vanishes into the mass of minds tumbling over each other, carrying information from place to place, passing it around at the speed of light.

There are special kinds of information that seem to be chemotactic. As soon as a trace is released, receptors at the back of the neck are caused to tremble, there is a massive convergence of motile minds flying upwind on a gradient of surprise, crowding around the source. It is an infiltration of intellects, an inflammation.

There is nothing to touch the spectacle. In the midst of what seems a collective derangement of minds in total disorder, with bits of information being scattered about, torn to shreds, disintegrated, reconstituted, engulfed, in a kind of activity that seems as random and agitated as that of bees in a disturbed part of the hive, there suddenly emerges, with the purity of a slow phrase of music, a single new piece of truth about nature. . . .

There is something like aggression in the activity, but it differs from other forms of aggressive behavior in having no sort of destruction as the objective. While it is going on, it looks and feels like aggression: get at it, uncover it, bring it out, grab it, halloo! It is like a primitive running hunt, but there is nothing at the end of it to be injured. More probably, the end is a sigh. But then, if the air is right and the science is going well, the sigh is immediately interrupted, there is a yawping new question, and the wild, tumbling activity begins once more, out of control all over again.

Lewis Thomas, *The Lives of a Cell: Notes of a Biology Watcher,* Viking Press, Inc., New York, 1974.

Science and Human Values

The raw materials of science are our observations of the phenomena of the natural universe. Science—unlike art, religion, or philosophy—is limited to what is observable and measurable and, in this sense, is rightly categorized as materialistic. Hunches are abandoned, hypotheses superseded, theories revised—and occasionally, shattered—but the observations endure, and, moreover, they are used over and over again, sometimes in wholly new ways. It is for this reason that scientists stress and seek objectivity. In the arts, by contrast, the emphasis is on subjectivity —experience as filtered through the individual consciousness.

Because of this emphasis on objectivity, value judgments cannot be made in science in the way that such judgments are made in philosophy, religion, and the arts, and indeed in our daily lives. Whether something is good or beautiful or right in a moral sense, for example, cannot be determined by scientific methods. Such judgments, even though they may be supported by a broad consensus, are not subject to scientific testing.

(a)

(b)

I–20 *In science, great stress is placed on objectivity. As a consequence, scientists, speaking as scientists, refrain from making value judgments. Thus, for example, they would not identify any organism as the ugliest or the most beautiful. As a private individual, however, a scientist may have his or her own opinions about* (a) *wart hogs and* (b) *butterflies and flowers.*

At one time, sciences, like the arts, were pursued for their own sake, for pleasure and excitement and satisfaction of the insatiable curiosity with which we are both cursed and blessed. In the twentieth century, however, the sciences have spawned a host of giant technological achievements—the hydrogen bomb, polio vaccine, pesticides, indestructible plastics, nuclear energy plants, perhaps even ways to manipulate our genetic heritage—but have not given us any clues about how to use them wisely. Moreover, science, as a result of these very achievements, appears enormously powerful. It is thus little wonder that there are many people who are angry at science, as one would be angry at an omnipotent authority who apparently has the power to grant one's wishes but who refuses to do so.

The reason that science cannot and does not solve the problems we want it to is inherent in its nature. Most of the problems we now confront can be solved only by value judgments. For example, science gave us nuclear weapons and can give us predictions as to the extent of the biological damage that their use might cause. Yet it cannot help us, as citizens, in weighing the risk of damage from a nuclear exchange against the desire for a strong national defense.

Science has produced the knowledge that makes possible not only the construction of kidney dialysis machines but also the surgical replacement of a diseased kidney by a healthy one. There are, however, many more patients who need kidney transplant operations than there are healthy kidneys available for transplanting. Scientific methods cannot help us decide who should be on the waiting list for a kidney transplant and who should remain dependent on repeated dialysis treatments. Similarly, scientists can predict the possible extent of damage to the plants and animals of a particular area from the use of pesticides; they can also predict the reduction in food crops or the increase in malaria that would occur were pesticides prohibited. But, scientists, in their capacity as scientists, cannot make the choice as to whether we should or should not use pesticides.

It is one of the ironies of this so-called "age of science and materialism" that probably never before have ordinary individual men and women, including scientists, been confronted with so many moral and ethical dilemmas. In this text, we shall discuss some of the dilemmas that have grown out of the achievements of modern science and technology. Our greater concern, however, is to provide you with the biological knowledge necessary to understand the relevant data as you make your own value judgments regarding the problems that confront us now and that will do so in the future.

Science as Process

You are fortunate to be studying biology now, in what many consider to be its "golden age." New ideas and unexpected discoveries have opened up exciting frontiers in many different areas—cell biology, genetics, immunology, neurobiology, evolution, ecology, to mention just a few. Because there is so much to tell, most biology texts, and this one is no exception, tend to stress what is known at the present time, rather than what is not known or how we came to know what we do. This tendency, although understandable, distorts the nature of biology and, indeed, of science in general.

A modern science is not a static accumulation of facts organized in a particular way but a somewhat amorphous body of knowledge that constantly grows, developing new bulges and unpredictable appendages. It may also suddenly change its entire shape—as biology did in the nineteenth century with the realization of the quantity and diversity of evidence supporting evolution. Science is not information contained within textbooks or libraries or information-retrieval centers, but rather it is a dynamic process taking place in the minds of living scientists. In our enthusiasm for telling you what biologists have learned thus far about living organisms—their history, their properties, and their activities—do not let us convince you that all is known. Many questions are still unanswered. More important, many good questions have not yet been asked. Perhaps you will be the one to ask them.

You may have been persuaded to study biology because of current environmental problems or because of a desire to know more about the mechanisms of your own body or an interest in genetic engineering or a career in medicine—in short, because it is "relevant." The study of biology is, indeed, pertinent to many aspects of our day-to-day existence, but do not make this your main focus as you embark on the study of biology. Above all other considerations, study biology because it is "irrelevant"—that is, study it for its own sake, because, like art and music and literature, it is an adventure for the mind and nourishment for the spirit.

QUESTIONS

1. What is the essential difference between Darwin's theory of evolution and that of Lamarck?

2. The chief predator of an English species of snail is the song thrush. Snails that inhabit woodland floors have dark shells, whereas those that live on grass have yellow shells, which are less clearly visible against the lighter background. Explain, in terms of Darwinian principles.

3. The phrase "chance and necessity" has been used to describe the Darwinian theory of evolution. Relate this to the fact that

snails living on grass do not have green shells, but there are, for example, green frogs and green insects.

4. The largest terrestrial organisms known are the giant sequoias (Figure I–17) of North America, which dwarf even an elephant. The largest animals, members of the whale family, are all aquatic. Can you explain why this should be the case?

5. When scientists report new findings, they are expected to reveal their methods and raw data as well as their conclusions. Why is such reporting considered essential?

SUGGESTIONS FOR FURTHER READING

BATES, MARSTON, and PHILIP S. HUMPHREY (eds.): *The Darwin Reader*, Charles Scribner's Sons, New York, 1956.*

A collection of Darwin's writings, including The Autobiography, *and excerpts from* The Voyage of the Beagle, The Origin of Species, The Descent of Man, *and* The Expression of the Emotions. *Darwin was a fine writer, and you can discover here the wide range of his interests and concerns at different periods of his life.*

BRONOWSKI, J.: *The Ascent of Man*, Little, Brown and Company, Boston, 1973.*

An informal and illuminating history of the sciences, originally prepared as a television series. The emphasis is on science's relation to human culture. Well designed and illustrated.

DARWIN, CHARLES: *The Origin of Species by Means of Natural Selection, or The Preservation of Favored Races in the Struggle for Life*, W. W. Norton & Company, New York, 1975.*

Darwin's "long argument." Every student of biology should, at the very least, browse through this book to catch its special flavor and to begin to understand its extraordinary force.

DARWIN, CHARLES: *The Voyage of the Beagle*, Doubleday & Company, Inc., Garden City, N.Y., 1962.*

Darwin's own chronicle of the expedition on which he made the discoveries and observations that eventually led him to his

theory of evolution. The sensitive, eager, young Darwin that emerges from these pages is very unlike the solemn image many of us have formed of him from his later portraits.

LEWIN, ROGER: *Thread of Life: The Smithsonian Looks at Evolution*, W. W. Norton & Company, New York, 1982.

An absorbing account of the historical development of evolutionary thought, with current applications from biochemistry, paleontology, and geology. Well written and beautifully illustrated with a rich assortment of color photographs.

MAYR, ERNST: *The Growth of Biological Thought: Diversity, Evolution, and Inheritance*, Harvard University Press, Cambridge, Mass., 1982.*

This is the first of two volumes on the history of biology and its major ideas, written by one of the leading figures in the study of evolution. The introductory chapters provide an outstanding analysis of the philosophy and methodology of the biological sciences. This book, like Darwin's masterpiece, should at least be sampled by every serious student of biology.

MOOREHEAD, ALAN: *Darwin and the Beagle*, Harper & Row, Publishers, Inc., New York, 1969.*

A delightful narrative of Darwin's journey, beautifully illustrated with contemporary or near-contemporary drawings, paintings, and lithographs.

---

* Available in paperback.

PART **1**

Biology of Cells

SECTION **1**

The Unity of Life

This explosion in the sky, a supernova in astronomers' terms, was caused by the death of a star. About 170,000 years ago, the star, then 10 million years old, exhausted its fuel. During its life span, thermonuclear reactions, such as those now taking place in our own sun, had turned hydrogen to helium, and helium to carbon and oxygen, which themselves fused to even heavier elements. Once twenty times the size of the sun, the star cooled and, under the force of gravity, exploded inward. In this way, in the death of stars, all of the atoms of which our planet and its inhabitants are formed had their beginnings. This supernova, the first to be recorded in 383 years, was initially seen on February 24, 1987, by astronomers at a remote observatory in Chile.

CHAPTER 1

Atoms and Molecules

Our universe began, according to current theory, with an explosion that filled all space, with every particle of matter hurled away from every other particle. The temperature at the time of the explosion—some 10 to 20 billion years ago—was about 100,000,000,000 degrees Celsius (10^{11} °C). At this temperature, not even atoms could hold together; all matter was in the form of subatomic, elementary particles. Moving at enormous velocities, even these particles had fleeting lives. Colliding with great force, they annihilated one another, creating new particles and releasing more energy.

As the universe cooled, two types of stable particles, previously present only in relatively small amounts, began to assemble. (By this time, several hundred thousand years after the "big bang" is believed to have taken place, the temperature had dropped to a mere 2500°C, about the temperature of a white-hot wire in an incandescent light bulb.) These particles—protons and neutrons—are very heavy as subatomic particles go. Held together by forces that are still incompletely understood, they formed the central cores, or nuclei, of atoms. These nuclei, with their positively charged protons, attracted small, light, negatively charged particles—electrons—which moved rapidly around them. Thus, atoms came into being.

It is from these atoms—blown apart, formed, and re-formed over the course of several billion years—that all the stars and planets of our universe are formed, including our particular star and planet. And it is from the atoms present on this planet that living systems assembled themselves and evolved. Each atom in our own bodies had its origin in that enormous explosion 10 to 20 billion years ago. You and I are flesh and blood, but we are also stardust.

This text begins where life began, with the atom. At first, the universe aside, it might appear that lifeless atoms have little to do with biology. Bear with us, however. A closer look reveals that the activities we associate with being alive depend on combinations and exchanges between atoms, and the force that binds the electron to the atomic nucleus stores the energy that powers living systems.

ATOMS

All matter, including the most complex living organisms, is made up of combinations of **elements**. Elements are, by definition, substances that cannot be broken down into other substances by ordinary chemical means. The smallest particle of an element is an **atom**. There are 92 naturally occurring elements, each differing from the others in the structure of its atoms (Table 1–1).

The atoms of each different element have a characteristic number of positively charged particles, called protons, in their nuclei. For example, an atom of hydrogen, the lightest of the elements, has 1 proton in its nucleus; an atom of carbon has 6 protons in its nucleus. The number of protons in the nucleus of a

1–1 *"From so simple a beginning,"* Darwin wrote in Origin of Species, *"endless forms most beautiful and most wonderful have been, and are being evolved."* Among them is this oceanic worm, about one-tenth of a centimeter in length. Its tentacles, which are green, can be withdrawn into its body. The yellow mass is its digestive gland, and the adjacent coiling orange-yellow tube is its gut. Like the water that surrounds it, this tiny worm, formally known as* Poeobius meseres, *is made up of atoms created in the death of stars.*

TABLE 1-1   **Atomic Structure of Some Familiar Elements**

ELEMENT	SYMBOL	NUCLEUS		NUMBER OF ELECTRONS
		NUMBER OF PROTONS	NUMBER OF NEUTRONS*	
Hydrogen	H	1	0	1
Helium	He	2	2	2
Carbon	C	6	6	6
Nitrogen	N	7	7	7
Oxygen	O	8	8	8
Sodium	Na	11	12	11
Phosphorus	P	15	16	15
Sulfur	S	16	16	16
Chlorine	Cl	17	18	17
Potassium	K	19	20	19
Calcium	Ca	20	20	20

* In most common isotope.

particular atom is called its **atomic number.** The atomic number of hydrogen is therefore 1, and the atomic number of carbon is 6.

Outside the nucleus of an atom are negatively charged particles, the electrons, which are attracted by the positive charge of the protons. The number of electrons in an atom equals the number of protons in its nucleus. The electrons determine the chemical properties of atoms, and chemical reactions involve changes in the numbers and energy of these electrons.

Atoms also contain neutrons, which are uncharged particles of about the same weight as protons. These, too, are found in the nucleus of the atom, where they seem to have a stabilizing effect. The **atomic weight** of an element is essentially equal to the number of protons plus neutrons in the nuclei of its atoms. The atomic weight of carbon is 12, whereas that of hydrogen, which contains no neutrons, is 1. Electrons are so light by comparison to protons and neutrons that their weight is usually disregarded. When you weigh yourself, only about 30 grams—approximately 1 ounce—of your total weight is made up of electrons.

Isotopes

All atoms of a particular element have the same number of protons in their nuclei. Sometimes, however, different atoms of the same element contain different numbers of neutrons. These atoms, which therefore differ from one another in their atomic weights but not in their atomic numbers, are known as **isotopes** of the element. For example, three different isotopes of hydrogen exist (Table 1–2). The common form of hydrogen, with its one proton, has an atomic weight of 1 and is

TABLE 1-2   **Isotopes of Hydrogen**

ISOTOPE		ATOMIC NUMBER	ATOMIC WEIGHT	NUMBER OF PROTONS	NUMBER OF NEUTRONS	NUMBER OF ELECTRONS
NAME	SYMBOL					
Hydrogen	^1H	1	1	1	0	1
Deuterium	^2H	1	2	1	1	1
Tritium	^3H	1	3	1	2	1

1–2 *The dating of fossils by determining the relative proportions of different isotopes in the nearby or surrounding volcanic rocks is an important tool in tracing the course of evolutionary history. About 50 million years ago, this perch (Mioplosus), unable to swallow or dislodge the herring (Knightia) it had voraciously attacked, suffocated and sank to the bottom of the lake in which it lived. Lake sediments built up year after year and eventually dried, preserving the bones in limestone. These fossils were found in the Green River Formation in southwestern Wyoming in an area that is now a dry, rocky basin but for millions of years was covered by a lake surrounded by lush vegetation.*

symbolized as ^{1}H, or simply H. A second isotope of hydrogen, known as deuterium, contains one proton and one neutron and so has an atomic weight of 2; this isotope is symbolized as ^{2}H. Tritium, ^{3}H, a third, extremely rare isotope, has one proton and two neutrons and so has an atomic weight of 3. The chemical behavior of the two heavier isotopes is essentially the same as that of ordinary hydrogen—all three isotopes have only one electron each, and it is the electrons that determine chemical properties.

Most elements have several isotopic forms. The differences in weight, although very small, are sufficiently great that they can be detected with modern laboratory apparatus. Moreover, many, but not all, of the less common isotopes are radioactive. This means that the nucleus of the atom is unstable and emits energy as it changes to a more stable form. The energy given off by the nucleus of a radioactive isotope may be in the form of rapidly moving subatomic particles, electromagnetic radiation, or both. It can be detected with a Geiger counter or on photographic film.

Isotopes have a number of important uses in biological research and in medicine. They can be used, for example, to determine the age of fossils and of the rocks in which fossils are found (Figure 1–2). Each type of radioactive isotope emits energy and changes into another kind of isotope at a characteristic and fixed rate. As a result, the relative proportions of different isotopes in a rock sample give a good indication of how long ago that rock was formed.

Another use of radioactive isotopes is as "tracers." Since isotopes of the same element all have the same chemical properties, a radioactive isotope will behave in an organism just as its more common nonradioactive isotope does. As a result, biologists have been able to use isotopes of a number of elements—especially hydrogen, carbon, nitrogen, oxygen, and phosphorus—to trace the course of many essential processes in living organisms.

Isotopes play a role in the treatment of many forms of cancer, and they also have a number of diagnostic uses in medicine. For example, an isotope of the element thallium, which is unreactive in the human body, can be used to identify blocked blood vessels in persons with symptoms of heart disease. The isotope is first injected into the bloodstream. Then, while the patient exercises on a treadmill, the movement of the radioactive isotope is detected by a Geiger counter that is connected to a computer. The result is a "picture" of the distribution of the isotope in the heart muscle. If a blood vessel is blocked by fatty deposits, the isotope cannot penetrate the region of heart muscle supplied by that blood vessel. This procedure, which has no known side effects, provides an extremely reliable indication of the presence or absence of a common type of heart disease (see page 71).

The Signs of Life

What do we mean when we speak of "the evolution of life," or "life on other planets," or "when life began"? Actually, there is no simple definition. Life does not exist in the abstract; there is no "life," only living things. Moreover, there is no single, simple way to draw a sharp line between the living and the nonliving. There are, however, certain properties that, taken together, distinguish animate (that is, living) objects from inanimate ones.

(a) Living things are highly organized, as in this cross section of a stem of a sycamore sapling. This stem reflects the complicated organization of many different kinds of atoms into molecules and of molecules into complex structures. Such complexity of form, which is never found in inanimate objects of natural origin, makes possible the specialization of different parts of a living organism for different functions.

(a)

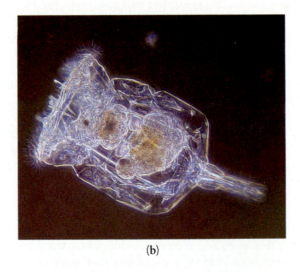

(b)

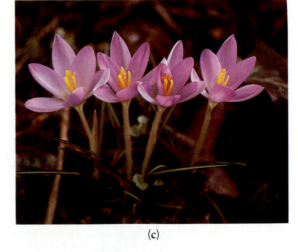

(c)

(b) Living things are homeostatic, which means simply "staying the same." Although they constantly exchange materials with the outside world, they maintain a relatively stable internal environment quite unlike that of their surroundings. Even this tiny, apparently fragile animal, a rotifer, has an internal chemical composition that differs from its changing environment.

(c) Living things reproduce themselves. They make more of themselves, generation after generation, with astonishing fidelity (and yet, as we shall see, with just enough variation to provide the raw material for evolution). Flowers, the familiar symbols of spring and romance, are the reproductive structures of the largest and most diverse group of plants.

(d) Living organisms grow and develop. Growth and development are the processes by which, for example, a single living cell, the fertilized egg, becomes a tree, or an elephant, or as shown here, a newborn zebra.

(e) Living things take energy from the environment and change it from one form to another. The pro-

(d)

(e)

(f)

(g)

cesses of energy conversion are highly specialized and remarkably efficient. This American bald eagle has converted chemical energy stored in its body to energy of motion used in catching a salmon. After the eagle has eaten and digested the salmon, the chemical energy stored in the body of the salmon will be available for the eagle's use.

(f) Living organisms respond to stimuli. When this fish jostled the tentacles hanging from the rim of the jellyfish's bell-shaped body, the response—injection

of a paralyzing substance into the fish's body—was immediate. As the fish lies helpless, attached to sticky strands from the tentacles, the fluttering mouth of the jellyfish responds to its presence.

(g) Living things are adapted. Moles, for instance, live underground in tunnels shoveled out by their large forepaws. Their eyes are small and almost sightless. Their noses, with which they sense the worms and other small invertebrates that make up their diet, are fleshy and enlarged.

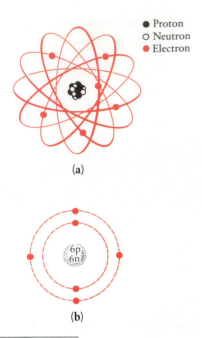

● Proton
○ Neutron
● Electron

(a)

(b)

1-3 *Two models of the carbon atom:* **(a)** *a planetary model and* **(b)** *a Bohr model.*

Models of Atomic Structure

The concept of the atom as the indivisible unit of the elements is almost 200 years old; however, our ideas about its structure have undergone many changes and may well undergo further changes in the future. These ideas, or hypotheses, are usually presented in the form of models, as are many other scientific hypotheses.

The earliest model of the atom, emphasizing its indivisibility, portrayed the atom as a sphere like a billiard ball. When it was realized that electrons could be removed from the atom, the billiard-ball model gave way to the plum-pudding model, in which the atom was represented as a solid, positively charged mass with negatively charged particles, the electrons, embedded in it. Subsequently, however, physicists found that an atom is, in fact, mostly empty space. The distance from electron to nucleus, experiments indicated, is about 1,000 times greater than the diameter of the nucleus; the electrons are so exceedingly small that the space is almost entirely empty. Thus the more familiar planetary model of the atom came into being, in which the electrons were depicted as moving in orbits around the nucleus (Figure 1–3a). Later, it was replaced by the Bohr model, named after physicist Niels Bohr. The Bohr model (Figure 1–3b) emphasized the fact that different electrons of an atom have different amounts of energy and are at different distances from the nucleus. As we shall see, neither the planetary model nor the Bohr model gives an accurate "picture" of an atom, and they have been superseded by another model, the orbital model (see Figure 1–6). However, the Bohr model can help us to understand certain properties of atoms that are of great importance in the chemistry of living systems.

ELECTRONS AND ENERGY

The distance of an electron from the nucleus is determined by the amount of **potential energy** (often called "energy of position") the electron possesses. The greater the amount of energy possessed by the electron, the farther it will be from the nucleus. Thus, an electron with a relatively small amount of energy is found close to the nucleus and is said to be at a low **energy level;** an electron with more energy is farther from the nucleus, at a higher energy level.

An analogy may be useful. A boulder resting on flat ground neither gains nor loses potential energy. If, however, you change its position by pushing it up a hill, you increase its potential energy. As long as it sits on the peak of the hill, the rock once more neither gains nor loses potential energy. If it rolls down the hill, however, potential energy is converted to energy of motion and released (Figure 1–4a). Similarly, water that has been pumped up to a water tank for storage has potential energy that will be released when the water runs back down.

The electron is like the boulder, or the water, in that an input of energy can move it to a higher energy level—farther away from the nucleus. As long as it

1-4 **(a)** *The energy used to push a boulder to the top of a hill (less the heat energy produced by friction between the boulder and hill) becomes potential energy, stored in the boulder as it rests at the top of the hill. This potential energy is converted to kinetic energy (energy of motion) as the boulder rolls downhill.*

(b) *When an atom, such as the hydrogen atom diagrammed here, receives an input of energy, an electron may be boosted to a higher energy level. The electron thus gains potential energy, which is released when the electron returns to its previous energy level.*

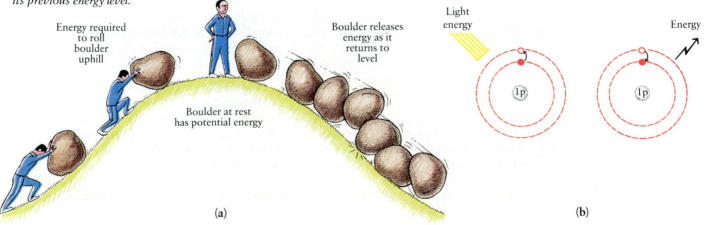

Energy required
to roll
boulder
uphill

Boulder at rest
has potential energy

Boulder releases
energy as it
returns to
level

Light
energy

Energy

(a)

(b)

remains at the higher energy level, it possesses the added energy. And, just as the rock is likely to roll, and the water to run, downhill, the electron also tends to go to its lowest possible energy level.

It takes energy to move a negatively charged electron farther away from a positively charged nucleus, just as it takes energy to push a rock up a hill. However, unlike the rock on the hill, the electron cannot be pushed partway up. With an input of energy, an electron can move from a lower energy level to any one of several higher energy levels, but it cannot move to an energy state somewhere in between. For an electron to move from one energy level to a higher one, it must absorb a discrete amount of energy, equal to the difference between the two particular energy levels. When the electron returns to its original energy level, that same amount of energy is released (Figure 1–4b). The discrete amount of energy involved in the transition between two energy levels is known as a quantum. Thus the study of electron movements is known as quantum mechanics, and the term "quantum jump," which has invaded our everyday discourse, refers to an abrupt, discontinuous movement from one level to another.

In the green cells of plants and algae, the radiant energy of sunlight raises electrons to a higher energy level. In a series of reactions, which will be described in Chapter 10, these electrons are passed "downhill" from one energy level to another until they return to their original energy level. During these transitions, the radiant energy of sunlight is transformed into the chemical energy on which life on earth depends.

The Arrangement of Electrons

At a given energy level, an electron moves around the nucleus at almost the speed of light. The electron is so small and moves so rapidly that it is theoretically impossible to determine, at any given moment, both its precise location and the exact amount of energy it possesses. As a result of this difficulty, the current model of atomic structure describes the pattern of the electron's motion rather than its position. The volume of space in which the electron will be found 90 percent of the time is defined as its **orbital.**

In any atom, the electrons at the lowest energy level—the first energy level—occupy a single spherical orbital, which can contain a maximum of two electrons (Figure 1–6a). Thus, for instance, hydrogen's single electron moves about the nucleus—90 percent of the time—within this single spherical orbital. Similarly, the two electrons of helium (atomic number 2) move within the single spherical orbital of the first energy level.

Atoms of higher atomic number than helium have more than two electrons. Since the first energy level is filled by two electrons, the additional electrons must occupy higher energy levels, farther from the nucleus. At the second energy level, there are four orbitals, each of which can hold a maximum of two electrons (Figure 1–6b). Thus, the second energy level can contain a total of eight electrons—and so can the third energy level of elements through atomic number 20 (calcium).

1–5 The leaves of these corn plants contain chlorophyll, which gives them their green color. When a packet of light energy —a photon—strikes a molecule of chlorophyll, electrons in the molecule are raised to higher energy levels. As each electron returns to its previous energy level, a portion of the energy released is captured in the bonds of carbon-containing molecules.

1–6 The most accurate representation of our knowledge of atomic structure is provided by orbital models. (a) The two electrons at the first energy level of an atom occupy a single spherical orbital. The nucleus is at the intersection of the axes.

(b) At the second energy level, there are four orbitals, each containing two electrons. One of these orbitals is spherical and the other three are dumbbell-shaped. The axes of the dumbbell-shaped orbitals are perpendicular to one another. The orbitals are shown individually in this diagram. In reality, the spherical orbital of the second energy level surrounds the orbital of the first energy level, and portions of the dumbbell-shaped orbitals pass through the two spherical orbitals. The orbitals influence one another and determine the overall shape of the atom.

First energy level:
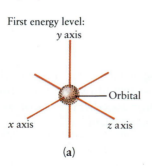
(a)

Second energy level:
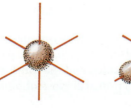
(b)

The way an atom reacts chemically is determined by the number and arrangement of its electrons. An atom is most stable when all of its electrons are at their lowest possible energy levels. Therefore, the electrons of an atom fill the energy levels in order—the first is filled before the second, the second before the third, and so on. Moreover, an atom in which the outermost energy level is completely filled with electrons is more stable than one in which the outer energy level is only partially filled. For example, helium (atomic number 2) has two electrons at the first energy level, which means that its outer energy level (in this case, also its lowest energy level) is completely filled. Helium is therefore extremely stable and tends to be unreactive. Similarly, neon (atomic number 10) has two electrons at the first energy level and eight at the second energy level; both energy levels are completely filled, and neon is unreactive. Helium, neon, and argon (atomic number 18) are called the "noble" gases because of their disdain for reacting with other elements.

In the atoms of most elements, however, the outer energy level is only partially filled (Table 1–3). These atoms tend to interact with other atoms in such a way that after the reaction both atoms have completely filled outer energy levels. Some atoms lose electrons; others gain electrons; and, in many of the most important chemical reactions that occur in living systems, atoms share their electrons with each other.

TABLE 1–3   **Electron Arrangements in Some Familiar Elements**

ELEMENT	ATOMIC NUMBER	NUMBER OF ELECTRONS IN EACH ENERGY LEVEL*			
		FIRST	SECOND	THIRD	FOURTH
Hydrogen (H)	1	1	—	—	—
Helium (He)	2	2	—	—	—
Carbon (C)	6	2	4	—	—
Nitrogen (N)	7	2	5	—	—
Oxygen (O)	8	2	6	—	—
Neon (Ne)	10	2	8	—	—
Sodium (Na)	11	2	8	1	—
Phosphorus (P)	15	2	8	5	—
Sulfur (S)	16	2	8	6	—
Chlorine (Cl)	17	2	8	7	—
Argon (Ar)	18	2	8	8	—
Potassium (K)	19	2	8	8	1
Calcium (Ca)	20	2	8	8	2

* The first energy level can hold a maximum of 2 electrons; the second energy level can hold a maximum of 8 electrons, as can the third energy level of the elements through atomic number 20 (calcium). In elements of higher atomic number, the third energy level has additional, inner orbitals that can hold a maximum of 10 more electrons.

BONDS AND MOLECULES

When atoms interact with one another, resulting in filled outer energy levels, new, larger particles are formed. These particles, consisting of two or more atoms, are known as **molecules,** and the forces that hold them together are known as **bonds.** There are two principal types of bonds: ionic and covalent.

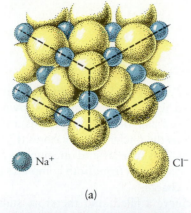

Na$^+$ Cl$^-$

(a)

(b)

1–7 (a) *Oppositely charged ions, such as the sodium and chloride ions depicted here as spheres, attract one another. Table salt is crystalline NaCl, a latticework of alternating Na$^+$ and Cl$^-$ ions held together by their opposite charges. Such bonds between oppositely charged ions are known as ionic bonds.*

(b) *The regularity of the latticework is reflected in the structure of salt crystals, magnified here about 14 times.*

Ionic Bonds

For many atoms, the simplest way to attain a completely filled outer energy level is either to gain or to lose one or two electrons. For example, chlorine (atomic number 17) needs one electron to complete its outer energy level (see Table 1–3). By contrast, sodium (atomic number 11) has a single electron in its outer energy level. This electron is strongly attracted by the chlorine atom and jumps from the sodium to the chlorine. As a result of this transfer, both atoms have outer energy levels that are completely filled, and all the electrons are at the lowest possible energy levels. In the process, however, the original atoms have become electrically charged. Such charged atoms are known as **ions.** The chlorine atom, having accepted an electron from sodium, now has one more electron than proton and is a negatively charged chloride ion: Cl$^-$. Conversely, the sodium ion has one less electron than proton and is positively charged: Na$^+$.

Because of their charges, positive and negative ions attract one another. Thus the sodium ion (Na$^+$) with its single positive charge is attracted to the chloride ion (Cl$^-$) with its single negative charge. The resulting substance, sodium chloride (NaCl), is ordinary table salt (Figure 1–7). Similarly, when a calcium atom (atomic number 20) loses two electrons, the resulting calcium ion (Ca^{2+}) can attract and hold two Cl$^-$ ions. Calcium chloride is identified in chemical shorthand as CaCl$_2$, with the subscript 2 indicating that two chloride ions are present for each ion of calcium.

Bonds that involve the mutual attraction of ions of opposite charge are known as **ionic bonds.** Such bonds can be quite strong, but, as we shall see in the next chapter, many ionic substances break apart easily in water, producing free ions. Small ions such as Na$^+$ and Cl$^-$ make up less than 1 percent of the weight of most living matter, but they play crucial roles. Potassium ion (K$^+$) is the principal positively charged ion in most organisms, and many essential biological processes occur only in its presence. Calcium ion (Ca^{2+}), K$^+$, and Na$^+$ are all involved in the production and propagation of the nerve impulse. In addition, Ca^{2+} is required for the contraction of muscles and for the maintenance of a normal heartbeat. Magnesium ion (Mg^{2+}) forms a part of the chlorophyll molecule, the molecule in green plants and algae that traps radiant energy from the sun.

1–8 *Ostriches, racing across the Etosha Pan in Namibia. Although these birds, the largest in the world, cannot fly, they are extremely graceful runners. Speeds of 40 to 60 kilometers per hour are not unusual. The movements of ostriches—like those of all complex animals—are the result of muscle contractions triggered by nerve impulses. Sodium, potassium, and calcium ions are involved in producing and propagating nerve impulses, and calcium ions are required for the contraction of muscle fibers.*

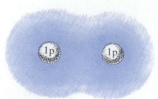

Hydrogen molecule (H$_2$)

1–9 *In a molecule of hydrogen, each atom shares its single electron with the other atom. As a result, both atoms effectively have a filled first energy level, containing two electrons—a highly stable arrangement. This type of bond, in which electrons are shared, is known as a covalent bond.*

Covalent Bonds

Another way for an atom to complete its outer energy level is by sharing electrons with another atom. Bonds formed by shared pairs of electrons are known as **covalent bonds.** In a covalent bond, the shared pair of electrons forms a new orbital (called a molecular orbital) that envelops the nuclei of both atoms (Figure 1–9). In such a bond, each electron spends part of its time around one nucleus and part of its time around the other. Thus the sharing of electrons both completes the outer energy level and neutralizes the nuclear charge.

Atoms that need to gain electrons to achieve a filled, and therefore stable, outer energy level have a strong tendency to form covalent bonds. Thus, for example, a hydrogen atom forms a single covalent bond with another hydrogen atom. It can also form a covalent bond with any other atom that needs to gain an electron to complete its outer energy level.

Of extraordinary importance in living systems is the capacity of carbon atoms to form covalent bonds. A carbon atom has four electrons in its outer energy level (see Table 1–3). It can share each of those electrons with another atom, forming covalent bonds with as many as four other atoms (Figure 1–10). The covalent bonds formed by a carbon atom may be with different atoms (most frequently hydrogen, oxygen, and nitrogen) or with other carbon atoms. As we shall see in Chapter 3, this tendency of carbon atoms to form covalent bonds with other carbon atoms gives rise to the large molecules that form the structures of living organisms and that participate in essential life processes.

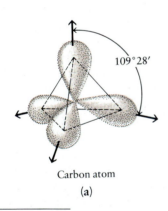

Carbon atom

(a)

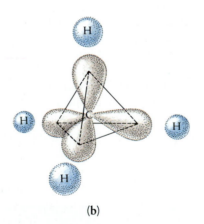

(b)

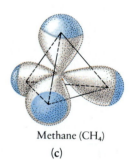

Methane (CH$_4$)

(c)

1–10 **(a)** *When a carbon atom forms covalent bonds with four other atoms, the electrons in its outer energy level form new orbitals. These new orbitals, which are all the same shape, are oriented toward the*

four corners of a tetrahedron. Thus the four orbitals are separated as far as possible. **(b)** *When a carbon atom reacts with four hydrogen atoms, each of the electrons in its outer energy level forms a covalent*

bond with the single electron of one hydrogen atom, producing a methane molecule **(c)**. *Each pair of electrons moves in a new, molecular orbital. The molecule has the shape of a tetrahedron.*

Polar Covalent Bonds

The atomic nuclei of different elements have different degrees of attraction for electrons. Factors that determine the strength with which a nucleus attracts its outer electrons include the number of protons it contains, the closeness of the outer electrons to the nucleus, and the number of other, "shielding" electrons between the nucleus and the outer electrons. In covalent bonds formed between atoms of different elements, the electrons are not shared equally between the atoms involved—instead, the shared electrons tend to spend more time around the nucleus with the greater attraction. The atom around which the electrons

Slightly positive charge — Slightly negative charge

Hydrogen chloride (HCl)

1–11 *In a polar molecule, such as hydrogen chloride (HCl), the shared electrons tend to spend more time around one atom, in this case, the chlorine atom, than around the other atom. As a result, the atom that attracts the electrons more strongly (chlorine) has a slightly negative charge, and the atom that attracts the electrons less strongly (hydrogen) has a slightly positive charge.*

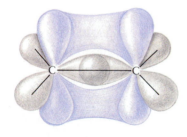

1–12 *An orbital model of a carbon-carbon double bond. One pair of electrons occupies the inner orbital between the two carbon atoms. The other pair of electrons occupies the outer orbital, which has two phases, one above the plane and one below. This creates a rigid bond, about which the atoms cannot rotate. Each of the two smaller orbitals (shown in gray) extending from each carbon atom contains one electron and can form a covalent bond with another atom. Thus two covalent bonds can be formed at each end of this structure.*

spend more time has a slightly negative charge; the other atom has a slightly positive charge, since the electrons spend less time around it and thus its nuclear charge is not entirely neutralized (Figure 1–11).

Covalent bonds in which electrons are shared unequally are known as **polar covalent bonds,** and the molecules containing these bonds are said to be **polar molecules.** Such molecules often contain oxygen atoms, to which electrons are strongly attracted. The polar properties of many oxygen-containing molecules have very important consequences for living things. For example, many of the special properties of water (H_2O), upon which life depends, derive largely from its polar nature, as we shall see in the next chapter.

Ionic, polar covalent, and covalent bonds actually may be considered different versions of the same type of bond. The differences depend on the differing attractions of the combining atoms for electrons. In a wholly nonpolar covalent bond, the electrons are shared equally; such bonds can exist only between identical atoms, H_2, Cl_2, O_2, and N_2, for example. In polar covalent bonds, electrons are shared unequally, and in ionic bonds, there is an electrostatic attraction between the negatively and positively charged ions as a result of their having previously gained or lost electrons.

Double and Triple Bonds

There are various ways in which atoms can participate in covalent bonds and fill their outer energy levels. Oxygen, for example, has six electrons in its outer energy level. Four of these electrons are grouped into two pairs and are generally unavailable for covalent bonding; the other two electrons are unpaired, and each can be shared with another atom in a covalent bond. In the water molecule (H_2O), one of these electrons participates in a covalent bond with one hydrogen atom, and the other in a covalent bond with a different hydrogen atom. Two single bonds are formed, and all three atoms have filled outer energy levels.

The bonding situation is different in another familiar substance, carbon dioxide (CO_2). In this molecule, the two available electrons of each oxygen atom participate with two electrons of a *single* carbon atom in the formation of *two* covalent bonds. Each oxygen atom is joined to the central carbon atom by two pairs of electrons (four electrons). Such bonds are called **double bonds,** and they are symbolized in a structural formula by two lines connecting the atomic symbols: O=C=O. Carbon atoms can form double or even triple bonds (in which three pairs of electrons are shared) with each other as well as with other atoms, and so the variety of kinds of molecules that carbon can form is very large.

Electrons shared in double and triple bonds form orbitals that differ in shape from the orbitals filled by single electron pairs. For instance, when four single bonds satisfy the electron requirements of carbon, they will be directed toward the four corners of a tetrahedron that has the carbon atom at its center, as we saw in Figure 1–10. When two bonds are replaced by a double bond, the remaining single bonds form the arms of a Y with the double bond as its leg (Figure 1–12). When two double bonds are made by a single carbon atom, as in carbon dioxide, the three bonded atoms lie in a straight line.

The symmetry of the carbon dioxide molecule has an important consequence. The double covalent bonds in carbon dioxide, like all covalent bonds between nonidentical atoms, are polar. However, because the molecule is perfectly symmetrical, the electrons are pulled in opposite directions by the two oxygen atoms, cancelling out the unequal distribution of charge. As a result, the carbon dioxide molecule is nonpolar. Similarly, the symmetry of the methane molecule (see Figure 1–10) produces a nonpolar molecule, even though the individual bonds between the atoms are polar.

Single bonds are flexible, leaving atoms free to rotate in relation to one another. Double and triple bonds hold the atoms relatively rigid in relation to one another. The presence of double bonds in a molecule can make a significant difference in its properties. For example, both fats and oils are composed of carbon and hydrogen atoms covalently bonded together, but in fats the bonds are all single and in oils some of the bonds are double. The rigidity caused by these double bonds prevents the oil molecules from packing together, and, as a consequence, oils are liquids at room temperature. In fats, by contrast, the molecules can bend and twist, fitting closely together in a solid structure at room temperature.

CHEMICAL REACTIONS

Chemical reactions—exchanges of electrons among atoms—can be compactly described by chemical equations. For example, the equation for the formation of sodium chloride is

$$Na^+ + Cl^- \longrightarrow NaCl$$

The arrow in the equation designates "forms" or "yields" and shows the direction of chemical change. Like algebraic equations, chemical equations "balance"; that is, there are the same number of atoms in the products of the reaction as in the original reactants. To take a slightly more complex example, hydrogen gas can combine with oxygen gas to produce water. Hydrogen gas is H_2, and oxygen gas is O_2. However, we know that each molecule of water contains two atoms of hydrogen and one of oxygen, and therefore the proportions must be two to one:

$$2H_2 + O_2 \longrightarrow 2H_2O$$

Two molecules of H_2 plus one molecule of O_2 yield two molecules of water. The equation for a chemical reaction thus tells the kinds of atoms that are present, their proportions, and the direction of the reaction.

A substance consisting of molecules that contain atoms of two or more different elements, held together in a definite and constant proportion by chemical bonds, is known as a chemical **compound.** Examples of chemical compounds include water (H_2O), sodium chloride (NaCl), carbon dioxide (CO_2), methane (CH_4), and glucose ($C_6H_{12}O_6$).

Types of Reactions

The multitude of chemical reactions that occur in both the animate and the inanimate worlds can be classified into a few general types. One type of reaction is a simple combination, represented by the expression:

$$A + B \longrightarrow AB$$

Examples of this type of reaction are the combination of sodium ions and chloride ions to form sodium chloride and the combination of hydrogen gas with oxygen gas to produce water.

A reaction may also take the form of a dissociation:

$$AB \longrightarrow A + B$$

For example, the earlier equation showing the formation of water can be reversed:

$$2H_2O \longrightarrow 2H_2 + O_2$$

This means that water molecules yield hydrogen and oxygen gases.

A reaction may also involve an exchange, taking the form:

$$AB + CD \longrightarrow AD + CB$$

An example of such an exchange occurs when the chemical compounds sodium hydroxide (NaOH) and hydrochloric acid (HCl) react, producing table salt and water:

$$NaOH + HCl \longrightarrow NaCl + H_2O$$

As we pursue our study of living organisms, we shall encounter numerous examples of these three general types of chemical reactions.

THE BIOLOGICALLY IMPORTANT ELEMENTS

Of the 92 naturally occurring elements, only six make up some 99 percent of all living tissue (Table 1–4). These six elements are carbon, hydrogen, nitrogen, oxygen, phosphorus, and sulfur, conveniently remembered as CHNOPS. These are not the most abundant of the elements of the earth's surface. Why, as life assembled and evolved from stardust, were these of such importance? One clue is that the atoms of all of these elements need to gain electrons to complete their outer energy levels (see Table 1–3). Thus they generally form covalent bonds. Because these atoms are small, the shared electrons in the bonds are held closely to the nuclei, producing very stable molecules. Moreover, with the exception of hydrogen, atoms of these elements can all form bonds with two or more atoms, making possible the formation of the large and complex molecules essential for the structures and functions of living systems.

TABLE 1–4 **Atomic Composition of Three Representative Organisms**

ELEMENT	HUMAN	ALFALFA	BACTERIUM
Carbon	19.37%	11.34%	12.14%
Hydrogen	9.31	8.72	9.94
Nitrogen	5.14	0.83	3.04
Oxygen	62.81	77.90	73.68
Phosphorus	0.63	0.71	0.60
Sulfur	0.64	0.10	0.32
CHNOPS total:	97.90%	99.60%	99.72%

LEVELS OF BIOLOGICAL ORGANIZATION

One of the fundamental principles of biology is that living things obey the laws of physics and chemistry. Organisms are made up of the same chemical components —atoms and molecules—as nonliving things. This does not mean, however, that organisms are "nothing but" the atoms and molecules of which they are composed. As we have seen, there are recognizable differences between living and nonliving systems. To understand the basis of these differences, let us consider the most thoroughly studied of all living things, the bacterium *Escherichia coli.**

---

* Biologists use a binomial ("two-name") system for designating organisms. Every different kind of organism has a unique two-part name. The first part of the name refers to the genus (plural, genera) to which the organism belongs; the second part, in combination with the first part, refers to the particular species, a subdivision of the genus category. In this name, for example, *Escherichia* denotes the genus, while *coli* designates a particular kind, or species, of *Escherichia,* distinguished from all others by certain characteristics. By convention, with the second mention of a scientific binomial, it is permissible to abbreviate the first (genus) name. This is fortunate, particularly when dealing with such names as *Escherichia.*

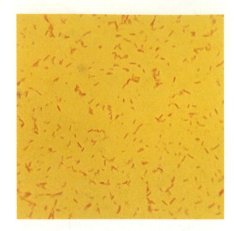

1-13 *Cells of* Escherichia coli, *as photographed through a light microscope. They have been stained with a dye that adheres to their surface, making them easier to see. Although these cells, magnified 450 times, are minute, their structure is quite complex and they exhibit all of the properties that characterize a living system.*

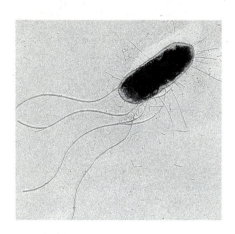

1-14 *An* E. coli *cell, magnified 11,280 times by the electron microscope. The short, rigid structures extending from the surface are used by the cell to attach to a food source or, when exchanging genetic information, to another* E. coli *cell. The longer, more flexible structures, which rotate at about 40 revolutions per second, propel the cell through the surrounding medium. When they rotate in a counterclockwise direction, the cell moves forward; when they rotate in a clockwise direction, it tumbles through the water and then, with the return of counterclockwise rotation, moves forward again.*

The atoms constituting this bacterium (see Table 1–4) are combined with each other in very specific ways. Much of the hydrogen and oxygen is present in the form of water, which accounts for most of the weight of *E. coli*. In addition to water, each bacterium contains about 5,000 different kinds of macromolecules (very large molecules). Some of these play structural roles, others regulate cell function, and nearly 1,000 of them are involved in decoding the genetic information. Some of the macromolecules interact with water to form a delicate, pliant film that encloses all the other atoms and molecules of which *E. coli* is composed. So enclosed, they constitute, remarkably, a cell, a living entity.

An *E. coli* cell is very small, appearing no bigger than a hyphen even when magnified by the most powerful light microscope, but it possesses some astonishing capacities. Like other living things, it can transform energy, taking molecules from the medium and using them to fuel its processes of growth and reproduction. It can exchange genetic information with other *E. coli* cells. It can move, propelling itself by the rotation of thin, flexible fibers attached to a structure that resembles—but considerably predates—an automobile transmission. The direction of motion is not random; *E. coli*, small as it is, has a number of different sensing devices, enabling it to detect and move toward nutrients and away from noxious substances.

E. coli is one of the most common of microscopic organisms. Its preferred residence is the human intestinal tract, where it lives in close association with the cells that form the lining of that tract. These human cells resemble *E. coli* in many important ways: they contain about the same proportions of the same six kinds of atoms, and, as in *E. coli*, these atoms are organized into macromolecules. However, the human cells are also very different from *E. coli*. For one thing, they are much larger; for another, they are much more complex. Most important, they are not independent entities, like the *E. coli* cells; each is part of a larger organism. Individual cells are specialized for particular functions that are subservient to the function of the organism as a whole. Each intestinal lining cell lives for only a few days; the organism, with any luck, will live for many decades.

E. coli, the cells of its human host, and other microorganisms living in the intestinal tract all interact with one another. Usually this takes place so uneventfully that we are unaware of the interactions, but occasionally we are reminded of the delicate balance. For example, many of us have had the experience of taking an antibiotic to cure one type of infection and ending up with another infection—usually caused by a type of yeast cell. What has happened is that the antibiotic has killed not only the bacteria causing our initial infection but also *E. coli* and the other normal inhabitants of our intestinal tract. Yeast cells are not susceptible to the antibiotic, and so they take over the territory, in much the same way that certain species of plants will quickly take over any patch of countryside from which the usual vegetation is removed.

E. coli and the cells with which it interacts illustrate what are known as levels of organization. The first level of organization with which biologists are usually concerned is subatomic—the particles that form atoms. The organization of these particles into atoms represents a second level, and the organization of atoms into molecules represents a third level. Although each level consists of components from the preceding level, the new organization of the components at a given level results in the emergence of new properties that are quite different from those of the preceding level. For example, hydrogen and oxygen are gases at ordinary temperatures, yet water—composed of hydrogen and oxygen—is a liquid with properties very different from those of either gas.

At a fourth level of organization, the most remarkable property of all emerges—life, in the form of a cell. Other properties emerge when individual, specialized cells are organized at a still higher level, in a multicellular organism. Organized in

1–15 *A gallery of cells.* (a) Amoeba proteus, *a single-celled organism named for*
Proteus, *a Greek god capable of changing*
his shape at will. Extensions of the cell,
known as pseudopodia, enable amoebas
to move and to capture prey.

(b) This simple organism, called Pan-
dorina, *is made up of 32 cells, most of*
which are visible here, held together by a
jellylike substance, whose outlines you can
also see. Each of these cells can survive
independently of the others. To reproduce,
each cell divides, producing a new cell
inside, and then the parent colony breaks
apart.

(c) The embryo of a sea urchin at the
two-celled stage. Within each cell is a
nucleus that carries all the genetic infor-
mation needed for every cell in the mature
sea urchin.

(d) These cells are from the cerebral
cortex of a human brain—the most highly
organized structure on earth. The actions
of the cells of the cerebral cortex and the
interconnections among them are respon-
sible for consciousness, intelligence,
dreams, and memory.

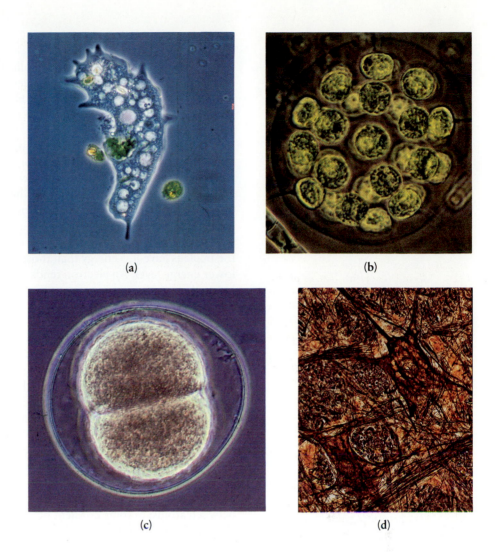

(a)   (b)

(c)   (d)

one way, the cells form a liver; in another way, the intestinal tract; in yet another,
the human brain, which represents an extraordinary level of organizational
complexity. Yet it, in turn, is only part of a larger entity whose characteristics are
different from those of the brain, although they depend on the characteristics of
the brain. Nor is the individual organism the ultimate level of biological order.
Living organisms interact with each other, and finally, groups of living organisms
are themselves part of an even vaster system of organization. This ultimate level of
organization, the **biosphere,** involves not only the great diversity of plants and
animals and microorganisms and their interactions with each other but also the
physical characteristics of the environment and the planet earth itself.

The organization of this text parallels the levels of biological organization. In
Part 1, we are beginning with atoms and molecules, and we will move on to
examine the structure and activities of the living cell. In Part 2, our focus will be on
individual organisms as we first consider their diversity and then, in more detail,
the essential characteristics of plants and animals. Our view will expand further in
Part 3, as we look at the interactions of organisms with each other. Over long
periods of time these interactions give rise to evolutionary change; on a shorter
time scale, they determine the organization of the communities of living organisms we find around us.

SUMMARY

Matter is composed of atoms, the smallest units of chemical elements. Atoms are made up of smaller particles. The nucleus of an atom contains positively charged protons and (except for hydrogen, ^1H) neutrons, which have no charge. The atomic number of an atom is equal to the number of protons in its nucleus. The atomic weight of an atom is the sum of the number of protons and neutrons in its nucleus. The chemical properties of an atom are determined by its electrons—small, negatively charged particles found outside the nucleus. The number of electrons in an atom equals the number of protons and thus the atomic number.

The nuclei of different isotopes of the same element contain the same number of protons but different numbers of neutrons. Thus the isotopes of an element have the same atomic number but different atomic weights.

The electrons of an atom have differing amounts of energy. Electrons closer to the nucleus have less energy than those farther from the nucleus and thus are at a lower energy level. An electron tends to occupy the lowest available energy level, but with an input of energy, it can be boosted to a higher energy level. When the electron returns to a lower energy level, energy is released.

The chemical behavior of an atom is determined by the number and arrangement of its electrons. An atom is most stable when all of its electrons are at their lowest possible energy levels and those energy levels are completely filled with electrons. The first energy level can hold two electrons, the second energy level can hold eight electrons, and so can the third energy level of the small atoms of greatest interest in biology. Chemical reactions between atoms result from the tendency of atoms to reach the most stable electron arrangement possible.

Particles consisting of two or more atoms are known as molecules, which are held together by chemical bonds. Two common types of bonds are ionic and covalent. Ionic bonds are formed by the mutual attraction of particles of opposite electric charge; such particles, formed when an electron jumps from one atom to another, are known as ions. In covalent bonds, pairs of electrons are shared between atoms; in some covalent bonds, known as polar covalent bonds, pairs of electrons are shared unequally, giving the molecule regions of positive and negative charge. Covalent bonds in which two atoms share two pairs of electrons (four electrons) are known as double bonds, and those in which they share three pairs of electrons (six electrons) are known as triple bonds.

Chemical reactions—exchanges of electrons among atoms—can be represented by chemical equations. Three general types of chemical reactions are (1) the combination of two or more substances to form a different substance, (2) the dissociation of a substance into two or more substances, and (3) the exchange of atoms among two or more substances. Substances that consist of the atoms of two or more different elements, in definite and constant proportions, are known as chemical compounds.

Living things are made up of the same chemical and physical components as nonliving things, and they obey the same chemical and physical laws. Six elements (CHNOPS) make up 99 percent of all living matter. The atoms of all of these elements are small and form tight, stable covalent bonds. With the exception of hydrogen, they can all form covalent bonds with two or more atoms, giving rise to the complex molecules that characterize living systems.

The properties of a complex molecule depend upon the organization of the atoms within the molecule. Similarly, the properties of a living cell depend upon the organization of molecules within the cell, and the properties of a multicellular organism depend upon the organization of the cells within its body. The ultimate level of biological organization, the biosphere, results from the interactions of the plants, animals, and microorganisms of the earth with each other and with physical factors in the environment.

QUESTIONS

1. Describe the three types of particles of which atoms are composed. What is the atomic number of an atom? The atomic weight?

2. For each of the following isotopes, determine the number of protons and neutrons in the nucleus: (a) ^{11}C, ^{12}C, ^{14}C; (b) ^{31}P, ^{32}P, ^{33}P; (c) ^{32}S, ^{35}S, ^{38}S.

3. Consider the isotopes of phosphorus listed in Question 2. Would you expect all three of these isotopes to exhibit the same chemical properties in a living organism? Why or why not?

4. Although no model of the atom gives us an exact "picture," different models can help us to understand important characteristics of atoms. What characteristic of the atom was stressed by the planetary model? What important characteristic of electrons is emphasized by the Bohr model? What additional information about electrons is provided by the orbital model?

5. The street lights in many cities contain bulbs filled with sodium vapor. When electrical energy is passed through the bulb, a brilliant yellow light is given off. What is happening to the sodium atoms to cause this?

6. What is the difference between an energy level and an orbital? How many electrons can the first energy level of an atom hold? The second energy level? The third energy level?

7. Determine the number of protons, the number of neutrons, the number of energy levels, and the number of electrons in the outermost energy level in each of the following atoms: oxygen, nitrogen, carbon, sulfur, phosphorus, chlorine, potassium, and calcium.

8. How many electrons does each of the atoms in Question 7 need to share, gain, or lose to acquire a completed outer energy level?

9. Magnesium has an atomic number of 12. How many electrons are in its first energy level? Its second energy level? Its third energy level? How would you expect magnesium and chlorine to interact? Write the formula for magnesium chloride.

10. Explain the differences between ionic, covalent, and polar covalent bonds. What tendency of atoms causes them to interact with each other, forming bonds?

11. Molecules that contain polar covalent bonds typically have regions of positive and negative charge and thus are polar. However, some molecules containing polar covalent bonds are nonpolar. Explain how this is possible.

12. Knowing that chemical reactions have to be balanced, fill the appropriate numbers into the underlined spaces (*hint:* from 1 to 6 in all cases):

(a) ____H_2CO_3 $\longrightarrow$ ____H_2O + ____CO_2
     Carbonic
     acid

(b) ____H_2 + ____N_2 $\longrightarrow$ ____NH_3
                       Ammonia

(c) ____$NaOH$ + ____H_2CO_3 $\longrightarrow$ ____Na_2CO_3 + ____H_2O
     Sodium   Sodium
     hydroxide   carbonate

(d) ____CH_3OH + ____O_2 $\longrightarrow$ ____CO_2 + ____H_2O
     Methyl
     alcohol

(e) ____O_2 + ____$C_6H_{12}O_6$ $\longrightarrow$ ____H_2O + ____CO_2
                Glucose

13. What six elements make up the bulk of living tissue? What characteristics do the atoms of these six elements share?

Water

In this chapter and the next, we are going to examine the molecules of which living things are composed. By far the most abundant of these molecules is water, which makes up 50 to 95 percent of the weight of any functioning living system.

Life on this planet began in water, and today, wherever liquid water is found, life is also present. There are one-celled organisms that eke out their entire existence in no more water than can cling to a grain of sand. Some kinds of algae are found only on the melting undersurfaces of polar ice floes. Certain bacteria can tolerate the near-boiling water of hot springs. In the desert, plants race through an entire life cycle—seed to flower to seed—following a single rainfall. In the tropical rain forest, the water cupped in the leaves of a plant forms a microcosm in which a myriad of small organisms are born, spawn, and die.

Water is the most common liquid on earth. Three-fourths of the surface of the earth is covered by water. In fact, if the earth's land surface were absolutely smooth, all of it would be 2.5 kilometers* under water. But do not mistake "common" for "ordinary"; water is not in the least an ordinary liquid. Compared with other liquids it is, in fact, quite extraordinary. If it were not, it is unlikely that life on earth could ever have evolved.

---

* A metric table with English equivalents is provided in Appendix A.

2–1 *The first living systems came into being, according to present hypotheses, in the warm primitive seas, and for many organisms, ourselves included, each new individual begins life bathed and cradled in water. These are salamander larvae.*

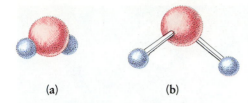

(a)   **(b)**

2-2 *The structure of the water molecule (H₂O) can be depicted in several different ways.* **(a)** *In the space-filling model, the oxygen atom is represented by the red sphere and the hydrogen atoms by the blue spheres. Because of its simplicity, this model is often used as a convenient symbol of the water molecule.*

(b) *The ball-and-stick model emphasizes that the atoms are joined by covalent bonds; it also gives some indication of the geometry of the molecule. A more accurate description of the molecule's shape is provided by the orbital model in Figure 2–3a.*

THE STRUCTURE OF WATER

In order to understand why water is so extraordinary and how, as a consequence, it can play its unique and crucial role in relation to living systems, we must look again at its molecular structure. Each water molecule is made up of two atoms of hydrogen and one atom of oxygen (Figure 2–2). Each of the hydrogen atoms is linked to the oxygen atom by a covalent bond; that is, the single electron of each hydrogen atom is shared with the oxygen atom, which also contributes an electron to each bond.

The water molecule as a whole is neutral in charge, having an equal number of electrons and protons. However, the molecule is polar (page 33). Because of the very strong attraction of the oxygen nucleus for electrons, the shared electrons of the covalent bonds spend more time around the oxygen nucleus than they do around the hydrogen nuclei. As a consequence, the region near each hydrogen nucleus is a weakly positive zone. Moreover, the oxygen atom has four additional electrons in its outer energy level. These electrons are paired in two orbitals that are not involved in covalent bonding to hydrogen. Each of these orbitals is a weakly negative zone. Thus, the water molecule, in terms of its polarity, is four-cornered, with two positively charged "corners" and two negatively charged ones (Figure 2–3a).

When one of these charged regions comes close to an oppositely charged region of another water molecule, the force of attraction forms a bond between them, which is known as a **hydrogen bond.** Hydrogen bonds are found not only in water but also in many large molecules, where they help maintain structural stability. They are, however, very specific. A hydrogen bond can form only between a hydrogen atom that is covalently bonded to an atom that has a strong attraction for electrons and a strongly electron-attracting atom in another molecule. In the molecules found in living systems, hydrogen bonding typically occurs between a hydrogen atom covalently bonded to either oxygen or nitrogen and the oxygen or nitrogen atom of another molecule. In water, a hydrogen bond forms between a negative "corner" of one water molecule and a positive "corner" of another. Every water molecule can establish hydrogen bonds with four other water molecules (Figure 2–3b).

Any single hydrogen bond is significantly weaker than either a covalent or an ionic bond. Moreover, it has an exceedingly short lifetime; on an average, each hydrogen bond in liquid water lasts approximately 1/100,000,000,000th of a second. But, as one is broken, another is made. All together, the hydrogen bonds have considerable strength, causing the water molecules to cling together as a liquid under ordinary conditions of temperature and pressure.

Now let us look at some of the consequences of these attractions among water molecules, especially as they affect living systems.

2-3 *The polarity of the water molecule and its consequences.* **(a)** *As shown in this model, four orbitals branch off from the oxygen nucleus of a water molecule. Two of the orbitals are formed by the shared electrons bonding the hydrogen atoms to the oxygen atom. They have a slightly positive charge. The other two orbitals have a slightly negative charge.*

(b) *As a result of these positive and negative zones, each water molecule can form hydrogen bonds (dashed lines) with four other water molecules. Under ordinary conditions of pressure and temperature, the hydrogen bonds are continually breaking and re-forming in a shifting pattern. Thus water is a liquid.*

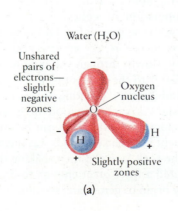

Water (H₂O)

Unshared
pairs of
electrons—
slightly
negative
zones

Oxygen
nucleus

Slightly positive
zones

(a)

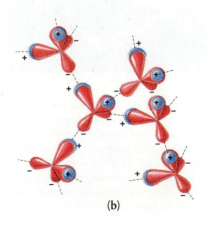

(b)

2-4 *This remarkable photo shows a kingfisher, just as it breaks the surface of the water in a dive for food. Note the many droplets surrounding the diving bird and the continuous sheet formed by the surface of the water. These are both results of the surface tension of water.*

CONSEQUENCES OF THE HYDROGEN BOND

Surface Tension

Look at water dripping from a faucet. Each drop clings to the rim and dangles for a moment by a thread of water; then just as the tug of gravity breaks it loose, its outer surface is drawn taut, to form a sphere as the drop falls free. Gently place a needle or a razor blade flat on the surface of the water in a glass. Although the metal is denser than water, it floats. Look at a pond in spring or summer; you will see water striders and other insects walking on its surface almost as if it were solid. These phenomena are all the result of **surface tension.** Surface tension is a result of the cohesion, or clinging together, of the water molecules. (**Cohesion** is, by definition, the holding together of molecules of the same substance. **Adhesion** is the holding together of molecules of different substances.)

The only liquid with a surface tension greater than that of water is mercury. Atoms of mercury are so greatly attracted to one another that they tend not to adhere to anything else. Water, however, because of its negative and positive charges, adheres strongly to any other charged molecules and to charged surfaces. The "wetting" capacity of water—that is, its ability to coat a surface—results from its polar structure, as does its cohesiveness.

Capillary Action and Imbibition

If you hold two dry glass slides together and dip one corner in water, the combination of cohesion and adhesion will cause water to spread upward between the two slides. This is **capillary action.** Capillary action similarly causes water to rise in very fine glass tubes, to creep up a piece of blotting paper, or to move slowly through the minute spaces between soil particles and so become available to the roots of plants.

Imbibition ("drinking up") is the capillary movement of water molecules into substances such as wood or gelatin, which swell as a result. The pressures developed by imbibition can be astonishingly great. It is said that stone for the ancient Egyptian pyramids was quarried by driving wooden pegs into holes drilled in the rock face and then soaking the pegs with water. The swelling of the wood created a force great enough to break the stone slab free. Seeds imbibe water as they begin to germinate, swelling and bursting their seed coats (Figure 2–5).

2-5 *The germination of seeds begins with changes in the seed coat that permit a massive uptake of water. The embryo and surrounding structures then swell, bursting the seed coat. In this acorn, photographed on the forest floor, the embryonic root has emerged through the tough outer layers of the fruit.*

Resistance to Temperature Change

If you go swimming in the ocean or a lake on one of the first hot days of summer, you will quickly be aware of a striking difference between the air temperature and the water temperature. This difference occurs because a greater input of energy is required to raise the temperature of water than to raise the temperature of air. The amount of heat a given amount of a substance requires for a given increase in temperature is its **specific heat** (also called heat capacity). One calorie* is defined as the amount of heat that will raise the temperature of 1 gram (1 milliliter or 1 cubic centimeter) of water 1°C. The specific heat of water is about twice the specific heat of oil or alcohol; that is, approximately 0.5 calorie will raise the temperature of 1 gram of oil or alcohol 1°C. It is four times the specific heat of air or aluminum and 10 times that of iron. Only liquid ammonia has a higher specific heat (Table 2–1).

Heat is a form of energy—the **kinetic energy,** or energy of motion, of molecules. Molecules are always moving; they vibrate, rotate, and shift position in relation to other molecules. Heat, which is measured in calories, reflects the *total* kinetic energy in a collection of molecules; it includes both the magnitude of the molecular movements and the mass and number of moving molecules present. By contrast, temperature, which is measured in degrees, reflects the *average* kinetic energy of the molecules. Thus, heat and temperature are not identical. For example, a lake may have a lower temperature than does a bird flying over it, but the lake contains more heat because it has many more molecules in motion.

The high specific heat of water is a consequence of hydrogen bonding. The hydrogen bonds in water tend to restrict the movement of the molecules. In order for the kinetic energy of water molecules to increase sufficiently for the temperature to rise 1°C, it is necessary first to rupture a number of the hydrogen bonds holding the molecules together. When you heat a pot of water, much of the heat energy added to the water is used in breaking the hydrogen bonds between the water molecules. Only a relatively small amount of heat energy is therefore available to increase molecular movement.

What does the high specific heat of water mean in biological terms? It means that for a given rate of heat input, the temperature of water will rise more slowly than the temperature of almost any other material. Conversely, the temperature will drop more slowly as heat is removed. Because so much heat input or heat loss is required to raise or lower the temperature of water, organisms that live in the oceans or large bodies of fresh water live in an environment where the temperature is relatively constant. Also, the high water content of terrestrial plants and animals helps them to maintain a relatively constant internal temperature. This constancy of temperature is critical because biologically important chemical reactions take place only within a narrow temperature range.

Vaporization

Vaporization—or evaporation, as it is more commonly called—is the change from a liquid to a gas. Water has a high **heat of vaporization.** At water's boiling point (100°C at a pressure of 1 atmosphere), it takes 540 calories to change 1 gram of liquid water into vapor, almost 60 times as much as for ether and twice as much as for ammonia (Table 2–2).

Hydrogen bonding is also responsible for water's high heat of vaporization. Vaporization comes about because some of the most rapidly moving molecules of a liquid break loose from the surface and enter the air. The hotter the liquid, the more rapid the movement of its molecules and, hence, the more rapid the rate of

TABLE 2–1 Comparative Specific Heats (The quantity of heat, in calories, required to raise the temperature of 1 gram through 1°C)

SUBSTANCE	SPECIFIC HEAT
Liquid ammonia	1.23
Water	1.00
Ethyl alcohol (ethanol)	0.60
Sugar (sucrose)	0.30
Chloroform	0.24
Salt (NaCl)	0.21
Glass	0.20
Iron	0.10
Lead	0.03

TABLE 2–2 Comparative Heats of Vaporization (The quantity of heat, in calories, required to convert 1 gram of liquid to 1 gram of gas)

LIQUID	HEAT REQUIRED
Water (at 0°C)	596
Water (at 100°C)	540
Hydrofluoric acid	360
Ammonia	295
Ethyl alcohol (ethanol)	236.5
Nitric acid	115
Carbon dioxide	72.2
Chlorine	67.4
Ether	9.4

* Nutritional calories are actually kilocalories (kcal); 1 kilocalorie equals 1,000 calories.

2–6 *Ammonia is very similar to water in its molecular structure, and biologists have speculated about whether it might substitute for water in life processes. The ammonia molecule (NH₃) is made up of hydrogen atoms covalently bonded to nitrogen, which, like the oxygen in the water molecule, retains a slight negative charge. However, because there are three hydrogens to one nitrogen, the charge difference between the positive and negative zones in the ammonia molecule is not as great as in the water molecule, and the hydrogen bonds formed by ammonia are slightly weaker than those formed by water. Moreover, the 3:1 ratio of hydrogen to nitrogen makes it difficult for ammonia molecules to form an interlocking network. As a consequence, ammonia does not have the cohesive power of water and evaporates much more quickly. Perhaps this is why no form of life based on ammonia has been found, although NH₃ may have been very common in the primitive atmosphere.*

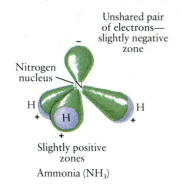

Unshared pair of electrons— slightly negative zone

Nitrogen nucleus

H   H

Slightly positive zones

Ammonia (NH₃)

"Ammonia! Ammonia!"

[*Drawing by R. Grossman, © 1962 The New Yorker Magazine, Inc.*]

evaporation. But, whatever the temperature, so long as a liquid is exposed to air that is less than 100 percent saturated with the vapor of that liquid, evaporation will take place, right down to the last drop.

In order for a water molecule to break loose from its fellow molecules—that is, to vaporize—the hydrogen bonds have to be broken. This requires heat energy. As a consequence, when water evaporates, as from the surface of your skin or a leaf, the escaping molecules carry a great deal of heat away with them. Thus evaporation has a cooling effect. Evaporation from the surface of a land-dwelling plant or animal is one of the principal ways in which these organisms "unload" excess heat and so stabilize their temperatures.

Freezing

Water exhibits another peculiarity when it undergoes the transition from a liquid to a solid (ice). In most liquids, the **density**—that is, the weight of the material in a given volume—increases as the temperature drops. This greater density occurs because the individual molecules are moving more slowly and so the spaces between them decrease. The density of water also increases as the temperature drops, until it nears 4°C. Then the water molecules come so close together and are moving so slowly that *every one* of them can form hydrogen bonds simultaneously with four other molecules—something they could not do at higher temperatures. However, the geometry of the water molecule is such that, as the temperature drops below 4°C, the molecules must move slightly apart from each other to maintain the maximum number of hydrogen bonds in a stable structure. At 0°C, the freezing point of water, this creates an open latticework (Figure 2–7) that is the most stable structure for an ice crystal. Thus water as a solid takes up more volume than water as a liquid. Ice is less dense than liquid water and therefore floats in it.

This increase in volume has occasional disastrous effects on water pipes but, on the whole, turns out to be enormously beneficial for life forms. If water continued to contract as it froze, ice would be heavier than liquid water. As a result, lakes and ponds and other bodies of water would freeze from the bottom up. Once ice began to accumulate on the bottom, it would tend not to melt, season after season. Spring and summer might stop the freezing process, but laboratory experiments have shown that if ice is held to the bottom of even a relatively shallow tank, water can be boiled on the top without melting the ice. Thus if water did not expand when it froze, it would continue to freeze from the bottom up, year after year, and never melt again. Eventually, the body of water would freeze solid and any life in it would be destroyed. By contrast, the layer of floating ice that actually forms tends to protect the organisms in the water. The ice layer effectively insulates the liquid water beneath it, keeping its temperature at or above the freezing point of water.

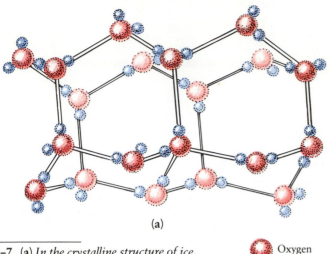

(a)

(b)

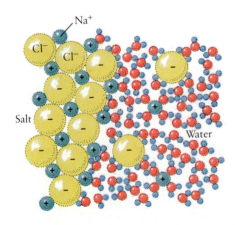

● Oxygen

● Hydrogen

2-7 **(a)** *In the crystalline structure of ice, each water molecule is hydrogen-bonded to four other water molecules in a three-dimensional open latticework. The bond angles in some of the water molecules are distorted as they link up in a hexagonal arrangement. This arrangement, shown here in a small section of the latticework, is repeated throughout the crystal and is responsible for the beautiful patterns seen in snowflakes and frost. The water molecules are actually farther apart in ice than they are in liquid water.*

(b) *When water freezes in the cracks and crevices of rock, the force created by its expansion splits the rock. Over long periods of time, this process breaks up masses of rock and contributes to the formation of soil.*

2-8 *Because of the polarity of water molecules, water can serve as a solvent for ionic substances and polar molecules. This diagram shows sodium chloride (NaCl) dissolving in water as the water molecules cluster around the individual sodium and chloride ions, separating them from one another. Notice the difference between the way the water molecules are arranged around the sodium ions and the way they are arranged around the chloride ions.*

The melting point of water is 0°C, the same temperature as the freezing point. To make the transition from a solid to a liquid, water requires 79.7 calories per gram, a quantity known as the **heat of fusion.** As ice melts, it draws this much heat from its surroundings, thereby cooling the surroundings. The heat energy absorbed by the ice breaks the hydrogen bonds of the latticework. Conversely, as water freezes, it releases the same amount of heat into its surroundings. In this way, ice and snow also serve as temperature stabilizers, particularly during the transition periods of fall and spring. Moderation of sudden changes in temperature gives organisms time to make seasonal adjustments essential to survival.

The presence of dissolved substances in water lowers the temperature at which water freezes, which is why salt is thrown on icy sidewalks and used in ice-cream freezers. The "hardening" process in several species of winter-hardy plants, by which they prepare themselves for cold weather, includes the breakdown of starch (which is insoluble in the fluids of the plant cell) into simple sugars (which are soluble). Freshwater fish, whose body fluids are salty compared to the pond or lake in which they live, do not freeze when the temperature of water is at or near 0°C. However, logically speaking, saltwater fish, whose body fluids are less salty than the ocean water surrounding them, should freeze at the below-zero temperatures of Arctic water. They do not, however, and animal physiologists investigating this phenomenon have discovered that at least one species, the ghost fish, produces a complex protein named, appropriately, antifreeze protein. This protein, which is secreted into the bloodstream, appears to interfere with the formation of the crystalline structure of ice. Recently, studies of several species of terrestrial frogs that spend the winter hibernating beneath leaf litter have revealed that their body fluids contain a high concentration of glycerol, one of the ingredients sometimes used in automobile antifreeze.

WATER AS A SOLVENT

Many substances within living systems are found in aqueous solution. (A **solution** is a uniform mixture of the molecules of two or more substances. The substance present in the greatest amount—usually a liquid—is called the **solvent,** and the substances present in lesser amounts are called **solutes.**) The polarity of water molecules is responsible for water's capacity as a solvent. The polar water molecules tend to separate ionic substances, such as sodium chloride (NaCl), into their constituent ions. As shown in Figure 2–8, the water molecules cluster around and segregate the charged ions.

The Seasonal Cycle of a Lake

As we have seen, water increases in density as its temperature drops, until it reaches 4°C, the temperature of maximum density. Water either colder or warmer than 4°C is less dense and floats above water at 4°C. As a result, the water of temperate-zone lakes is stratified in the summer and winter but undergoes considerable mixing in the fall and spring. The stratifications of summer and winter enable lake-dwelling organisms to avoid life-threatening temperature extremes, while the mixing that occurs in fall and spring provides nutrients and oxygen to organisms at all levels of the lake.

In the summer, the top layer of water, called the epilimnion, is heated by the sun and the surrounding air, becoming warmer than the lower layers. Since it becomes less dense as it becomes warmer, this water remains at the surface. Only the water in the epilimnion circulates. In the middle layer, there is an abrupt drop in temperature, known as the thermocline. Since the water in this layer is progressively more dense, it does not mix with the lighter water above. The water of the middle layer effectively cuts off the circulation of oxygen from the surface into the third layer, the hypolimnion. As the organisms of the hypolimnion gradually use up the available oxygen, the summer stagnation results (a).

In the fall, the temperature of the epilimnion drops until it is the same as that of the hypolimnion. The warmer water in the middle layer then rises to the surface, producing the fall overturn (b). Aided by the fall winds, water begins to circulate throughout the lake (c); oxygen is returned to the depths, and nutrients released by the activities of bottom-dwelling bacteria are carried to the upper layers of the lake.

As winter deepens, the surface water cools below 4°C, becoming lighter as it expands. This water remains on the surface and, in many areas, freezes. The result is winter stratification (d).

In the spring, as the ice melts and the water on the surface warms to 4°C, it sinks to the bottom, producing the spring overturn (e). Another thorough mixing of the water in the lake follows, comparable to that shown in (c).

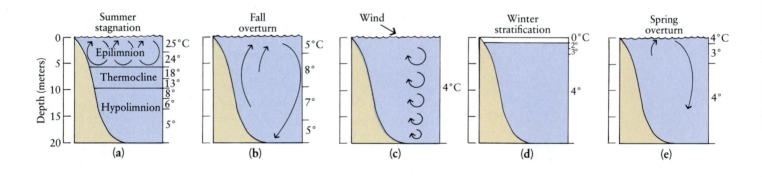

Many of the covalently bonded molecules important in living systems—such as sugars—have regions of partial positive and negative charge. (Such polar regions arise, as you might expect, in the neighborhood of covalently bonded atoms whose nuclei exert differing degrees of attraction for electrons.) These molecules therefore attract water molecules and also dissolve in water. Polar molecules that readily dissolve in water are often called **hydrophilic** ("water-loving"). Such molecules slip into aqueous solution easily because their partially charged regions attract water molecules as much as or more than they attract each other. The polar water molecules thus compete with the attraction between the solute molecules themselves.

Molecules, such as fats, that lack polar regions tend to be very insoluble in water. The hydrogen bonding between the water molecules acts as a force to exclude the nonpolar molecules. As a result of this exclusion, nonpolar molecules tend to cluster together in water, just as droplets of fats tend to coalesce, for example, on the surface of chicken soup. Such molecules are said to be **hydrophobic** ("water-fearing"), and the clusterings are known as hydrophobic interactions.

We will encounter these properties of hydrophilic and hydrophobic molecules again in later chapters. These weak forces—hydrogen bonds and hydrophobic forces—play crucial roles in shaping the architecture of large, biologically important molecules and, as a consequence, in determining their properties.

IONIZATION OF WATER: ACIDS AND BASES

In liquid water, there is a slight tendency for a hydrogen atom to jump from the oxygen atom to which it is covalently bonded to the oxygen atom to which it is hydrogen-bonded (Figure 2–9). In this reaction, two ions are produced: the hydronium ion (H_3O^+) and the hydroxide ion (OH^-). In any given volume of pure water, a small but constant number of water molecules will be ionized in this way. The number is constant because the tendency of water to ionize is offset by the tendency of the ions to reunite; thus even as some molecules are ionizing, an equal number of others are forming, a state known as dynamic **equilibrium.**

2–9 *When water ionizes, a hydrogen nucleus (that is, a proton) shifts from the oxygen atom to which it is covalently bonded to the oxygen atom to which it is hydrogen-bonded. The resulting ions are the negatively charged hydroxide ion and the positively charged hydronium ion. In this diagram, the large spheres represent oxygen and the small spheres represent hydrogen.*

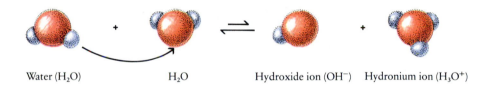

Water (H_2O) H_2O Hydroxide ion (OH^-)   Hydronium ion (H_3O^+)

Although the positively charged ion formed when water ionizes is the hydronium ion (H_3O^+), rather than the hydrogen ion (H^+), by convention the ionization of water is expressed by the equation:

$$HOH \rightleftharpoons H^+ + OH^-$$

The arrows indicate that the reaction goes in both directions. The fact that the arrow pointing toward HOH is longer indicates that, at equilibrium, most of the H_2O is not ionized. As a consequence, in any sample of pure water, only a small fraction exists in ionized form.

In pure water, the number of H^+ ions exactly equals the number of OH^- ions. This is necessarily the case since neither ion can be formed without the other when only H_2O molecules are present. However, when an ionic substance or a substance with polar molecules is dissolved in water, it may change the relative numbers of H^+ and OH^- ions. For example, when hydrogen chloride (HCl) dissolves in water, it is almost completely ionized into H^+ and Cl^- ions; as a result, an HCl solution (hydrochloric acid) contains more H^+ ions than OH^- ions. Conversely, when sodium hydroxide (NaOH) dissolves in water, it forms Na^+ and OH^- ions; thus, in a solution of sodium hydroxide in water, there are more OH^- ions than H^+ ions.

A solution acquires the properties we recognize as acidic when the number of H^+ ions exceeds the number of OH^- ions; conversely, a solution is basic (alkaline) when the number of OH^- ions exceeds the number of H^+ ions. Thus, an **acid** is a substance that causes an increase in the relative number of H^+ ions in a solution, and a **base** is a substance that causes an increase in the relative number of OH^- ions.

Strong and Weak Acids and Bases

Strong acids and bases are substances, like HCl and NaOH, that ionize almost completely in water, resulting in relatively large increases in the concentrations of H^+ and OH^- ions, respectively. Weak acids and bases, by contrast, are those that ionize only slightly, resulting in relatively small increases in the concentration of H^+ or OH^- ions.

Because of the strong tendency of H^+ and OH^- ions to combine and the weak tendency of water to ionize, the concentration of OH^- ions will always decrease as the concentration of H^+ ions increases (as, for example, when HCl is added to water), and vice versa. If HCl is added to a solution containing NaOH, the following reaction will take place:

$$H^+ + Cl^- + Na^+ + OH^- \longrightarrow H_2O + Na^+ + Cl^-$$

In other words, if an acid and a base of comparable strength are added in equivalent amounts, the solution will not have an excess of either H^+ or OH^- ions.

Many of the acids important in living systems owe their acidic properties to a group of atoms called the carboxyl group, which includes one carbon atom, two oxygen atoms, and a hydrogen atom (symbolized as —COOH). When a substance containing a carboxyl group is dissolved in water, some of the —COOH groups dissociate to yield hydrogen ions:

$$—COOH \rightleftharpoons —COO^- + H^+$$

Thus compounds containing carboxyl groups are hydrogen-ion donors, or acids. They are weak acids, however, because, as indicated by the arrows, the —COOH ionizes only slightly.

Among the most important bases in living systems are compounds that contain the amino group (—NH_2). This group has a weak tendency to accept hydrogen ions, thereby forming —NH_3^+:

$$—NH_2 + H^+ \rightleftharpoons —NH_3^+$$

As hydrogen ions are removed from solution by the amino group, the relative concentration of H^+ ions decreases and the relative concentration of OH^- ions increases. Groups, such as —NH_2, that are weak hydrogen-ion acceptors are thus weak bases.

The pH Scale

Chemists express degrees of acidity by means of the **pH scale.** The symbol "pH" is derived from the French *pouvoir hydrogène* ("hydrogen power"). It stands for the negative logarithm of the concentration of hydrogen ions in moles per liter. Although this sounds complicated, in practice it is relatively simple. As you may recall from your mathematics courses, the logarithm is the exponential power to which a specified number (commonly 10) must be raised to equal a given number. For example, the logarithm of 100 is 2, since 100 equals 10^2 (that is, 10×10). The logarithm of 1/100 is -2, since 1/100 equals 10^{-2} (that is, $1/10 \times 1/10$). The numbers whose logarithms are of interest to us are the concentrations of hydrogen ions in solutions, expressed in moles per liter.

A **mole** is the amount of an element equivalent to its atomic weight expressed in grams, or the amount of a substance equivalent to its molecular weight expressed in grams. (The **molecular weight** of a substance is the sum of the atomic weights of the atoms constituting the molecule.) Thus, a mole of atomic hydrogen (atomic weight 1) is 1 gram of hydrogen atoms; a mole of atomic oxygen (atomic weight 16) is 16 grams of oxygen atoms; and a mole of water (molecular weight 18) is 18 grams of water molecules. The most interesting thing about the mole is that a mole—of any substance—contains the same number of particles as any other mole. This number, known as Avogadro's number, is 6.02×10^{23}. Thus, a mole of water molecules (18 grams) contains exactly the same number of molecules as a mole of hydrogen chloride molecules (36.5 grams). Use of the mole in specifying quantities of substances involved in chemical reactions makes it possible for us to consider comparable numbers of reacting particles.

The ionization that occurs in a liter of pure water results in the formation, at equilibrium, of 1/10,000,000 mole of hydrogen ions (and, as we noted earlier, of exactly the same quantity of hydroxide ions). In decimal form, this concentration of hydrogen ions is written as 0.0000001 mole per liter. This same concentration of hydrogen ions can be written even more conveniently in exponential form as 10^{-7} mole per liter. The logarithm is the exponent, -7, and the negative logarithm is 7; in terms of the pH scale, it is referred to simply as pH 7 (see Table 2–3). At pH 7, the concentrations of free H^+ and OH^- are exactly the same, as they are in pure water. This is a neutral state. Any pH below 7 is acidic, and any pH above 7 is basic. The lower the pH number, the higher the concentration of hydrogen ions. Thus pH 2 means 10^{-2} mole of hydrogen ions per liter of water, or 1/100 mole per liter (0.01 mole per liter)—which is, of course, a much larger figure than 1/10,000,000 (0.0000001). Since the pH scale is logarithmic, a difference of one pH unit represents a tenfold difference in the concentration of hydrogen ions; for example, a solution at pH 3 has 1,000 times as many H^+ ions as a solution at pH 6.

We can now define "acid" and "base" more fully:

1. An acid is a substance that causes an increase in the number of H^+ ions and a decrease in the number of OH^- ions in a solution. Most acids are hydrogen-ion donors, but some acids function by removing OH^- ions from the solution. A solution with a pH below 7 (with more than 10^{-7} mole of H^+ ions per liter) is acidic.

2. A base is a substance that causes a decrease in the number of H^+ ions and an increase in the number of OH^- ions in a solution. Some bases, such as NaOH, donate OH^- ions to the solution; others, such as the $-NH_2$ group, are hydrogen-ion acceptors, removing H^+ ions from the solution. A solution with a pH above 7 (with less than 10^{-7} mole of H^+ ions per liter) is basic.

TABLE 2–3 **The pH Scale**

			CONCENTRATION OF H+ IONS (MOLES PER LITER)		pH	CONCENTRATION OF OH− IONS (MOLES PER LITER)
Increasing H+ / Decreasing OH−	Acidic	1.0	$= 10^0$	0	10^{-14}
		0.1	$= 10^{-1}$	1	10^{-13}
		0.01	$= 10^{-2}$	2	10^{-12}
		0.001	$= 10^{-3}$	3	10^{-11}
		0.0001	$= 10^{-4}$	4	10^{-10}
		0.00001	$= 10^{-5}$	5	10^{-9}
		0.000001	$= 10^{-6}$	6	10^{-8}
	Neutral	0.0000001	$= 10^{-7}$	7	$10^{-7} = 0.0000001$
Decreasing H+ / Increasing OH−	Basic		10^{-8}	8	$10^{-6} = 0.000001$
			10^{-9}	9	$10^{-5} = 0.00001$
			10^{-10}	10	$10^{-4} = 0.0001$
			10^{-11}	11	$10^{-3} = 0.001$
			10^{-12}	12	$10^{-2} = 0.01$
			10^{-13}	13	$10^{-1} = 0.1$
			10^{-14}	14	$10^0 = 1.0$

Acid Rain

The average pH of normal rainfall is about 5.6 (mildly acidic), a result of the combination of carbon dioxide with water vapor to produce carbonic acid. In the 1920s, however, the pH of rain and snow in Scandinavia began to drop, and by the 1950s, similar phenomena were observed elsewhere in Europe and in the northeastern United States. As more data were collected, it was found that, in certain geographic areas, the average annual pH of precipitation was between 4.0 and 4.5. Occasional storms would release rain with a pH as low as 2.1, which is extremely acidic.

The low pH was traced primarily to two acids found in the rainfall: sulfuric (H_2SO_4) and nitric (HNO_3), both of which ionize almost completely in aqueous solution, releasing hydrogen ions. These acids are formed when the gaseous oxides of sulfur and nitrogen react with water vapor and other gases in the air. Sulfur and nitrogen oxides are released into the atmosphere by some natural processes (for example, volcanic eruptions), but far greater quantities are released as a result of human activities. Sulfur oxides are produced by the combustion of high-sulfur coal and oil and by the smelting of sulfur-containing ores. Nitrogen oxides are by-products of gasoline combustion in automobile engines and of some generating processes for electricity.

That sulfur oxides could be damaging to vegetation was evident as early as the turn of the century, when a large copper smelter was opened in a mountainous area in Tennessee. Within a few years, all vegetation had been killed in the formerly luxuriant forest surrounding the smelter. The solution devised for this problem—still used today—was to build very tall smokestacks, so the wind would carry the pollutants away from the immediate area. It was assumed that they would be so widely dispersed that they would be rendered harmless. By the 1960s, accumulating evidence indicated that sulfur oxides released from tall smokestacks are transported hundreds or thousands of miles by the prevailing winds (generally west to east in the Northern Hemisphere) and are then returned to the earth in rain and snow. Nitrogen oxides released from automobiles are also carried off by the wind. What was once a local problem has become an international problem, in which the pollutants respect no boundaries.

The biological consequences of acid rain depend in part on the characteristics of the soil and underlying rock on which it falls. In areas where the principal rock is limestone (calcium carbonate), the buffering action of the H_2CO_3–HCO_3^- system (see page 52) can generally prevent acidification of the soil, lakes, and streams. In other areas, where the soil and bodies of water do not contain such a natural buffer, the pH drops gradually but steadily as a result of repeated additions of acid rain. The drop in pH is often quite sudden and extreme when the spring melt brings an infusion of acid accumulated in the winter snows. Although the low pH resulting from the spring melt is usually temporary, it can be particularly devastating to salamanders and frogs, many of which lay their eggs in small ponds and puddles formed by the melt water.

Lakes in mountainous regions are especially vulnerable to acid rain. A 1977 Cornell University study of the lakes at high elevations (above 600 meters) in the western Adirondack Mountains of New York found that 51 percent had a pH below 5.0, of which 90 percent were devoid of fish life. By contrast, a similar study performed between 1929 and 1937 found that only 4 percent of the lakes were acidic and without fish. More recent studies indicate increasingly acidic conditions in lakes at lower levels in the Adirondacks (which were previously unaffected) and in many lakes of the Cascade Mountains in the Pacific Northwest. The effects of low pH on fish include the depletion of calcium in their bodies, leading to weakened and deformed bones; the failure of many eggs to hatch and deformed fish from those that do hatch; and the clogging of the gills by aluminum, which is released from the soil by acid.

The effects of acid rain on plants depend on both the species and the soil conditions. Among the observed effects are reduced germination of seeds, a decrease in the number of seedlings that mature, reduced growth, and lowered resistance to disease. If the soil is not adequately buffered, essential nutrients are leached from it and are thus unavailable to the plants. Recently, it has become apparent that the forests of the eastern United States, from Maine through Georgia, are in serious decline. Detailed studies reveal a dramatic slowing of growth during the past 20 years, and in some locations at high altitude,

(a) This map, based on estimates made by the Environmental Protection Agency, shows the sensitivity of different areas of the continental United States to acid rain. It takes into account such factors as major sources of sulfur and nitrogen oxides, weather patterns, altitude, and soil characteristics.

(b) In the mountains of New England, red spruce trees are dying at a high rate. Only their skeletons remain, standing silent sentinel over the forest.

(c) In a forest in Vermont, a student sets up an apparatus to collect rainwater that will be tested for acidity.

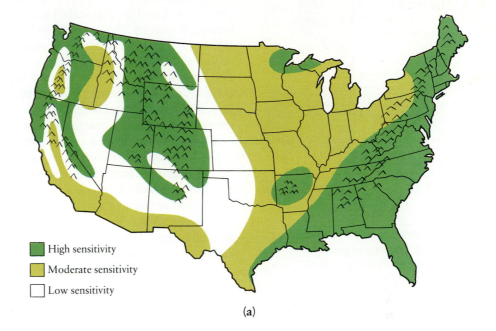

High sensitivity

Moderate sensitivity

Low sensitivity

(a)

trees are dying in large numbers and reproduction of some plants has come to a halt. At the present time, it is not known whether the decline is a result of acid rain, other pollutants, disease, subtle climate changes, or, most likely, some combination of these factors.

The accumulating evidence indicates that acid rain is one of the most serious worldwide pollution problems confronting us today. The potential consequences of its effects on biological systems are immense: lowered crop yields, decreased timber production, the need for greater amounts of increasingly expensive fertilizer to compensate for nutrient leaching, the loss of important freshwater fishing areas, and, possibly, of the eastern forests as well. The monetary and social costs of allowing the conditions that create acid rain to continue (or even to increase) are potentially very great, as are the costs of available processes to remove the sulfur and nitrogen oxides at the source, before they enter the air.

Scientists from many fields are presently engaged in research to gain a greater understanding of the causes and effects of acid rain and the likely consequences of proposed solutions. Although scientists can provide information on which decisions can be based, the choices that lie ahead are essentially social and economic, to be made through political processes.

(b)

(c)

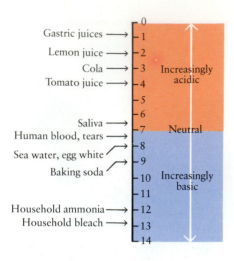

Gastric juices → 0, 1
Lemon juice → 2
Cola → 3   Increasingly acidic
Tomato juice → 4
5
6
Saliva → 7   Neutral
Human blood, tears → 8
Sea water, egg white → 9
Baking soda → 10   Increasingly basic
11
Household ammonia → 12
Household bleach → 13
14

2–10 *pH values of various common solutions. A difference of one pH unit reflects a tenfold difference in the H^+ ion concentration. Cola, for instance, is 10 times as acidic as tomato juice, and gastric juices are about 100 times more acidic than cola drinks.*

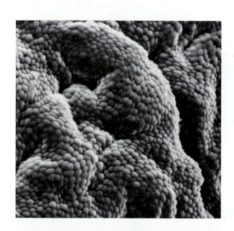

2–11 *Surface of the lining of the stomach, as shown in a scanning electron micrograph (magnified approximately 185 times). The numerous indentations are the openings into gastric pits, in which acid-secreting cells are located. Mucus, also secreted by cells of the stomach lining, coats the surface of the stomach and protects it from the acid.*

Buffers

Solutions more acidic than pH 1 or more basic than pH 14 are possible, but these are not included in the scale because they are almost never encountered in biological systems (Figure 2–10). In fact, almost all the chemistry of living things takes place at pH's between 6 and 8. Notable exceptions are the chemical processes in the stomach of humans and other animals, which take place at a pH of about 2 (Figure 2–11). Human blood, for instance, maintains an almost constant pH of 7.4 despite the fact that it is the vehicle for a large number and variety of nutrients and other chemicals being delivered to the cells, as well as for the removal of wastes, many of which are acids and bases.

The maintenance of a constant pH—an example of homeostasis (see page 26)—is important because the pH greatly influences the rate of chemical reactions. Organisms resist strong, sudden changes in the pH of blood and other fluids by means of **buffers,** which are combinations of H^+-donor and H^+-acceptor forms of weak acids or bases.

Buffers help maintain constant pH by their tendency to combine with H^+ ions and thus remove them from solution as the H^+ ion concentration begins to rise and to release them as it falls. The capacity of a buffer system to resist changes in pH is greatest when the concentrations of its H^+-donor and H^+-acceptor forms are equal. As the concentration of one form increases and that of the other decreases, the buffer becomes less effective. A variety of buffers function in living systems, each most effective at the particular pH at which its H^+-donor and H^+-acceptor concentrations are equal.

The major buffer system in the human bloodstream is the acid-base pair H_2CO_3–HCO_3^-. The weak acid H_2CO_3 (carbonic acid) dissociates into H^+ and bicarbonate ions as shown in the equation below.

$$H_2CO_3 \rightleftharpoons H^+ + HCO_3^-$$
$$\text{H}^+ \text{ donor} \qquad \text{H}^+ \text{ acceptor}$$

The H_2CO_3–HCO_3^- buffer system resists the changes in pH that might result from the addition of small amounts of acid or base by "soaking up" the acid or base. For example, if a small amount of H^+ is added to the system, it combines with the H^+ acceptor HCO_3^- to form H_2CO_3. This reaction removes the added H^+ and maintains the pH near its original value. If a small amount of OH^- is added, it combines with the H^+ to form H_2O; more H_2CO_3 tends to ionize to replace the H^+ as it is used.

Control of the pH of the blood is rendered even "tighter" by the fact that the H_2CO_3 is in equilibrium with dissolved carbon dioxide (CO_2) in the blood:

$$H_2O + CO_2 \rightleftharpoons H_2CO_3$$

As the arrows indicate, the two reactions are in equilibrium, and the equilibrium favors the formation of CO_2; in fact, the ratio is about 100 to 1 in favor of CO_2 formation.

Dissolved CO_2 in the blood is, in turn, in equilibrium with the CO_2 in the lungs. By changing your rate of breathing, you can change the HCO_3^- concentration in the blood and thus adjust the pH of your internal fluids.

Obviously, if the blood should be flooded with a very large excess of acid or base, the buffer would fail, but normally it is able to adjust continuously and very rapidly to the constant small additions of acid or base that normally occur in body fluids.

2–12 *The water cycle.*

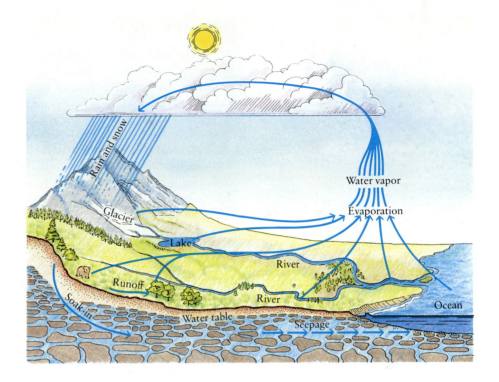

THE WATER CYCLE

Most of the water on earth—almost 98 percent—is in liquid form, in the oceans, lakes, and streams. Of the remaining 2 percent, some is frozen in polar ice and glaciers, some is in the soil, some is in the atmosphere in the form of vapor, and some is in the bodies of living organisms.

Water is made available to land organisms by processes powered by the sun. Solar energy evaporates water from the oceans, leaving the salt behind. Water is also evaporated, but in much smaller amounts, from moist soil surfaces, from the leaves of plants, and from the bodies of other organisms. These molecules—now water vapor—are carried up into the atmosphere by air currents. Eventually they fall to the earth's surface again as snow or rain. Most of the water falls on the oceans, since these cover most of the earth's surface. The water that falls on land is pulled back to the oceans by the force of gravity. Some of it, reaching low ground, forms ponds or lakes and streams or rivers, which pour water back into the oceans.

Some of the water that falls on the land percolates down through the soil until it reaches a zone of saturation. In the zone of saturation, all pores and cracks in the rock are filled with water (groundwater). The upper surface of the zone of saturation is known as the water table. Below the zone of saturation is solid rock, through which the water cannot penetrate. The deep groundwater, moving extremely slowly, eventually also reaches the ocean, thereby completing the water cycle.

As we have seen in this chapter, water, essential for life, is a most extraordinary substance. The earth's supply of water is the permanent possession of our planet, held to its surface by the force of gravity. Through the movements of the water cycle, it is perpetually available to living organisms.

SUMMARY

Water, the most common liquid on the earth's surface and the major component, by weight, of all living things, has a number of remarkable properties. These properties are a consequence of its molecular structure and are responsible for water's "fitness" for its roles in living systems.

Water is made up of two hydrogen atoms and one oxygen atom held together by covalent bonds. The water molecule is polar, with two weakly negative zones and two weakly positive zones. As a consequence, weak bonds form between water molecules. Such bonds, which link a somewhat positively charged hydrogen atom that is part of one molecule to a somewhat negatively charged oxygen atom that is part of another molecule, are known as hydrogen bonds. Each water molecule can form hydrogen bonds with four other water molecules. Although individual bonds are weak and constantly shifting, the total strength of the bonds holding the molecules together is very great.

Because of the hydrogen bonds holding the water molecules together (cohesion), water has a high surface tension and a high specific heat (the amount of heat that a given amount of the substance requires for a given increase in temperature). It also has a high heat of vaporization (the heat required to change a liquid to a gas) and a high heat of fusion (the heat required to change a solid to a liquid). Just before water freezes, it expands; thus ice has a lower density and a larger volume than liquid water. As a result, ice floats in water.

The polarity of the water molecule is responsible for water's adhesion to other polar substances and hence its tendency for capillary movement. Similarly, water's polarity makes it a good solvent for ions and polar molecules. Molecules that dissolve readily in water are known as hydrophilic. Water molecules, as a consequence of their polarity, actively exclude nonpolar molecules from solution. Molecules that are excluded from aqueous solution are known as hydrophobic.

Water has a slight tendency to ionize, that is, to separate into H^+ ions (actually H_3O^+, hydronium ions) and OH^- ions. In pure water, the number of H^+ ions and OH^- ions is equal at 10^{-7} mole per liter. A solution that contains more H^+ ions than OH^- ions is acidic; one that contains more OH^- ions than H^+ ions is basic. The pH scale reflects the proportion of H^+ ions to OH^- ions. An acidic solution has a pH lower than 7.0; a basic solution has a pH higher than 7.0. Almost all of the chemical reactions of living systems take place within a narrow range of pH around neutrality. Organisms maintain this narrow pH range by means of buffers, which are combinations of H^+-donor and H^+-acceptor forms of weak acids or bases.

Through the water cycle, the water above, on, and below the earth's surface is recirculated. As a result, it is continuously available to living organisms.

QUESTIONS

1. (a) Sketch the water molecule and label the zones of positive and negative charge. (b) What are the major consequences of the polarity of the water molecule? (c) How are these effects important to living systems?

2. The trick with the razor blade (see page 42) works better if the blade is a little greasy. Why?

3. Surfaces such as glass or raincoat cloth can be made "nonwettable" by application of silicone oils or other substances that cause water to bead up instead of spread flat. What do you suppose is happening, in molecular terms, when a surface becomes nonwettable?

4. Can you explain maple sugar production in terms of its value to the sugar maple tree?

5. Generally, coastal areas have more moderate temperatures (not as cold in winter, nor as hot in summer) than inland areas at the same latitude. What reasonable explanation can you give for this phenomenon?

6. What is vaporization? Describe the changes that take place in water as it vaporizes. What is heat of vaporization? Why does water have an unusually high heat of vaporization?

7. As we have seen, the digestive processes in the human stomach take place at a pH of about 2. When the food being digested reaches the small intestine, sodium bicarbonate ($NaHCO_3$) is released from the pancreas into the small intestine. What effect would you expect this to have on the pH of the partially digested food mass?

CHAPTER 3

Organic Molecules

3–1 *In the process of photosynthesis, carbon from carbon dioxide in the atmosphere is incorporated into organic molecules by plants. These molecules provide the energy that powers living systems and are also used to build the larger structural molecules of which living organisms are composed. Some 300 million years ago, conditions on the earth were such that the carbon-containing dead bodies of vast numbers of organisms did not decay but were instead converted into coal and petroleum. Coal deposits are rich with the fossilized remains of plants that lived at the time, such as the leaves of the fern* Alethopteris *and the branch of the giant horsetail* Calamites *shown here.*

In this chapter, we present some of the types of **organic molecules**—molecules containing carbon—that are found in living things. As you will see, the molecular drama is a grand spectacular with, literally, a cast of thousands; a single bacterial cell contains some 5,000 different kinds of molecules, and an animal or plant cell has about twice that many. These thousands of molecules, however, are composed of relatively few elements (CHNOPS). Similarly, relatively few kinds of molecules play the major roles in living systems. As we noted previously, water makes up from 50 to 95 percent of a living system, and small ions such as K^+, Na^+, and Ca^{2+} account for no more than 1 percent. Almost all the rest, chemically speaking, is composed of organic molecules.

Four different kinds of organic molecules are found in large quantities in organisms. These four are **carbohydrates** (composed of sugars), **lipids** (nonpolar molecules, many of which contain fatty acids), **proteins** (composed of amino acids), and **nucleotides** (complex molecules that play key roles in energy exchanges and that can also combine to form very large molecules known as nucleic acids). All of these molecules—carbohydrates, lipids, proteins, and nucleotides—contain carbon, hydrogen, and oxygen. In addition, proteins contain nitrogen and sulfur, and nucleotides, as well as some lipids, contain nitrogen and phosphorus.

It has been said that it is necessary only to be able to recognize about 30 molecules for a working knowledge of the biochemistry of cells. Two of these are the sugars glucose and ribose; another is a fatty acid; 20 are the biologically important amino acids; and five are nitrogenous bases, nitrogen-containing molecules that are key constituents of nucleotides. If you bear with us, you will find that you readily learn to recognize the players and their roles and to distinguish the stars from the members of the chorus. Consider this, if you will, an introduction to the principal characters; the plot begins to unfold in Chapter 4.

THE CENTRAL ROLE OF CARBON

The Carbon Backbone

As you will recall from Chapter 1, a carbon atom has six protons and six electrons, two electrons in its first energy level and four in its second energy level. Thus carbon can form four covalent bonds with as many as four different atoms. Methane (CH_4), which is natural gas, is an example (Figure 1–10, page 32). Even more important, in terms of carbon's biological role, carbon atoms can form bonds with each other. Ethane, for example, contains two carbons; propane, three; butane, four; and so on, forming long chains (Figure 3–2). In general, an organic molecule derives its overall shape from the arrangement of the carbon atoms that form the backbone, or skeleton, of the molecule. The shape of the molecule, in turn, determines many of its properties and its function within living systems.

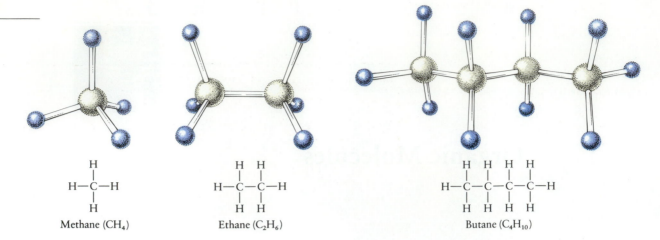

Methane (CH$_4$) Ethane (C$_2$H$_6$) Butane (C$_4$H$_{10}$)

3-2 *Ball-and-stick models and structural formulas of methane, ethane, and butane. In the models, the gray spheres represent carbon atoms and the smaller blue spheres represent hydrogen atoms. The sticks in the models—and the lines in the structural formulas—represent covalent bonds, each of which consists of a pair of shared electrons. Note that every carbon atom forms four covalent bonds.*

In the molecules shown in Figure 3–2, every carbon bond that is not occupied by another carbon atom is taken up by a hydrogen atom. Such compounds, consisting of only carbon and hydrogen, are known as **hydrocarbons.** Structurally, they are the simplest kind of organic compounds. Although most hydrocarbons are derived from the remains of organisms that died millions of years ago, they are relatively unimportant in living organisms. They are, however, of great economic importance; the liquid fuels upon which we depend—gasoline, diesel fuel, and heating oil—are all composed of hydrocarbons.

Functional Groups

The specific chemical properties of an organic molecule derive principally from groups of atoms known as **functional groups.** These groups are attached to the carbon skeleton, replacing one or more of the hydrogens that would be present in a hydrocarbon. An —OH (hydroxyl) group is an example of a functional group.* When one hydrogen and one oxygen are bonded covalently, one outer electron of the oxygen is left over, unpaired and unshared; it can be shared with a similarly available outer electron of a carbon atom, thereby forming a covalent bond with the carbon. A compound with a hydroxyl group in place of one or more of the hydrogens in a hydrocarbon is known as an alcohol. Thus methane (CH$_4$), with the replacement of one hydrogen atom by a hydroxyl group, becomes methanol, or wood alcohol (CH$_3$OH), a pleasant-smelling, poisonous compound noted for its ability to cause blindness and death. Ethane similarly becomes ethanol, or grain alcohol (C$_2$H$_5$OH), which is present in all alcoholic beverages. Glycerol, C$_3$H$_5$(OH)$_3$, contains, as its formula indicates, three carbon atoms, five hydrogen atoms, and three hydroxyl groups.

Table 3–1 illustrates the functional groups that will be of greatest interest to us in our exploration of living systems. A knowledge of functional groups makes it easy to recognize particular molecules and to predict their properties. For example, the carboxyl group (—COOH), mentioned in the previous chapter, is a functional group that gives a molecule the properties of an acid. Alcohols, with their polar hydroxyl groups, tend to be soluble in water, for instance, whereas hydrocarbons, such as butane, with only nonpolar functional groups (such as methyl groups), are highly insoluble in water. Aldehyde groups are often associated with pungent odors and tastes. Smaller molecules with aldehyde groups, such as formaldehyde, have unpleasant odors, whereas larger ones, such as the chemicals that give vanilla, apples, cherries, and almonds their distinctive flavors, tend to be pleasing to the human sensory apparatus.

As you can see, most of the functional groups in Table 3–1 are polar and so have regions of positive and negative charge in aqueous solution. Thus they confer water-solubility and local electric charge to the molecules that contain them.

---

* —OH, the functional group, is called hydroxyl; OH$^-$, the ion, is called hydroxide.

TABLE 3–1 **Some Biologically Important Functional Groups**

GROUP	NAME	BIOLOGICAL SIGNIFICANCE
—OH	Hydroxyl	Polar, thus water-soluble; forms hydrogen bonds
—C$\diagup^{O}_{\diagdown OH}$	Carboxyl	Weak acid (hydrogen donor); when it loses a hydrogen ion, it becomes negatively charged: $-C\diagup^{O}_{\diagdown O^-}$ $+ H^+$
—N$\diagup^{H}_{\diagdown H}$	Amino	Weak base (hydrogen acceptor); when it accepts a hydrogen ion, it becomes positively charged: $-\overset{\displaystyle H}{\underset{\displaystyle H}{N^+}}-H$
$\overset{\displaystyle H}{\underset{}{-C}}=O$	Aldehyde	Polar, thus water-soluble; characterizes some sugars
$\diagdown C=O$	Ketone (or carbonyl)	Polar, thus water-soluble; characterizes other sugars
$-\overset{\displaystyle H}{\underset{\displaystyle H}{C}}-H$	Methyl	Hydrophobic (insoluble in water)
$-\overset{\displaystyle O}{\overset{\displaystyle \|}{\underset{\displaystyle OH}{P}}}-OH$	Phosphate	Acid (hydrogen donor); in solution, usually negatively charged: $-\overset{\displaystyle O}{\overset{\displaystyle \|}{\underset{\displaystyle O^-}{P}}}-O^- + 2H^+$

Some of the polar functional groups tend to become fully ionized, depending on the pH of the solution. Many functional groups participate directly in the chemical reactions of greatest interest in biological systems.

The Energy Factor

Covalent bonds—the bonds commonly found in organic molecules—are strong, stable bonds consisting of electrons moving in orbitals about two or more atomic nuclei. These bonds have different characteristic strengths, depending on the configurations of the orbitals. You will recall from the last chapter that molecules are always in motion—vibrating, rotating, and shifting position in relation to other molecules. The atoms within molecules are also in motion—vibrating and, often, rotating about the axes of their bonds. If this motion becomes great enough (that is, if the atoms possess enough kinetic energy), the bond will "break" and the atoms will become separated from each other. Bond strengths are conventionally expressed in terms of the energy, in kilocalories per mole, that must be supplied to break the bond under standard conditions of temperature and pressure (Figure 3–3).

Kilocalories per mole

93.4	$\diagdown$N$\underline{\quad 0.102 \text{ nm} \quad}$H	
98.8	$\diagdown$C$\underline{\quad 0.109 \quad}$H	
171	$\diagdown$C$\underline{\quad 0.123 \quad}$=O	
147	$\diagdown$C$\underline{\quad 0.127 \quad}$=N$\diagup$	
147	$\diagdown$C$\underline{\quad 0.133 \quad}$=C$\diagup$	
84	$\diagdown$C$\underline{\quad 0.143 \quad}O\diagup$	
69.7	$\diagdown$C$\underline{\quad 0.148 \quad}N\diagup$	
83.1	$\diagdown$C$\underline{\quad 0.154 \quad}C\diagup$	

3–3 *A chemical bond is a force holding atoms together. The strength of the bond is measured in terms of the energy required to break it. The figures at the left indicate the number of kilocalories that will break the bonds between the pairs of atoms shown. The lines connecting the atoms represent the bonds; the figures above the lines represent the characteristic center-to-center distances between the atoms, expressed in nanometers (1 nanometer, abbreviated nm, equals 10^{-9} meter). Double lines indicate double bonds, which, as you can see, hold the atoms closer together and are stronger.*

Why Not Silicon?

Silicon (atomic number 14) is more abundant than carbon (atomic number 6). As you can tell from its atomic number, silicon also requires four electrons to complete its outer energy level. Why then is it found so rarely in living systems? Because silicon atoms are larger than carbon atoms, the distance between two silicon atoms is much greater than the distance between two carbon atoms. As a result, the bonds between the more tightly held carbon atoms are almost twice as strong as those between silicon atoms. Thus carbon can form long stable chains and silicon cannot.

Carbon's capacity to form double bonds is also crucial to its central role in biology. As we saw in Chapter 1, a carbon atom can combine with two oxygen atoms by means of two double bonds; the carbon dioxide molecule, all its electron requirements satisfied, floats in air as a gas, free and independent. It also dissolves readily in water and so is available to living systems. In silicon dioxide, by contrast, a silicon atom forms single bonds to two oxygen atoms, leaving two unpaired electrons on the silicon atom and one on each oxygen. As a consequence, the silicon atom needs to gain two electrons to fill its outer energy level, and each oxygen atom needs to gain one electron. Thus the unpaired electrons are readily

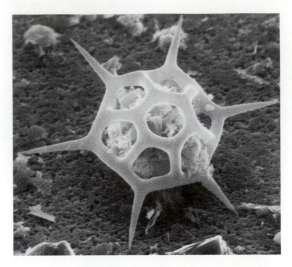

This delicate skeleton of a microorganism, from Narragansett Bay, Rhode Island, is composed of silicon dioxide. The material within the skeleton is organic debris.

shared with the unpaired electrons on neighboring SiO_2 molecules, forming, eventually, grains of sand, rocks, or with biological intervention, the shells of microscopic marine organisms.

When a covalent bond breaks, atoms (or, in some cases, groups of atoms) are released, and each atom usually takes its own electrons with it. This results in atoms whose outer energy levels are only partially filled with electrons. For example, when the atoms of a methane molecule are vibrating and rotating so rapidly that the four carbon-hydrogen bonds break, one carbon atom and four hydrogen atoms are produced—and each of these atoms needs to gain electrons to complete its outer energy level. Thus, the atoms tend to form new covalent bonds quite rapidly, restoring the stable condition of filled outer energy levels. Whether the new bonds that form are identical to those that were broken or are different depends on a number of factors—the temperature, the pressure, and, most important, what other atoms are available in the immediate vicinity.

Chemical reactions in which new combinations are formed always involve a change in electron configurations and therefore in bond strengths. Depending on the relative strengths of the bonds broken and the bonds formed in the course of a chemical reaction, energy will either be released from the system or will be taken up by it from the surroundings. Consider, for example, the burning of methane, represented by the following equation:

$$CH_4 + 2O_2 \longrightarrow CO_2 + 2H_2O$$

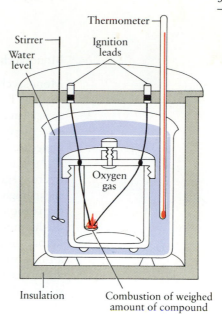

3–4 *A calorimeter is used to measure the amount of energy stored in an organic compound. A known quantity of the compound is ignited electrically. As it burns, the rise in the temperature of the surrounding water is measured. Using the specific heat of water and the known weight of water in the calorimeter, one can then calculate the number of calories released by the burning of the sample.*

This reaction, which can be set in motion by a spark, is often the cause of explosions in coal mines; when it occurs, it releases energy in the form of heat. The amount of energy released can be measured quite precisely, as shown in Figure 3–4. It turns out to be 213 kilocalories per mole of methane. This can be expressed by a simple equation:

$$\Delta H° = -213 \text{ kcal/mole}$$

The Greek letter delta (Δ) stands for change, H for heat, and the superscript ° indicates that the reaction occurs under certain standard conditions of temperature and pressure. The minus sign indicates that energy has been released.

Similarly, changes in energy occur in the chemical reactions that take place in organisms. However, as we shall see in Section 2, living systems have evolved strategies for minimizing not only the energy required to set a reaction in motion but also the proportion of energy released as heat. These strategies involve, among other factors, specialized protein molecules known as **enzymes,** which are essential participants in the chemical reactions of living systems. (The word "strategy" in its ordinary meaning is a deliberate plan to achieve a specified goal. Biologists use it to mean a group of related traits, evolved by organisms under the influence of natural selection, that solves particular problems encountered by living systems.)

CARBOHYDRATES: SUGARS AND POLYMERS OF SUGARS

Carbohydrates are the primary energy-storage molecules in most living things. In addition, they form a variety of structural components of living cells; the walls of young plant cells, for example, are about 40 percent cellulose, which is the most common organic compound in the biosphere.

Carbohydrates are formed from small molecules known as **sugars.** There are three principal kinds of carbohydrates, classified according to the number of sugar molecules they contain. **Monosaccharides** ("single sugars"), such as ribose, glucose, and fructose, contain only one sugar molecule. **Disaccharides** consist of two sugar molecules linked covalently. Familiar examples are sucrose (table sugar), maltose (malt sugar), and lactose (milk sugar). **Polysaccharides,** such as cellulose and starch, contain many sugar molecules linked together. Large molecules, such as polysaccharides, that are made up of similar or identical subunits are known as **polymers** ("many parts"), and the subunits are called **monomers** ("single parts").

Monosaccharides: Ready Energy for Living Systems

Monosaccharides are organic compounds composed of carbon, hydrogen, and oxygen. They can be described by the formula $(CH_2O)_n$, where n may be as small as 3, as in $C_3H_6O_3$, or as large as 8, as in $C_8H_{16}O_8$ (Figure 3–5). These proportions gave rise to the term carbohydrate ("hydrate of carbon") for sugars and the larger molecules formed from sugar subunits.

As you can see by studying Figure 3–5, monosaccharides are characterized by hydroxyl groups and an aldehyde or ketone group. These functional groups make sugars highly soluble in aqueous solution and, in molecules containing more than five carbon atoms, lead to an internal reaction that dramatically changes the shape of the molecule. When these monosaccharides are in solution, the aldehyde or ketone group has a tendency to react with one of the hydroxyl groups, producing a ring structure. In glucose, for example, the aldehyde group on the first carbon atom reacts with the hydroxyl group on the fifth carbon atom, producing a six-membered ring structure, as shown in Figure 3–6. When the ring forms, it may close in one of two ways, with the hydroxyl group now on the first carbon positioned either above or below the plane of the ring. The form in which the hydroxyl group is below the plane is known as alpha-glucose, and the form in which it is above the plane is known as beta-glucose. As we shall see, this small difference between the alpha and beta forms of glucose can lead to very significant differences in the properties of larger molecules formed by living systems from glucose.

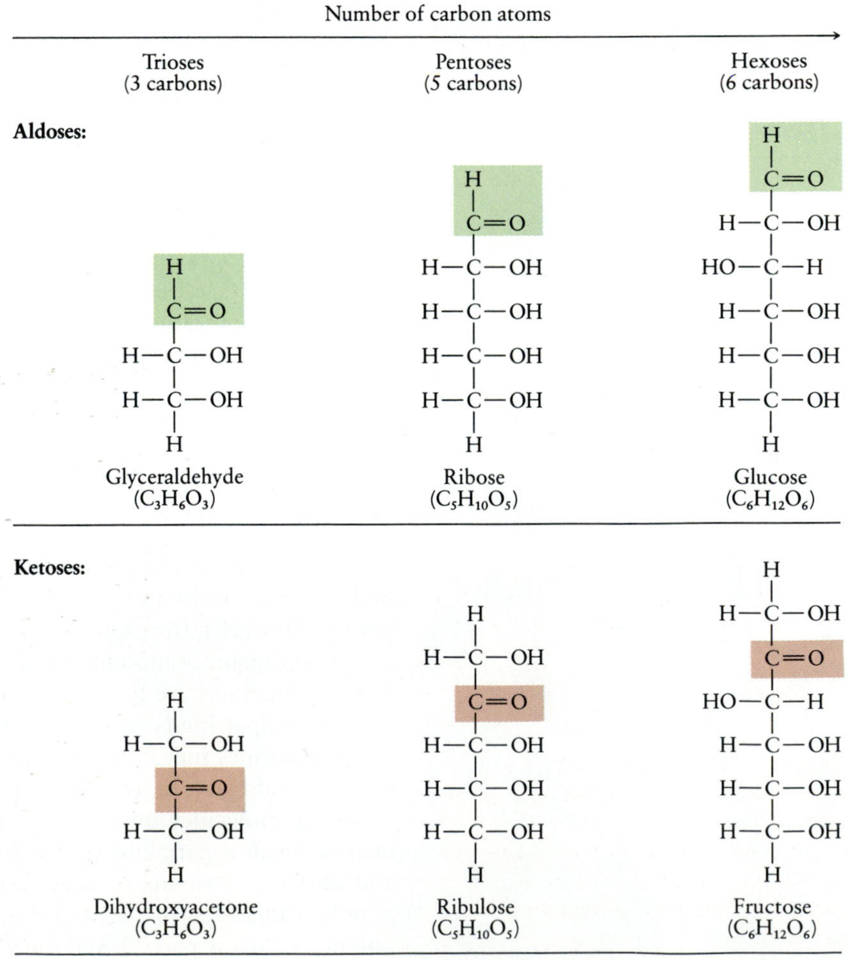

3–5 *Two different ways of classifying monosaccharides: according to the number of carbon atoms and according to the functional groups, indicated here in color. Glyceraldehyde, ribose, and glucose contain, in addition to hydroxyl groups, an aldehyde group, indicated in green; they are called aldose sugars (aldoses). Dihydroxyacetone, ribulose, and fructose each contain a ketone group, indicated in brown, and are called ketose sugars (ketoses).*

Number of carbon atoms

	Trioses (3 carbons)	Pentoses (5 carbons)	Hexoses (6 carbons)

Aldoses: ... Glyceraldehyde ($C_3H_6O_3$), Ribose ($C_5H_{10}O_5$), Glucose ($C_6H_{12}O_6$)

Ketoses: ... Dihydroxyacetone ($C_3H_6O_3$), Ribulose ($C_5H_{10}O_5$), Fructose ($C_6H_{12}O_6$)

⁶CH₂OH

alpha-Glucose

$$(CH_2O)_n + nO_2 \longrightarrow (CO_2)_n + (H_2O)_n$$

Like hydrocarbons, monosaccharides can be burned, or oxidized, to yield carbon dioxide and water:

$$(CH_2O)_n + nO_2 \longrightarrow (CO_2)_n + (H_2O)_n$$

This reaction, like the burning of methane, releases energy, and the amount of energy released as heat can be calculated by burning sugar molecules in a calorimeter. The same amount of energy is released—although not nearly so wastefully—when the equivalent amount of carbohydrate is oxidized in a living cell as when it is burned in a calorimeter.

This statement comparing the oxidation of food molecules with that of fuel molecules is not a metaphor; it is a fact. For example, the energy cost of transporting a kilogram of body weight the distance of a kilometer is 0.95 kilocalorie for a pigeon, 0.73 kcal for a person, and 0.83 kcal for a Cadillac.

A principal energy source for humans and other vertebrates is the monosaccharide glucose. It is in this form that sugar is generally transported in the animal body. A patient receiving an intravenous feeding in a hospital is getting glucose dissolved in a salt solution that approximates the ionic composition of body fluids. The dissolved glucose is carried through the bloodstream to the cells of the body where the energy-releasing reactions are carried out. As measured in a calorimeter, the oxidation of a mole of glucose releases 673 kilocalories:

$$C_6H_{12}O_6 + 6O_2 \longrightarrow 6CO_2 + 6H_2O$$
$$\Delta H^\circ = -673 \text{ kcal}$$

Disaccharides: Transport Forms

Although glucose is the common transport sugar for vertebrates, sugars are often transported in other organisms as disaccharides. Sucrose, commonly called cane sugar, is the form in which sugar is transported in plants from the photosynthetic cells (mostly in the leaves), where it is produced, to other parts of the plant body. Sucrose is composed of the monosaccharides glucose and fructose. Sugar is transported through the blood of many insects in the form of another disaccharide, trehalose, which consists of two glucose units linked together. Another common disaccharide is lactose, a sugar that occurs only in milk. Lactose is made up of glucose combined with another monosaccharide, galactose.

3–6 *In aqueous solution, the six-carbon sugar glucose exists in two different ring structures, alpha and beta, that are in equilibrium with each other. The molecules pass through the straight-chain form to get from one structure to the other. The sole difference in the two ring structures is the position of the hydroxyl group attached to carbon atom 1; in the alpha form it is below the plane of the ring, and in the beta form it is above the plane.*

3–7 *A male calliope hummingbird drinking sugary syrup—nectar—from the flower of an Oregon grape. Many animals have sensitive detection mechanisms for sugar and apparently find its taste pleasurable. In the course of consuming sugary plant products, animals obtain not only a rich energy supply but also other essential nutrients, such as plant proteins, lipids, vitamins, and minerals.*

Representations of Molecules

As we saw in Chapters 1 and 2, chemists have developed various models to represent the structures of atoms and molecules. Each of these models is a way of organizing a particular set of scientific data and of focusing attention on particular characteristics of atoms and molecules.

Because the properties of a molecule depend on its three-dimensional characteristics, physical models are often the most useful. For example, ball-and-stick models of the kind shown in Figure 3–2 emphasize the geometry of a molecule and, in particular, the bonds between atoms. But these models fail to suggest the overall shape of the molecule created by the movement of electrons within their orbitals.

A closer approximation of molecular shape is provided by space-filling models. Each atom is represented by the edge of its outermost orbitals. Space-filling models are misleading, however, in that molecules do not fill space in the same way that we think of a table or a rock as filling space. The atoms that make up molecules consist mostly of empty space. If the perimeter of the outer orbitals of the electrons in an oxygen atom were the size of the perimeter of the Astrodome in Houston, the nucleus would be a ping-pong ball in the center of the stadium. What "fills" the space in molecules are regions of charge, associated with the movements of the electrons around the nuclei. One molecule "sees" another molecule in terms of these regions of charge. As a consequence, for instance, a protein that transports glucose molecules into the living cell will not transport fructose molecules because of the differences in the shape of the regions of charge. All the intricate biochemistry that goes on in the cell is based on this ability of molecules to "recognize" one another.

Ball-and-stick and space-filling models are often used in the laboratory, but they are less useful on paper because it is necessary to see them from all angles to see all of the atoms and their bonds. The most accurate two-dimensional representations of molecular structure are orbital models, such as those shown in Figure 2–3 (page 41). For molecules containing more than a few atoms, however, orbital models become extremely complicated. Thus, when representing complex molecules, such as those found in living systems, chemists usually use molecular formulas or structural formulas. A molecular formula indicates the number of atoms of each kind within the molecule, while a structural formula shows how the atoms are bonded to one another.

Space-filling models of the sugars glucose and fructose. The gray spheres, almost completely hidden at the center of each molecule, represent the carbon atoms. The red spheres at the surface of each molecule represent oxygen atoms, while the blue spheres represent hydrogen atoms.

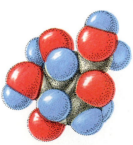 

Glucose   Fructose

Glucose, for example, has 6 carbon atoms, 12 hydrogen atoms, and 6 oxygen atoms. Its molecular formula is $C_6H_{12}O_6$. However, fructose also contains 6 carbons, 12 hydrogens, and 6 oxygens and has a similar structure—a chain of carbon atoms to which hydrogen and oxygen atoms are attached. The differences between glucose and fructose are determined by which carbon atoms the other atoms are attached to. The molecules can therefore be distinguished by their structural formulas:

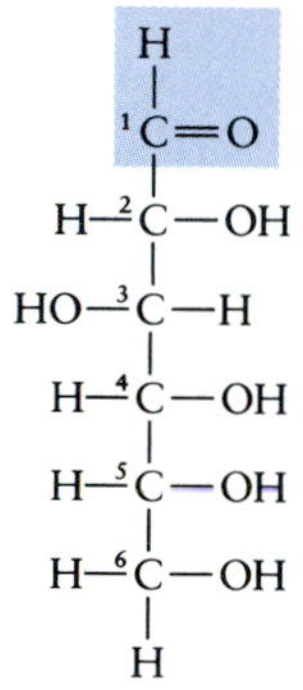

Glucose
(chain)

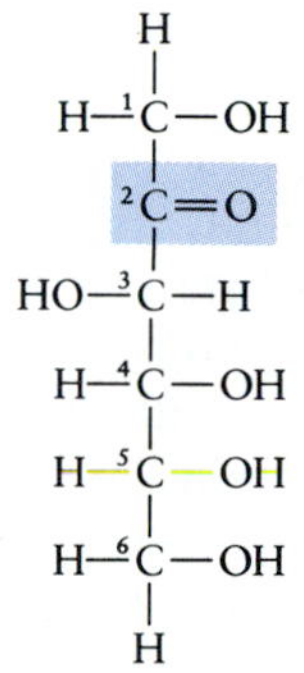

Fructose
(chain)

Or, in the ring forms:

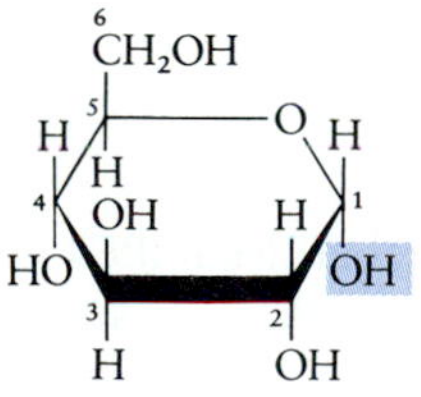

alpha-Glucose

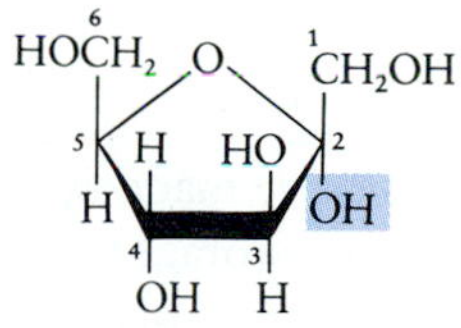

alpha-Fructose

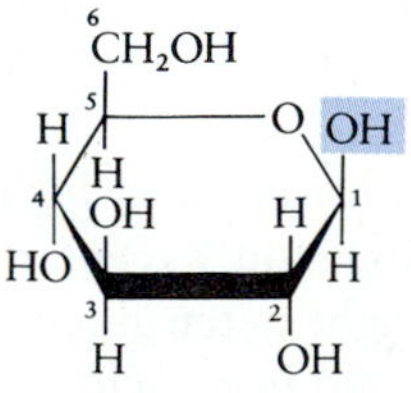

beta-Glucose

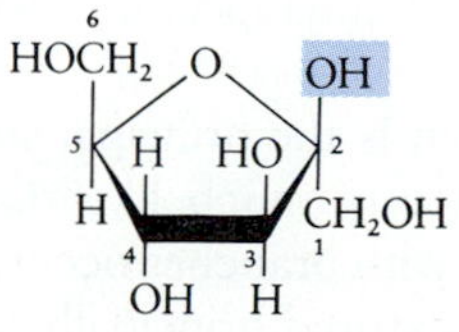

beta-Fructose

The lower edges of the rings are made thicker to hint at a three-dimensional structure. By convention, the carbon atoms at the intersections of the links in an organic ring structure are "understood" and not labeled. Although it is not necessary to number the carbon atoms, doing so often makes it easier to interpret the structural formula.

Notice that the reaction leading to the formation of the ring structures of fructose involves the ketone group, which is on carbon 2, and the hydroxyl group on carbon 5. The result is a five-membered ring, with an overall shape quite unlike that of the six-membered ring formed by glucose. The hydroxyl group whose position above or below the plane of the ring determines whether the molecule is alpha-fructose or beta-fructose is on carbon 2.

In sugars that form ring structures in solution, the positions of the hydroxyl groups not involved in ring formation are easily determined: —OH groups that appear on the left side of the structural formula for the straight-chain form go above the plane of the ring, and —OH groups that appear on the right side of that structural formula go below the plane of the ring.

Although structural formulas do not give us exact information about the shapes of the regions of charge that are so critical in biological reactions, they do give us more information than is apparent at first glance. You will find them a convenient tool as we examine the molecules involved in the structures and processes of living systems.

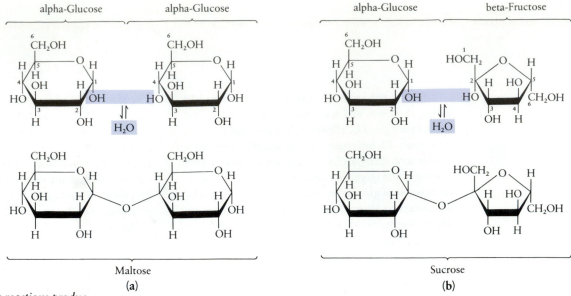

Maltose

(a)

Sucrose

(b)

3–8 *The condensation reactions producing two common disaccharides.* **(a)** *Maltose is a disaccharide made up of two alpha-glucose units, joined in what is known as a 1 ⟶ 4 linkage (the bonding between the two rings involves the 1-carbon of one glucose subunit and the 4-carbon of the other).* **(b)** *Sucrose is a disaccharide formed from an alpha-glucose unit and a beta-fructose unit, joined in a 1 ⟶ 2 linkage (the bonding between the two rings involves the 1-carbon of glucose and the 2-carbon of fructose). In order to represent this bond on paper, we must rotate the structural formula for beta-fructose 180° (right to left), which has the disconcerting effect of turning everything upside down to our eyes. In the three-dimensional world in which the molecules actually exist, however, formation of this 1 ⟶ 2 linkage creates no problems.*

As you can see, the condensation reactions producing these disaccharides involve the removal of a molecule of water. Splitting them back into their constituent monosaccharides requires the addition of a water molecule (hydrolysis).

In the synthesis of a disaccharide molecule from two monosaccharide molecules, a molecule of water is removed in the process of forming the new bond between the two monosaccharides (Figure 3–8). This type of chemical reaction, which occurs in the synthesis of most organic polymers from their subunits, is known as **condensation.** Thus, only the free monomers of carbohydrates actually have a CH_2O ratio because of the removal of two atoms of hydrogen and one of oxygen every time such a bond is formed.

When a disaccharide is split into its monosaccharide units, which happens when it is used as an energy source, the molecule of water is added again. This splitting is known as **hydrolysis,** from *hydro,* meaning "water," and *lysis,* meaning "breaking apart." Hydrolysis is an energy-releasing reaction. The hydrolysis of sucrose, for example, releases 5.5 kilocalories per mole. Conversely, the formation of sucrose from glucose and fructose requires an energy input of 5.5 kilocalories per mole of sucrose.

Storage Polysaccharides

Polysaccharides are made up of monosaccharides linked together in long chains. Some of them are storage forms of sugar. **Starch,** for instance, is the principal food storage form in most plants. A potato, for example, contains starch produced from the sugar formed in the green leaves of the plant; the sugar is transported underground and accumulated there in a form suitable for winter storage, after which it will provide for new growth in the spring. Starch occurs in two forms, amylose and amylopectin. Both consist of glucose units linked together (Figure 3–9).

Glycogen is the principal storage form for sugar in higher animals. Glycogen has a structure much like that of amylopectin except that it is more highly branched, with branches occurring every eight to ten glucose units. In vertebrates, glycogen is stored principally in the liver and in muscle tissue. When there is an excess of glucose in the bloodstream, the liver forms glycogen. When the concentration of glucose in the blood drops, the hormone glucagon, produced by the pancreas, is released into the bloodstream; glucagon stimulates the liver to hydrolyze glycogen to glucose, which then enters the bloodstream.

Formation of polysaccharides from monosaccharides requires energy. However, when the cell needs energy, these polysaccharides can be hydrolyzed, releasing monosaccharides that can, in turn, be oxidized to provide energy for cellular work.

Amylose
(a)

A branch point in amylopectin
(b)

(c)

(d)

(e)

3–9 *In plants, sugars are stored in the form of starch. Starch is composed of two different types of polysaccharides, amylose (a) and amylopectin (b). A single molecule of amylose may contain 1,000 or more alpha-glucose units with carbon 1 of one glucose ring linked to carbon 4 of the next in a long, unbranched chain, which coils to form a helix (c). A molecule of amylopectin may contain from 1,000 to 6,000 alpha-glucose units; short chains containing about 24 to 36 alpha-glucose units periodically branch off from the main chain.*

(d) Starch molecules, perhaps because of their helical nature, tend to cluster into granules. In this scanning electron micrograph of a single storage cell of a potato, the spherical and egg-shaped objects are starch granules. They are magnified about 1,000 times.

(e) Glycogen, which is the common storage form for sugar in vertebrates, resembles amylopectin in its general structure except that each branch contains only 16 to 24 alpha-glucose units. The dark granules in this liver cell, magnified about 55,000 times, are glycogen. When glucose is needed, it is provided by the hydrolysis of glycogen.

Structural Polysaccharides

A major function of molecules in living systems is to form the structural components of cells and tissues. The principal structural molecule in plants is **cellulose**. In fact, half of all the organic carbon in the biosphere is contained in cellulose. Wood is about 50 percent cellulose, and cotton is nearly pure cellulose.

Cellulose molecules form the fibrous part of the plant cell wall. The cellulose fibers, embedded in a matrix of other kinds of polysaccharides, form an external envelope around the plant cell. When the cell is young, this envelope is flexible and stretches as the cell grows, but it becomes thicker and more rigid as the cell matures. In some plant tissues, such as the tissues that form wood and bark, the cells eventually die, leaving only their tough outer walls.

Cellulose is a polymer composed of monomers of glucose, just as starch and glycogen are. Starch and glycogen can be readily utilized as fuels by almost all kinds of living systems, but only a few microorganisms—certain bacteria, protozoa, and fungi—can hydrolyze cellulose. Cows and other ruminants, termites, and cockroaches can use cellulose for energy only because of microorganisms that inhabit their digestive tracts.

To understand the differences between structural polysaccharides, such as cellulose, and energy-storage polysaccharides, such as starch or glycogen, we must look again at the glucose molecule. You will remember that the molecule is basically a chain of six carbon atoms and that when it is in solution, as it is in the cell, it assumes a ring form. The ring may close in either of two ways (see Figure 3–6). One ring form is known as alpha, and the other as beta. The alpha and beta forms are in equilibrium, with a certain number of molecules changing from one form to the other all the time, going through the open-chain structure to reach the other form. Starch and glycogen are both made up entirely of alpha units.

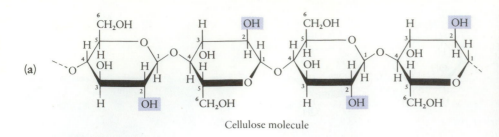

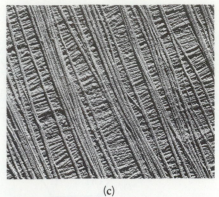

Cellulose molecule

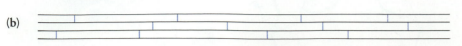

(b)

Model of cross-linked cellulose molecules

(c)

3–10 (a) *Cellulose consists of beta-glucose monomers, joined in 1 ⟶ 4 linkages. (Note that the structural formulas for alternating beta-glucose units have been rotated 180° to show the bonding.) In cellulose, the —OH groups (indicated in color), which project from both sides of the chain, form hydrogen bonds with neighboring —OH groups, resulting in* the formation of bundles of cross-linked parallel chains (b). *By contrast, in the starch molecule (Figure 3–9), most of the —OH groups capable of forming hydrogen bonds face toward the exterior of the helix, making it more readily soluble in the surrounding water.*

(c) *The wall of a young plant cell is about 40 percent cellulose. Each of the* microfibrils you can see here (magnified about 30,000 times) is a bundle of hundreds of cellulose strands, and each strand is a chain of beta-glucose monomers (a). *The microfibrils, as strong as an equivalent amount of steel, are embedded in other polysaccharides, one of which is pectin.*

Cellulose, however, consists entirely of beta units (Figure 3–10). This slight difference has a profound effect on the three-dimensional structure of the molecules, which align in parallel, forming crystalline cellulose microfibrils. As a result, cellulose is impervious to the enzymes that so successfully break down the storage polysaccharides.

Chitin, which is a major component of the exoskeletons of arthropods, such as insects and crustaceans, and also of the cell walls of many fungi, is a tough, resistant, modified polysaccharide (Figure 3–11). At least 900,000 different species of organisms can synthesize chitin, and it has been estimated that the individuals belonging to a single species of crab produce several million tons of chitin a year.

3–11 (a) *Chitin is a polymer consisting of repeated modified monosaccharides. As you can see, the monomer is a six-carbon sugar, like glucose, in which a nitrogen-containing group has replaced the —OH group on carbon 2.* (b) *A cicada molting. The relatively hard outer coverings, or* exoskeletons, of insects contain chitin. Because exoskeletons do not grow as the insect grows, they must be molted periodically. The discarded exoskeleton is at the top, above the insect, which is drying out and waiting for its new exoskeleton to harden.

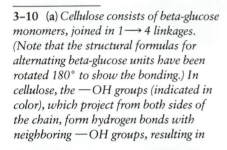

Chitin

(a)

(b)

LIPIDS

Lipids are a general group of organic substances that are insoluble in polar solvents, such as water, but that dissolve readily in nonpolar organic solvents, such as chloroform, ether, and benzene. Typically, lipids serve as energy-storage molecules—usually in the form of fats or oils—and for structural purposes, as in the case of phospholipids, glycolipids, and waxes. Some lipids, however, play major roles as chemical "messengers," both within and between cells.

Fats and Oils: Energy in Storage

Unlike many plants, such as the potato, animals have only a limited capacity to store carbohydrates. In vertebrates, sugars in excess of what can be stored as glycogen are converted into fats. Some plants also store food energy as oils, especially in seeds and fruits. Fats and oils contain a higher proportion of energy-rich carbon-hydrogen bonds (see Figure 3–3) than carbohydrates do and, as a consequence, contain more chemical energy. On the average, fats yield about 9.3 kilocalories per gram* as compared to 3.79 kcal per gram of carbohydrate, or 3.12 kcal per gram of protein. Also, because fats are nonpolar, they do not attract water molecules and hence are not "weighted down" by them, as glycogen is. Taking into account the water factor, fats store six times as much energy, gram for gram, as glycogen, which is undoubtedly why in the course of evolution they came to play a major role in energy storage.

An example of the value of this concentrated energy storage is provided by hummingbirds. A male ruby-throated hummingbird has a fat-free weight of 2.5 grams (about 1/10 ounce). It migrates every fall from Florida to Yucatan, some 2,000 kilometers. Before doing so, it accumulates 2.0 grams of body fat, an amount almost equal to its original weight. However, if it were to carry the same energy reserves in the form of glycogen, it would have to carry 5 grams, twice its own fat-free weight.

A fat molecule consists of three molecules of fatty acid joined to one glycerol molecule. Glycerol, as we noted previously, is a three-carbon alcohol that contains three hydroxyl groups. A fatty acid consists of a long hydrocarbon chain that terminates in a carboxyl group (—COOH); the nonpolar chain is hydrophobic, whereas the carboxyl group gives one portion of the molecule the properties of an acid. As with the disaccharides and polysaccharides, each bond between glycerol and a fatty acid is formed by the removal of a molecule of water (condensation), as shown in Figure 3–12. Fat molecules, which are also known as triglycerides, are said to be neutral because they contain no polar groups. As you would expect, they are extremely hydrophobic.

Fatty acids, which are seldom found in cells in a free state (that is, not as part of another molecule), consist of chains containing an even number of carbon atoms, typically between 14 and 22. About 70 different fatty acids are known. They differ in their chain lengths, in whether the chain contains any double bonds (as in oleic acid) or not (as in stearic acid), and in the position in the chain of any double bonds (see Figure 3–12). A fatty acid, such as stearic acid, in which there are no double bonds is said to be **saturated** because the bonding possibilities are complete for all the carbon atoms of the chain (that is, each carbon atom has formed bonds to four other atoms). A fatty acid, such as oleic acid, that contains carbon atoms joined by double bonds is said to be **unsaturated** because those carbon atoms have the potential to form additional bonds with other atoms.

Unsaturated fats, which tend to be oily liquids, are more common in plants than in animals; examples are olive oil, peanut oil, and corn oil. Animal fats, such as butter and lard, contain saturated fatty acids and usually have higher melting temperatures.

---

* 1,000 grams = 1 kilogram = 2.2 pounds, so oxidation of a pound of fat would yield about 4,200 kilocalories, more than the 24-hour requirement for a moderately active adult.

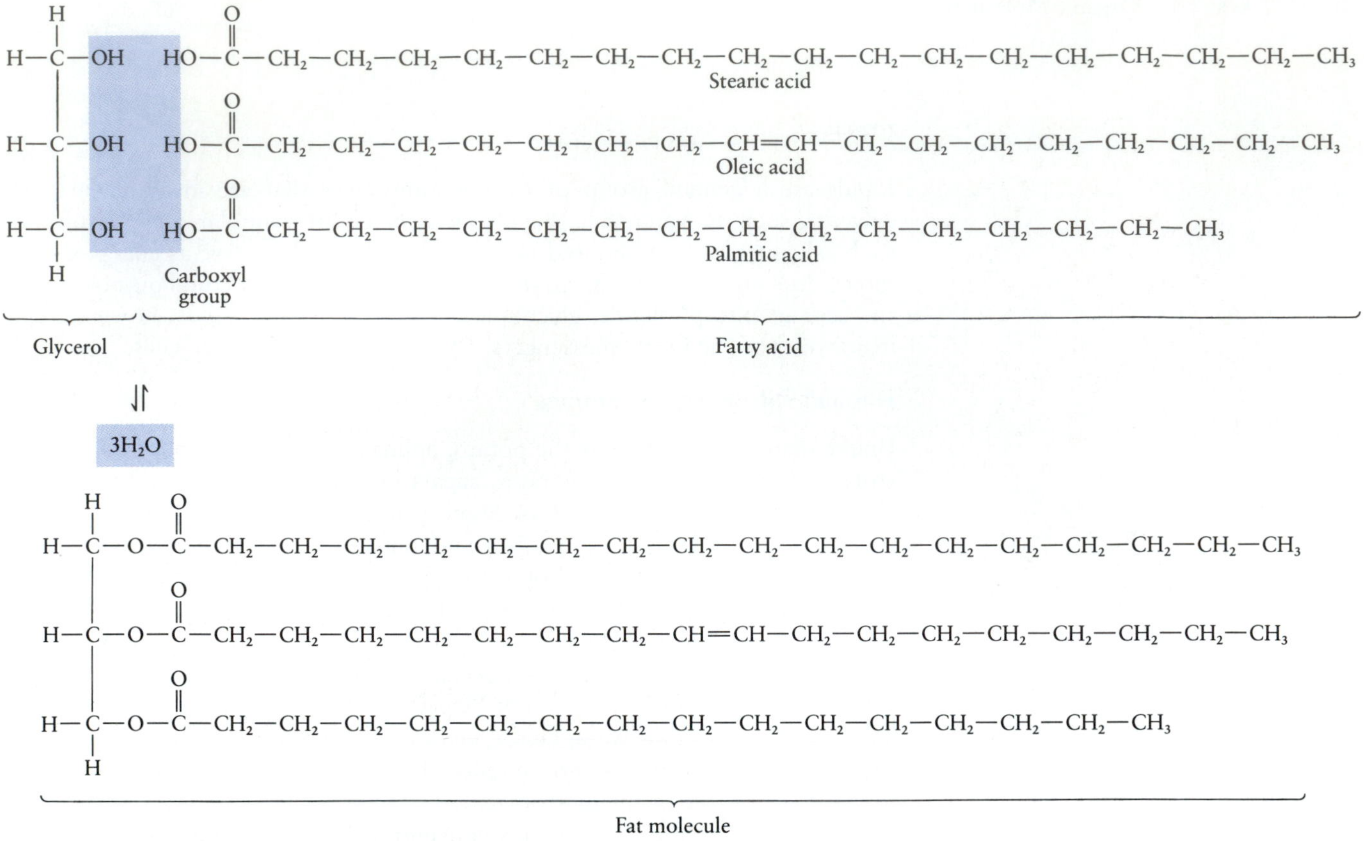

H
H—C—OH   HO—C—CH₂—CH₂—CH₂—CH₂—CH₂—CH₂—CH₂—CH₂—CH₂—CH₂—CH₂—CH₂—CH₂—CH₂—CH₂—CH₃

$$H-C-OH \quad HO-\overset{O}{\overset{\|}{C}}-CH_2-CH_2-CH_2-CH_2-CH_2-CH_2-CH_2-CH_2-CH_2-CH_2-CH_2-CH_2-CH_2-CH_2-CH_2-CH_2-CH_3$$
Stearic acid

$$H-C-OH \quad HO-\overset{O}{\overset{\|}{C}}-CH_2-CH_2-CH_2-CH_2-CH_2-CH_2-CH_2-CH=CH-CH_2-CH_2-CH_2-CH_2-CH_2-CH_2-CH_2-CH_3$$
Oleic acid

$$H-C-OH \quad HO-\overset{O}{\overset{\|}{C}}-CH_2-CH_2-CH_2-CH_2-CH_2-CH_2-CH_2-CH_2-CH_2-CH_2-CH_2-CH_2-CH_2-CH_2-CH_3$$
H   Palmitic acid

Carboxyl group

Glycerol   Fatty acid

$\Updownarrow$

3H₂O

$$H-C-O-\overset{O}{\overset{\|}{C}}-CH_2-CH_2-CH_2-CH_2-CH_2-CH_2-CH_2-CH_2-CH_2-CH_2-CH_2-CH_2-CH_2-CH_2-CH_2-CH_2-CH_3$$

$$H-C-O-\overset{O}{\overset{\|}{C}}-CH_2-CH_2-CH_2-CH_2-CH_2-CH_2-CH_2-CH=CH-CH_2-CH_2-CH_2-CH_2-CH_2-CH_2-CH_2-CH_3$$

$$H-C-O-\overset{O}{\overset{\|}{C}}-CH_2-CH_2-CH_2-CH_2-CH_2-CH_2-CH_2-CH_2-CH_2-CH_2-CH_2-CH_2-CH_2-CH_2-CH_3$$
H

Fat molecule

3–12 *A fat molecule consists of three fatty acids joined to a glycerol molecule (hence the term "triglyceride"). The long hydrocarbon chains of which the fatty acids are composed terminate in carboxyl (—COOH) groups, which become covalently bonded to the glycerol molecule. Each bond is formed when a molecule of water (color) is removed (condensation). The physical properties of a fat—such as its melting point—are determined by the lengths of its fatty acid chains and by whether the chains are saturated or unsaturated. Three different fatty acids are shown here. Stearic acid and palmitic acid are saturated, and oleic acid is unsaturated, as you can see by the double bond in its structure.*

Sugars, Fats, and Calories

As we noted earlier, when carbohydrates are taken into the body in excess of the body's energy requirements, they are stored temporarily as glycogen or, more permanently, as fats. Conversely, when the energy requirements of the body are not met by its immediate intake of food, glycogen and, subsequently, fat are broken down to fill these requirements. Whether or not the body uses up its own storage molecules has nothing to do with the molecular form in which the energy comes into the body. It is simply a matter of whether these molecules, as they are broken down, release sufficient numbers of calories.

Insulators and Cushions

In general, fat stored in fat cells can be mobilized for energy when caloric intake is less than caloric expenditures. Some types of fat, however, seem to be protected from such mobilization. Large masses of fatty tissue, for example, surround mammalian kidneys and serve to protect these precious organs from physical shock. For reasons that are not understood, these fat deposits remain intact even at times of starvation. Another mammalian characteristic is a layer of fat under the skin, which serves as thermal insulation. This layer is particularly well developed in seagoing mammals.

Among humans, females characteristically have a thicker layer of subdermal ("under-the-skin") fat than males. This capacity to store fat, although not much admired in our present culture, was undoubtedly very valuable 10,000 or more years ago. At that time, as far as we know, there was no other reserve food supply, and this extra fat not only nourished the woman but, more important, the unborn child and the nursing infant, whose ability to fast without damage is much less than that of the adult. Thus many of us are strenuously dieting off what millennia of evolution have given us the capacity to accumulate.

3-13 *This harp seal, resting on an ice floe in the Gulf of St. Lawrence, is well insulated by a thick layer of fat under the skin, which serves the same function that a wet suit serves for a diver.*

Phospholipids and Glycolipids

Lipids, especially phospholipids and glycolipids, also play extremely important structural roles. Like fats, both phospholipids and glycolipids are composed of fatty acid chains attached to a glycerol backbone. In the **phospholipids,** however, the third carbon of the glycerol molecule is occupied not by a fatty acid but by a phosphate group (Figure 3–14) to which another polar group is usually attached. Phosphate groups are negatively charged. As a result, the phosphate end of the molecule is hydrophilic, whereas the fatty acid portions are hydrophobic. The consequences are shown in Figure 3–15. As we shall see in Chapter 5, this arrangement of phospholipid molecules, with their hydrophilic heads exposed and their hydrophobic tails clustered together, forms the structural basis of cellular membranes.

In the **glycolipids** ("sugar lipids"), the third carbon of the glycerol molecule is occupied not by a phosphate group but by a short carbohydrate chain. Depending on the particular glycolipid, this chain may contain anywhere from 1 to 15 monosaccharide monomers. Like the phosphate head of a phospholipid, the carbohydrate head of a glycolipid is hydrophilic, and the fatty acid tails are, of course, hydrophobic. In aqueous solution, glycolipids behave in the same fashion as phospholipids, and they are also important components of cellular membranes.

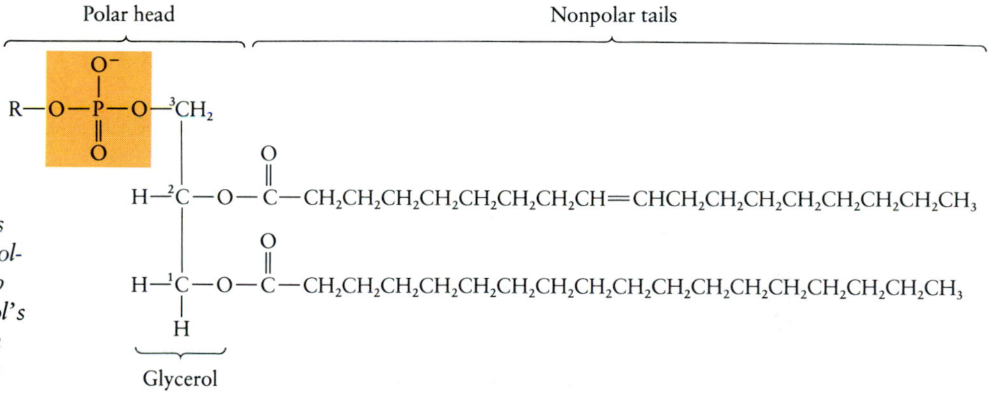

3-14 *A phospholipid molecule consists of two fatty acids linked to a glycerol molecule, as in a fat, and a phosphate group (indicated by color) linked to the glycerol's third carbon. It also usually contains an additional chemical group, indicated by the letter R. The fatty acid "tails" are nonpolar and therefore insoluble in water (hydrophobic); the polar "head" containing the phosphate and R groups is soluble (hydrophilic).*

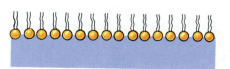

(a)

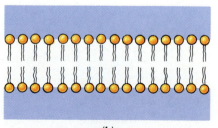

(b)

3-15 (a) *Because phospholipids have water-soluble heads and water-insoluble tails (Figure 3–14), they tend to form a thin film on a water surface with their tails extending above the water.* (b) *Surrounded by water, they spontaneously* *arrange themselves in two layers with their hydrophilic (water-loving) heads extending outward and their hydrophobic (water-fearing) tails inward. This arrangement forms the structural basis of cellular membranes.*

Waxes

Waxes are also a form of structural lipid. They form protective coatings on skin, fur, feathers, on the leaves and fruits of land plants (Figure 3–16), and on the exoskeletons of many insects.

Cholesterol and Other Steroids

Cholesterol belongs to an important group of compounds known as the **steroids** (Figure 3–17). Although steroids do not resemble the other lipids structurally, they are grouped with them because they are insoluble in water. All the steroids have four linked carbon rings, like cholesterol, and several of them, like cholesterol, have a tail. In addition, many of them have the —OH functional group, which makes them alcohols.

Cholesterol is found in cell membranes (with the exception of bacterial cells); about 25 percent (by dry weight) of the cell membrane of a red blood cell is cholesterol. It is also a major component of the myelin sheath, the lipid membrane that wraps around fast-conducting nerve fibers, speeding the nerve impulse. Cholesterol is synthesized in the liver from saturated fatty acids and is also obtained in the diet, principally in meat, cheese, and egg yolks. High concentrations of cholesterol in the blood are associated with atherosclerosis, in which cholesterol is found in fatty deposits on the interior lining of diseased blood vessels (see essay).

Sex hormones and the hormones of the adrenal cortex (the outer portion of the adrenal glands, which lie atop the kidneys) are also steroids. These hormones are formed from cholesterol in the ovaries, testes, adrenal cortex, and other glands that produce them. Prostaglandins are a group of lipids with hormonelike actions, which are derived from fatty acids. Both the steroid hormones and the prostaglandins will be discussed more fully in Section 6.

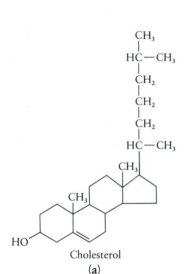

Cholesterol

(a)

Testosterone

(b)

3–17 *Two examples of steroids. (a) The cholesterol molecule consists of four carbon rings and a hydrocarbon chain. (b) Testosterone, a male sex hormone synthesized from cholesterol by cells in the testes, also has the characteristic four-ring structure but lacks the hydrocarbon tail.*

PROTEINS

Proteins are among the most abundant organic molecules; in most living systems they make up 50 percent or more of the dry weight. Only plants, with their high cellulose content, are less than half protein. There are many different protein molecules: enzymes; hormones; storage proteins, such as those in the eggs of birds and reptiles and in seeds; transport proteins, such as hemoglobin; contractile proteins of the sort found in muscle; immunoglobulins (antibodies); membrane proteins; and many different types of structural proteins (Table 3–2). In their functions, their diversity is overwhelming. In their structure, however, they all follow the same simple blueprint: they are all polymers of amino acids, arranged in a linear sequence.

Regulation of Blood Cholesterol

Although cholesterol plays essential roles in the animal body, it is also a principal villain in heart disease. Deposits containing cholesterol can narrow the arteries carrying blood to the heart muscle, and people with unusually large amounts of cholesterol in their blood have a high risk of heart attacks. How does the body regulate cholesterol levels? What goes wrong to cause elevated levels? How does cholesterol cause heart attacks? Given the fact that heart disease is the major cause of death in this country, these questions are not only of biological interest but are also of importance to just about every one of us.

The key organ in cholesterol regulation is the liver, which not only synthesizes needed cholesterol from saturated fatty acids but also degrades excess cholesterol circulating in the blood—as a result, for example, of a diet rich in milk, cheese, and egg yolks. Cholesterol is transported to and from body cells, including those of the liver, by way of the bloodstream. Like other lipids, however, it is insoluble in water, and thus in plasma, the fluid portion of the blood. It is carried in particles consisting of a cholesterol interior and a lipid "wrapper" that has water-soluble proteins embedded in its outer surface. These large complexes exist in two principal forms: low-density lipoproteins (LDLs) and high-density lipoproteins (HDLs). LDLs function as the delivery trucks of the system, carrying dietary cholesterol and newly synthesized cholesterol to various destinations in the body, including both the liver and the hormone-synthesizing organs. HDLs, however, function more like garbage trucks, carrying excess cholesterol on a one-way trip to the liver for degradation and excretion.

Normally the system is in balance, and the liver synthesizes or degrades cholesterol depending on the body's current needs and the amount of circulating cholesterol. It can, however, be thrown out of balance by a number of factors. If, for example, the dietary intake of cholesterol is high, the liver becomes swamped and cannot degrade all of the excess. If the dietary intake of saturated fats is high, even in the absence of a high intake of cholesterol itself, the liver increases its synthesis of cholesterol. Current evidence indicates that the liver monitors the level of cholesterol in the blood through its uptake of LDLs, for which cell surfaces have specialized receptors. If these receptors are absent or damaged, liver cells continue

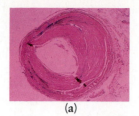 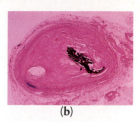

(a) (b)

In one type of heart disease, atherosclerosis, cholesterol and other fatty substances accumulate in the walls of the coronary arteries, which supply the heart muscle. This accumulation triggers abnormal growth and the production of fibrous tissues by the cells of the walls. (a) A cross section of a coronary artery in which moderate atherosclerosis has developed. Fatty deposits have formed, and the space left for blood flow is significantly decreased. (b) A coronary artery in which the deposits have become so great that only a very narrow channel remains open. Such a narrow channel can be completely blocked by a blood clot. The result is a heart attack and the death of the heart muscle supplied by the artery.

to synthesize and export cholesterol in the form of LDLs, even when blood cholesterol levels are high.

When the quantities of circulating LDLs are greater than can be taken up by the liver and hormone-synthesizing organs, they are taken up by the cells lining the arteries supplying the heart. This can ultimately lead to total blockage of an artery and thus to a heart attack.

Heart disease often runs in families, suggesting that hereditary factors are involved in some cases. In one type of hereditary heart disease, the cells of the body have no LDL receptors. Individuals with this disease have six to eight times the normal amount of cholesterol in their blood, usually have their first heart attack in childhood, and die of heart disease in their early twenties. Other families seem to be protected against heart disease, apparently because the individuals' bodies synthesize large quantities of HDLs, ensuring that all excess cholesterol makes a speedy one-way trip to the liver. For most of us, however, the degree of risk depends on our behavior: whether or not we exercise regularly, which seems to increase HDL levels and thus protect against cholesterol buildup; whether or not we smoke cigarettes, which seems to decrease HDL levels; and the quantities of cholesterol and saturated fats that we consume.

Protein molecules are large, often containing several hundred amino acids. Thus the number of different amino acid sequences, and therefore the possible variety of protein molecules, is enormous—about as enormous as the number of different sentences that can be written with our own 26-letter alphabet. Organisms, however, have only a very small fraction of the proteins that are theoretically possible. The single-celled bacterium *Escherichia coli*, for example, contains 600 to 800 different kinds of proteins at any one time, and the cell of a plant or animal has several times that number. In a complex organism, there are at least several thousand different proteins, each with a special function and each, by its unique chemical nature, specifically fitted for that function.

TABLE 3-2 **Biological Functions of Proteins**

TYPES OF PROTEINS*	EXAMPLES
Structural proteins	Collagen, silk, virus coats, microtubules
Regulatory proteins	Insulin, ACTH, growth hormones
Contractile proteins	Actin, myosin
Transport proteins	Hemoglobin, myoglobin
Storage proteins	Egg white, seed protein
Protective proteins in vertebrate blood	Antibodies, complement
Membrane proteins	Receptors; membrane-transport proteins; antigens
Toxins	Botulism toxin, diphtheria toxin
Enzymes	Sucrase, pepsin

* Many of the proteins listed here will be discussed in other sections of the book, particularly in Section 6.

Amino Acids: The Building Blocks of Proteins

Every amino acid has the same fundamental structure, which consists of a central carbon atom bonded to an amino group ($-NH_2$), to a carboxyl group ($-COOH$), and to a hydrogen atom (Figure 3–18a). In every amino acid there is also another atom or group of atoms (designated as $-R$) bonded to the central carbon. As we saw in Chapter 2 (page 48), the amino group is a weak base and the carboxyl group is a weak acid. Depending on the pH of the surrounding solution, a free amino acid may be uncharged, negatively charged (if the $-COOH$ group is ionized to $-COO^-$ and H^+), or positively charged (if the $-NH_2$ group has acquired a hydrogen ion, becoming $-NH_3^+$).

A large variety of different amino acids is theoretically possible, but only 20 different kinds are used to build proteins (Figure 3–18b). And it is always the same 20, whether in a bacterial cell, a plant cell, or a cell in your own body. The only differences in these 20 amino acids lie in their side ($-R$) groups. In eight of the molecules, the side group consists of short chains or rings of carbon and hydrogen atoms; as you would expect, such groups are nonpolar and thus hydrophobic. The side groups in seven of the amino acids have polar regions; in acidic or basic solution, these regions can become charged. The remaining five amino acids have side groups that are either weak acids or weak bases; depending on the particular side group and the pH of the solution, they may be negatively or positively charged.

In yet another example of a condensation reaction, the amino "head" of one amino acid can be linked to the carboxyl "tail" of another by the removal of a

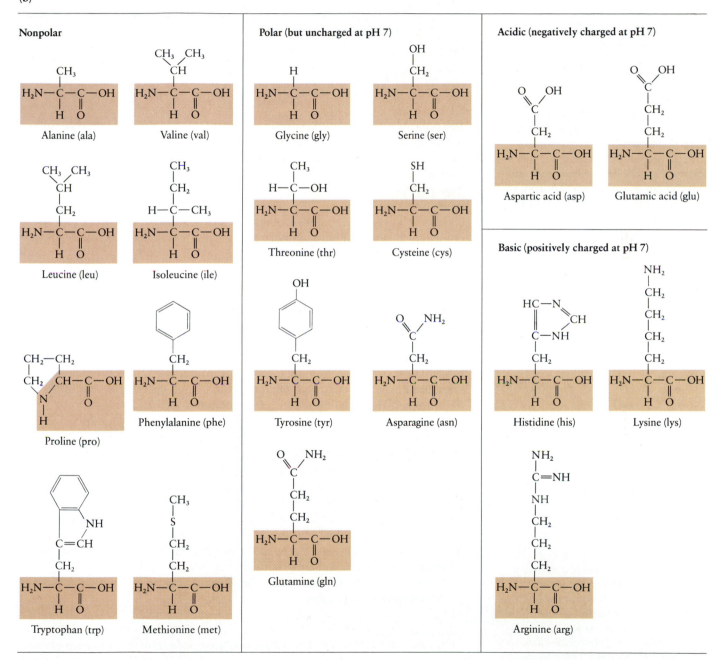

3–18 (a) *Every amino acid contains an amino group (—NH₂) and a carboxyl group (—COOH) bonded to a central carbon atom. A hydrogen atom and a side group are also bonded to the same carbon atom. This basic structure is the same in all amino acids. The "R" stands for the* side group, *which is different in each kind of amino acid.* (b) *The 20 different kinds of amino acids used in making proteins. As you can see, the essential structure is the same in all 20 molecules, but the side groups differ. These groups may be nonpolar (with no difference in charge* between one zone and another), polar but with the charges balancing one another out so that the side group as a whole is uncharged, negatively charged, or positively charged. The nonpolar side groups are not soluble in water, whereas the polar and charged side groups are water-soluble.

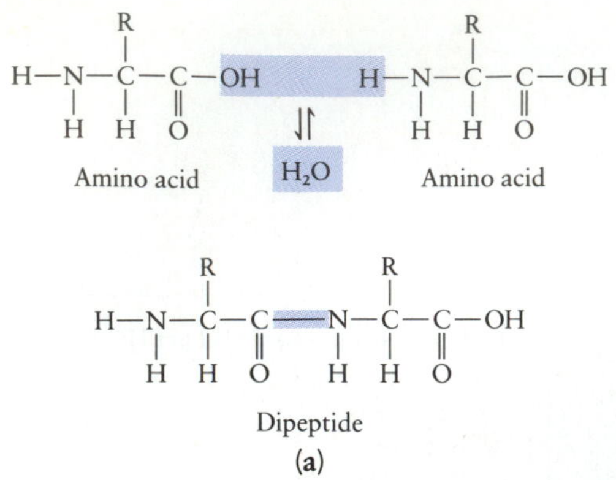

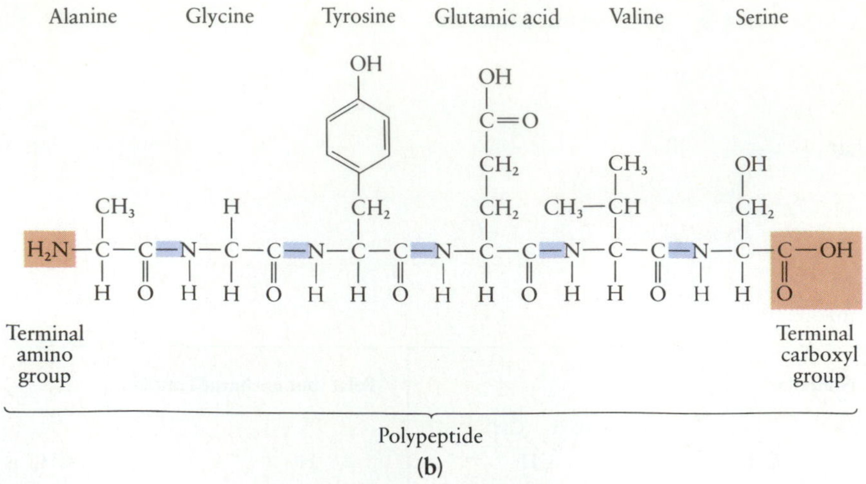

Alanine   Glycine   Tyrosine   Glutamic acid   Valine   Serine

3–19 (a) *A peptide bond is a covalent bond formed by condensation.* (b) *Polypeptides are polymers of amino acids linked together by peptide bonds, with the amino group of one acid joining the carboxyl group of its neighbor. The polypeptide chain shown here contains only six amino acids, but some chains may contain as many as 1,000 linked amino acid monomers.*

molecule of water (Figure 3–19a). The covalent linkage that is formed is known as a **peptide bond,** and the molecule that is formed by the linking of many amino acids is called a **polypeptide** (Figure 3–19b). The sequence of amino acids in the polypeptide chain determines the biological character of the protein molecule; even one small variation in the sequence may alter or destroy the way in which the protein functions.

In order to assemble amino acids into proteins, a cell must have not only a large enough quantity of amino acids but also enough of every kind. This fact is of great importance in human nutrition (see essay).

The Levels of Protein Organization

In a living system, a protein is assembled in a long polypeptide chain, one amino acid at a time. In this process (to be described in some detail in Chapter 15), the amino group of one amino acid is linked to the carbonyl* of another, like a line of boxcars. The linear sequence of amino acids, which is dictated by the hereditary information in the cell for that particular protein, is known as the **primary structure** of the protein. Each different protein has a different primary structure. The primary structure of one protein is shown in Figure 3–20.

As the chain is assembled, interactions begin to take place among the various amino acids along the chain. Linus Pauling and coworker Robert Corey discovered that hydrogen bonds could form between the slightly positive amino hydrogen of one amino acid and the slightly negative carbonyl oxygen of another amino acid. They elucidated two structures that could result from these hydrogen bonds. One of these they called the alpha helix, because it was the first to be discovered, and the second, the beta pleated sheet. These structures are shown in Figure 3–21. Biochemists refer to the regular, repeated configurations caused by hydrogen bonding between atoms of the polypeptide backbone as the **secondary structure** of a protein. Proteins that exist for most of their length in a helical or pleated-sheet form are known as fibrous proteins, and they play important structural roles in organisms.

3–20 *Primary structure of a relatively small protein, human adrenocorticotropic hormone (ACTH). This was one of the first proteins for which the primary structure was determined. As you can see, it consists of a single polypeptide chain containing 39 amino acids. This hormone, secreted by the pituitary gland, stimulates production of cortisol and related steroid hormones by the adrenal cortex.*

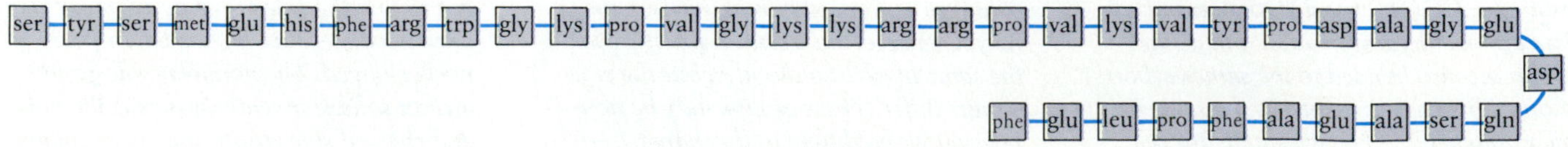

---

* When a peptide bond forms, the OH of the carboxyl group and an H of the amino group split out to form a water molecule. All that remains of the carboxyl group is the $>C=O$ group, which, in this context, we refer to as "carbonyl" (see Table 3–1).

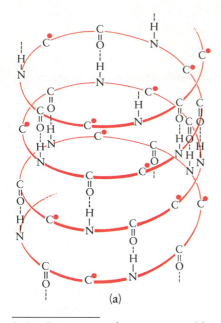

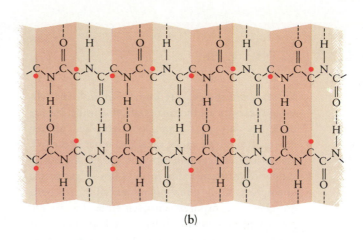

(b)

3–21 *Protein secondary structures.* (a) *The alpha helix. The helix is held in shape by hydrogen bonds, indicated by the dashed lines. The hydrogen bonds form between the oxygen atom of the carbonyl group in one amino acid and the hydrogen atom of the amino group in another amino acid that occurs four amino acids farther along the chain. The R groups, not shown in this diagram, are attached to the carbons indicated by the red dots. The R groups extend out from the helix.* (b) *The beta pleated sheet. The pleats are formed by hydrogen bonding between atoms of the backbone of the polypeptide; the R groups, which are attached to the carbons indicated by the red dots, extend above and below the folds of the pleat.*

Other forces, which involve the nature of the R groups in the individual amino acids, are also at work on the polypeptide chain, and these counteract the formation of the hydrogen bonds just described. For instance, an R group such as that of isoleucine is so bulky that it interrupts the turn of the helix, making the hydrogen bonding impossible. When the —SH portion of the R group of a cysteine encounters the same portion of another cysteine, two hydrogen atoms may split out, resulting in the formation of a covalent bond between the sulfur atoms of the two amino acids. This bond, known as a disulfide bridge, locks the molecule in that position. R groups with unlike charges are attracted to each other, and those with like charges are mutually repelled. As the molecule is twisted and turned in solution, the hydrophobic R groups tend to cluster together in the interior of the molecule and the hydrophilic R groups tend to extend outward into the aqueous solution. Hydrogen bonds form, linking together segments of the amino acid backbone. The intricate three-dimensional structure that results from these interactions among R groups is called the **tertiary structure** of a protein. Figure 3–22a shows the various types of bonds that are involved in forming the tertiary structure.

● = Amino acid
N = Nonpolar (hydrophobic) interactions
S — S = Disulfide bridges
– – – = Hydrogen bonds
P⁺, P⁻ = Polar (hydrophilic) groups

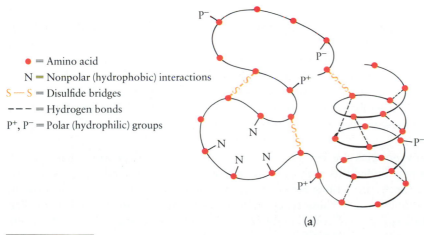

(a)

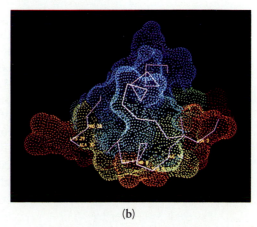

(b)

3–22 (a) *Types of bonds that stabilize the tertiary structure of a protein molecule. These same types of bonds also stabilize the structure of protein molecules that consist of more than one polypeptide chain.* (b) *A computer-generated model of the insulin molecule, which consists of*

two short polypeptide chains, folded together in an intricate three-dimensional structure. The two lines represent the backbones of the chains, and the dots represent the atoms on the surface of the molecule that are accessible to the surrounding solvent. In the insulin mole-

cule, as in all molecules, atoms are constantly vibrating and rotating. The atoms shown in red and orange are most likely to undergo slight shifts of position in the insulin crystal, whereas those shown in green and blue are least likely to shift position.

Amino Acids and Nitrogen

Like fats, amino acids are formed within living cells using sugars as starting materials. But while fats are made up only of carbon, hydrogen, and oxygen atoms, all available in the sugar and water of the cell, amino acids also contain nitrogen. Most of the earth's supply of nitrogen exists in the form of gas in the atmosphere. Only a few organisms, all microscopic, are able to incorporate nitrogen from the air into compounds—ammonia, nitrites, and nitrates—that can be used by living systems. Hence, the proportion of the earth's nitrogen supply available to the living world is very small.

Plants incorporate the nitrogen in ammonia, nitrites, and nitrates into carbon-hydrogen compounds to form amino acids. Animals are able to synthesize some of their amino acids, using ammonia as a nitrogen source. The amino acids they cannot synthesize, the so-called essential amino acids, must be obtained either directly or indirectly from plants. For adult human beings, the essential amino acids are lysine, tryptophan, threonine, methionine, phenylalanine, leucine, valine, and isoleucine.

People who eat meat usually get enough protein and the correct balance of amino acids. People who are vegetarians, whether for philosophical, esthetic, or economic reasons, have to be careful that they get enough protein and, in particular, the essential amino acids.

Until recently, agricultural scientists concerned with the world's hungry people concentrated on developing plants with a high caloric yield. Increasing recognition of the role of plants as a major source of amino acids for human populations has led to emphasis on the development of high-protein strains of food plants and of plants with essential amino acids, such as "high-lysine" corn.

Another approach to the right balance of amino acids is to combine certain foods. Beans, for instance, are likely to be deficient in tryptophan and in the sulfur-containing amino acids, but they are a good-to-excellent source of isoleucine and lysine. Rice is deficient in isoleucine and lysine but provides an adequate amount of the other essential amino acids. Thus rice and beans in combination make just about as perfect a protein menu as eggs or steak, as some nonscientists seem to have known for quite a long time.

In many proteins, the tertiary structure produces an intricately folded, globular shape for the molecule as a whole; these proteins are called globular proteins. Enzymes—proteins that regulate chemical reactions in living systems—are globular proteins, as are the membrane receptors for an enormous variety of molecules. Antibodies, important components of the immune system, are also globular proteins. As we shall see in subsequent chapters, the three-dimensional structures of all of these molecules are of critical importance in determining their biological functions.

Many proteins are composed of more than one polypeptide chain. These chains may be held to each other by hydrogen bonds, disulfide bridges, hydrophobic forces, attractions between positive and negative charges, or, most often, by a combination of these types of interactions. Such proteins are often called multimeric; a protein containing two polypeptide chains is termed a dimer, one containing three chains is a trimer, and one containing four chains is a tetramer. The hormone insulin, for example, is a dimer; it is composed of two polypeptide chains (Figure 3–22b). This level of organization of proteins, which involves interaction of two or more polypeptides, is called a **quaternary structure.**

The secondary, tertiary, and quaternary structures of a protein all depend on the primary structure—the sequence of amino acids—and on the local chemical environment.

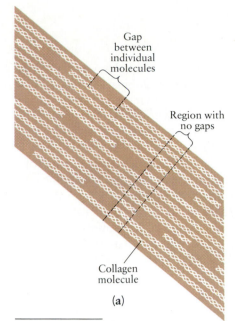

Gap between individual molecules

Region with no gaps

Collagen molecule

(a)

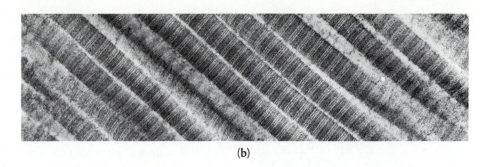

(b)

3–23 *Collagen molecules are packed together in fibrils that are a major constituent of skin, tendon, ligament, cartilage, and bone. Within an individual fibril (a), the collagen molecules are arranged in a staggered pattern, with gaps between individual molecules. This arrangement strengthens the fibrils, making them resistant to shearing forces. In electron micrographs of collagen fibrils, such as (b), a striated pattern is observed because the stain used in preparing the specimen is concentrated in the gaps between the molecules, causing the regions with no gaps to appear as lighter bands. These fibrils are magnified 23,500 times.*

Structural Uses of Proteins

Fibrous Proteins

In general, fibrous proteins have a regular, repeated sequence of amino acids and therefore a regular, repetitious structure. An example is collagen, which makes up about one-third of all the protein in vertebrates. The basic collagen molecule is composed of three very long polypeptides—about 1,000 amino acids per chain. These three polypeptides, which are made up of repeating groups of amino acids, are held together by hydrogen bonds linking amino acids of different chains into a tight coil. The molecules can coil so tightly because every third amino acid is glycine, the smallest of the amino acids. The collagen molecules are packed together to form fibrils (Figure 3–23), which are, in turn, associated into larger fibers.

Collagen is actually a family of proteins. Different types of collagen molecules contain polypeptides with slightly different amino acid sequences. The larger structures formed from the different types of molecules perform a variety of functions in the body. Consider a cow: Tendons, which link muscle to bone, are made up of collagen fibers in parallel bundles; thus arranged, they are very strong but do not stretch. The cow's hide, by contrast, is made up of collagen fibrils arranged in an interlacing network laid down in sheets. Even its corneas—the transparent coverings of the eyeballs—are composed of collagen. When collagen is boiled in water, the polymers are dispersed into shorter chains, which we know as gelatin.

Other fibrous proteins include keratin (Figure 3–24), silk, and elastin, present in the elastic tissue of ligaments.

3–24 *The fibrous protein keratin is found in all vertebrates. It is the chief component of scales, wool, nails, and feathers. (a) The horn of a rhinoceros consists of tightly packed strands of keratin. Solid rhino horn is used for dagger handles, and powdered horn as an aphrodisiac. A single horn can net a poacher far more than the average annual wage in many parts of Africa. (b) A feather, such as this spectacularly colored peacock feather, is made up of a shaft to which thousands of barbs—each with many tiny barbules—are attached.*

(a)

(b)

α tubulin

β tubulin

Soluble
tubulin
dimer

Microtubule

(a)

(b)

3–25 (a) *Microtubules are hollow tubes,
so small that they cannot be visualized by
a light microscope. They are composed of
subunits, each of which is a globular protein. The subunits are of two types, alpha
tubulin and beta tubulin, which first come
together to form a soluble dimer. The
dimers then self-assemble into insoluble
hollow tubules. (b) Among their many
functions, microtubules make up the
internal structure of cilia, the small, hair-
like appendages found on the surface of
many eukaryotic cells, such as the protist
Dileptus. The organism is magnified
1,000 times.*

Globular Proteins

Some structural proteins are globular. For example, microtubules, which function
in a variety of ways inside the cell, are made up of globular proteins. These
proteins associate to form long, hollow tubes—so long that their entire length can
seldom be traced in a single microscopic section. Microtubules play a critical role
in cell division, as we shall see in Chapter 7. They also participate in the internal
skeleton that stiffens parts of the cell body and also seem to function as a kind of
scaffolding for cellular construction projects. For example, the formation of a
new cell wall in a plant can be predicted by the appearance at the site of large
numbers of microtubules; when a plant cell wall is forming or growing and
cellulose fibrils are being laid down outside the cell membrane, microtubules can
be detected inside the cell, aligned in the same direction as the fibrils outside.

Chemical analysis shows that each microtubule consists of a very large number
of subunits. There are two types of subunits, each of which is a globular protein
formed from one polypeptide chain. Because of their complementary configurations, the two subunits fit together, forming approximately dumbbell-shaped
dimers. The dimers assemble themselves into tubules (Figure 3–25), adding on
length as required. When their job is over, they separate. The way in which the cell
controls the assembly and disassembly of microtubules is the subject of a great
deal of current research.

Hemoglobin: An Example of Specificity

Fibrous proteins, like polysaccharides, are usually molecules with a relatively
small variety of monomers in a repetitive sequence. Many globular proteins, by
contrast, have extremely complex, irregular amino acid sequences, as complex
and irregular as the sequence of letters in a sentence on this page. Just as these
sentences make sense (if they do) because the letters are the right ones and in the
right order, the proteins make sense, biologically speaking, because their amino
acids are the right ones in the right order.

Hemoglobin, for example, is a protein that is manufactured and carried in the
red blood cells. Its molecules have the special property of being able to combine
loosely with oxygen, collecting it in the lungs and releasing it in the tissues. The
hemoglobin molecule has a quaternary structure that consists of four polypeptide
chains, each of which is combined with an iron-containing group known as **heme.**
In heme, an iron atom is held by nitrogen atoms that are part of a larger structure
known as a porphyrin ring (Figure 3–26). Hemoglobin has two identical alpha
chains and two identical beta chains, each with a unique primary structure
containing about 150 amino acids, for a total of about 600 amino acids in all
(Figure 3–27).

Sickle cell anemia is a disease in which the hemoglobin molecules are defective.
When oxygen is removed from them, these molecules change shape and combine
with one another to form stiffened rodlike structures. Red blood cells containing
large proportions of such hemoglobin molecules become stiff and deformed,
taking on the characteristic sickle shape (Figure 3–28). The deformed cells may
clog the smallest blood vessels (capillaries). This causes blood clots and deprives
vital organs of their full supply of blood, resulting in pain, intermittent illness, and,
in many cases, a shortened life span.

Analysis of the hemoglobin molecules reveals that the only difference between
normal and sickle cell hemoglobin is that in a precise location in each beta chain,
one glutamic acid is replaced by one valine. In the quaternary structure of the
molecule, this particular location is on the outer surface, and valine, unlike
glutamic acid, contains a nonpolar R group. The result is a hydrophobic, "sticky"
region that can interact with hydrophobic regions on neighboring hemoglobin
molecules, producing the observed clumping. When one considers that this

3–26 *The heme group of hemoglobin. It contains an iron atom (Fe) held in a porphyrin ring. The porphyrin ring consists of four nitrogen-containing rings, which are numbered in the diagram. Each heme group is attached to a long polypeptide chain that wraps around it. The oxygen molecule is held flat against the heme.*

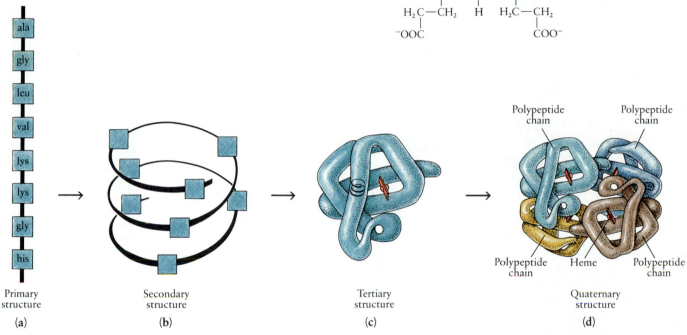

Primary structure
(a)

Secondary structure
(b)

Tertiary structure
(c)

Quaternary structure
(d)

3–27 *Levels of organization in the hemoglobin molecule. (a) The sequence of the amino acids in each chain is its primary structure. (b) The helical form assumed by any part of the chain as a consequence of hydrogen bonding between nearby* $>$C$=$O *and* —NH *groups is its secondary structure. (c) The folding of the chains in three-dimensional shapes is the tertiary structure, and (d) the combination of the four chains into a single functional molecule is the quaternary structure. The outside of the molecule and the hole through the middle are lined by charged amino acids, and the uncharged amino acids are packed inside. Each of the four chains surrounds a heme group (red), which can hold a single oxygen molecule. A hemoglobin molecule is therefore capable of transporting four oxygen molecules.*

3–28 *Scanning electron micrographs of (a) human red blood cells containing normal hemoglobin, and (b) a red blood cell containing the abnormal hemoglobin associated with sickle cell anemia. When the oxygen concentration in the blood is low, the abnormal hemoglobin molecules stick together, distorting the shape of the cells. As a result, the cells cannot pass readily through the capillaries. These cells are magnified about 7,000 times.*

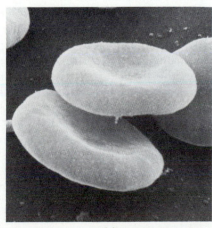

(a)

(b)

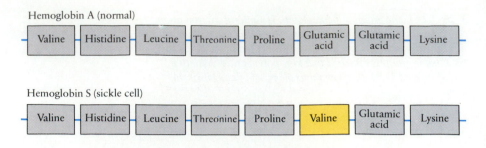

Hemoglobin A (normal)

| Valine | Histidine | Leucine | Threonine | Proline | Glutamic acid | Glutamic acid | Lysine |

Hemoglobin S (sickle cell)

| Valine | Histidine | Leucine | Threonine | Proline | Valine | Glutamic acid | Lysine |

3–29 *An example of the remarkable precision of the "language" of proteins. Portions of the beta chains of the hemoglobin A (normal) molecule and the hemoglobin S (sickle cell) molecule are shown. The entire structural difference between the normal molecule and the sickle cell molecule (literally, a life-and-death difference) consists of one change in the sequence of each beta chain: one glutamic acid is replaced by one valine.*

difference of two amino acids in a total of almost 600 can make such a profound difference in the properties of the molecule as a whole—indeed, can be the difference between life and death—one begins to get an idea of the precision and the importance of the arrangement of amino acids in a particular sequence in a protein. In living systems, which must perform many different activities simultaneously, the specificity of function that results from the structural precision of different protein molecules is of crucial importance.

NUCLEOTIDES

The information dictating the structures of the enormous variety of protein molecules found in living organisms is encoded in and translated by molecules known as **nucleic acids.** Just as proteins consist of long chains of amino acids, nucleic acids consist of long chains of nucleotides. A nucleotide, however, is a more complex molecule than an amino acid. As shown in Figure 3–30, it consists of three subunits: a phosphate group, a five-carbon sugar, and a **nitrogenous base**—a molecule that has the properties of a base and contains nitrogen.

The sugar subunit of a nucleotide may be either ribose or deoxyribose, which contains one less oxygen atom than ribose (Figure 3–31). Ribose is the sugar subunit in the nucleotides that form **ribonucleic acid (RNA),** and deoxyribose is the subunit in the nucleotides that form **deoxyribonucleic acid (DNA).** Five different nitrogenous bases are found in the nucleotides that are the building blocks of nucleic acids. Two of these, adenine and guanine, have a two-ring structure and are known as **purines** (Figure 3–32a). The other three, cytosine, thymine, and uracil, have a single-ring structure and are known as **pyrimidines** (Figure 3–32b). Adenine, guanine, and cytosine are found in both DNA and RNA, while thymine is found only in DNA and uracil only in RNA. As we shall see in Chapter 8, adenine and ribose sugar are also found in the nucleotides that are essential participants in the chemical reactions occurring within living systems.

Although their chemical components are very similar, DNA and RNA play very different biological roles. DNA is the primary constituent of the chromosomes of the cell and is the carrier of the genetic message. The function of RNA is to transcribe the genetic message present in DNA and translate it into proteins. The discovery of the structure and function of these molecules is undoubtedly the greatest triumph thus far of the molecular approach to the study of biology. In Section 3, we shall trace the events leading to the key discoveries and shall consider in some detail the marvelous processes—the details of which are still being worked out—by which these molecules perform their functions. First, however, we must turn our attention to the living cell—its origins, its structure, and the activities by which it maintains itself as an entity distinct from the nonliving world surrounding it.

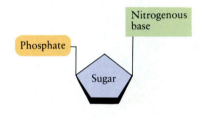

3–30 *A nucleotide is made up of three different subunits: a phosphate group, a five-carbon sugar, and a nitrogenous base. As we shall see in Chapter 14, nucleotides can be linked together in long chains by condensation reactions involving the hydroxyl groups of the phosphate and sugar subunits.*

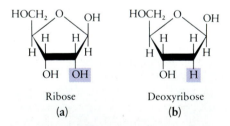

Ribose
(a)

Deoxyribose
(b)

3–31 *The sugar subunit of a nucleotide may be either (a) ribose or (b) deoxyribose. As you can see, the structural difference between the two sugars is slight. In ribose, carbon 2 bears a hydrogen atom above the plane of the ring and a hydroxyl group below the plane; in deoxyribose, the hydroxyl group on carbon 2 is replaced by a hydrogen atom.*

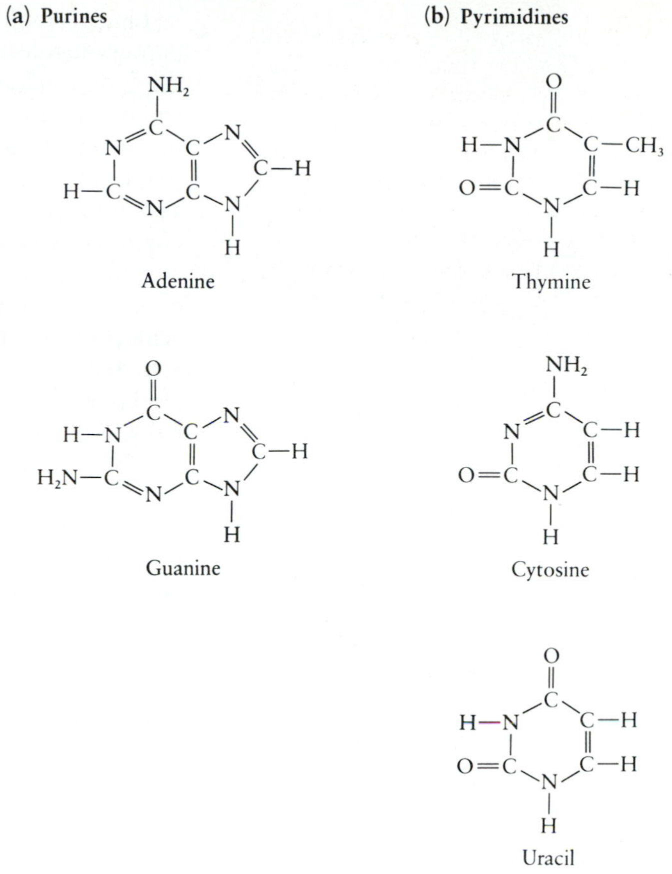

3–32 *The five nitrogenous bases of the nucleotides that form nucleic acids.* **(a)** *Adenine and guanine, the purines, occur in both DNA and RNA, as does cytosine, one of the pyrimidines* **(b)***. Thymine, also a pyrimidine, is found in DNA but not in RNA, and uracil, a third pyrimidine, is found in RNA but not in DNA. As we shall see in Section 3, the sequence of these simple molecules within long chains of nucleic acids is responsible for the transmission and translation of hereditary information, generation after generation.*

(a) Purines

Adenine

Guanine

(b) Pyrimidines

Thymine

Cytosine

Uracil

SUMMARY

The chemistry of living organisms is, in essence, the chemistry of organic compounds—that is, compounds containing carbon. Carbon is uniquely suited to this central role by the fact that it is the lightest atom capable of forming multiple covalent bonds. Because of this capacity, carbon can combine with carbon and other atoms to form a great variety of strong and stable chain and ring compounds. Organic molecules derive their three-dimensional shapes primarily from their carbon skeletons. Many of their specific properties, however, are dependent on functional groups. A general characteristic of all organic compounds is that they release energy when oxidized. Among the major types of organic molecules important in living systems are carbohydrates, lipids, proteins, and nucleotides.

Carbohydrates serve as a primary source of chemical energy for living systems. The simplest carbohydrates are the monosaccharides ("single sugars"), such as glucose and fructose. Monosaccharides can be combined to form disaccharides ("two sugars"), such as sucrose, and polysaccharides (chains of many monosaccharides). The polysaccharides starch and glycogen are storage forms for sugar, whereas cellulose, another polysaccharide, is an important structural material in plants. Disaccharides and polysaccharides are formed by condensation reactions in which monosaccharide units are covalently bonded with the removal of a molecule of water. They can be broken apart again by hydrolysis, with the addition of a water molecule.

Lipids are hydrophobic organic molecules that, like carbohydrates, play important roles in energy storage and as structural components. Compounds in this group include fats and oils, phospholipids, glycolipids, waxes, and cholesterol and other steroids. Fats are the chief energy-storing lipids. A fat molecule consists of one molecule of glycerol bonded to three fatty acids. Fats are designated as unsaturated or saturated depending on whether or not their fatty acids contain any double bonds. Unsaturated fats, which tend to be oily liquids, are more commonly found in plants.

Phospholipids are major structural components of cellular membranes. Phospholipids consist of one unit of glycerol, two fatty acids (instead of the three fatty acids present in fats), and a phosphate group to which another polar group may be attached. Because of their hydrophilic "heads" and hydrophobic "tails," phospholipids spontaneously orient in water to form films and clusters that are the basis of membrane structure. Glycolipids, which consist of one unit of glycerol, two fatty acids, and a short carbohydrate chain attached to the third carbon of the glycerol, are also important components of cellular membranes.

Proteins are very large molecules composed of long chains of amino acids; these are known as polypeptide chains. The 20 different amino acids used in making proteins vary according to the properties of their side (R) groups. From these relatively few amino acids, an extremely large variety of different kinds of protein molecules can be synthesized, each of which has a highly specific function in living systems.

The sequence of amino acids is known as the primary structure of the protein. Depending on the amino acid sequence, the molecule may take on any of a variety of forms. Hydrogen bonds between $\diagup C{=}O$ and $\diagup NH$ groups tend to fold the chain into a repeating secondary structure, such as the alpha helix or the beta pleated sheet. Interactions between the R groups of the amino acids may result in further folding into a tertiary structure, which is often an intricate, globular form. Two or more polypeptides may interact to form a quaternary structure.

In fibrous proteins, the long molecules interact with other similar, or identical, long polypeptide chains to form cables or sheets. Collagen and keratin are fibrous proteins that play a variety of structural roles. Globular proteins may also serve structural purposes. Microtubules, which are important cell components, are composed of repeating units of globular proteins assembled helically into a hollow tubule. Other globular proteins have regulatory, transport, and protective functions.

Because of the variety of amino acids, proteins can have a high degree of specificity. An example is hemoglobin, the oxygen-carrying molecule of the blood, which is composed of four (two pairs of) polypeptide chains, each attached to an iron-containing (heme) group. Substitution of one amino acid for another in one of the pairs of chains alters the surface of the molecule, producing a serious and sometimes fatal disease known as sickle cell anemia.

Nucleotides are complex molecules consisting of a phosphate group, a five-carbon sugar, and a nitrogenous base. They are the building blocks of the nucleic acids deoxyribonucleic acid (DNA) and ribonucleic acid (RNA), which transmit and translate the genetic information. Nucleotides also play key roles in the energy exchanges accompanying chemical reactions within living systems.

QUESTIONS

1. Distinguish among the following: hydrocarbon/carbohydrate; glucose/fructose/sucrose; monomer/polymer; glycogen/starch/cellulose; saturated/unsaturated; phospholipid/glycolipid; polysaccharide/polypeptide; peptide bond/disulfide bridge/hydrophobic interaction; primary structure/secondary structure/tertiary structure/quaternary structure; heme/hemoglobin; nitrogenous base/nucleotide/nucleic acid.

2. Identify the functional groups in the compounds below. Which of these is hydrophilic? Hydrophobic?

(a) CH_3COOH
Major component
of vinegar

(b) HCOOH
Active ingredient
in an ant's sting

(c) CH_2—CH_2
       | |
      OH   OH
Automobile
antifreeze

(d) H—C=O
         |
         H
Preservative used
for biological
specimens

(e) CH_3—C—CH_3
              ||
              O
Nail-polish
remover

(f) NH_2

Used in manufacture
of commercial dyes

3. Draw a structural formula for (a) a monosaccharide; (b) a fatty acid; (c) an amino acid.

4. Butyric acid, $CH_3CH_2CH_2COOH$, gives rancid butter its odor and flavor. Draw its structural formula.

5. Many of the synthetic reactions in living systems take place by condensation. What is a condensation reaction? What types of molecules undergo condensation reactions to form disaccharides and polysaccharides? To form fats? To form proteins?

6. Disaccharides and polysaccharides, as well as lipids and proteins, can be broken down by hydrolysis. What is hydrolysis? What two types of products are released when a polysaccharide such as starch is hydrolyzed? How are these products important for the living cell?

7. What do we mean when we say that some polysaccharides are "energy-storage" molecules and that others are "structural" molecules? Give an example of each. In what sense should any polysaccharide be regarded as an "energy-storage" molecule?

8. Plants usually store energy reserves as polysaccharides, whereas, in most animals, lipids are the principal form of energy storage. Why is it advantageous for animals to have their energy reserves stored as lipids rather than as polysaccharides? (Think about the differences in "life-style" between plants and animals.) What kinds of storage materials would you expect to find in seeds?

9. Sketch the arrangement of phospholipids when they are surrounded by water.

10. In pioneer days, soap was made by boiling animal fat with lye (potassium hydroxide). The bonds linking the fatty acids to the glycerol molecule were hydrolyzed, and the potassium hydroxide reacted with the fatty acid to produce soap. A typical soap available today is sodium stearate. In water, it ionizes to produce sodium ions (Na^+) and stearate ions:

$$CH_3(CH_2)_{16}C \overset{\displaystyle =O}{\underset{\displaystyle O^-}{}}$$

Explain how soap functions to trap and remove particles of dirt and grease.

11. Silk is a protein in which polypeptide chains are arranged in a beta pleated sheet. In these chains, the peptide sequence glycine-serine-glycine-alanine-glycine-alanine occurs repeatedly. (a) Draw the structural formula for this hexapeptide, and show the peptide bonds in color. (b) Explain how a peptide bond is formed.

Cells: An Introduction

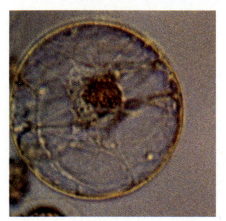

⊢ 500 μm ⊣

4–1 *A single living cell from the leaf of a poplar tree. Made up mostly (more than 95 percent) of only four kinds of atoms, it is nevertheless capable of carrying out a wide variety of intricate chemical reactions. This cell's capacity to exhibit the properties, such as energy conversion, homeostasis, reproduction, and growth, that are characteristic of living organisms depends on the complex organization of its constituent parts.*

The short, straight line at the bottom of this micrograph and those that follow provide a reference for size; a micrometer, abbreviated μm, is 1/1,000,000 meter. The same system is used to indicate distances on a road map.

In the last three chapters, we have progressed from subatomic particles through atoms and molecules to complex macromolecules, such as proteins and nucleic acids. At each level of organization, new properties appear. For instance, water, as we have seen, is not the sum of the properties of elemental hydrogen and oxygen; it is something more and also something different. In proteins, amino acids become organized into polypeptides, and polypeptide chains are arranged in a new level of organization, the tertiary or quaternary structure of the complete protein molecule. Only at this level of organization do the complex properties of the protein emerge, and only then can the molecule assume its function.

The characteristics of living systems, like those of atoms or of molecules, do not emerge gradually as the degree of organization increases. They appear quite suddenly and specifically, in the form of the living cell—something that is more than and different from its constituent atoms and molecules. No one knows exactly when or how this new level of organization—the living cell—first came into being. However, increasing knowledge of the history of our planet and the results of numerous laboratory experiments provide evidence for the hypothesis that living cells spontaneously self-assembled from molecules present in the primitive seas.

THE FORMATION OF THE EARTH

About 5 billion years ago, cosmologists calculate, the star that is our sun came into being. According to current theory, it formed, like other stars, from an accumulation of particles of dust and hydrogen and helium gases whirling in space among the older stars.

The immense cloud that was to become the sun condensed gradually as the hydrogen and helium atoms were pulled toward one another by the force of gravity, falling into the center of the cloud and gathering speed as they fell. As the cluster grew denser, the atoms moved more rapidly. More atoms collided with each other, and the gas in the cloud became hotter and hotter. As the temperature rose, the collisions became increasingly violent until the hydrogen atoms collided with such force that their nuclei fused, forming additional helium atoms and releasing nuclear energy. This thermonuclear reaction is still going on at the heart of the sun and is the source of the energy radiated from its glowing surface.

The planets, according to current theory, formed from the remaining gas and dust moving around the newly formed star. At first, particles would have collected at random, but as each mass grew larger, other particles began to be attracted by the gravity of the largest masses. The whirling dust and forming spheres continued to revolve around the sun until finally each planet had swept its own path clean, picking up loose matter like a giant snowball. The orbit nearest the sun was swept

4-2 *The tremendous amounts of energy released by the thermonuclear reactions at the heart of the sun give rise to an envelope of extremely hot gases surrounding its surface. The layer of gases may extend as far as 64,000 kilometers from the surface of the sun—a distance about five times greater than the diameter of the earth.*

by Mercury, the next by Venus, the third by earth, the fourth by Mars, and so on out to Neptune and Pluto, the most distant of the planets. The planets, including earth, are calculated to have come into being about 4.6 billion years ago.

During the time earth and the other planets were being formed, the release of energy from radioactive materials kept their interiors very hot. When earth was still so hot that it was mostly liquid, the heavier materials collected in a dense core whose diameter is about half that of the planet. As soon as the supply of stellar dust, stones, and larger rocks was exhausted, the planet ceased to grow. As earth's surface cooled, an outer crust, a skin as thin by comparison as the skin of an apple, was formed. The oldest known rocks in this layer have been dated by isotopic methods as about 4.1 billion years old.

Only 50 kilometers below its surface, the earth is still hot—a small fraction of it is even still molten. We see evidence of this in the occasional volcanic eruption that forces lava (molten rock) through weak points in the earth's skin, or in the geyser, which spews up boiling water that has trickled down to the earth's interior.

The biosphere is the part of the planet within which life exists. It forms a thin film on the outermost layer, extending only about 8 or 10 kilometers up into the atmosphere and about as far down into the depths of the sea.

THE BEGINNING OF LIFE

Until very recently, the earliest fossil organisms known were a mere 600 million years old, and for a long time after the publication of *The Origin of Species,* biologists regarded the earliest events in the history of life as chapters that would probably remain forever closed to scientific investigation.

Two developments, however, have greatly improved our long-distance vision. The first was the formulation of a testable hypothesis about the events preceding life's origins. This hypothesis generated questions for which answers could be sought experimentally. The results of the initial experimental tests led to the formulation of further hypotheses and to additional experiments, a process that continues today as scientists in many laboratories explore the question of life's origins. The second development was the discovery of fossilized cells more than 3 billion years old.

The Question of Spontaneous Generation

Most of the early biologists, from the time of Aristotle, believed that simple living things, such as worms, beetles, frogs, and salamanders, could originate spontaneously in dust or mud, that rodents formed from moist grain, and that plant lice condensed from a dewdrop. In the seventeenth century, Francesco Redi performed a famous experiment in which he put out decaying meat in a group of wide-mouthed jars—some with lids, some covered by a fine veil, and some open—and demonstrated that maggots arose only where flies were able to lay their eggs.

By the nineteenth century, no scientist continued to believe that complex organisms arose spontaneously. The advent of microscopy, however, led a vigorous renewal of belief in the spontaneous generation of very simple organisms. It was necessary only to put decomposing substances in a warm place for a short time and tiny "live beasts" appeared under the lens, before one's very eyes. By 1860, the controversy had become so spirited that the Paris Academy of Sciences offered a prize for experiments that would throw new light on the question. The prize was claimed in 1864 by Louis Pasteur, who devised experiments to show that microorganisms appeared only as contaminants from the air and not "spontaneously" as his opponents claimed. In his experiments he used swan-necked flasks, which permitted the entrance of oxygen, thought to be necessary for life, but which, in their long, curving necks, trapped bacteria, fungal spores, and other microbial life and thereby protected the contents of the flasks from contamination. He showed that if the liquid in the flask was boiled (which killed microorganisms already present) and the neck of the flask was allowed to remain intact, no microorganisms would appear. Only if the curved neck of the flask was broken off, permitting contaminants to enter the flask, did microorganisms appear. (Some of his original flasks, still sterile, remain on display at the Pasteur Institute in Paris.)

"Life is a germ, and a germ is Life," Pasteur proclaimed at a brilliant "scientific evening" at the Sorbonne before the social elite of Paris. "Never will the doctrine of spontaneous generation recover from the mortal blow of this simple experiment."

In retrospect, Pasteur's well-planned experiments were so decisive because the broad question of whether or not spontaneous generation had *ever* occurred was reduced to the simpler question of whether or not it occurred under the specific conditions claimed for it. Pasteur's experiments answered *only* the latter question, but the results were so dramatic that for many years very few scientists were able to entertain the possibility that under quite different conditions, when the earth was very young, some form of "spontaneous generation" might indeed have taken place. The question of the origin of the first living systems remained unasked until well into the twentieth century.

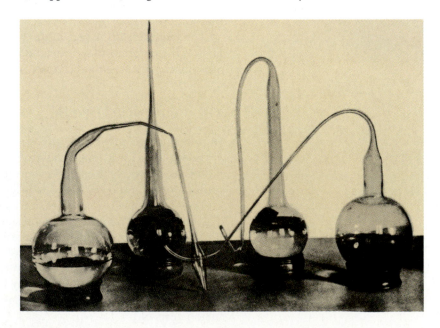

Pasteur's swan-necked flasks, which he used to counter the argument that spontaneous generation failed to occur in sealed vessels because air was excluded. These flasks permitted the entrance of oxygen, thought to be essential for life, but their long, curving necks trapped spores of microorganisms and thereby protected the liquids in the flasks from contamination.

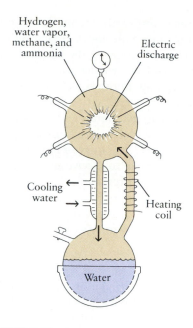

4–3 *Miller's experiment. Conditions believed to have existed on the primitive earth were simulated in the apparatus diagrammed here. Methane (CH_4) and ammonia (NH_3) were continuously circulated between a lower "ocean," which was heated, and an upper "atmosphere," through which an electric discharge was transmitted. At the end of 24 hours, about half of the carbon originally present in the methane gas was converted to amino acids and other organic molecules. This was the first test of Oparin's hypothesis.*

The testable hypothesis was offered by the Russian biochemist A. I. Oparin. According to Oparin, the appearance of life was preceded by a long period of what is sometimes called chemical evolution. The identity of the substances, particularly gases, present in the primitive atmosphere and in the seas during this period is a matter of controversy. There is general agreement, however, on two critical issues: (1) Little or no free oxygen was present, and (2) the four elements—hydrogen, oxygen, carbon, and nitrogen—that make up more than 95 percent of living tissues were available in some form in the atmosphere and waters of the primitive earth.

In addition to these raw materials, energy abounded on the young earth. There was heat energy, both boiling (moist) heat and baking (dry) heat. Water vapor spewed out of the primitive seas, cooled in the upper atmosphere, collected into clouds, fell back on the crust of the earth, and steamed up again. Violent rainstorms were accompanied by lightning, which provided electrical energy. The sun bombarded the earth's surface with high-energy particles and ultraviolet light, another form of energy. Radioactive elements within the earth released their energy into the atmosphere. Oparin hypothesized that under such conditions organic molecules were formed from the atmospheric gases and collected in a thin soup in the earth's seas and lakes. Because there was no free oxygen present to react with and degrade these organic molecules to simple substances such as carbon dioxide (as would happen today), they tended to persist. Some of these molecules might have become locally more concentrated by the drying up of a lake or by the adhesion of the molecules to a solid surface.

Oparin published his hypothesis in 1922, but at that time biochemists were so convinced by Pasteur's demonstration disproving spontaneous generation (see essay) that the scientific community ignored his ideas. In the 1950s, the first test of Oparin's hypothesis was performed by Stanley Miller, then a graduate student at the University of Chicago (Figure 4–3). Experiments of this sort, now repeated many times, have shown that almost any source of energy—lightning, ultraviolet radiation, or hot volcanic ash—would have converted molecules believed to have been present on the earth's surface into a variety of complex organic compounds. With various modifications in the experimental conditions and in the mixture of gases placed in the reaction vessel, almost all of the common amino acids have been produced, as well as the nucleotides that are the essential components of DNA and RNA.

These experiments have not proved that such organic compounds were formed spontaneously on the primitive earth, only that they could have formed. The accumulated evidence is nevertheless very great, and most biochemists now

4–4 *Volcanic eruptions, such as the one shown here off the coast of Iceland, occurred frequently on the restless surface of the young earth. The intense fields of electrical, thermal, and shock-wave energy generated by such eruptions of steam and lava could have been a major factor in the formation of organic molecules. This eruption, which took place in 1963, resulted in the birth of the island of Surtsey.*

believe that, given the conditions existing on the young earth, chemical reactions producing amino acids, nucleotides, and other organic molecules were inevitable.

As the concentrations of such molecules increased, bringing them into closer proximity to each other, they would have been subject to the same chemical forces that act on organic molecules today. As we saw in the last chapter, small organic molecules react with each other, typically in condensation reactions, to form larger molecules; moreover, such forces as hydrogen bonds and hydrophobic interactions cause these molecules to assemble themselves into more complex aggregates. In modern chemical systems—either in the laboratory or in the living organism—the more stable molecules and aggregates tend to survive and the least stable are transitory. Similarly, the compounds and aggregates that had the greatest chemical stability under the prevailing conditions on the primitive earth would have tended to survive. Hence a form of natural selection played a role in chemical evolution as well as in the biological evolution that was to follow.

The First Cells

From a biochemical perspective, three characteristics distinguish the living cell from other chemical systems: (1) the capacity to replicate itself, generation after generation; (2) the presence of enzymes, the complex proteins that are essential for the chemical reactions on which life depends; and (3) a membrane that separates the cell from the surrounding environment and enables it to maintain a distinct chemical identity. Which of these characteristics appeared first—and made possible the development of the others—remains an open question. However, as we shall see in Chapter 18, recently discovered functions of RNA suggest that the starting point may well have been the self-assembly of RNA molecules from nucleotides produced by chemical evolution.

In other studies, simulating conditions during the earth's first billion years, Sidney W. Fox and his coworkers at the University of Miami have produced membrane-bound protein structures that can carry out a few chemical reactions analogous to those of living cells. These structures are produced through a series of chemical reactions, beginning with dry mixtures of amino acids. When the mixtures are heated at moderate temperatures, polymers (known as thermal proteinoids) are formed, each of which may contain as many as 200 amino acid monomers. When these polymers are placed in an aqueous salt solution and maintained under suitable conditions, they spontaneously form proteinoid microspheres (Figure 4–5). The microspheres grow slowly by the addition of

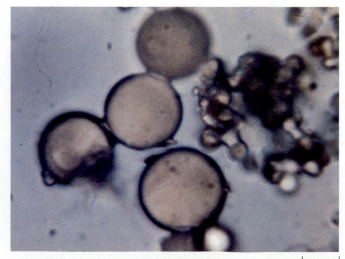

4–5 *Proteinoid microspheres, which form spontaneously when thermal proteinoids are maintained in aqueous solution under suitable conditions. As you can see, the microspheres are separated from the surrounding solution by a membrane that appears to be two-layered.*

1μm

proteinoid material from the solution and eventually bud off smaller microspheres. These microspheres are not living cells. Their formation, however, suggests the kinds of processes that could have given rise to self-sustaining protein entities, separated from their environment and capable of carrying out the chemical reactions necessary to maintain their physical and chemical integrity.

It is not known when the first living cells appeared on earth, but we can establish some sort of time scale. The earliest fossils found so far (Figure 4–6), which resemble present-day bacteria, have been dated at 3.4 and 3.5 billion years—about 1.1 billion years after the formation of the earth itself. Although the fossils are so small that their structure can be made visible only by electron microscopy, they are sufficiently complex that it is clear that some little aggregation of chemicals had moved through the twilight zone separating the living from the nonliving millions of years before.

Why on Earth?

On the basis of astronomical studies and the explorations carried out by unmanned space vehicles, it appears that earth alone among the planets of our solar system supports life. The conditions on earth are ideal for living systems based on carbon-containing molecules. A major factor is that earth is neither too close to nor too distant from the sun. The chemical reactions on which life—at least as we know it—depends require liquid water, and they virtually cease at very low temperatures. At high temperatures, the complex chemical compounds essential for life are too unstable to survive.

Earth's size and mass are also important factors. Planets much smaller than earth do not have enough gravitational pull to hold a protective atmosphere, and any planet much larger than earth is likely to have so dense an atmosphere that light from the sun cannot reach its surface. The earth's atmosphere blocks out many of the most energetic radiations from the sun, which are capable of breaking the covalent bonds between carbon atoms. It does, however, permit the passage of visible light, which made possible one of the most significant steps in the evolution of complex living systems.

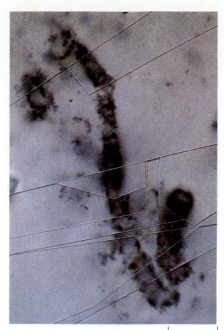

├─── 50 μm ───┤

4–6 *This microfossil of a filament of bacteria-like cells was found in Western Australia in a deposit of a flintlike rock known as black chert. Dated at 3.5 billion years of age, it is one of the oldest fossils known.*

HETEROTROPHS AND AUTOTROPHS

The energy that produced the first organic molecules came from a variety of sources on the primitive earth and in its atmosphere—heat, ultraviolet radiations, and electrical disturbances. When the first primitive cells or cell-like structures evolved, they required a continuing supply of energy to maintain themselves, to grow, and to reproduce. The manner in which these cells obtained energy is currently the subject of lively discussion.

Modern organisms—and the cells of which they are composed—can meet their energy needs in one of two ways. **Heterotrophs** are organisms that are dependent upon outside sources of organic molecules for both their energy and their small building-block molecules. (*Hetero* comes from the Greek word meaning "other," and *troph* comes from *trophos*, "one that feeds.") All animals and fungi, as well as many single-celled organisms, are heterotrophs. **Autotrophs,** by contrast, are "self-feeders." They do not require organic molecules from outside sources for energy or to use as small building-block molecules; they are, instead, able to synthesize their own energy-rich organic molecules from simple inorganic substances. Most autotrophs, including plants and several different types of single-celled organisms, are **photosynthetic,** meaning that the energy source for their synthetic reactions is the sun. Certain groups of bacteria, however, are **chemosynthetic;** these organisms capture the energy released by specific inorganic reactions to power their life processes, including the synthesis of needed organic molecules.

Both heterotrophs and autotrophs seem to be represented among the earliest microfossils. It has long been postulated that the first living cell was an extreme heterotroph. As the primitive heterotrophs increased in number, according to this hypothesis, they began to use up the complex molecules on which their existence depended and which had taken millions of years to accumulate. As the supply of these molecules decreased, competition began. Under the pressure of this competition, cells that could make efficient use of the limited energy sources now available were more likely to survive and reproduce than cells that could not. In the course of time, other cells evolved that were able to synthesize organic molecules out of simple inorganic materials.

Recent discoveries, however, have raised the possibility that the first cells may have been either chemosynthetic or photosynthetic autotrophs rather than heterotrophs. First, several different groups of chemosynthetic bacteria have been found that would have been well-suited to the conditions prevailing on the young earth (Figure 4–7). Some of these bacteria are the inhabitants of swamps, while others have been found in deep ocean trenches in areas where gases escape from fissures in the earth's crust. There is evidence (to be discussed in Chapter 21) that these bacteria are the surviving representatives of very ancient groups of unicellular organisms. Second, organic molecules that are, in plants, the chemical precursors of chlorophyll have been produced in experiments analogous to that performed by Miller. When these molecules are mixed with simple organic molecules in an oxygen-free environment and illuminated, primitive photosynthetic reactions occur. These reactions resemble the reactions that occur in some types of photosynthetic bacteria.

Although biologists are presently unable to resolve the question of whether the earliest cells were heterotrophs or autotrophs, it is certain that without the evolution of autotrophs, life on earth would soon have come to an end. In the more than 3.5 billion years since life first appeared on earth, the most successful autotrophs (that is, those that have left the most offspring and diverged into the greatest variety of forms) have been those that evolved a system for making direct use of the sun's energy in the process of photosynthesis. With the advent of photosynthesis, the flow of energy in the biosphere came to assume its dominant modern form: radiant energy from the sun channeled through photosynthetic autotrophs to all other forms of life.

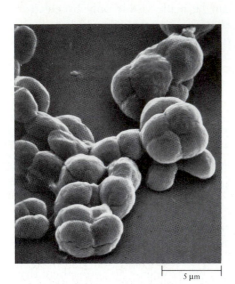

5 μm

4–7 *Methanogens, such as the cells shown here, are chemosynthetic bacteria that produce methane from carbon dioxide and hydrogen gas. They can live only in the absence of oxygen—a condition prevailing on the young earth but occurring today only in isolated environments, such as the muck and mud at the bottom of swamps.*

PROKARYOTES AND EUKARYOTES

The cell theory, as we noted in the Introduction, is one of the foundations of modern biology. This theory states simply that (1) all living organisms are composed of one or more cells; (2) the chemical reactions of a living organism, including its energy-releasing processes and its biosynthetic reactions, take place within cells; (3) cells arise from other cells; and (4) cells contain the hereditary information of the organisms of which they are a part, and this information is passed from parent cell to daughter cell. All available evidence indicates that there is an unbroken continuity between modern cells—and the organisms they compose—and the first primitive cells that appeared on earth.

All cells share two essential features. One is an outer membrane, the **cell membrane** (also known as the **plasma membrane**), that separates the cell from its external environment. The other is the genetic material—the hereditary information—that directs a cell's activities and enables it to reproduce, passing on its characteristics to its offspring.

The organization of the genetic material is one of the characteristics that distinguish two fundamentally distinct kinds of cells, **prokaryotes** and **eukaryotes.**

In prokaryotic cells, the genetic material is in the form of a large, circular molecule of DNA, with which a variety of proteins are loosely associated. This molecule is known as the **chromosome.** In eukaryotic cells, by contrast, the DNA is linear, forming a number of distinct chromosomes; moreover, it is tightly bound to special proteins known as **histones,** which are an integral part of the chromosome structure. Within the eukaryotic cell, the chromosomes are surrounded by a double membrane, the **nuclear envelope,** that separates them from the other cell contents in a distinct **nucleus** (hence the name, *eu*, meaning "true," and *karyon*, meaning "nucleus" or "kernel"). In prokaryotes ("before a nucleus"), the chromosome is not contained within a membrane-bound nucleus, although it is localized in a distinct region known as the **nucleoid.**

The remaining components of a cell (that is, everything within the cell membrane except the nucleus or nucleoid and its contents) constitute the **cytoplasm.** The cytoplasm contains a large variety of molecules as well as formed bodies called **organelles.** These specialized structures carry out particular functions within the cell. Both prokaryotes and eukaryotes contain very small organelles called **ribosomes,** on which protein molecules are assembled. In addition, eukaryotes contain a variety of more complex organelles, which are often enclosed within membranes.

The cell membrane of prokaryotes is surrounded by an outer **cell wall** that is manufactured by the cell itself. Some eukaryotic cells, including plant cells and fungi, have cell walls, although their structure is different from that of prokaryotic cell walls. Other eukaryotic cells, including those of our own bodies and of other animals, do not have cell walls. Another feature distinguishing eukaryotes and prokaryotes is size: eukaryotic cells are usually larger than prokaryotic cells.

Modern prokaryotes include the bacteria (Figure 4–8) and the cyanobacteria (Figure 4–9, on the next page), a group of photosynthetic prokaryotes that were formerly known as the blue-green algae.* According to the fossil record, the

---

4–8 *Cells of* Escherichia coli, *the heterotrophic prokaryote that is the most thoroughly studied of all living organisms. The genetic material (DNA) is in the lighter-appearing area in the center of each cell; this region, which is not enclosed by a membrane, is known as the nucleoid. The small, dense bodies in the cytoplasm are ribosomes. The two cells in the center have just finished dividing and have not yet separated completely.*

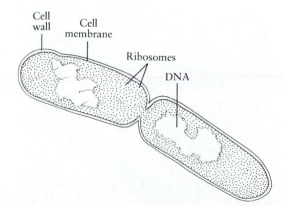

---

* The Latin term *alga*, plural *algae*, means "seaweed." Algae is a general term aplied to eukaryotic single-celled photosynthetic organisms and to many simple multicellular forms. Until recently, it was also applied to some prokaryotes.

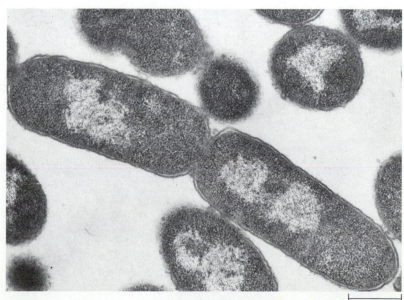

0.5 μm

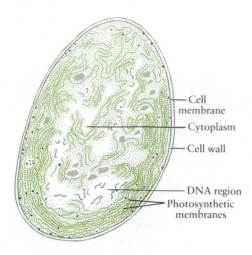

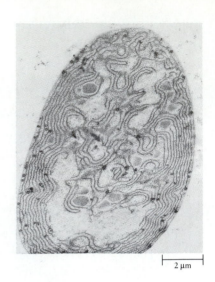

2 µm

4-9 *Electron micrograph and diagram of a photosynthetic prokaryotic cell, the cyanobacterium* Anabaena azollae. *In addition to the genetic material, this cell contains a series of membranes in which chlorophyll and other photosynthetic pigments are embedded.* Anabaena *synthesizes its own energy-rich organic compounds in chemical reactions powered by the radiant energy of the sun.*

Cell membrane
Cytoplasm
Cell wall
DNA region
Photosynthetic membranes

earliest living organisms were comparatively simple cells, resembling present-day prokaryotes. Prokaryotes were the only forms of life on this planet for almost 2 billion years until eukaryotes evolved. (The evolutionary relationship between prokaryotes and eukaryotes—quite an interesting subject—will be explored in Chapter 22.)

Figure 4–10 gives an example of a single-celled photosynthetic eukaryote, the alga *Chlamydomonas*. It is a common inhabitant of freshwater ponds and aquariums. These organisms are small and bright green (because of their chlorophyll), and they move very quickly with a characteristic darting motion. Being photosynthetic, they are usually found near the water's surface.

The Origins of Multicellularity

The first multicellular organisms, as far as can be told by the fossil record, made their appearance a mere 750 million years ago (Figure 4–11). The major groups of multicellular organisms—such as the fungi, the plants, and the animals—are thought to have evolved from different types of single-celled eukaryotes.

4-10 *Electron micrograph of* Chlamydomonas, *a photosynthetic eukaryotic cell, which contains a membrane-bound ("true") nucleus and numerous organelles. The most prominent organelle is the single, irregularly shaped chloroplast that fills most of the cell. It is surrounded by a double membrane and is the site of photosynthesis. Other membrane-bound organelles, the mitochondria, provide energy for cellular functions, including the flicking movements of the two flagella (one of which is visible in the micrograph). These movements propel the cell through the water. The organism's food reserves are in the form of starch granules. The cytoplasm is surrounded by a cell membrane, outside of which is a cell wall composed of polysaccharides.*

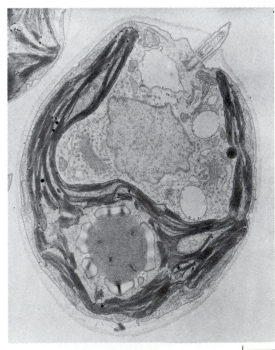

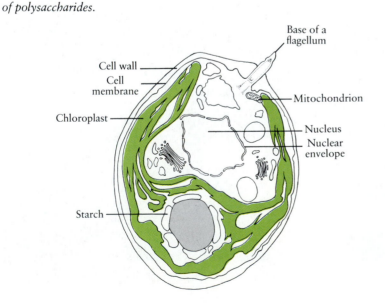

Base of a flagellum
Cell wall
Cell membrane
Chloroplast
Mitochondrion
Nucleus
Nuclear envelope
Starch

1 µm

4–11 *The clockface of biological time. Life first appears relatively early in the earth's history, before 6:00 A.M. on a 24-hour time scale. The first multicellular organisms do not appear until the early evening of that 24-hour day, and Homo, the genus to which humans belong, is a late arrival—at about 30 seconds to midnight.*

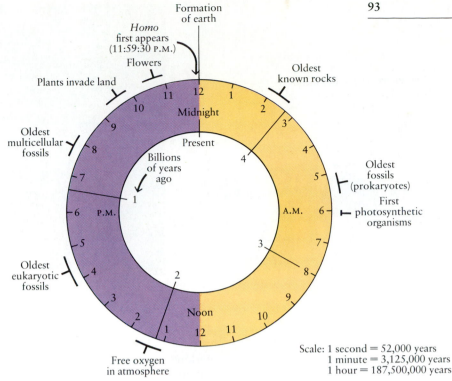

Scale: 1 second = 52,000 years
1 minute = 3,125,000 years
1 hour = 187,500,000 years

4–12 *Electron micrograph of cells from the leaf of a corn plant. The nucleus can be seen on the right side of the central cell. The granular material within the nucleus is chromatin; it contains DNA associated with histone proteins. Note the many mitochondria and chloroplasts, all enclosed by membranes. The vacuole and cell wall are characteristic of plant cells but are generally not found in animal cells. As you can see, this cell closely resembles* Chlamydomonas, *shown in Figure 4–10.*

The cells of modern multicellular organisms closely resemble those of single-celled eukaryotes. They are bound by a cell membrane identical in appearance to the cell membrane of a single-celled eukaryote. Their organelles are constructed according to the same design. The cells of multicellular organisms differ from single-celled eukaryotes in that each type of cell is specialized to carry out a relatively limited function in the life of the organism. However, each remains a remarkably self-sustaining unit.

Notice how similar a cell from the leaf of a corn plant (Figure 4–12) is to *Chlamydomonas*. This plant cell is also photosynthetic, supplying its own energy needs from sunlight. However, unlike the alga, it is part of a multicellular organism and depends on other cells for water, minerals, protection from desiccation (drying out), and other necessities.

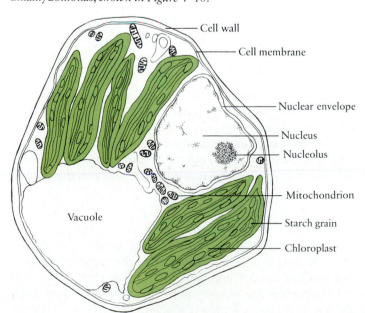

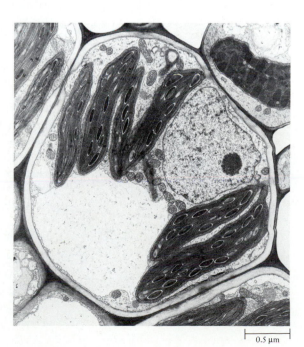

0.5 μm

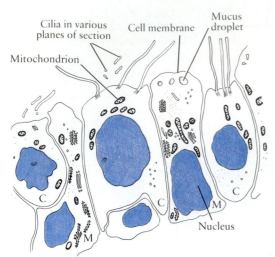

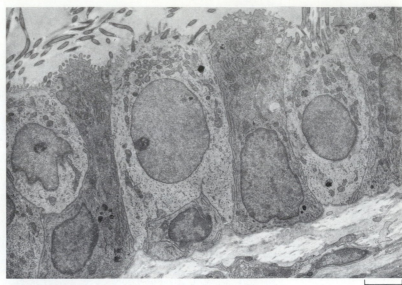

2.5 μm

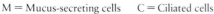

4-13 *Cells from the surface of the trachea (windpipe) of a bat. The free surface of the larger cell is covered with cilia, which are essentially the same in structure as the flagella of Chlamydomonas. (When they are fewer and longer, they are usually called flagella, whereas when shorter and more numerous they are called cilia.) Next to the ciliated cells are cells that secrete mucus onto the cell surface. Mucus currents, swept by cilia, remove foreign particles from the surface of the trachea. Note the mitochondria located near the base of the cilia.*

The human body, made up of trillions of individual cells, is composed of at least 200 different types of cells, each specialized for its particular function but all working as a cooperative whole. Figure 4–13 shows cells from an animal trachea (windpipe). These cells are epithelial cells, the sort of cells that line all the internal and external body surfaces of animals. The epithelial cells of the trachea are part of an elaborate organ system involved in delivering oxygen to other cells in the body.

As we discussed in the Introduction, prokaryotes, protists, fungi, plants, and animals constitute the five major categories, or kingdoms, into which organisms are classified in this text. The prokaryotes (kingdom Monera) are essentially unicellular, although in some types the cells form clusters, threads, or chains; this kingdom includes chemosynthetic, photosynthetic, and heterotrophic forms. Protists are a diverse group of eukaryotic one-celled organisms and some simple multicellular ones; the protists include both heterotrophs and photosynthetic autotrophs. The fungi, plants, and animals are all multicellular eukaryotic organisms. All animals and fungi are heterotrophs, whereas all plants, with a few curious exceptions (such as Indian pipe and dodder, which are parasites) are photosynthetic autotrophs. Within the multicellular plant body, however, some of the cells are photosynthetic, such as the cells of a leaf, and some are heterotrophic, such as the cells of a root. The photosynthetic cells supply the heterotrophic cells of the plant with sucrose.

VIEWING THE CELLULAR WORLD

In the three centuries since Robert Hooke first observed the structure of cork through his simple microscope (see page 10), a wealth of knowledge has been accumulated both about the structure of cells and their component parts and about the dynamic processes that characterize the living cell. This knowledge, which we shall begin to examine in the next chapter, has generally come in bursts, following the development of new and better techniques for studying the cell and its contents.

Types of Microscopes

Unaided, the human eye has a resolving power of about 1/10 millimeter, or 100 micrometers (Table 4–1). Resolving power is a measure of the capacity to distinguish objects from one another; it is the minimum distance that must be between two objects for them to be perceived as separate objects. For example, if you look at two lines that are less than 100 micrometers apart, you will see a single,

TABLE 4-1 **Measurements Used in Microscopy**

1 centimeter (cm) = 1/100 meter = 0.4 inch*
1 millimeter (mm) = 1/1,000 meter = 1/10 cm
1 micrometer (μm)† = 1/1,000,000 meter = 1/10,000 cm
1 nanometer (nm) = 1/1,000,000,000 meter = 1/10,000,000 cm
1 angstrom (Å)‡ = 1/10,000,000,000 meter = 1/100,000,000 cm
or
1 meter = 10^2 cm = 10^3 mm = 10^6 μm = 10^9 nm = 10^{10} Å

* A metric-to-English conversion table is found in Appendix A.
† Micrometers were formerly known as microns (μ), and nanometers as millimicrons (mμ).
‡ The angstrom is not an accepted measurement in the International System of Units; in the past, however, it was widely used in microscopy, and you will occasionally encounter it in your reading.

somewhat thickened line. Similarly, two dots less than 100 micrometers apart look like a single blurry dot. Conversely, if you look at two lines (or two dots) that are 120 micrometers apart, you can easily distinguish them from each other.

Most eukaryotic cells are between 10 and 30 micrometers in diameter—some 3 to 10 times below the resolving power of the human eye—and prokaryotic cells are smaller still. In order to distinguish individual cells, to say nothing of examining the structures of which they are composed, we must use instruments that provide greater resolution. Most of our current knowledge of cell structure has been gained with the assistance of three different types of instruments: the light microscope, the transmission electron microscope, and the scanning electron microscope (Figure 4–14).

The best light microscopes have a resolving power of 0.2 micrometer, or 200 nanometers, and so improve on the naked eye about 500 times. It is theoretically

4–14 *A comparison of* (**a**) *the light microscope,* (**b**) *the transmission electron microscope, and* (**c**) *the scanning electron microscope. The light microscope is shown upside-down, to emphasize its similarities with the electron microscopes. The focusing lenses in the light microscope are glass or quartz; those in the electron microscopes are magnetic coils. In both the light microscope and the transmission electron microscope, the illuminating beam passes through the specimen; in the scanning electron microscope, it is deflected from the surface.*

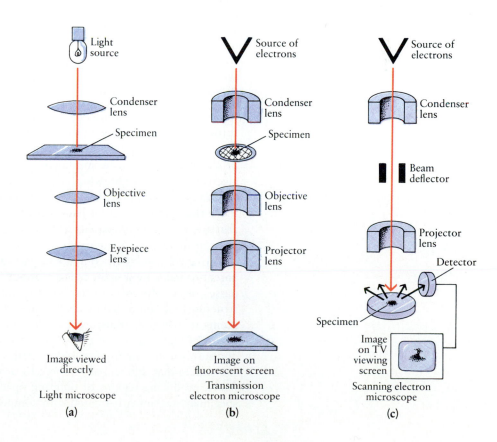

Light source — Condenser lens — Specimen — Objective lens — Eyepiece lens — Image viewed directly

Light microscope
(a)

Source of electrons — Condenser lens — Specimen — Objective lens — Projector lens — Image on fluorescent screen

Transmission electron microscope
(b)

Source of electrons — Condenser lens — Beam deflector — Projector lens — Detector — Specimen — Image on TV viewing screen

Scanning electron microscope
(c)

impossible to build a light microscope that will do better than this. The limiting factor is the wavelength of light, which ranges from about 0.4 micrometer for violet light to about 0.7 micrometer for red light. With the light microscope, we can distinguish the larger structures within eukaryotic cells and can also distinguish individual prokaryotic cells. We cannot, however, visualize the internal structure of prokaryotic cells or distinguish between the finer structures of eukaryotic cells.

Notice that resolving power and magnification are two different things. If you take a picture through the best light microscope of two lines that are less than 0.2 micrometer, or 200 nanometers, apart, you can enlarge that photograph indefinitely, but the two lines will continue to blur together. By using more powerful lenses, you can increase magnification, but this will not improve resolution.

With the transmission electron microscope, resolving power has been increased about 1,000 times over that provided by the light microscope. This is achieved by using "illumination" of a much shorter wavelength, consisting of electron beams instead of light rays. Areas in the specimen that permit the transmission of more electrons—"electron-transparent" regions—show up bright, and areas that scatter electrons away from the image—"electron-opaque" regions—are dark. Transmission electron microscopy at present affords a resolving power of about 0.2 nanometer, roughly 500,000 times greater than that of the human eye. This is about twice the diameter of a hydrogen atom.

Although the resolving power of the scanning electron microscope is only about 10 nanometers, this instrument has become a valuable tool for biologists. In scanning electron microscopy, the electrons whose imprints are recorded come from the surface of the specimen, rather than from a section through it. The electron beam is focused into a fine probe, which is rapidly passed back and forth over the specimen; complete scanning from top to bottom usually takes a few seconds. Variations in the surface of the specimen affect the pattern in which the electrons are scattered from it; holes and fissures appear dark, and knobs and ridges are light. The scattered electrons are amplified and transmitted to a television monitor, producing a visual image of the specimen. Scanning electron microscopy provides vivid three-dimensional representations of cells and cellular structures that compensate, in part, for its limited resolution.

Preparation of Specimens

In both the light microscope and the transmission electron microscope, the formation of an image with perceptible contrast requires that different parts of the cell differ in their transparency to the beam of illumination—either light rays or electrons. Parts of the specimen that readily permit the passage of light or of electrons appear bright, whereas parts that block the passage of the illuminating beam appear dark. In the scanning electron microscope, areas that appear light are those that deflect electrons back into the image; parts that appear dark are those that are not well illuminated by the electron beam or that deflect electrons away from the detector.

Living cells and their component parts are, however, almost completely transparent to light. By weight, cells are about 70 percent water, through which light passes easily. Moreover, water and the much larger molecules that form cellular structures are composed of small atoms of low atomic weight (CHNOPS). These atoms are relatively transparent to electrons, which are strongly deflected only by atoms of high atomic weight, such as those of heavy metals. To create sufficient contrast for the light microscope, cells must be treated with dyes or other substances that differentially adhere to or react with specific subcellular components, producing regions of differing opacity. For the electron microscope, the specimens are similarly treated with compounds of heavy metals.

4–15 *Rabbit sperm cells, as seen in* (**a**) *a light micrograph,* (**b**) *a transmission electron micrograph, and* (**c**) *a scanning electron micrograph. Note the dramatic increase in the resolution of ultrastructural detail in the electron micrographs.*

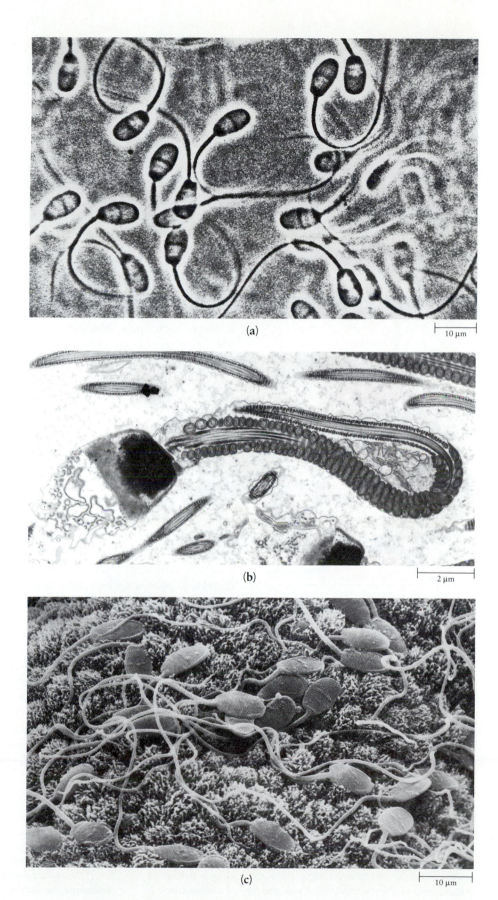

(a)

10 μm

(b)

2 μm

(c)

10 μm

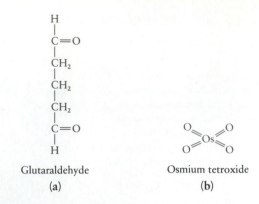

4–16 (a) *Glutaraldehyde and* (b) *osmium tetroxide, two compounds frequently used as fixatives in preparing specimens for microscopic examination. Note that glutaraldehyde has two aldehyde functional groups; each of these groups can react with a different protein molecule, linking the molecules together. The oxygen atoms of osmium tetroxide react with lipids, leading to a similar cross-linking.*

Glutaraldehyde

(a)

Osmium tetroxide

(b)

After a specimen has been treated with a staining substance, all of the stain that has not adhered to specific structures must be washed away. Cells, however, are quite fragile and any kind of rough treatment disrupts their structure. To solve this problem, biological specimens are generally "fixed" before staining. This procedure involves treatment with compounds that bind cell structures in place, usually through the formation of additional covalent bonds between molecules. Aldehydes, for instance, react with the amino groups of protein molecules, linking adjacent protein molecules together in a fairly rigid structure. Osmium tetroxide, a compound frequently used in preparing specimens for electron microscopy, interacts with lipids, binding the molecules together. Fixation has the added advantage of making cells more permeable to staining substances and to the solutions used to wash away excess stain.

Fixation and staining procedures are usually carried out with groups of cells as in, for example, a piece of liver tissue. Such specimens are not transparent to an illuminating beam—they are simply too thick to allow the passage of light rays or electrons. Before examination under the light microscope or the transmission electron microscope, they must be sliced into sections so thin that the unstained regions are transparent. After fixation and staining, the specimens are usually embedded in wax or a plastic resin to provide enough firmness to allow a clean slicing.

With the scanning electron microscope, as we noted earlier, electrons do not pass through the sample but are instead deflected from its surface. Usually, the surface of the specimen is first coated with metal, a process known as shadowing (Figure 4–17). Often, the organic material of the original specimen is removed by chemical treatment, leaving only a metallic replica of the surface, which is reinforced with a carbon film. Depending on its thickness, the replica can be examined under either the transmission electron microscope or the scanning electron microscope.

In addition to these rather drastic treatments, specimens for the electron microscope must also be dehydrated, either by chemical methods or by freeze-dry methods similar to those used in preparing freeze-dried coffee. This step is necessitated by the properties of the electrons forming the illuminating beam. If they pass through a chamber containing gaseous molecules, the electrons are deflected by the molecules and cannot be focused into a beam. Thus all air must be evacuated from the inner, working chamber of an electron microscope, creating a vacuum. If the sample were not first dehydrated, water molecules would evaporate from the sample into the chamber, destroying the vacuum and the focused electron beam.

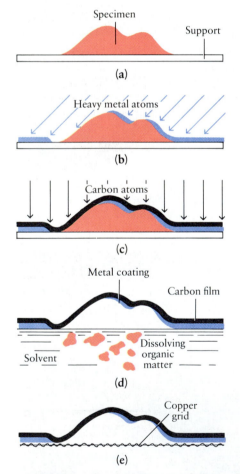

4–17 *"Shadowing" and the preparation of a replica.* (a) *The specimen is placed on a support and* (b) *is "shadowed" by heavy metal atoms evaporated from a heated filament at the side. Because the atoms are deposited from an angle, the metal coating is thicker on raised areas of the specimen.* (c) *A uniform film of carbon atoms is deposited from above, reinforcing and strengthening the replica,* (d) *which is then floated to the surface of a solvent that dissolves away the organic material. The completed replica* (e) *is washed and picked up on a copper grid.*

If the replica is thin enough, it can be examined under the transmission electron microscope as well as under the scanning electron microscope.

The procedures required to prepare most specimens for either the light microscope or the electron microscope usually result in the death of the constituent cells. Moreover, they raise serious questions about whether the structures we see in micrographs are "real" or are distortions introduced by the preparation process. One way of reducing the likelihood that microscopic observations are in error is to prepare similar samples using different techniques. If a feature appears repeatedly in different types of preparation, the probability is great that it exists within the living cell. Another approach has been the development of a variety of new techniques for viewing living cells.

Observation of Living Cells

When light waves are emitted from a coherent source (such as a laser), the waves are in phase, that is, the peaks and troughs of the waves match (Figure 4–18a). This has the effect of reinforcing the waves and creating a greater amplitude, which we perceive as increased brightness. When light waves are out of phase, however, they interfere with one another, reducing amplitude and brightness (Figure 4–18b).

As light waves pass from one material through another, they are bent, or diffracted, and their paths are slightly changed. This diffraction also alters the phase relationships of the light waves, resulting in varying amounts of interference. The amount of interference produced when light passes through the different structures of a cell is not great and provides little detectable contrast when the cell is viewed through an ordinary light microscope—which is, of course, why cells must be stained. In phase-contrast and differential-interference microscopes, however, specially designed optical systems enhance the small amounts of interference, providing greater contrast. The resolution of these microscopes is limited, as in ordinary light microscopes, but they do provide a different perspective on the living cell, revealing features difficult to detect with other systems.

Another technique frequently used with living cells is dark-field microscopy. The illuminating beam strikes the specimen from the side and the lens system detects light reflected from the sample, which appears as a bright object against a dark background. Features of the cell that are invisible in other micrographs often come into sharp relief in dark-field micrographs.

At the present time, work is progressing rapidly on other microscopic techniques. For example, video cameras are being coupled with light microscopes, producing a display on a screen and a videotape record. By adjusting the controls, as you do on a television set, the background "noise" can be reduced, contrast improved, and particular features enhanced. Video techniques, as applied to the study of the living cell, are in their infancy, but they are generating great excitement as they reveal previously unseen processes within the cell.

4–18 *The brightness of light depends on the amplitude of the light waves. (a) When two light waves are in phase, they reinforce one another, resulting in greater amplitude and brightness. (b) When the waves are completely out of phase, as shown here, they cancel each other out and no light is perceived.*

Waves that have passed through different structures within the unstained cell are partially out of phase. The resulting interference produces slight differences in contrast that can be enhanced by special optical systems, such as those of phase-contrast and differential-interference microscopes.

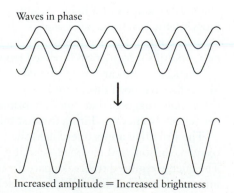

Waves in phase

Waves out of phase

Increased amplitude = Increased brightness

Decreased amplitude = Darkness

(a)

(b)

4-19 *Four views of a living fibroblast, a type of connective tissue cell that can be propagated in the laboratory in a culture dish. This cell has been photographed through (a) a conventional light microscope, (b) a phase-contrast microscope, (c) a differential-interference microscope, and (d) a dark-field microscope.*

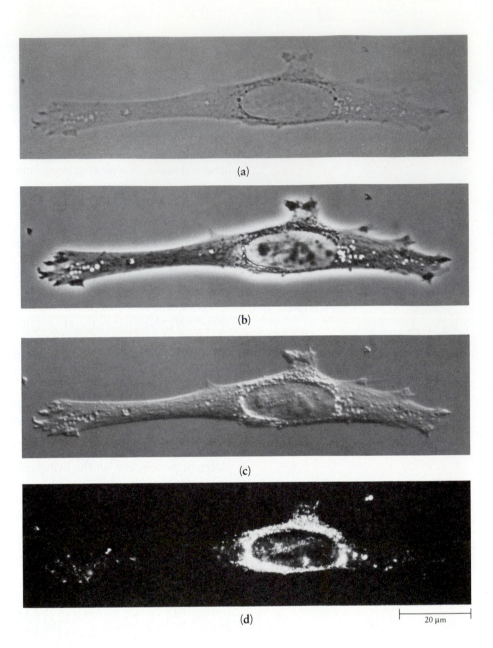

(a)

(b)

(c)

(d)   20 μm

SUMMARY

The properties associated with living systems emerge at the cellular level of organization. One of the fundamental principles of biology is the cell theory, which states that (1) all living organisms consist of one or more cells; (2) the chemical reactions of a living organism, including energy-releasing processes and biosynthetic reactions, take place within cells; (3) cells arise from other cells; and (4) cells contain the hereditary information of the organisms of which they are a part, and this information is passed from parent cell to daughter cell.

The age of the earth is estimated at 4.6 billion years. Microfossils of bacteria-like cells have been discovered that are 3.5 billion years old. The complexity of these cells suggests that the first primitive cells arose very early in the earth's existence—sometime during the first billion years.

The primitive atmosphere held the raw materials of living matter—hydrogen, oxygen, carbon, and nitrogen—combined in water vapor and other simple gases. The energy required to break apart the molecules of these gases and re-form them into more complex molecules was present in heat, lightning, radioactive elements, and high-energy radiation from the sun. Laboratory experiments have shown that under these conditions, the types of organic molecules characteristic of living systems can be formed. Other experiments have suggested the kinds of processes by which aggregations of organic molecules could have formed cell-like structures, separated from their environment by a membrane and capable of maintaining their structural and chemical integrity.

The earliest cells may have been heterotrophs (organisms that depend on outside sources for their energy-rich organic molecules) or autotrophs (organisms that can make their own organic molecules from inorganic substances). The first autotrophs may have been chemosynthetic (using the energy released by specific inorganic reactions to synthesize their own organic molecules) or photosynthetic (using the sun's energy to power their synthetic reactions). With the advent of photosynthesis, the flow of energy through the biosphere assumed its dominant modern form—radiant energy from the sun is captured by photosynthetic autotrophs and channeled through them to heterotrophic organisms. Modern heterotrophs include fungi and animals, as well as many types of single-celled organisms; modern autotrophs include other types of single-celled organisms and, most important, the green plants.

There are two fundamentally distinct types of cells—prokaryotes, which include only the bacteria and cyanobacteria, and eukaryotes, which include the protists, fungi, plants, and animals. Prokaryotic cells lack membrane-bound nuclei and most of the organelles found in eukaryotic cells. Prokaryotes were the only form of life on earth for almost 2 billion years, and then, about 1.5 billion years ago, eukaryotic cells evolved. Multicellular organisms, which are composed of eukaryotic cells specialized to perform particular functions, evolved comparatively recently—only about 750 million years ago.

Because of the small size of cells and the limited resolving power of the human eye, microscopes are required to visualize cells and subcellular structures. The three principal types are the light microscope, the transmission electron microscope, and the scanning electron microscope. Specimens to be studied using a conventional light microscope or a transmission electron microscope must be fixed, stained, dehydrated (for the electron microscope), embedded, and sliced into thin sections. Surface replicas are generally prepared for study with the scanning electron microscope. Concern about the distortions that may be introduced by these preparation procedures has led to the development of other microscopic techniques. The special optical systems of phase-contrast, differential-interference, and dark-field microscopes make it possible to study living cells. An important new development is the use of video cameras with microscopes.

QUESTIONS

1. Distinguish among the following: heterotroph/autotroph; chemosynthetic autotroph/photosynthetic autotroph; prokaryote/eukaryote; light microscope/transmission electron microscope/scanning electron microscope.

2. Why would energy sources have been necessary for the synthesis of simple organic molecules on the primitive earth?

3. Although there is some uncertainty as to the exact mixture of gases that constituted the early atmosphere, there is general agreement that free oxygen was not present. What properties of oxygen would have made chemical evolution unlikely in an atmosphere containing O_2?

4. A key event in the origin of life was the formation of a membrane that separated the contents of primitive cells from their surroundings. Why was this so critical?

5. Some scientists think that other planets in our galaxy may well contain some form of life. If you were seeking such a planet, what characteristics would you look for?

6. Return to Chapter 2 and add approximate scale markers to Figures 2–1, 2–5, and 2–11.

7. What are the advantages and disadvantages of studying cells with the transmission electron microscope and the scanning electron microscope? With special optical microscopes such as phase-contrast, differential-interference, and dark-field?

How Cells Are Organized

There are many, many different kinds of cells. In a drop of pond water, you are likely to find a variety of protists, and in even a small pond, there are probably several hundred different kinds of protists, plus a variety of prokaryotes. Our own tissues and organs are constructed of at least 200 different and distinct types of somatic ("body") cells. Plants are composed of cells that appear quite different from those of our bodies, and insects have many cells of kinds not found either in plants or in vertebrates. Thus, the first remarkable fact about cells is their diversity.

The second, even more remarkable, fact is their similarity. Every cell is a self-contained and at least partially self-sufficient unit, surrounded by a membrane that controls the passage of materials into and out of the cell. This makes it possible for the cell to differ biochemically and structurally from its surroundings. All cells also have an information and control center in which the genetic material is localized. As we noted in the last chapter, this region in prokaryotic cells is known as the nucleoid; in eukaryotic cells, it is the nucleus. Many eukaryotic cells also have a variety of internal structures, the organelles, which are similar or identical from one cell to another through a wide range of cell types. And, all cells are composed of the same remarkably few kinds of atoms and molecules.

CELL SIZE AND SHAPE

Most of the cells that make up a plant or animal body are between 10 and 30 micrometers in diameter. A principal restriction on cell size is that imposed by the relationship between volume and surface area. As Figure 5–1 shows, as volume decreases, the ratio of surface area to volume increases rapidly. Materials—such as oxygen, carbon dioxide, ions, food molecules, and waste products—entering and leaving the cell must move through its membrane-bound surface. These substances are the raw materials and products of the cell's **metabolism,** which is the total of all of the chemical activities in which it is engaged. The more active the cell's metabolism, the more rapidly materials must be exchanged with the environment if the cell is to continue to function. In smaller cells, the ratio of surface area to volume is higher than in larger cells, and thus proportionately greater quantities of materials can move into, out of, and through smaller cells in a given period of time. A larger cell, by contrast, requires the exchange of greater quantities of materials in order to meet the needs of the larger volume of living matter—and yet, the larger the cell, the smaller the ratio of surface area to volume.

A second limitation on cell size appears to involve the capacity of the nucleus, the cell's control center, to provide enough copies of the information needed to regulate the processes occurring in a large, metabolically active cell. The exceptions seem to "prove" the rule. In certain large, complex one-celled protists—the ciliates, of which *Paramecium* is an example—each cell has two or more nuclei, the additional ones apparently copies of the original. Other organisms, such as the

5–1 *The single 4-centimeter cube, the eight 2-centimeter cubes, and the sixty-four 1-centimeter cubes all have the same total volume. As the volume is divided up into smaller units, however, the total amount of surface area increases, as does the ratio of surface area to volume. For example, the sixty-four 1-centimeter cubes have four times the total surface area of the single 4-centimeter cube, and the ratio of surface area to volume in each 1-centimeter cube is four times that in the 4-centimeter cube. Similarly, smaller cells have a higher ratio of surface area to volume than larger cells. This means not only more membrane surface through which materials can move into or out of the cell but also less living matter to be serviced and shorter distances through which materials must move within the cell.*

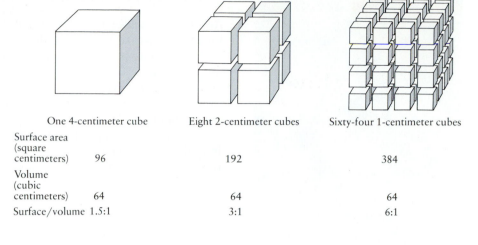

	One 4-centimeter cube	Eight 2-centimeter cubes	Sixty-four 1-centimeter cubes
Surface area (square centimeters)	96	192	384
Volume (cubic centimeters)	64	64	64
Surface/volume	1.5:1	3:1	6:1

slime molds, are made up, in effect, of one giant cell but have thousands of nuclei. Such organisms are also, frequently, very thin and spread out, thereby avoiding the problem of the ratio of surface area to volume.

It is not surprising that the most metabolically active cells are usually small. The relationship between cell size and metabolic activity is nicely illustrated by egg cells. Many egg cells are very large. A frog's egg, for instance, is 1,500 micrometers in diameter. Some egg cells are several centimeters across—for example, the cell, or yolk, of a chicken's egg. Most of this mass consists of stored nutrients for the developing embryo. When the egg cell is fertilized and begins to be active metabolically, many nuclear divisions occur and the cell divides many times before there is any actual increase in volume or mass. Thus the total mass is cut into cellular units small enough for efficient transfer and control processes.

Like drops of water and soap bubbles, cells have a tendency to be spherical. As we have seen, however, cells often have other shapes. This is because of cell walls, found in plants, fungi, and many one-celled organisms; because of attachments to and pressure from other neighboring cells or surfaces (as in intestinal epithelial cells); or because of arrays of microtubules and other structural filaments within the cell (Figure 5–2).

5–2 *(a) Numerous fine strands of cytoplasm (axopods) extend radially from the body of the protist* Actinosphaerium. *Each axopod is bounded by an extension of the cell membrane and contains many microtubules, arranged longitudinally, which stiffen and extend the axopods. (b) An axopod in cross section. The microtubules are arranged in two interlocking spirals that form a twelvefold pattern.*

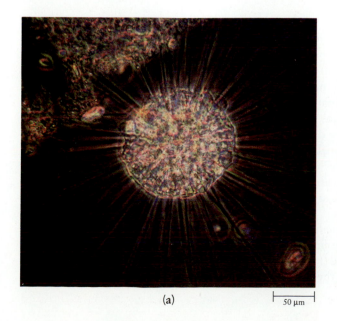

(a) ⊢ 50 µm ⊣

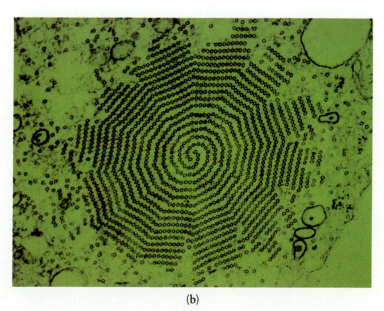

(b)

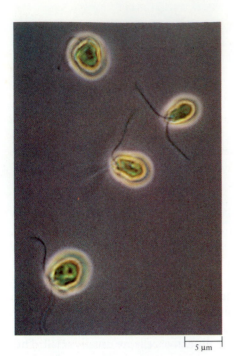

|_____| 5 μm

5–3 *Cells of the green alga* Chlamydomonas. *Note the pair of flagella with which each cell propels itself through the water.*

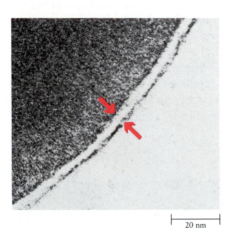

|_____| 20 nm

5–4 *Electron micrograph showing a cross section of the cell membrane of a human red blood cell. The cell membrane is indicated by the arrows. The "molecular sandwich" structure of the membrane is believed to consist of two electron-opaque (dark) layers of phospholipid molecules arranged with their hydrophobic tails pointing inward, forming the electron-transparent (light) inner "filling," with globular proteins embedded throughout. The darker material at the left of the micrograph is hemoglobin, which fills the red blood cell.*

SUBCELLULAR ORGANIZATION

Antony van Leeuwenhoek discovered the protists some 300 years ago. "This was for me," he wrote, "among all the marvels that I have discovered in nature, the most marvelous of all." As Leeuwenhoek and his successors observed the thousands of living creatures that they found "all alive in a drop of water," they were able to see, but only barely, structures within them. They interpreted these structures as miniature hearts and stomachs and lungs, in other words, tiny organs, or organelles.

Modern microscopic techniques have confirmed that eukaryotic cells do indeed contain a multitude of structures. They are not, of course, organs such as those found in multicellular organisms, but they are in some ways comparable: they are specialized in form and function to carry out particular activities required in the cellular economy. Just as the organs of multicellular animals work together in organ systems, the organelles of cells engage in a number of cooperative and interdependent functions.

Every cell must carry out essentially the same processes—acquire and assimilate food, eliminate wastes, synthesize new cellular materials, and in many cases, be able to move and to reproduce. Just as the various organs of your body have a structure that suits them for the specific functions they carry out (the kidney for elimination of wastes from the blood, the intestine for food absorption, and so on), so all cells have an internal architecture that includes organelles suited to the functions they perform. It is important to realize that a cell is not a random assortment of parts but a dynamic, integrated entity.

Also, remember that although we can look at only one structure or process at a time, most activities of a cell go on simultaneously and influence one another. *Chlamydomonas,* for instance, is swimming, photosynthesizing, absorbing nutrients from the water, building its cell wall, making proteins, converting sugars to starch (or vice versa), and oxidizing food molecules for energy, all at the same time. It is also likely to be orienting itself in the sunlight, it is probably preparing to divide, it is possibly "looking" for a mate, and it is undoubtedly carrying out at least a dozen or more other important activities, many of which may be still unknown.

CELL BOUNDARIES

The Cell Membrane

All cells, as we stated previously, are basically very similar. They all have DNA as the genetic material, they perform the same types of chemical reactions, and they are all surrounded by an external cell membrane that conforms to the same general design in both prokaryotic and eukaryotic cells. The living matter bounded by the membrane consists, in eukaryotes, of the nucleus and the cytoplasm, which contains the organelles.

A cell can exist as a distinct entity because of the cell membrane, which regulates the passage of materials into and out of the cell. The cell membrane (also called, as we noted previously, the plasma membrane) is only about 7 to 9 nanometers thick and cannot be resolved by the light microscope. With the electron microscope, it can be visualized as a continuous, thin double line (Figure 5–4).

The eukaryotic cell membrane is formed from a phospholipid bilayer, that is, a double layer of phospholipid molecules arranged with their hydrophobic fatty acid tails pointing inward (Figure 5–5). Cholesterol molecules are embedded in the hydrophobic interior of the bilayer, in which numerous protein molecules are also suspended. These proteins, known as integral membrane proteins, generally span the bilayer and protrude on either side; the portions embedded in the bilayer have hydrophobic surfaces, whereas the surfaces of those portions that extend beyond the bilayer are hydrophilic.

The two surfaces of the cell membrane differ considerably in chemical composition. The two layers of the bilayer generally have different concentrations of specific types of lipid molecules. In many types of cells, the outer layer is particularly rich in glycolipid molecules. The carbohydrate chains of these molecules are, like the phosphate heads of the phospholipid molecules, exposed on the surface of the membrane; the hydrophobic fatty acid tails are within the membrane. The protein composition of the two layers also differs. The integral membrane proteins have a definite orientation within the bilayer, and the portions extending on either side are completely different in both amino acid composition and tertiary structure. On the cytoplasmic side of the membrane, additional protein molecules, known as peripheral membrane proteins, are bound to some of the integral proteins protruding from the bilayer. On the outside of the membrane, short carbohydrate chains are covalently linked to the protruding proteins. These chains, along with the carbohydrate chains of the glycolipids, form a carbohydrate coat on the outer surface of the membranes of many types of cells. The carbohydrates are thought to play a role in the adhesion of cells to one another and in the "recognition" of molecules that interact with the cell (such as hormones, antibodies, and viruses).

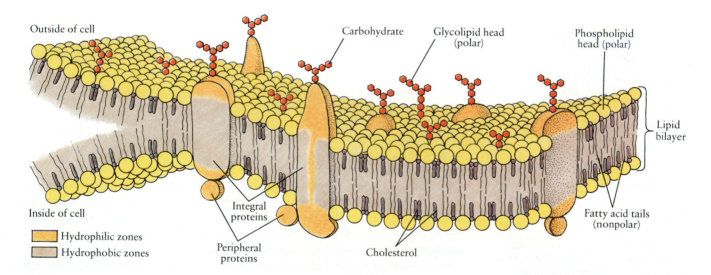

5–5 Model of a cell membrane, as determined from electron micrographs and biochemical data. The basic membrane structure is formed from a network of phospholipid molecules, in which cholesterol molecules and large protein molecules are embedded. The phospholipid molecules are arranged in a bilayer with their hydrophobic tails pointing inward and their hydrophilic phosphate heads pointing outward. Cholesterol molecules nestle among the hydrophobic tails.

The proteins embedded in the bilayer are known as integral membrane proteins. On the cytoplasmic face of the membrane, peripheral membrane proteins are bound to some of the integral proteins. The portion of a protein molecule's surface that is within the lipid bilayer is hydrophobic; the portion of the surface that is exposed outside the bilayer is hydrophilic. It is believed that pores with hydrophilic surfaces pass through some of the protein molecules.

Interspersed among the phospholipid molecules of the outer layer of the bilayer are glycolipid molecules. Their carbohydrate chains and the carbohydrate chains attached to the proteins protruding on the outside of the membrane are thought to be involved in the adhesion of cells to each other and with the "recognition" of molecules at the membrane surface.

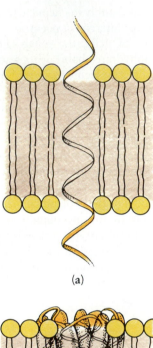

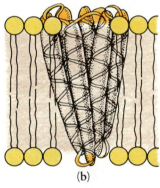

5–6 *Two principal configurations that have been determined for membrane proteins are (a) an alpha helix, and (b) a globular tertiary structure, formed by repeating segments of alpha helix that zig-zag through the membrane. The helical segments are linked by irregular, hydrophilic segments of the polypeptide chain that extend on either side of the membrane.*

Although many of the integral proteins appear to be anchored in place, either by peripheral proteins or by cytoplasmic protein filaments that are concentrated near the membrane, the structure of the bilayer is generally quite fluid. The lipid molecules and at least some of the protein molecules can move laterally within it, forming different patterns that vary from time to time and place to place. Consequently, this widely accepted model of membrane structure is known as the **fluid-mosaic model.**

Recent studies have revealed the detailed structure of a number of membrane proteins. Two basic configurations have been identified among the integral proteins studied thus far (Figure 5–6). One is a relatively simple rodlike structure, consisting of an alpha helix embedded in the hydrophobic interior of the membrane, with less regular, hydrophilic portions extending on either side of the membrane. These hydrophilic portions are often extensively folded into an intricate tertiary structure. The other configuration is found in large, globular molecules with complex tertiary or quaternary structures that result from repeated "passes" through the membrane. The portions of these proteins embedded in the hydrophobic interior of the bilayer consist of segments of tightly coiled alpha helix. In globular proteins formed from a single polypeptide chain, these regular helical segments alternate with segments of the polypeptide chain that have an irregular structure. The irregular segments, which are hydrophilic, are exposed on either side of the membrane, while the helical segments zig-zag back and forth through the membrane. Although the embedded surfaces in contact with the lipid bilayer are always hydrophobic, the interior portions of some of the globular proteins are apparently hydrophilic, creating "pores" through which certain polar substances can cross the membrane.

The cell membrane of bacterial cells is much the same in basic composition as the cell membrane of eukaryotic cells, except that, with a few exceptions, bacterial cell membranes do not contain cholesterol. In eukaryotes, all the membranes of a cell, including those surrounding the various organelles, also have the same general structure. There are, however, differences in the types of lipids and, particularly, in the number and types of proteins and carbohydrates, which vary from membrane to membrane and also from place to place on the same membrane. These differences give the membranes of different types of cells and of the different organelles unique properties that can be correlated with differences in function. Most membranes are about 40 percent lipid and 60 percent protein, though there is considerable variation. The proteins, which are extremely diverse structurally, perform a variety of essential functions. Some are enzymes, regulating particular chemical reactions; others are receptors, involved in the recognition and binding of signalling molecules, such as hormones; and still others are transport proteins, playing critical roles in the movement of substances across the membrane. As we shall see repeatedly in the course of this text, discoveries concerning the structure and function of specific membrane proteins are shedding new light on a diversity of processes, ranging from the navigation of bacterial cells to photosynthesis to the transmission of the nerve impulse.

The Cell Wall

A principal distinction between plant and animal cells is that plant cells are surrounded by a cell wall. The wall is outside the membrane and is constructed by the cell. As a plant cell divides, a thin layer of gluey material forms between the two new cells; this becomes the **middle lamella** (Figure 5–7a). Composed of pectins (the compounds that make jellies gel) and other polysaccharides, it holds adjacent cells together. Next, on either side of the middle lamella, each plant cell constructs its primary cell wall. The primary wall contains cellulose molecules bundled together in microfibrils that are laid down in a matrix of gluey polymers.

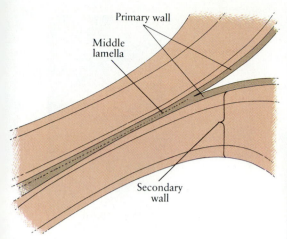

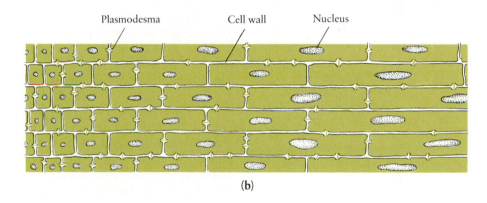

(a)

1 μm

5–7 (a) *Electron micrograph of two adjacent cell walls of tracheids, cells through which water is conducted in plants. You can see the middle lamella, the primary walls, and the layered secondary walls, deposited inside the primary wall. The cells, which are from the wood of a ground hemlock, have died. The electron-transparent areas in the top left and lower right portions of the micrograph represent empty space, once filled by the living matter of the cells.*

(b) *Growth of plant cells is limited by the rate at which the cell walls expand. The walls control both the rate of growth and its direction; they do not expand in all directions but elongate in a single dimension. Cells at the left are newly formed; cells farther to the right are older and have started to elongate. Plasmodesmata (singular, plasmodesma) are channels connecting adjacent cells.*

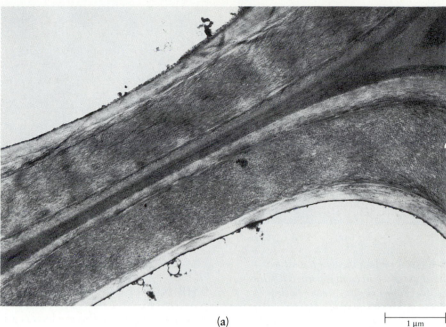

(b)

As you can see in Figure 3–10c on page 66, successive layers of cellulose microfibrils are oriented at right angles to one another in the completed cell wall. (Those of you familiar with building materials will note that the cell wall thus combines the structural features of both fiberglass and plywood.)

In plants, growth takes place largely by cell elongation. Studies have shown that the cell adds new materials to its walls throughout this elongation process. The cell, however, does not simply expand in all directions; its final shape is determined by the structure of its cell wall (Figure 5–7b).

As the cell matures, a secondary wall may be constructed. This wall is not capable of expansion, as is the primary wall. It often contains other molecules, such as lignin, that have stiffening properties. In such cells, the living material of the cell often dies, leaving only the outer wall, a monument to the cell's architectural abilities (see Figure I–12, page 10).

Cellulose-containing cell walls are also found in many algae. Fungi and prokaryotes also have cell walls, but they usually do not contain cellulose. The cell walls of fungi are composed principally of chitin (Figure 3–11a, page 66). Prokaryotic cell walls contain polysaccharides and complex polymers known as peptidoglycans, which are formed from amino acids and sugars. We shall examine the structure of prokaryotic cell walls in Chapter 21.

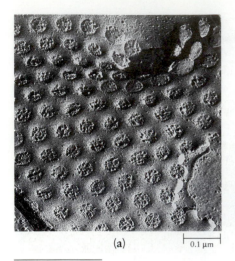

(a)   ⊢ 0.1 μm ⊣

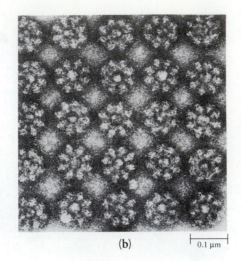

(b)   ⊢ 0.1 μm ⊣

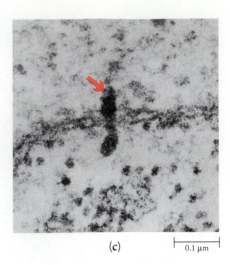

(c)   ⊢ 0.1 μm ⊣

5–8 (a) *A surface view of the nuclear envelope of a guinea pig sperm cell. Clearly visible on this surface are the nuclear pores. Biochemical studies and electron micrographs of sections through the plane of the envelope have revealed that the structure of each nuclear pore consists of eight protein-containing granules (b). The opening of the pore is a very narrow channel in the center of each octagonal array.*

(c) A granule of protein and RNA, indicated by the arrow, moves from the nucleus (top of the micrograph), through a nuclear pore, and into the cytoplasm (bottom). This cell is from the salivary gland of the midge Chironomus, *a delicate insect that resembles a mosquito but does not feed on humans or other mammals.*

THE NUCLEUS

In eukaryotic cells, the nucleus is a large, often spherical body, usually the most prominent structure within the cell. It is surrounded by the nuclear envelope, which is made up of two concentric membranes, each of which is a lipid bilayer. These two membranes are separated by a gap of about 20 to 40 nanometers. At frequent intervals, however, they are fused together, creating small **nuclear pores** through which materials pass between the nucleus and cytoplasm (Figure 5–8). The pores, which are surrounded by large protein-containing granules arranged in an octagonal pattern, form a narrow channel through the fused lipid bilayers.

The chromosomes are found within the nucleus. When the cell is not dividing, the chromosomes are visible only as a tangle of fine threads, called **chromatin.** The most conspicuous body within the nucleus is the **nucleolus.** There are typically two nucleoli per nucleus, although often only one is visible in a micrograph. As we shall see in Chapter 18, the nucleolus is the site at which ribosomal subunits are constructed. Viewed with the electron microscope, the nucleolus appears to be a collection of fine granules and tiny fibers (Figure 5–9). These are thought to be parts of ribosomal subunits and threads of chromatin.

5–9 *Electron micrograph of the nucleus of the alga* Chlamydomonas. *The dark body in the center of the nucleus is the nucleolus. The major components of ribosomes are produced in the nucleolus; you can see partially formed ribosomes around its periphery. Notice also the nuclear envelope with its nuclear pores, two of which are indicated by arrows. Above the nucleus is a portion of a chloroplast containing starch grains. A Golgi complex is to the upper left of the nucleus, and mitochondria are visible below.*

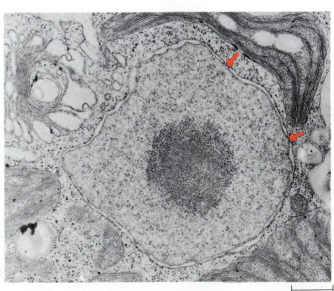

⊢ 0.5 μm ⊣

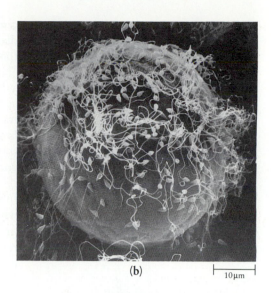

(a)   (b)   ├──┤ 10μm

5–10 (a) *Sea urchins, readily identified by their spiny body surfaces, are common inhabitants of rocky seashores.* (**b**) *This egg from a sea urchin is surrounded by sperm cells. Despite the great differences in size of egg and sperm, both contribute equally to the hereditary characteristics of the individual. Since the nucleus is approximately the same size in both cells, the early microscopists postulated that this part of the cell must be the carrier of the hereditary information. Sea urchin eggs and sperm cells have been used in many studies because sea urchins are relatively easy to obtain and fertilization, which is external, can be easily observed in the laboratory.*

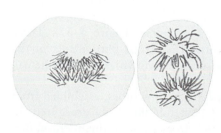

5–11 *Drawings made by Walther Flemming in 1882 of chromosomes in dividing cells of salamander larvae. Flemming's observations were dependent upon the development of new cytological staining techniques.*

The Functions of the Nucleus

Our current understanding of the role of the nucleus in the life of the cell began with some early microscopic observations. One of the most important of these observations was made more than a hundred years ago by a German embryologist, Oscar Hertwig, who was observing the eggs and sperm of sea urchins (Figure 5–10). Sea urchins produce eggs and sperm in great numbers. The eggs are relatively large and so are easy to observe. They are fertilized in the open water, rather than internally, as is the case with land-dwelling vertebrates such as ourselves. Watching the eggs being fertilized under his microscope, Hertwig observed that only a single sperm cell was required. Further, when the sperm cell penetrated the egg, its nucleus was released and fused with the nucleus of the egg. This observation, confirmed by other scientists and in other kinds of organisms, was important in establishing the fact that the nucleus is the carrier of the hereditary information: the only link between father and offspring is the nucleus of the sperm.

Another clue to the importance of the nucleus came about as the result of the observations of Walther Flemming, also about 100 years ago. Flemming observed the "dance of the chromosomes" that takes place when eukaryotic cells divide (a process to be described in Chapter 7), and he painstakingly pieced together the sequence of events. (The fact that Hertwig and Flemming made their observations at about the same time was no coincidence; enormous improvements had just occurred in the light microscope and in techniques of microscopy.)

Since Flemming's time, a number of experiments have explored the role of the nucleus of the cell. In one simple experiment, the nucleus was removed from an amoeba by microsurgery. The amoeba stopped dividing and, in a few days, it died. If, however, a nucleus from another amoeba was implanted within 24 hours after the original one was removed, the cell survived and divided normally.

In the early 1930s, Joachim Hämmerling studied the comparative roles of the nucleus and the cytoplasm by taking advantage of some unusual properties of the marine alga *Acetabularia*. The body of *Acetabularia* consists of a single huge cell 2 to 5 centimeters in height. Individuals have a cap, a stalk, and a "foot," all of which are differentiated portions of the single cell. If the cap is removed, the cell will rapidly regenerate a new one. Different species of *Acetabularia* have different kinds of caps. *Acetabularia mediterranea*, for example, has a compact umbrella-shaped cap, and *Acetabularia crenulata* has a cap of petal-like structures.

Hämmerling took the "foot," which contains the nucleus, from a cell of *A. crenulata* and grafted it onto a cell of *A. mediterranea*, from which he had first

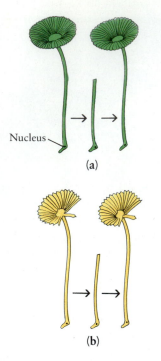

Nucleus

(a)

(b)

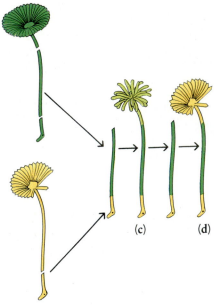

(c)   (d)

5-12 (a) *One species of* Acetabularia *has an umbrella-shaped cap, and* (b) *another has a ragged, petal-like cap. If the cap is removed, a new cap forms, similar in appearance to the amputated one. However, if the "foot" (containing the nucleus) is removed at the same time as the cap and a new nucleus from the other species is transplanted, the cap* (c) *that forms will have a structure with characteristics of both species. If this cap is removed, the next cap* (d) *that grows will be characteristic of the cell that donated the nucleus, not of the cell that donated the cytoplasm.*

removed the "foot" and the cap. The cap that then formed had a shape intermediate between those of the two species. When this cap was removed, the next cap that formed was completely characteristic of *A. crenulata* (Figure 5–12).

Hämmerling interpreted these results as meaning that certain cap-determining substances are produced under the direction of the nucleus. These substances accumulate in the cytoplasm, which is why the first cap that formed after nuclear transplantation was of an intermediate type. By the time the second cap formed, however, the cap-determining substances present in the cytoplasm before the transplant had been exhausted, and the form of the cap was completely under the control of the new nucleus.

We can see from these experiments that the nucleus performs two crucial functions for the cell. First, it carries the hereditary information that determines whether a particular cell will develop into (or be a part of) a *Paramecium*, an oak, or a human—and not just any *Paramecium*, oak, or human, but one that resembles the parent or parents of that particular, unique organism. Each time a cell divides, this information is passed on to the two new cells. Second, as Hämmerling's work indicated, the nucleus exerts a continuing influence over the ongoing activities of the cell, ensuring that the complex molecules that the cell requires are synthesized in the number and of the kind needed. The way in which the nucleus performs these functions will be described in Section 3.

THE CYTOPLASM

Not long ago, the cell was visualized as a bag of fluid containing enzymes and other dissolved molecules along with the nucleus, a few mitochondria, and occasional other organelles that could be seen by special microscopic techniques. With the development of electron microscopy, however, an increasing number of structures have been identified within the cytoplasm, which is now known to be highly organized and crowded with organelles. On page 112, the interior of a typical animal cell is shown in Figure 5–15; Figure 5–16 shows a corresponding view of a typical plant cell.

The Cytoskeleton

As we noted in the previous chapter, extremely thin sections are required for study with the transmission electron microscope. With the recent development, however, of the high-voltage electron microscope, which produces a beam of electrons with greater penetration, it has become possible to use much thicker specimens—in some cases, whole cells. The resulting visualization of the interior of the cell in three dimensions has revealed previously unsuspected interconnections among filamentous protein structures within the cytoplasm. These structures form an internal **cytoskeleton** that maintains the shape of the cell, enables it to move, anchors its organelles, and directs its traffic (Figure 5–13). Three different types of filaments have been identified as major participants in the cytoskeleton: **microtubules, actin filaments** (formerly known as microfilaments), and **intermediate filaments.**

Microtubules, as we saw on page 78, are long, hollow tubes, assembled from dimers of the globular proteins alpha and beta tubulin. They are about 22 nanometers in diameter, but their length varies. In many cells, the microtubules extend outward from an "organizing center" near the nucleus, ending near the cell surface (Figure 5–14a). As noted previously, microtubules play an important role in cell division and seem to provide a temporary scaffolding for the construction of other cellular structures. As we shall see later in this chapter, they are also key components of cilia and flagella, permanent structures used for locomotion by many types of cells.

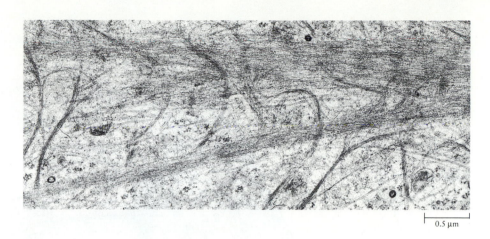

0.5 μm

5-13 *The three principal elements of the cytoskeleton are visible in this electron micrograph of an epithelial cell from a rat kangaroo. The thick bundles of relatively straight fibers running horizontally across the micrograph are actin filaments. Microtubules are the somewhat thicker individual fibers that resemble railroad tracks as seen in aerial photographs taken at high altitude. The bundles of fibers that curve vertically through the micrograph consist of intermediate filaments.*

Actin filaments are fine protein threads, averaging 6 nanometers in diameter, formed from molecules of the globular protein actin. Each filament consists of many globular actin molecules, linked together in a helical chain. Like microtubules, actin filaments can be readily assembled and disassembled by the cell, and they also play important roles in cell division and cell motility. In some cells, they are concentrated in bundles, known as stress fibers, near the cell membrane (Figure 5–14b).

Intermediate filaments, as their name implies, are intermediate in size between microtubules and actin filaments, with a diameter of 7 to 11 nanometers. Unlike microtubules and actin filaments, which consist of globular protein subunits, intermediate filaments are composed of fibrous proteins and cannot be as easily disassembled by the cell once they are formed. The specific protein forming intermediate filaments varies according to the cell type; in different types of epithelial cells, for example, these filaments are composed of different types of keratin. Each of the protein molecules making up an intermediate filament has a rodlike portion of constant length, with terminal regions that vary in length and amino acid composition. In many cells, the intermediate filaments radiate out from the nuclear envelope and are closely associated with the microtubules (Figure 5–14c); in epithelial cells, they are also anchored at specific points in the cell membrane. The function of intermediate filaments in the life of the cell is still poorly understood, but they are found in the greatest density in cells subject to mechanical stress.

5-14 *The distribution of elements of the cytoskeleton in whole cells is dramatically revealed by immunofluorescence microscopy. The cell is treated with specially prepared fluorescent antibodies to the protein of interest. The antibodies attach to the protein, and the pattern of their fluorescence indicates the location of the protein. These micrographs of epithelial cells from the rat kangaroo reveal (a) microtubules radiating from the cell center, (b) actin filaments, bundled together in stress fibers, and (c) intermediate filaments, extending throughout the cytoplasm.*

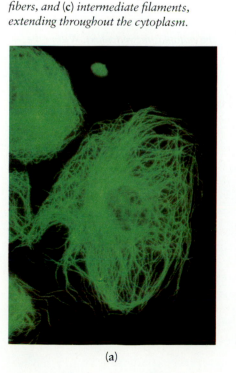

(a)

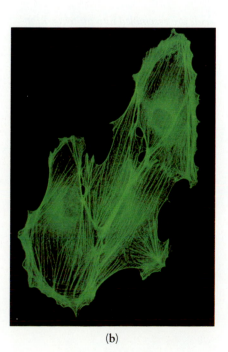

(b)

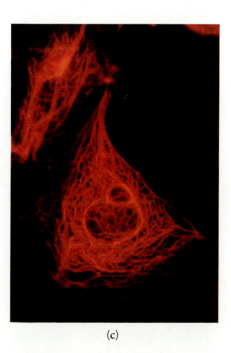

(c)

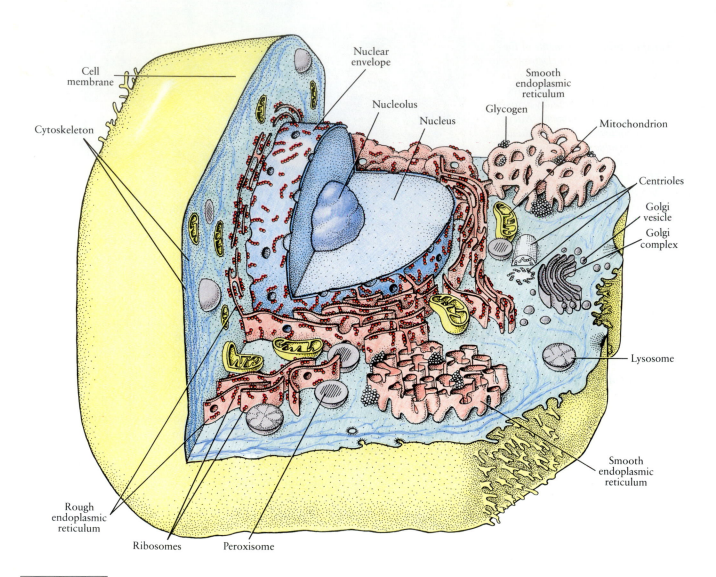

5–15 *A representative animal cell, as interpreted from electron micrographs. Like all cells, this one is bounded by a cell membrane (the plasma membrane), which acts as a selectively permeable barrier to the surrounding environment. All materials that enter or leave the cell, including food, wastes, and chemical messages, must pass through this barrier.*

Within the membrane is found the cytoplasm, which contains the enzymes and other solutes of the cell. The cytoplasm is traversed and subdivided by an elaborate system of membranes, the endoplasmic reticulum, a portion of which is shown here. In some areas, the endoplasmic reticulum is covered with ribosomes, the special structures on which amino acids are assembled into proteins. Ribosomes are also found elsewhere in the cytoplasm.

Golgi complexes are packaging centers for molecules synthesized within the cell. Lysosomes and peroxisomes are vesicles in which a number of different types of molecules are broken down to simpler constituents that can either be used by the cell or, in the case of waste products, safely removed from it. The mitochondria are the sites of the chemical reactions that provide energy for cellular activities.

The largest body in the cell is the nucleus. It is surrounded by a double membrane, the nuclear envelope, the outer membrane of which is continuous with the endoplasmic reticulum. Within the nuclear envelope is a nucleolus, the site where the ribosomal subunits are formed, and the chromatin, which is the material of the chromosomes in an extended form.

An elaborate, highly structured network of protein filaments, the cytoskeleton, pervades the cytoplasm. Among its components are microtubules, which have a rodlike appearance, and intermediate filaments, threadlike structures concentrated near the cell membrane. Other elements of the cytoskeleton are too fine to be visible at this magnification. The filaments of the cytoskeleton maintain the cell's shape, anchor its organelles, and direct the intracellular molecular traffic.

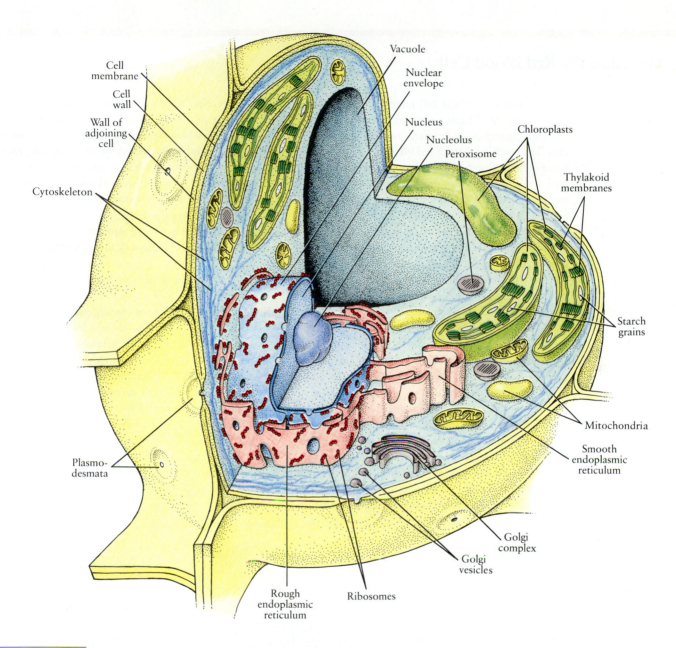

Cell membrane

Cell wall

Wall of adjoining cell

Cytoskeleton

Vacuole

Nuclear envelope

Nucleus

Nucleolus

Peroxisome

Chloroplasts

Thylakoid membranes

Starch grains

Plasmodesmata

Mitochondria

Smooth endoplasmic reticulum

Golgi complex

Golgi vesicles

Rough endoplasmic reticulum

Ribosomes

5–16 *A relatively young plant cell, as interpreted from electron micrographs. Like the animal cell, it is bounded by a cell membrane. Surrounding the cell membrane is a cellulose-containing cell wall. Plasmodesmata, which are channels through the cell walls, provide a cytoplasmic connection between adjacent cells.*

The most prominent structure in many plant cells is a large vacuole, filled with a solution of salts and other substances. In mature plant cells, the vacuole often occupies the bulk of the cell, and the other cel-

lular contents are squeezed into a narrow region next to the cell membrane. As we shall see in the next chapter, the vacuole plays a key role in keeping the cell wall stiff and the plant body crisp.

Chloroplasts, the large organelles in which photosynthesis takes place, are generally concentrated near the surface of the cell. Molecules of chlorophyll and the other substances involved in the capture of light energy from the sun are located in the thylakoid membranes within the chloroplasts.

Like the animal cell, the living plant cell contains a prominent nucleus, extensive endoplasmic reticulum, and many ribosomes and mitochondria. Especially numerous in the growing plant cell are Golgi complexes, which play an important role in the assembly of materials for the expanding cell wall. The orientation of cellulose microfibrils as they are added to the cell wall is determined by the orientation of microtubules in portions of the cytoskeleton close to the cell membrane.

Spectrin and the Red Blood Cell

As the human red blood cell matures, it synthesizes large quantities of hemoglobin and then extrudes its nucleus, organelles, and other cytoplasmic constituents. The resulting mature red blood cell is essentially a bag of hemoglobin, enclosed by the cell membrane. And yet it has a definite shape, resembling a doughnut in which the center has been compressed but not removed (see Figure 3–28a, page 79). Moreover, as it is carried along in the bloodstream, the cell can twist, turn, bend, and fold as it makes its way through the smallest of the blood vessels, the capillaries. Many of the capillaries are so narrow that a single red blood cell can barely squeeze through to deliver its precious load of oxygen to the body tissues.

Until recently, the question of how the red blood cell maintains its shape, springing back unchanged from its contortions in the capillaries, remained a mystery. Studies of the cell membrane and the cytoskeleton have now revealed that the secret lies in several different kinds of protein molecules that together form a supporting meshwork just inside the cell membrane. The major component of this meshwork is a protein known as spectrin, which constitutes about 30 percent of all the protein associated with the red blood cell membrane. Each spectrin molecule consists of two long polypeptide chains, loosely intertwined

with each other to form a dimer. Another important component of the meshwork is actin, the globular protein from which the actin filaments of the cytoskeleton are formed. As shown in the diagram, each spectrin dimer is anchored at one end to the cell membrane by a peripheral protein known as ankyrin. The other end of each spectrin dimer is linked to another dimer by a short filament of actin monomers and an actin-spectrin link protein. The result is a secure, yet flexible framework for the red blood cell.

It is hard to know whether to consider spectrin a peripheral protein of the cell membrane or an additional constituent of the cytoskeleton. This difficulty in definition—which is, of course, of no consequence to the living red blood cell—underlines once more the structural and functional integration of all the components of a living cell. Although the spectrin meshwork appears to be a unique feature of the highly specialized red blood cell, proteins related to spectrin have now been identified in a number of other cell types. It is hoped that studies of these spectrin-like proteins will reveal, in more general terms that apply to a wide range of cell types, how the cytoskeleton interacts with the proteins of the cell membrane to give each type of cell its particular overall structure and shape.

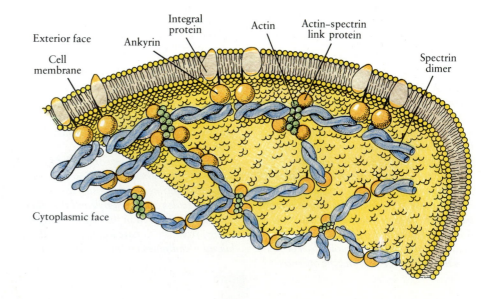

The supporting framework of the mature red blood cell. Each spectrin dimer is attached at one end to a peripheral membrane protein known as ankyrin; another spectrin dimer is attached to a second, nearby molecule of ankyrin. The ankyrin molecules, in turn, are linked to integral membrane proteins. At the opposite end, each spectrin dimer is attached to another spectrin dimer by way of a short filament of globular actin monomers and still another protein, the actin-spectrin link protein. The resulting meshwork, securely anchored to the cell membrane, is flexible yet strong, enabling the red blood cell to move efficiently through the capillaries.

Exterior face

Cell membrane

Ankyrin

Integral protein

Actin

Actin–spectrin link protein

Spectrin dimer

Cytoplasmic face

A dense network of wisplike fibers interconnects all of the other structures within the cytoplasm. These fibers, which are formed from accessory proteins of the cytoskeleton, link the cytoskeletal filaments together in specific ways. Although the resulting network gives the cell a highly ordered three-dimensional structure, it is neither rigid nor permanent. The cytoskeleton is a dynamic framework, changing and shifting according to the activities of the cell.

Vacuoles and Vesicles

In addition to organelles and the cytoskeleton, the cytoplasm of many cells, especially plant cells, contains **vacuoles**. A vacuole is a space in the cytoplasm filled with water and solutes; it is surrounded by a single membrane, known in plant cells as the **tonoplast**. Immature plant cells characteristically have many vacuoles, but as a plant cell matures, the numerous smaller vacuoles coalesce into one large, central, fluid-filled vacuole that then becomes a major supporting element of the cell (Figure 5–17). This vacuole also increases the size of the cell, including the amount of surface exposed to the environment, with a minimal investment in structural materials by the cell.

Vesicles, which are found in all metabolically active eukaryotic cells, have the same general structure as vacuoles. They are distinguished by size, function, and composition. Vesicles are usually less than 100 nanometers in diameter, whereas vacuoles are larger. One of the principal functions of vesicles is transport; as we shall see, vesicles participate in the transport of materials both within the cell and into and out of the cell.

Ribosomes

Ribosomes are the most numerous of the cell's many organelles. A rapidly growing *E. coli* cell has approximately 15,000 ribosomes, and a eukaryotic cell may have many times that number. Ribosomes, which are not enclosed by a membrane, are of similar construction in both prokaryotic and eukaryotic cells; the ribosomes of eukaryotic cells, however, are somewhat larger than those of prokaryotes. As we noted in Chapter 3, proteins are chains of amino acids, assembled in a specific sequence. The ribosomes are the sites at which this assembly takes place, a process that will be explored in some detail in Chapter 15. The more protein a cell is making, the more ribosomes it has.

The way in which the ribosomes are distributed in the eukaryotic cell is related to the way the newly synthesized proteins are utilized. Some proteins, such as collagen, digestive enzymes, hormones, or mucus, are released outside the cell, sometimes carrying out their functions at a great distance, on a cellular scale, from their source. Other proteins are essential components of cellular membranes. And, still others—hemoglobin and some enzymes, for example—are used within the cytoplasm. In cells, such as immature red blood cells, that are making cytoplasmic proteins for their own use, the ribosomes are distributed throughout the cytoplasm. In cells that are making new membrane material or proteins that are to be exported from the cell, large numbers of ribosomes are found attached to a complex system of internal membranes, the endoplasmic reticulum.

Endoplasmic Reticulum

The **endoplasmic reticulum** is a network of interconnecting flattened sacs, tubes, and channels found in eukaryotic cells. The amount of endoplasmic reticulum in a cell is not fixed but increases or decreases depending on the cell's activity.

There are two general categories of endoplasmic reticulum, rough (with ribosomes attached) and smooth (without ribosomes), which are, however, continuous with each other. Rough endoplasmic reticulum is present in all eukaryotic cells and predominates in cells making large amounts of proteins for export. It is

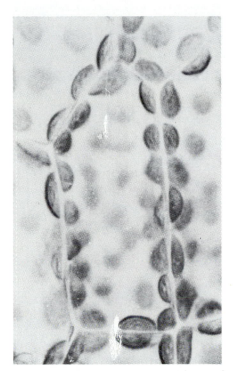

├─ 10 µm ─┤

5-17 *In this cell from the photosynthetic structure of a moss, the vacuole has expanded until it almost entirely fills the cell. The small amount of living cytoplasm, containing the chloroplasts, has been forced to the edges of the cell, up against the cell membrane.*

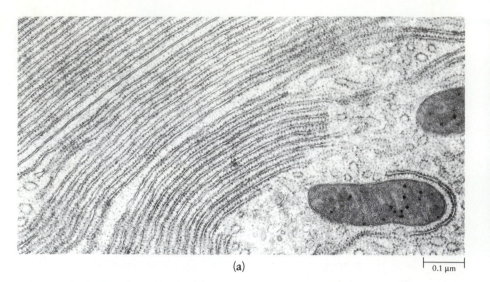

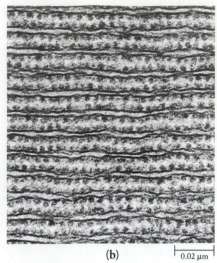

(a)   0.1 μm   (b)   0.02 μm

5–18 (a) *Rough endoplasmic reticulum, which fills most of this micrograph, is a system of membranes that separates the cell into channels and compartments and provides surfaces on which chemical activities take place. The dense objects on the membrane surfaces are ribosomes. This cell is from a pancreas, an organ extremely active in the synthesis of digestive enzymes, which are "exported" to the upper intestine, where most digestion takes place. In the lower right-hand corner of the micrograph are one mitochondrion and, above it, a portion of another.* (b) *Rough endoplasmic reticulum at higher magnification, showing the individual ribosomes. The compartments formed by the membranes of the endoplasmic reticulum are filled with newly synthesized proteins.* (c) *An interpretation of the rough endoplasmic reticulum based on electron micrographs.*

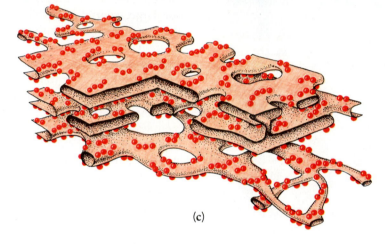

(c)

continuous with the outer membrane of the nuclear envelope, which also has ribosomes attached (see Figures 5–15 and 5–16). Rough endoplasmic reticulum often includes large, flattened sacs called cisternae. If cells engaged in protein synthesis are permitted to take up radioactive amino acids, the radioactive labels are first detected in the cytoplasm, then at the membrane of the rough endoplasmic reticulum, and then, slightly later, within its cisternae.

The synthesis of a protein destined for export from the cell, for incorporation into a specific organelle or its membrane, or for incorporation into the cell membrane itself begins in the cytoplasm with the synthesis of a "leader" of hydrophobic amino acids. This portion of the molecule, known as the **signal sequence,** is thought to direct the newly forming protein and the ribosomes that are participating in its synthesis to a specific region of the rough endoplasmic reticulum. The ribosomes attach to the endoplasmic reticulum, and the hydrophobic amino acids of the signal sequence assist in the transport of the protein through the lipid bilayer to the interior cavity, or **lumen,** of the endoplasmic reticulum. As synthesis of the protein proceeds, the growing polypeptide chain continues to move into the lumen. The newly synthesized protein molecule then moves from the rough endoplasmic reticulum through a special transitional endoplasmic reticulum, in which it is packaged in a transport vesicle destined for the Golgi complex. In the course of this progression from the endoplasmic reticulum to the Golgi complex, and subsequently to its ultimate destination in the cell, the protein molecule undergoes further processing. This processing includes cleavage of the signal sequence and, often, the addition of carbohydrate groups to the protein.

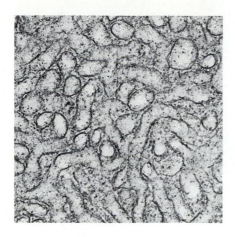

5-19 *Smooth endoplasmic reticulum from the testicle of an opossum. These membranes participate in the synthesis of the steroid hormone testosterone.*

Only cells specialized for the synthesis or metabolism of lipids—such as the gland cells that make steroid hormones—have large amounts of smooth endoplasmic reticulum. Smooth endoplasmic reticulum is also found in liver cells, where it appears to be involved in various detoxification processes (one of the many functions of the liver). For instance, in experimental animals fed large amounts of phenobarbital, the amount of smooth endoplasmic reticulum in the liver cells increases severalfold. A specialized transitional endoplasmic reticulum also seems to be active in the liver's breakdown of glycogen to glucose. As more of its functions are discovered, it seems likely that smooth endoplasmic reticulum actually represents a number of quite different specializations of endoplasmic reticulum, resembling one another only in their lack of ribosomes.

Golgi Complexes

Each **Golgi complex** consists of flattened, membrane-bound sacs stacked loosely on one another and surrounded by tubules and vesicles (Figure 5-20). The function of the Golgi complex is to accept vesicles from the endoplasmic reticulum, to modify the membranes and contents of the vesicles, and to incorporate the finished products in transport vesicles that deliver them to other parts of the cell and, especially, to the cell surface. Thus Golgi complexes serve as packaging and distribution centers. They are found in almost all eukaryotic cells. Animal cells usually contain 10 to 20 Golgi complexes, and plant cells may have several hundred.

One of the most critical products processed, packaged, and distributed by the Golgi complexes is new material for the membranes of the cell and its organelles. Membrane lipids and proteins, synthesized in the endoplasmic reticulum, are delivered to the Golgi complex in vesicles that fuse with it. Within the cisternae of the Golgi complex, the final assembly of carbohydrates with proteins (forming glycoproteins) and with lipids (forming glycolipids) occurs; as we noted previously, these carbohydrate combinations found on the surface of cell membranes are thought to play key roles in membrane function. Current evidence indicates that different stages in this chemical processing occur in different cisternae of the Golgi complex, and that materials are transported from one cisterna to the next via vesicles. After the chemical processing is completed, the new membrane material is packaged in vesicles that are targeted to the correct location, whether it

5-20 *Graphic interpretation and electron micrograph of a Golgi complex. A Golgi complex consists of four or more membrane-bound cisternae, arranged in a loose stack. Materials are packaged in membrane-enclosed vesicles at the Golgi complexes and are distributed within the cell or shipped to the cell surface. Note the vesicles pinching off from the edges of the flattened cisternae.*

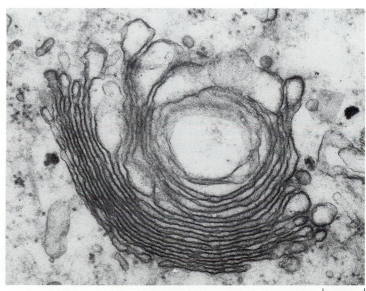

0.25 μm

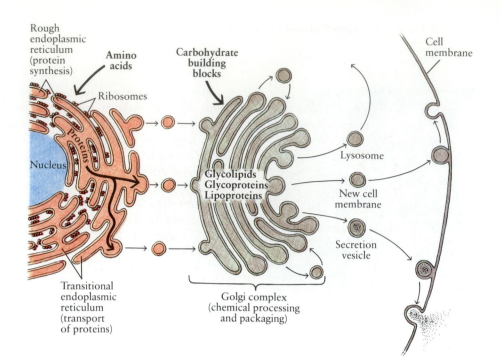

5–21 *Diagram illustrating the interaction of ribosomes, the endoplasmic reticulum, and the Golgi complex and its vesicles. These organelles work together in the synthesis, chemical processing, packaging, and distribution of macromolecules and new membrane material. As proteins are synthesized on the ribosomes, they are fed into the rough endoplasmic reticulum. They then move through a specialized transitional region of endoplasmic reticulum and are released in vesicles that fuse with the sacs of the Golgi complex. These vesicles incorporate in their membranes lipids newly synthesized in the endoplasmic reticulum. In the Golgi complex, carbohydrates are added to some of the proteins and lipids, producing glycoproteins and glycolipids; these macromolecules are common components of membranes. In some types of cells, lipids are added to other proteins in the Golgi complex, producing lipoproteins. Molecules destined for export from the cell also undergo chemical processing in the Golgi complex. Vesicles containing the finished molecules and macromolecules are released from the Golgi complex and move to other locations within the cell or to its exterior surface.*

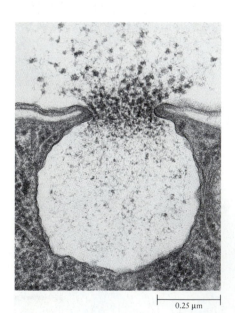

0.25 μm

5–22 *A secretion vesicle, formed by a Golgi complex of the protist* Tetrahymena furgasoni, *discharges mucus at the cell surface. Notice how the membrane enclosing the vesicle has fused with the cell membrane.*

be the cell membrane or the membrane of a particular organelle. In plant cells, Golgi complexes also bring together some of the components of the cell walls and export them to the cell surface where they are assembled.

In addition to their function in the assembly of cellular membranes, Golgi complexes have a similar function in the processing and packaging of materials that are released outside the cell. Figure 5–21 summarizes the way in which the ribosomes, the endoplasmic reticulum, and the Golgi complex and its vesicles interact to produce new material for the cell membrane and macromolecules for export.

Lysosomes

One type of relatively large vesicle commonly formed from the Golgi complex is the **lysosome**. Lysosomes are essentially membranous bags that enclose hydrolytic enzymes, thereby separating them from the rest of the cell; these enzymes are involved in breaking down proteins, polysaccharides, and lipids. If the lysosomes break open, the cell itself will be destroyed, since the enzymes they carry are capable of hydrolyzing all the major types of macromolecules found in a living cell. The tenderness and inflammation associated with rheumatoid arthritis and gout appear to be related to the escape of hydrolytic enzymes from lysosomes.

An example of the function of lysosomes is seen among white blood cells, which engulf bacteria in the human body. As the bacteria are taken up by the cell, they are wrapped in a membrane-enclosed sac, a vacuole. (This process is known as phagocytosis; we shall discuss it further in the next chapter.) When this occurs, the lysosomes within the cell fuse with the vacuoles containing the bacteria and release their hydrolytic enzymes. The bacteria are quickly digested. In similar fashion, the lysosomes of heterotrophic protists, such as *Paramecium*, *Didinium*, and amoebas, fuse with the phagocytic vacuoles containing food organisms. Hydrolytic enzymes released by the lysosomes into the vacuoles digest the contents. Why the enzymes do not destroy the membranes of the lysosomes that carry them is a pertinent question yet to be answered.

Peroxisomes

Another type of relatively large vesicle containing lytic enzymes is the **peroxisome**. Peroxisomes are vesicles in which purines (one of the two major categories of nitrogenous bases) and several other types of compounds are broken down by the cell. In plants, peroxisomes are also the site of a series of reactions that occur in

5–23 *Lysosomes and peroxisomes are vesicles within which different types of molecules are broken down. (**a**) In this portion of a cell from the adrenal gland, the dark oval bodies are lysosomes. They may contain 40 or more different hydrolytic enzymes. (**b**) A peroxisome. The crystalline material in the center is a peroxide-producing enzyme involved in the breakdown of purines. Another enzyme destroys the peroxide, preventing its escape into the cytoplasm.*

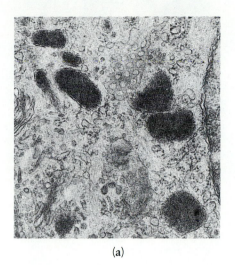

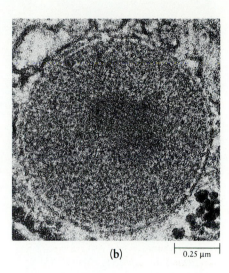

(a)   (b)   0.25 µm

sunlight when the cell contains relatively high concentrations of oxygen. These reactions and the breakdown of purines both produce hydrogen peroxide (H_2O_2), a compound that is extremely toxic to living cells. The peroxisomes, however, contain another enzyme that immediately breaks hydrogen peroxide into water and oxygen, preventing any damage to the cell.

Mitochondria

Mitochondria (singular, mitochondrion) are among the largest organelles in a cell. In the mitochondria, energy-yielding organic molecules are broken down and their energy is repackaged into smaller units, convenient for most cellular processes. The higher the energy requirements of a particular eukaryotic cell, the more mitochondria it is likely to have. A liver cell, for example, which has modest energy requirements, has about 2,500, making up about 25 percent of the volume of the cell, whereas a heart muscle cell has several times as many very large mitochondria. Mitochondria are often found clustered in areas in the cell where energy requirements are high.

Mitochondria vary in shape from almost spherical, to potato-shaped, to greatly elongated cylinders. As shown in Figure 5–24, they are always surrounded by two membranes, the inner one of which folds inward; these folds, known as **cristae,** are working surfaces for mitochondrial reactions. The more active a mitochondrion, the more cristae it is likely to have. In Chapter 9, we shall examine in more detail the structure of mitochondria and the processes that occur within these important organelles.

5–24 *A mitochondrion is surrounded by two membranes. The inner membrane folds inward to make a series of shelves, or cristae. The membrane forming these shelves plays a crucial role in the energy-releasing chemical reactions that occur in the mitochondria.*

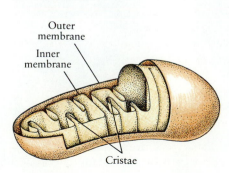

Outer membrane
Inner membrane
Cristae

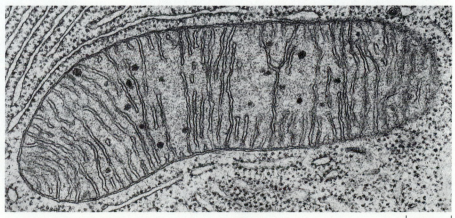

0.25 µm

5–25 (a) *A leucoplast from the embryo sac of a soybean. The embryo sac is the structure in the flower in which the egg is fertilized and the embryonic plant begins its development. The large round, clear bodies are starch granules, and the smaller, dark bodies are lipid droplets.*

(b) Chromoplast from a forsythia petal. The large, dark granules contain the orange and yellow pigments characteristic of certain flowers and fall leaves. To the left of the chromoplast are the faintly visible cell wall and two mitochondria. To the left and right of the micrograph are portions of two vacuoles.

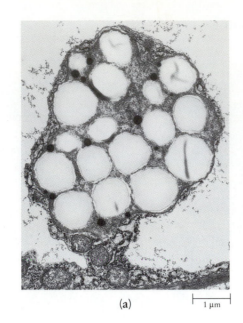
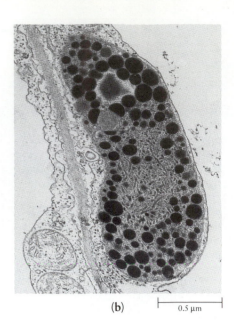

(a)   1 µm   (b)   0.5 µm

5–26 *Chloroplast development. (a) The immature plastid contains small crystalline structures (top). (b) In the presence of light, these structures begin to break up into elongated vesicles. (c) The vesicles flatten into membranes that are grouped together in stacks. These internal membranes are known as thylakoids, and the stacks they form as grana. (d) A mature chloroplast. Like the mitochondrion, it is surrounded by two membranes; in addition, it contains an elaborate internal membrane system in which the light-capturing reactions of photosynthesis take place.*

Chromoplasts and leucoplasts develop from plastids similar to (a).

Plastids

Plastids are membrane-bound organelles found only in the cells of plants and algae. They are surrounded by two membranes, like mitochondria, and have an internal membrane system that may be folded intricately. Mature plastids are of three types: leucoplasts, chromoplasts, and chloroplasts.

Leucoplasts (*leuco* means "white") store starch or, sometimes, proteins or oils. Leucoplasts are likely to be numerous in storage organs such as roots, as in a turnip, or tubers, as in a potato.

Chromoplasts (*chromo* means "color") contain pigments and are associated with the bright orange and yellow colors of fruits, flowers, fall leaves, and carrots.

Chloroplasts (*chloro* means "green") are the chlorophyll-containing plastids in which photosynthesis takes place. Like other plastids, they are surrounded by two membranes; the third, internal membrane of chloroplasts forms an elaborate series of internal compartments and work surfaces. The sequence of events that lead to the formation of a mature chloroplast are shown in Figure 5–26. In Chapter 10, we shall consider the structure of the chloroplast and its functional significance in more detail.

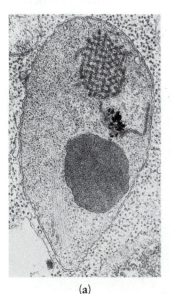
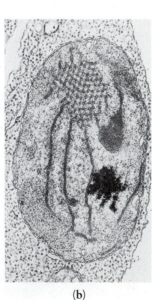
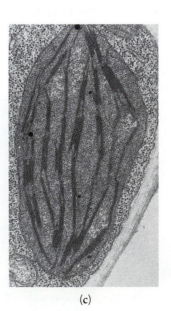
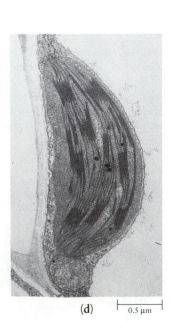

(a)   (b)   (c)   (d)   0.5 µm

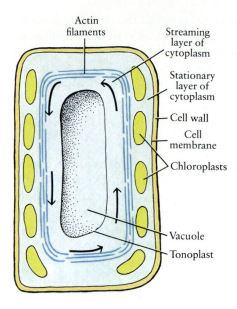

Actin
filaments

Streaming
layer of
cytoplasm

Stationary
layer of
cytoplasm

Cell wall

Cell
membrane

Chloroplasts

Vacuole

Tonoplast

5-27 *The role of actin in cytoplasmic streaming. In the cell of a green alga, the portion of the cytoplasm containing the chloroplasts is located against the cell membrane. Laid down across the chloroplasts are bundles of actin filaments that form distinct "tracks"; these are thought to direct the regular streaming movement that occurs in the portion of the cytoplasm located between the actin filaments and the vacuole. The outer portion of the cytoplasm, by contrast, is stationary.*

HOW CELLS MOVE

All cells exhibit some form of movement. Even plant cells, encased in a rigid cell wall, exhibit active cytoplasmic streaming (movement of cytoplasm within the cell) as well as chromosomal movements and changes in shape during cell division. As we saw in Figure 4–13 (page 94), cilia beat along the surface of the tracheal cells of animals. Embryonic cells migrate in the course of animal development. Differentiating and regenerating nerve cells send out axons, which are long, slender extensions that may be a meter or more in length. Amoebas pursue and engulf their prey, and little *Chlamydomonas* cells dart toward a light source. Two different molecular mechanisms of cellular movement have been identified: (1) assemblies of protein filaments, in which actin filaments (page 111) are present in large quantity, and (2) assemblies of microtubules in cilia and flagella.

Actin and Associated Proteins

As you will recall, the actin filaments of the cytoskeleton consist of helical chains composed of subunits of the globular protein actin. Actin filaments are present in a great variety of cells, including plant cells. They participate not only in the maintenance of cytoplasmic organization but also in cell motility and the internal movement of cellular contents. Interacting with the actin filaments in producing cellular motion are bundles of another protein, known as myosin. Additional proteins, thought to play regulatory roles, are associated with the actin and myosin molecules.

Actin, myosin, and their associated proteins are involved in a variety of different cellular processes. For example, they are found in the large, multinucleate organisms known as slime molds, which move like giant amoebas and exhibit vigorous cytoplasmic streaming. Actin has also been found in a true amoeba, *Amoeba proteus.* In such cells, which move by gradual changes of shape, the actin filaments are found concentrated in bundles or a meshwork near the moving edge. These filaments have also been shown to act as a sort of "purse string" in animal cells during cell division, pinching off the cytoplasm to separate the two daughter cells. In algal cells, actin filaments occur in bundles wherever cytoplasmic streaming is taking place (Figure 5–27). The way in which actin and its associated proteins bring about amoeboid movement and cytoplasmic streaming is currently under active investigation.

5-28 *A macrophage, a type of phagocytic white blood cell, on the move across the surface of a culture dish. Its motility is made possible by the interaction of actin filaments with another protein known as myosin. As we shall see in Chapter 39, macrophages perform a number of vital functions in the body's response to invading microorganisms. They are also one of the principal cell types attacked by the AIDS virus.*

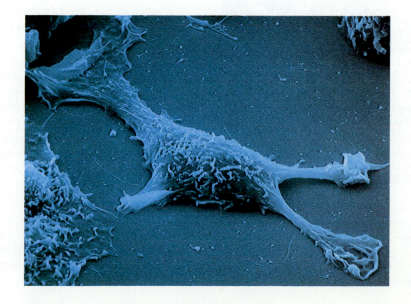

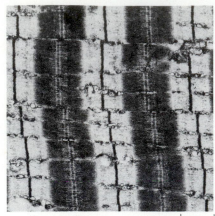

├─ 1 µm

5–29 *Contractile assemblies of protein filaments, principally actin and myosin, in a vertebrate skeletal muscle. Each unit is known as a sarcomere, approximately 18 of which are visible in this electron micrograph. When the muscle contracts, the distance from left to right in this micrograph shortens.*

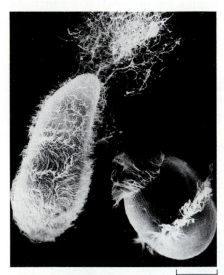

├─ 50 µm

5–30 *Two ciliates, one-celled protists distinguished by their many cilia. On the left,* Paramecium; *on the right,* Didinium. Didinium *is stalking* Paramecium. *Paramecium, in defense, has discharged a barrage of barbs (visible as a cloud at the top of the micrograph).* Didinium *is about to eject a bundle of slender, poisonous strands (not visible), which will paralyze* Paramecium *in a matter of seconds. In* Paramecium, *the cilia are distributed fairly evenly over the cell surface. In* Didinium, *they form two wreaths that circle the organism's barrel-shaped body.*

Actin and myosin are also the principal components of the elaborate contractile assemblies (Figure 5–29) found in the muscle cells of vertebrates and many other animals. This specialized organization of actin and myosin (to be described in detail in Chapter 42) makes possible the rapid, coordinated movements that give animals, ranging from insects to fishes to birds to race horses to ourselves, their great mobility.

Cilia and Flagella

Cilia and **flagella** are long, thin (0.2 micrometer) structures extending from the surface of many types of eukaryotic cells. They are essentially the same except for length (the names were given before their basic similarity was realized). When they are shorter and occur in larger numbers, they are more likely to be called cilia; when they are longer and fewer, they are usually called flagella. (Prokaryotic cells also have flagella, but they are so different in construction from those of eukaryotes that it would be useful if they had a different name. We shall examine their structure and mechanism of movement in Chapter 21.)

In one-celled protists and some small animals (such as a few types of flatworms), cilia and flagella are associated with movement of the organism. For example, one species of *Paramecium* has approximately 17,000 cilia, each about 10 micrometers long, which propel it through the water by beating in a coordinated fashion. Other protists, such as the members of the genus *Chlamydomonas*, have only two whiplike flagella, which protrude from the anterior end of the organism and move it through the water (see Figure 5–3, page 104). The motile power of the human sperm cell comes from its single powerful flagellum, or "tail."

Many of the cells that line the surfaces within our bodies are also ciliated. These cilia do not move the cells but, rather, serve to sweep substances across the cell surface. For example, cilia on the surface of cells of the respiratory tract beat upward, propelling a current of mucus that sweeps bits of soot, dust, pollen, tobacco tar—whatever foreign substances we have inhaled either accidentally or on purpose—to our throats, where they can be removed by swallowing. Human egg cells are propelled down the oviducts by the beating of cilia that line the inner surfaces of these tubes. Cilia and flagella are found extensively throughout the living world, on the cells of invertebrates, vertebrates, the sex cells of ferns and other plants, as well as on protists. Only a few large groups of eukaryotic organisms, such as red algae, fungi, flowering plants, and roundworms (nematodes), have no cilia or flagella on any cells.

Almost all eukaryotic cilia and flagella, whether on a *Paramecium* or a sperm cell, have the same internal structure. Nine pairs of fused microtubules form a ring that surrounds two additional, solitary microtubules in the center (Figure 5–31). Microtubules, you will recall, are composed of identical globular protein units assembled in a hollow helix. The movement of cilia and flagella comes from within the structures themselves; if cilia are removed from cells and placed in a medium containing energy-yielding chemicals, they beat or swim through the medium. The movement, according to the generally accepted hypothesis, is caused by each outer pair of microtubules moving tractor-fashion with respect to its nearest neighbor. The two "arms" that you can see on one of each pair of outer tubules (Figure 5–31a) have been shown to be enzymes involved in energy-releasing chemical reactions. Other proteins are involved in the formation of spokes connecting the nine pairs of outer microtubules to the central pair, and still other proteins form more widely spaced links, rather like the hoops of a barrel, connecting the nine outer pairs to each other. The spokes are thought to play a role in coordinating the tractorlike movements of the microtubules, whereas the links limit the amount of sliding possible and thus convert it into a bending motion.

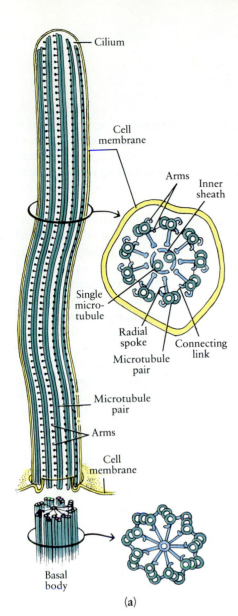

Cilium

Cell
membrane

Arms

Inner
sheath

Single
micro-
tubule

Radial
spoke

Connecting
link

Microtubule
pair

Microtubule
pair

Arms

Cell
membrane

Basal
body

(a)

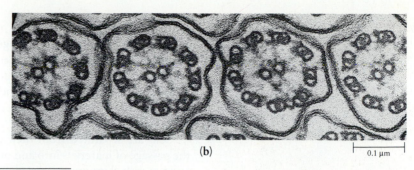

(b)

0.1 µm

5–31 **(a)** *Diagram of a cilium with its underlying basal body. Virtually all eukaryotic cilia and flagella, whether they are found on protists or on the surfaces of cells within our own bodies, have this same internal structure, which consists of an outer ring of nine pairs of microtubules surrounding two additional microtubules in the center. The "arms," the radial spokes, and the connecting links are formed from different types of protein. The basal bodies from which cilia and flagella arise have nine outer triplets, with* no microtubules in the center. The "hub" of the wheel in the basal body is not a microtubule, although it has about the same diameter.

(b) *Cross section of cilia from a gill cell of a mussel. The beating of cilia on the gills of mussels, clams, oysters, and other two-shelled mollusks sweeps water through the sievelike gills. Small organisms and particles of food are trapped in mucus on the gill surface and are then swept toward the mouth by the cilia.*

Basal Bodies and Centrioles

Underlying each cilium is a structure known as a **basal body,** which has the same diameter as a cilium, about 0.2 micrometer. It consists of microtubules arranged in nine triplets (rather than pairs) around the periphery. Unlike the cilium, it has no microtubules in the center, and none of the microtubules in the basal body have arms. Cilia and flagella arise from basal bodies. For instance, as a sperm cell takes form, a basal body moves near the cell membrane, and the sperm's flagellum arises from it through the assembly of microtubules.

Many types of eukaryotic cells contain **centrioles.** Centrioles, which typically occur in pairs, are small cylinders, about 0.2 micrometer in diameter, containing nine microtubule triplets (Figure 5–32). Their structure is identical to that of basal bodies; however, their distribution in the cell is different. Until recently, it appeared that their function was also different, which is why they are called by different names even though electron microscopy has revealed their identical structures. Centrioles usually lie in pairs with their long axes at right angles to one another in the region of the cytoplasm, near the nuclear envelope, from which the microtubules of the cytoskeleton radiate. They are found only in those groups of organisms that also have cilia or flagella (and, therefore, basal bodies). There is evidence that centrioles play a role in organizing a structure known as the spindle, which appears at the time of cell division and is involved in chromosome movements. The spindle, it has been shown, also contains numerous microtubules; thus centrioles and basal bodies both appear to be organizers of microtubules. However, as we shall see in Chapter 7, cells that have no centrioles—such as the cells of the flowering plants—are also able to organize microtubules into a spindle.

The discovery of the complex internal structure of cilia and flagella, basal bodies and centrioles, repeated over and over again throughout the living world, was one of the spectacular revelations of electron microscopy. For biologists, it is another glimpse down the long corridor of evolution, providing overwhelming evidence, once again, of the basic unity of earth's living things.

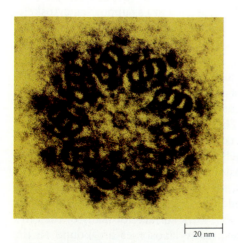

20 nm

5–32 *Cross section of a centriole from a cell of the fruit fly* Drosophila. *Centrioles are structurally identical to basal bodies.*

SUMMARY

Cells are the basic units of biological structure and function. The size of cells is limited by proportions of surface to volume; the greater a cell's surface area in proportion to its volume, the greater the quantity of materials that can move into and out of the cell in a given period of time. Cell size is also limited by the capacity of the nucleus to regulate cellular activities. Cells that are active metabolically are likely to be small.

Cells are separated from their environment by a cell membrane that restricts the passage of materials into and out of the cell and so protects the cell's structural and functional integrity. According to the fluid-mosaic model, cell membranes are formed from phospholipid bilayers in which cholesterol and protein molecules are embedded. The embedded protein molecules, which typically span the membrane, are known as integral membrane proteins. Different integral proteins perform different functions; some are enzymes, others are receptors, and still others are transport proteins. The two faces of the membrane differ in chemical composition. The cytoplasmic face is characterized by peripheral membrane proteins, attached to the integral proteins embedded in the bilayer. The exterior face of the membrane is characterized by short carbohydrate chains. Some of these chains are the hydrophilic heads of glycolipid molecules that are interspersed among the phospholipid molecules of the outer layer of the bilayer; other carbohydrate chains are covalently linked to the protruding portions of integral membrane proteins.

The cells of plants, most algae, fungi, and prokaryotes are further separated from the environment by a cell wall constructed by the cell itself. Cellulose is an important constituent of the cell walls of plants and many algae.

The nucleus of eukaryotic cells is separated from the cytoplasm by the nuclear envelope, which consists of two concentric lipid bilayers. Pores in the nuclear envelope provide channels through which molecules pass to and from the cytoplasm. The nucleus contains the hereditary material, the chromosomes, which, when the cell is not dividing, exist in an extended form called chromatin. The nucleolus, visible within the nucleus, is involved in the formation of ribosomes. Interacting with the cytoplasm, the nucleus helps to regulate the cell's ongoing activities.

The cytoplasm of the cell is a concentrated aqueous solution containing enzymes, many other dissolved molecules and ions, and also, in the case of eukaryotic cells, a variety of membrane-bound organelles with specialized functions in the life of the cell. The eukaryotic cytoplasm has a supporting cytoskeleton that includes three principal types of structures: microtubules, actin filaments, and intermediate filaments. The cytoskeleton maintains the shape of the cell, enables it to move, anchors its organelles, and directs its traffic. Vacuoles, which are bounded by a single membrane, are also present in the cytoplasm of many cells, particularly plant cells. The vacuoles of plant cells are storage reservoirs and play a role in cell support. Vesicles, present in all metabolically active eukaryotic cells, are usually smaller than vacuoles. They perform a variety of functions, of which transport is one of the most important.

Eukaryotic cells contain many organelles, most of which are not found in prokaryotic cells (see Table 5–1). The most numerous organelles (in both prokaryotes and eukaryotes) are the ribosomes, the sites of assembly of proteins. The cytoplasm of eukaryotic cells is subdivided by a network of membranes known as the endoplasmic reticulum, which serves as a work surface for many of the cell's biochemical activities. In eukaryotic cells, many ribosomes are bound to the surface of the endoplasmic reticulum, producing the rough endoplasmic reticulum. Rough endoplasmic reticulum is especially abundant in cells producing proteins for export. Smooth endoplasmic reticulum, which lacks ribosomes, is

TABLE 5-1 A Comparison of Cell Characteristics

KINGDOM	MONERA	PROTISTA	FUNGI	PLANTAE	ANIMALIA
Cell type	Prokaryotic	Eukaryotic	Eukaryotic	Eukaryotic	Eukaryotic
Cell membrane	Present	Present	Present	Present	Present
Cell wall	Noncellulose (polysaccharide and peptidoglycan)	Present in some forms, various types	Chitin and other noncellulose polysaccharides	Cellulose and other polysaccharides	Absent
Nuclear envelope	Absent	Present	Present	Present	Present
Chromosomes	Single, continuous DNA molecule	Multiple, consisting of DNA and histone proteins	Multiple, consisting of DNA and histone proteins	Multiple, consisting of DNA and histone proteins	Multiple, consisting of DNA and histone proteins
Ribosomes	Present (smaller)	Present	Present	Present	Present
Endoplasmic reticulum	Absent	Present	Present	Present	Present
Mitochondria	Absent	Usually present	Present	Present	Present
Plastids	Absent	Present in some forms	Absent	Present	Absent
Golgi complexes	Absent	Present	Present	Present	Present
Lysosomes	Absent	Often present	Often present	Similar structures (lysosomal compartments present)	Often present
Peroxisomes	Absent	Often present	Present in some forms	Often present	Often present
Vacuoles	Absent	Present	Present	Usually large single vacuole in mature cell	Small or absent
9 + 2 cilia or flagella	Absent	Often present	Absent	Absent (in flowering plants)	Often present
Centrioles	Absent	Often present	Absent	Absent (in flowering plants)	Present

abundant in cells specialized for lipid synthesis or metabolism. Golgi complexes, also composed of membranes, are processing and packaging centers for materials being moved through and out of the cell. Lysosomes, which contain hydrolytic enzymes, are involved in intracellular digestive activities in some cells. The enzymes for cellular reactions that produce hydrogen peroxide as a by-product are sequestered in the peroxisomes, along with an enzyme that breaks down the toxic peroxide into water and oxygen.

Mitochondria are membrane-bound organelles in which energy-yielding organic molecules are broken down and the released energy repackaged into smaller units. Plastids are membrane-bound organelles found only in photosynthetic organisms. Leucoplasts are storage compartments, chromoplasts contain pigments, and chloroplasts are the sites of photosynthesis in plants and algae.

Assemblies of protein filaments, principally actin and myosin, are associated with internal cellular movement, whereas cilia and flagella are associated with the external movement of cells or the movement of materials along cell surfaces. These whiplike appendages are found on the surface (yet within the cell membrane) of many types of eukaryotic cells. They have a highly characteristic 9 + 2 structure, with nine pairs of microtubules forming a ring surrounding two central microtubules. One of each pair of outer microtubules contains enzymes involved in the chemical reactions that release energy for ciliary motion.

Cilia and flagella arise from basal bodies, which are cylindrical structures containing nine microtubule triplets with no inner pair. Centrioles have the same internal structure as basal bodies and are found in those groups of organisms that also have cilia or flagella. They typically occur in pairs, lying near the nuclear envelope, and may play a role in the formation of the spindle during cell division.

QUESTIONS

1. Distinguish between the following: cell membrane/cell wall; nucleus/nucleolus; rough endoplasmic reticulum/smooth endoplasmic reticulum; lysosomes/peroxisomes; chloroplasts/mitochondria; cilia/flagella; basal body/centriole.

2. Describe the structure of the cell membrane. How do the two faces of the membrane differ? What is the functional significance of these differences?

3. (a) Sketch an animal cell. Include the principal organelles and label them. (b) Prepare a similar, labeled sketch of a plant cell. (c) What are the major differences between the animal cell and the plant cell?

4. Why is the secondary wall of a plant cell *inside* the primary cell wall? Where is the cell membrane in relation to the two cell walls?

5. What are the functions of the cytoskeleton? Describe the similarities and differences between microtubules, actin filaments, and intermediate filaments.

6. Explain the functions of each of the following structures: ribosomes, endoplasmic reticulum, vesicles, and Golgi complexes. How do they interact in the synthesis and delivery of new membrane material and in the export of proteins from the cell?

7. Use a ruler and the scale marker at the bottom of each micrograph on pages 104 and 123 to determine: (a) the thickness (roughly) of a cell membrane, (b) the diameter of a cilium, and (c) the diameter of a microtubule within a cilium. (This is how the sizes of cellular components are determined by microscopists.) Would a cilium be resolvable in a light microscope (that is, is its diameter more than 0.2 µm)?

8. (a) Sketch a cross section of a cilium. (b) Sketch a cross section of the basal body of a cilium. (c) What are the differences between the two structures?

9. On the basis of what you know of the functions of each of the structures in Table 5–1, what components would you expect to find most prominently in each of the following cell types: muscle cells, sperm cells, green leaf cells, red blood cells, white blood cells?

10. Two brothers were under medical treatment for infertility. Microscopic examination of their semen showed that the sperm were immotile and that the little "arms" were missing from the microtubular arrays. The brothers also had chronic bronchitis and other respiratory difficulties. Can you explain why?

How Things Get into and out of Cells

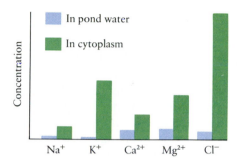

6–1 *The relative concentrations of different ions in pond water and in the cytoplasm of the green alga* Nitella. *Differences such as these indicate that cells regulate their exchanges of materials with the surrounding environment.*

One of the criteria by which we identify living systems is that living matter, although surrounded on all sides by nonliving matter with which it constantly exchanges materials, is different from that nonliving matter in the kinds and amounts of chemical substances it contains (Figure 6–1). Without this difference, of course, living systems would be unable to maintain the organization and structure on which their existence depends.

In all living systems, ranging from prokaryotes to the most complex multicellular eukaryotes, regulation of exchanges of substances between the living system and the nonliving world occurs at the level of the individual cell and is accomplished by the cell membrane. In multicellular organisms, the cell membrane has the additional task of regulating exchanges of substances among the various specialized cells that constitute the organism. Control of these exchanges is essential to protect each cell's integrity, to maintain those very narrow conditions of pH and ionic concentrations at which its metabolic activities can take place, and to coordinate the activities of the different cells. In addition to the cell membrane, which controls the passage of materials between the cell and its environment, internal membranes, such as those surrounding mitochondria, chloroplasts, and the nucleus, control the passage of materials among intracellular compartments (Figure 6–2). This makes it possible for the cell to maintain the specialized chemical environments necessary for the processes occurring in the different organelles.

Maintenance of the internal environment of the cell and its constituent parts requires that cell membranes perform a complex double function: they must keep

6–2 *Electron micrograph of a portion of a cell from the pancreas. The cell has a large, central nucleus with scattered chromatin, many mitochondria, large quantities of rough endoplasmic reticulum, and many small vesicles. Not only is the cell itself surrounded by a membrane and the nucleus by a double membrane system (the nuclear envelope), but its organelles are also surrounded by membranes. The membranes of the endoplasmic reticulum further divide the cell into membrane-bound compartments. Collectively, all of these different membranes regulate the movements of substances into and out of cells and restrict their passage from one part of the cell to another.*

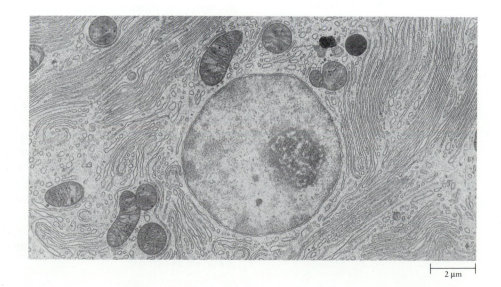

2 μm

certain substances out while letting others in, and, conversely, they must keep certain substances in while letting others out. The capacity of a membrane to accomplish this function depends not only on the physical and chemical properties that result from its lipid and protein structure, but also on the physical and chemical properties of the substances—ions, molecules, and aggregations of molecules—that interact with the membrane. Of the many kinds of molecules surrounding and contained within the cell, by far the most common is water. Further, the many other molecules and ions important in the life of the cell are carried in an aqueous solution. Therefore, let us begin our consideration of transport across cell membranes by looking again at water, focusing our attention this time on how it moves.

THE MOVEMENT OF WATER AND SOLUTES

In both the animate and inanimate worlds, water molecules move from one place to another because of differences in potential energy, usually referred to as the **water potential.** Water moves from a region where water potential is greater to a region where water potential is lower, regardless of the reason for the water potential. A simple example is water running downhill in response to gravity. Water at the top of a hill has more potential energy (that is, a greater water potential) than water at the bottom of a hill. As the water runs downhill, its potential energy is converted to kinetic energy; this, in turn, can be converted to mechanical energy doing useful work if, for example, a water mill is placed in the path of the moving water.

Pressure is another source of water potential. If we fill a rubber bulb with water and then squeeze, water will squirt out of the nozzle. Like water at the top of a hill, this water has been given a high water potential and will move to a lower one. Can we make the water that is running downhill run uphill by means of pressure? Obviously we can. But only so long as the water potential produced by the pressure exceeds the water potential produced by gravity.

In solutions, water potential is affected by the concentration of dissolved particles (solutes). As the concentration of solute particles (that is, the number of solute particles per unit volume of solution) increases, the concentration of water molecules (that is, the number of water molecules per unit volume of solution) must necessarily decrease, and vice versa. In the absence of other factors (such as pressure), the water potential of a solution is directly related to the concentration of water molecules—the higher the concentration of water molecules, the greater the water potential. Conversely, the higher the concentration of solute particles, the lower the water potential. Water molecules move from regions of higher water potential to regions of lower water potential, a fact of great importance for living systems.

The concept of water potential is a useful one because it enables us to predict the way that water will move under various combinations of circumstances. Measurements of water potential are usually made in terms of the pressure required to stop the movement of water—that is, the hydrostatic (water-stopping) pressure—under the particular circumstances. The unit usually used to measure this pressure is the atmosphere. One atmosphere is the average pressure of the air at sea level, about 1 kg/cm² (or 15 lb/in²).

Two mechanisms are involved in the movement of water and solutes: bulk flow and diffusion. In living systems, bulk flow moves water and solutes from one part of a multicellular organism to another part, whereas diffusion moves molecules and ions into, out of, and through cells. A particular instance of diffusion—that of water across a membrane that separates solutions of different concentration—is known as osmosis.

6–3 *Water at the top of a falls, like a boulder on a hilltop, has potential energy. The movement of water molecules as a group, as from the top of the falls to the bottom, is referred to as bulk flow.*

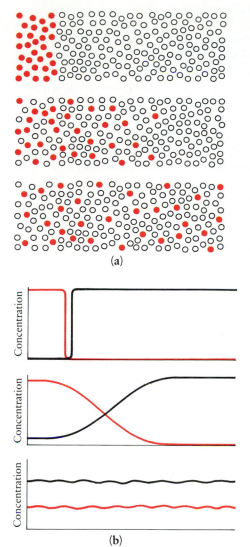

(a)

(b)

6–4 (a) *Diagram of the diffusion process.*
Diffusion is the result of the random
movement of individual molecules (or
ions), which produces a net movement
from a more concentrated to a less con-
centrated region. Notice that as one type
of molecule (indicated by color) diffuses to
the right, the other diffuses in the opposite
direction. The result will be an even distri-
bution of both types of molecules. Can
you see why the net movement of mole-
cules will slow down as equilibrium is
reached? (b) Graphs showing the concen-
tration gradients of dye and water.

Bulk Flow

Bulk flow is the overall movement of a fluid. The molecules move all together and in the same direction. For example, water runs downhill by bulk flow in response to the differences in water potential at the top and the bottom of a hill. Blood moves through your body by bulk flow as a result of the water potential (blood pressure) created by the pumping of your heart. Sap—an aqueous solution of sucrose and other solutes—moves by bulk flow from the leaves of a plant to other parts of the plant body.

Diffusion

Diffusion is a familiar phenomenon. If you sprinkle a few drops of perfume in one corner of a room, the scent will eventually permeate the entire room even if the air is still. If you put a few drops of dye in one end of a glass tank full of water, the dye molecules will slowly become evenly distributed throughout the tank. The process may take a day or more, depending on the size of the tank, the temperature, and the relative size of the dye molecules.

Why do the dye molecules move apart? If you could observe the individual dye molecules in the tank (Figure 6–4), you would see that each one of them moves individually and at random. Looking at any single molecule—at either its rate of motion or its direction of motion—gives you no clue at all about where the molecule is located with respect to the others. So how do the molecules get from one end of the tank to the other? Imagine a thin section through the tank, running from top to bottom. Dye molecules will move in and out of this section, some moving in one direction, some moving in the other. But you will see more dye molecules moving from the side of greater dye concentration. Why? Simply because there are more dye molecules at that end of the tank. If there are more dye molecules on the left, more dye molecules, moving at random, will move to the right, even though there is an equal probability that any one molecule of dye will move from right to left. Consequently, the *net* movement of dye molecules will be from left to right. Similarly, if you could see the movement of the individual water molecules in the tank, you would see that their *net* movement is from right to left.

Substances that are moving from a region of higher concentration of their own molecules to a region of lower concentration are said to be moving *down a gradient*. (A substance moving in the opposite direction, toward a higher concentration of its own molecules, moves *against a gradient*, which is analogous to being pushed uphill.) Diffusion occurs only down a gradient. The steeper the downhill gradient—that is, the larger the difference in concentration—the more rapid the diffusion. In our imaginary tank, there are two gradients; the dye molecules are moving down one of them, and the water molecules are moving down the other in the opposite direction. In each case, the molecules are moving from a region of higher potential energy to a region of lower potential energy.

What happens when all the molecules are distributed evenly throughout the tank? The even distribution does not affect the behavior of the molecules as individuals; they still move at random. And, since the movements are random, just as many molecules go to the left as to the right. But because there are now as many molecules of dye and as many molecules of water on one side of the tank as on the other, there is no *net* movement of either. There is, however, just as much random motion as before, provided the temperature has not changed. When the molecules have reached a state of equal distribution, that is, when there are no more gradients, they are said to be in dynamic equilibrium.

The essential characteristics of diffusion are (1) that each molecule or ion moves independently of the others and (2) that these movements are random. The net result of diffusion is that the diffusing substances become evenly distributed.

Cells and Diffusion

Water, oxygen, carbon dioxide, and a few other simple molecules diffuse freely across cell membranes. Carbon dioxide and oxygen, which are both nonpolar, are soluble in lipids and move easily through the lipid bilayer of the membrane. Despite their polarity, water molecules also move through the membrane without hindrance, apparently through hydrophilic apertures. These may be either permanent pores created by the tertiary structure of some of the integral membrane proteins or momentary openings resulting from movements of the lipid molecules. Other polar molecules, provided they are small enough, also diffuse through these openings. The permeability of the membrane to these solutes varies inversely with the size of the molecules, indicating that the apertures are small and that the membrane acts like a sieve in this respect.

Diffusion is also a principal way in which substances move within cells. One of the major factors limiting cell size is this dependence upon diffusion, which is essentially a slow process, except over very short distances. As you can see by studying Figure 6–4, the process becomes increasingly slower and less efficient as the distance "covered" by the diffusing molecules increases. The rapid spread of a substance through a large volume, such as perfume through the air of a room, is due not primarily to diffusion but rather to the circulation of air currents. Similarly, in many cells, the transport of materials is speeded by active streaming of the cytoplasm, a process in which actin filaments of the cytoskeleton play a key role (see Figure 5–27, page 121).

Efficient diffusion requires not only a relatively short distance but also a steep concentration gradient. Cells maintain such gradients by their metabolic activities, thereby hastening diffusion. For example, carbon dioxide is constantly produced as the cell oxidizes fuel molecules for energy. As a result, there is a higher concentration of carbon dioxide inside the cell than out. Thus a gradient is maintained between the inside of the cell and the outside, and carbon dioxide diffuses out of the cell down this gradient. Conversely, oxygen is used up by the cell in the course of its activities, so oxygen present in air or water or blood tends to move into the cells by diffusion, again down a gradient. Similarly, within a cell, molecules or ions are often produced at one place and used at another. Thus a concentration gradient is established between the two regions, and the substance diffuses down the gradient from the site of production to the site of use.

Countercurrent Exchange

Although diffusion is efficient only over short distances, it nevertheless plays a key role in the transport of substances into and out of multicellular organisms, as well as between different compartments within the organism. For example, oxygen enters the bloodstream of an animal by diffusing through cells that are in contact with the environment. Unless the animal is quite small or leads a sedentary existence, a high rate of diffusion is required to supply it with adequate oxygen. Organisms have developed a number of anatomical arrangements that maintain steep concentration gradients and thus maximize diffusion rates. One of the most common types of arrangement is found in fish gills. Gills are divided into filaments, which are further divided into a number of flat, densely packed lamellae—platelike structures containing many blood vessels (Figure 6–5). Water, containing dissolved oxygen, flows over the lamellae, and oxygen moves by diffusion into the blood vessels just below the thin surfaces. In the lamellae, the direction of blood flow is opposite to the direction of water flow. This countercurrent arrangement maintains a constant concentration gradient between the bloodstream and the water, and oxygen can diffuse in through the entire surface of the lamellae. As we shall see in Section 6, this principle of countercurrent exchange is used in several different anatomical systems.

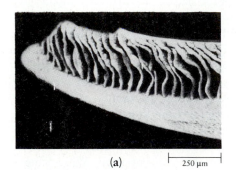

(a)   250 μm

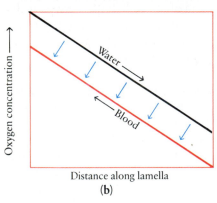

Distance along lamella
(b)

6–5 (a) *Micrograph of a gill filament from a trout, showing the densely packed lamellae, which contain the blood vessels. Water, carrying dissolved oxygen, flows between the lamellae in one direction; blood flows through them in the opposite direction. Thus the blood carrying the most oxygen (that is, the blood leaving the lamellae) meets the water carrying the most oxygen, and the blood carrying the least oxygen (the blood entering the lamellae) meets the water carrying the least oxygen. (b) In this way, a constant concentration gradient is maintained along the lamella, and the transfer of oxygen to blood by diffusion (blue arrows) takes place all across its surface.*

Sensory Responses in Bacteria: A Model Experiment

Concentration gradients are important not only in the diffusion of substances into, out of, and through cells but also, in the case of many single-celled organisms, in the movement of the cell itself through the surrounding medium. Bacterial cells, as we noted on page 36, are able to swim toward a food source or away from a noxious chemical. They accomplish this by moving along a concentration gradient, from a lower concentration of a particular type of molecule to a higher one, or vice versa. Such directed movements in bacteria are extremely sensitive and highly specific: the bacteria can sense only certain molecules and can sense them at very low concentrations. The sensory abilities are due to receptor sites in the cell membrane that detect the molecules in question.

If you watch flagellated bacteria swimming freely, you will see two types of movement. When the flagella are rotating, they drive the cell through the water in much the same way a propeller drives a boat. When the flagella stop, the cell tumbles wildly for perhaps a tenth of a second. Then the propellerlike motion begins again, and the cell moves off in a new direction. When the concentration of chemicals in the water is uniform, the cell tumbles often, changing direction every time. By contrast, when the cell is moving along a gradient, there are fewer tumbles, so the cell continues longer in the same direction.

How do bacterial cells "decide" to move in a particular direction? How do they know there is a concentration gradient? For many years, the most widely held hypothesis was that a bacterial cell could detect the difference in concentration between its front end and its rear end. However, when Daniel E. Koshland, Jr., of the University of California, calculated the concentrations of molecules to which a cell could respond, he began to question this concept. A bacterial cell is so small that, in a gradient steep enough to produce a strong response, the difference in concentration between the front end of the cell and the rear end would be only on the order of one molecule in 10,000. Also, no gradient would be exactly uniform. In short, the analytical task confronting the cell on its journey would seem virtually impossible.

Koshland then formulated an alternative hypothesis and, more important, figured out a way to test between the two. Koshland's hypothesis was that the bacterial cells were making a comparison not in space —between their front end and their rear end—but in time—from one microsecond to the next as they moved along the gradient. In order to choose between the alternatives, he formulated an ingeniously simple experiment. Using a strain of the common bacterium *Salmonella,* Koshland set up an apparatus with which he could transfer cells almost instantaneously from one liquid medium to another and compare their motility. First, he put *Salmonella* in a medium that contained no chemical attractants. The cells exhibited their normal tumble-and-run pattern of behavior. He transferred them to a new medium, also containing no chemical attractants. They did not change their pattern of movement. This part of the experiment— known as a control—showed that moving the cells, by itself, did not affect their motility.

Next, in the crucial part of the experiment, he placed the bacteria in a medium containing a uniform concentration of the amino acid serine, an attractant for *Salmonella.* The cells behaved just as they had when no attractant was present. Then he transferred the cells to a medium with a slightly higher concentration of serine. There was an immediate change: for a few seconds, the cells ran more than they tumbled. Then he transferred them to a medium with a lower concentration of serine; for a few seconds, they tumbled more and ran less. In other words, although the bacteria were actually moving from one uniform concentration to another, they behaved as if they were moving up or down a gradient. Koshland had tricked the bacteria into revealing their secret and so was able to choose between the alternative hypotheses. The bacteria were analyzing differences in time, not space.

This experiment is a minilesson in how scientists go about their business. They formulate a testable hypothesis and then they challenge it. The test of the hypothesis can take the form of a clever, well-designed experiment, as in this example, of accumulated observations, or of the analysis of reports made by other observers. However, two components are always necessary: the testable hypothesis and the data with which to test it.

Of course, many questions remain. Exactly how do the receptor sites on the cell membrane recognize particular substances? How does the cell "remember" the concentration from one moment to the next? How does the sensory response (the detection of the chemical) trigger the motor response (the movement of the flagella)? Here again is a characteristic of the scientific process: the answer to one question nearly always raises still more questions.

Osmosis: A Special Case of Diffusion

A membrane that permits the passage of some substances, while blocking the passage of others, is said to be **selectively permeable.** The movement of water molecules through such a membrane is a special case of diffusion, known as **osmosis.** Osmosis results in a net transfer of water from a solution that has higher water potential to a solution that has lower water potential. In the absence of other factors that influence water potential (such as pressure), the movement of water in osmosis will be from a region of lower solute concentration (and therefore of higher water concentration) into a region of higher solute concentration (lower water concentration). The presence of solute decreases the water potential and so creates a gradient of water potential along which water diffuses.

The diffusion of water is not affected by *what* is dissolved in the water, only by *how much* is dissolved—that is, the concentration of particles of solute (molecules or ions) in the water. The word **isotonic** was coined to describe two or more solutions that have equal numbers of dissolved particles per unit volume and therefore the same water potential. There is no net movement of water across a membrane separating two solutions that are isotonic to one another, unless, of course, pressure is exerted on one side. In comparing solutions of different concentration, the solution that has less solute (and therefore a higher water potential) is known as **hypotonic,** and the one that has more solute (a lower water potential) is known as **hypertonic.** (Note that *iso* means "the same"; *hyper* means "more"—in this case, more particles of solute; and *hypo* means "less"—in this case, fewer particles of solute.) In osmosis, water molecules diffuse from a hypotonic solution (or from pure water), through a selectively permeable membrane, into a hypertonic solution (Table 6–1).

TABLE 6–1   **The Direction of Water Movement in Osmosis**

WATER MOVES ACROSS A SELECTIVELY PERMEABLE MEMBRANE FROM	TO
Region of higher water potential	Region of lower water potential
Higher water concentration	Lower water concentration
Lower solute concentration	Higher solute concentration
Hypotonic solution (less solute)	Hypertonic solution (more solute)
Region of lower osmotic potential	Region of higher osmotic potential

Osmosis and Living Organisms

The osmotic movement of water across the selectively permeable cell membrane causes some crucial problems for living systems. These problems vary according to whether the cell or organism is hypotonic, isotonic, or hypertonic in relation to its environment. One-celled organisms that live in the seas, for example, are usually isotonic with the salty medium they inhabit, which is one way of solving the problem. The cells of most marine invertebrates are also isotonic with sea water. Similarly, the cells of vertebrate animals are isotonic with the blood and lymph that constitute the watery medium in which they live.

Many types of cells, however, live in a hypotonic environment. In all single-celled organisms that live in fresh water, such as *Paramecium,* the interior of the cell is hypertonic to the surrounding water; consequently, water tends to move into the cell by osmosis. If too much water were to move into the cell, it could dilute the cell contents to the point of interfering with function and could even eventually rupture the cell membrane. This is prevented by a specialized organelle

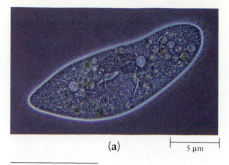

(a)

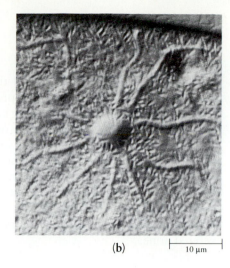

(b)   10 µm

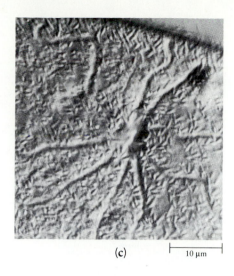

(c)   10 µm

6–6 A Paramecium *is hypertonic in relation to its environment, and hence water tends to move into the cell by osmosis. Excess water is expelled through its contractile vacuoles. (a) Phase-contrast micrograph of a living* Paramecium *showing the position of its two rosette-like contractile vacuoles. As revealed by the scanning electron microscope, (b) collecting tubules converge toward the vacuole, filling it. (c) Then it contracts, emptying outside the cell membrane by way of a small central pore. Actin filaments (page 121) are involved in the contraction of the vacuole.*

known as a contractile vacuole, which collects water from various parts of the cell and pumps it out with rhythmic contractions (Figure 6–6). As you might expect, this bulk transport process requires energy.

Osmotic Potential

The water potential on the two sides of a selectively permeable membrane will become equal if enough water moves from the hypotonic solution into the hypertonic solution to equalize the water concentrations (and therefore the solute concentrations)—that is, to make the solutions isotonic. If, however, physical barriers prevent the expansion of the hypertonic solution as water moves into it by osmosis, there will be increasing resistance as water molecules continue to move across the membrane. This resistance creates a buildup of pressure that gradually increases the water potential of the hypertonic solution, decreasing the gradient of water potential between the two solutions. As the pressure increases, the *net* flow of water molecules will slow and then cease as the gradient of water potential disappears. (Individual water molecules, of course, continue to move back and forth across the membrane, but these movements are in equilibrium and there is no net movement of water.) The pressure that is required to stop the osmotic movement of water into a solution is called the osmotic pressure. It is a measure of the **osmotic potential** of the solution—that is, of the tendency of water to move across a membrane into the solution.

The measurement of the osmotic potential of a solution is illustrated in Figure 6–7. The beaker contains distilled water, and within the tube is the solution. Across the mouth of the tube is a selectively permeable membrane; such a membrane is freely permeable to water but not to the particles (ions or molecules) in the solution. Water moving from the beaker through the membrane into the solution causes the solution to rise in the tube. As it does so, the pressure created by the force of gravity acting on the column of solution gradually increases, thus increasing the water potential of the solution. The solution rises in the tube until

6–7 Osmosis and the measurement of osmotic potential. (a) The tube contains a solution and the beaker contains distilled water. (b) The selectively permeable membrane permits the passage of water but not of solute. The diffusion of water into the solution causes the solution to rise in the tube until the tendency of water to move into a region of lower water concentration is counterbalanced by the pressure resulting from the force of gravity acting on the column of solution. This hydrostatic (water-stopping) pressure is proportional to the height, h, and density of the column of solution. (c) The pressure that must be applied to the piston to force the column of solution back to the level of the water in the beaker provides a quantitative measure of the osmotic potential of the solution—that is, of the tendency of water to diffuse across a membrane into the solution.

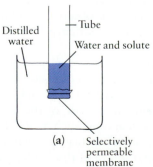

Distilled water Tube

Water and solute

(a) Selectively permeable membrane

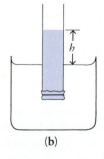

h

(b)

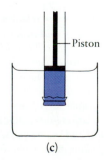

Piston

(c)

equilibrium is reached—that is, until the water potential on both sides of the membrane is equal. The amount of pressure that must then be applied to the piston to force the solution in the tube back to the level of the water in the beaker provides a quantitative measure of the osmotic potential.

The lower the water potential of a solution, the greater the tendency of water molecules to move into it by osmosis and, therefore, the greater its osmotic potential. Since solutes decrease the water potential of a solution, a higher solute concentration means a greater osmotic potential.

As we shall see in Chapter 37, the osmotic potential of solutions within the vertebrate kidney plays an important part in determining the composition of the excreted urine.

Turgor

Plant cells are usually hypertonic to their surrounding environment, and so water tends to diffuse into them. This movement of water into the cell creates pressure within the cell against the cell wall. The pressure causes the cell wall to expand and the cell to enlarge. The elongation that occurs as a plant cell matures (see Figure 5–7b, page 107) is a direct result of the osmotic movement of water into the cell.

As the plant cell matures, the cell wall stops growing. Moreover, mature plant cells typically have large central vacuoles that contain solutions of salts and other materials. (In citrus fruits, for example, they contain the acids that give the fruits their characteristic sour taste.) Because of these concentrated solutions, plant cells have a high osmotic potential—that is, water has a strong tendency to move into the cells. In the mature cell, however, the cell wall does not expand further. Its resistance to expansion results in an inward directed pressure, analogous to the pressure exerted by the depressed piston in Figure 6–7. This pressure, known as the wall pressure, prevents the net movement of additional water into the cell. Consequently, equilibrium of water concentration is not reached and water continues to "try" to move into the cell, maintaining a constant pressure on the cell wall from the inside (Figure 6–8). This internal pressure on the cell wall is known as **turgor,** and it keeps the cell walls stiff and the plant body crisp. When turgor is reduced, as a consequence of water loss, the plant wilts.

CARRIER-ASSISTED TRANSPORT

As we noted in Chapter 2, water and other polar or charged (hydrophilic) molecules exclude lipids and other hydrophobic molecules. Conversely, hydrophobic molecules exclude hydrophilic ones. This behavior of molecules, determined by the presence or absence of polar or charged regions, is of fundamental importance in the capacity of cellular membranes to regulate the passage of materials into and out of cells and organelles. As we have seen, cell membranes are formed from a lipid bilayer, the interior of which is filled by the hydrophobic tails of the lipid molecules. This interior lipid sea is a formidable barrier to ions and most hydrophilic molecules, but it does allow easy passage of hydrophobic molecules, such as steroid hormones. (It was, in fact, the observation that hydrophobic molecules diffuse readily across cell membranes that provided the first evidence of the lipid nature of the membrane.)

Most organic molecules of biological importance, however, have polar functional groups and are therefore hydrophilic; unlike carbon dioxide, oxygen, and water, they cannot move freely through the lipid barrier by simple diffusion. Similarly, the ions that are of crucial importance in the life of the cell cannot diffuse through the membrane. Although individual ions, such as sodium (Na^+) and chloride (Cl^-) are quite small, in aqueous solution they are surrounded by water molecules (see Figure 2–8, page 45); both the size of the resulting aggrega-

Movement of water

Solutes

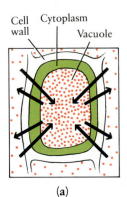
Cell wall   Cytoplasm   Vacuole

(a)

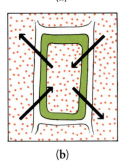

(b)

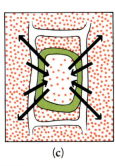

(c)

6–8 (a) A turgid plant cell. The central vacuole is hypertonic in relation to the fluid surrounding it and so gains water. The expansion of the cell is held in check by the cell wall. (b) A plant cell begins to wilt if it is placed in an isotonic solution, so that water pressure no longer builds up within the vacuole. (c) A plant cell in a hypertonic solution loses water to the surrounding fluid and so collapses, with its membrane pulling away from the cell wall. Such a cell is said to be plasmolyzed.

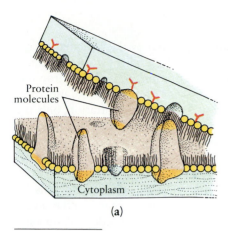

(a)

(b)

0.1 μm

6-9 (a) *The interior surface of cellular membranes can be revealed when specimens are prepared for the electron microscope by the freeze-fracture technique. In this procedure, the cell is frozen and then fractured with a sharp blow. The fracture line generally runs between the two lipid layers of a membrane, exposing the embedded proteins. A replica of the exposed surface is prepared for examination, using the procedures described on page 98.*

(b) The interior surface of the membrane of a red blood cell, prepared by the freeze-fracture technique. The arrow indicates the edge of the fracture line. The numerous particulate structures visible in the micrograph are integral membrane proteins, many of which are thought to be transport proteins.

tions and their charges prevent ions from slipping through the apertures that allow the passage of water molecules (see page 130). Transport of these aggregations and of all but the very smallest hydrophilic molecules depends upon integral membrane proteins that act as carriers, ferrying molecules back and forth.

The transport proteins of cell and organelle membranes are highly selective; a particular protein may accept one molecule while it excludes a nearly identical one. It is the configuration of the protein molecule—that is, its tertiary or, in some cases, quaternary structure—that determines what molecules it can transport. Although the protein typically undergoes temporary changes in configuration in the course of the transport process, it is not permanently altered. As we shall see in Chapter 8, enzymes are also highly selective in the molecules with which they interact and are not permanently altered by those interactions. Enzymes are often given names that end in "-ase," and to emphasize the similarities, transport proteins are sometimes referred to as permeases. Unlike enzymes, however, transport proteins do not necessarily produce chemical change in the molecules with which they interact.

Some transport proteins can move substances across the membrane only if there is a favorable concentration gradient; such carrier-assisted transport is known as **facilitated diffusion.** Other proteins can move molecules against a concentration gradient, a process known as **active transport.** Facilitated diffusion, like the simple diffusion discussed earlier, is a passive process, requiring no energy outlay by the cell; active transport, by contrast, requires the expenditure of cellular energy (Figure 6–10).

6-10 *Modes of transport through the cell membrane. In simple diffusion and facilitated diffusion, molecules or ions move down a concentration gradient. The potential energy of the concentration gradient drives these processes, which are, from the standpoint of the cell, passive. In active transport, by contrast, molecules or ions are moved against a concentration gradient. Energy, released by cellular reactions, is required to power active transport. Both facilitated diffusion and active transport require the presence of integral membrane proteins, specific for the substance being transported.*

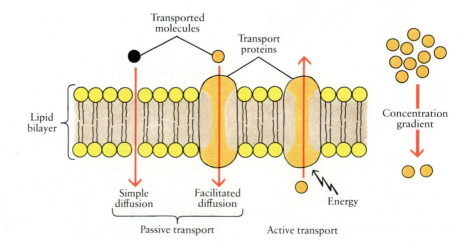

Transported molecules

Transport proteins

Lipid bilayer

Simple diffusion

Facilitated diffusion

Energy

Concentration gradient

Passive transport

Active transport

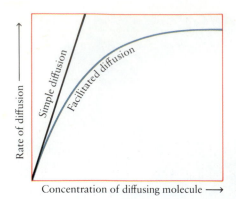

6–11 *A comparison of the rates of simple diffusion and facilitated diffusion across the cell membrane. In simple diffusion, the rate increases steadily as the concentration of diffusing molecules (or ions) increases. In facilitated diffusion, by contrast, the rate increases only as long as there are additional transport protein molecules, specific for the diffusing substance, available. When all of the protein molecules are in use, the rate levels off and does not increase further.*

6–12 *Mitochondria clustered near the surface of kidney cells of a bat. These cells are concerned with pumping out sodium ions against a concentration gradient. The mitochondria provide the energy for this active-transport process.*

Facilitated Diffusion

Both facilitated diffusion and simple diffusion are driven by the potential energy of a concentration gradient. Molecules move down the gradient from a region of higher concentration to a region of lower concentration. Ions and hydrophilic molecules, however, can move across the lipid barrier of a cell membrane, from the region of higher concentration to the region of lower concentration, only if a specific transport protein is available to allow them passage. The rate at which they can diffuse across the membrane depends not only upon the steepness of the concentration gradient but also upon the number of their specific transport protein molecules present in the membrane.

Glucose, for example, is a hydrophilic molecule that enters most cells by facilitated diffusion. Because glucose is rapidly broken down when it enters a cell, a steep concentration gradient is maintained between the inside and outside. However, when very large numbers of glucose molecules are present in the surrounding medium, the rate of entry does not increase beyond a certain point; it reaches a peak and then remains steady at that level (Figure 6–11). This limitation on the rate of entry is a result of the limited number of specific glucose-transporting protein molecules in the cell membrane.

Active Transport

Depending on the direction of the concentration gradient, a molecule may be transported across the cell membrane by either facilitated diffusion or active transport. In active transport, molecules or ions are moved against a concentration gradient, an energy-requiring process analogous to pushing a boulder up a hill. For example, glucose is transported into liver cells, where it is stored as glycogen, even though the concentration of glucose is higher inside the liver cells than in the bloodstream. This active-transport process presumably involves different membrane proteins than those used in facilitated diffusion.

The Sodium-Potassium Pump

One of the most important and best-understood active-transport systems is the sodium-potassium pump. Most cells maintain a differential concentration gradient of sodium ions (Na^+) and potassium ions (K^+) across the cell membrane: Na^+ is maintained at a lower concentration inside the cell, and K^+ is kept at a higher concentration. This concentration gradient is exploited by nerve cells to propagate electrical impulses, as you will learn in Chapter 41. The sodium-potassium pump requires energy made available by a molecule known as ATP (adenosine triphosphate), which is the form in which most of a cell's ready energy currency is carried. A measure of the importance of the sodium-potassium pump to the organism is that more than a third of the ATP used by a resting animal is consumed by this one ion-pumping mechanism.

The pumping of Na^+ and K^+ ions is accomplished by a transport protein thought to exist in two alternative configurations. One configuration has a cavity opening to the inside of the cell, into which Na^+ ions can fit; the other has a cavity opening to the outside, into which K^+ ions fit. As shown in Figure 6–13, Na^+ within the cell binds to the transport protein. At the same time, an energy-releasing reaction involving ATP results in the attachment of a phosphate group to the protein. This triggers its shift to the alternative configuration and the release of the Na^+ to the outside of the membrane. The transport protein is now ready to pick up K^+, which results in the release of the phosphate group from the protein, thus causing it to return to the first configuration and to release the K^+ to the inside of the cell. As you can see, this process will generate a gradient of Na^+ and K^+ ions across the membrane.

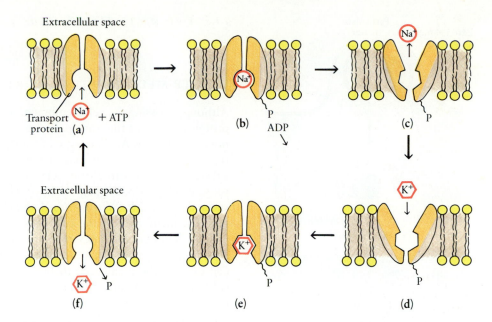

6–13 *A model of the sodium-potassium pump.* **(a)** *An Na⁺ ion in the cytoplasm fits precisely into the transport protein.* **(b)** *A chemical reaction involving ATP then attaches a phosphate group (P) to the protein, releasing ADP (adenosine diphosphate). This process results in* **(c)** *a change of shape that causes the Na⁺ to be released outside the cell.* **(d)** *A K⁺ ion in the extracellular space is bound to the transport protein* **(e)***, which in this form provides a better fit for K⁺ than for Na⁺.* **(f)** *The phosphate group is then released from the protein, inducing conversion back to the other shape, and the K⁺ ion is released into the cytoplasm. The protein is now ready once more to transport Na⁺ out of the cell.*

For clarity, only single ions are shown in this diagram. Quantitative studies have shown, however, that each complete pumping sequence transports three Na⁺ ions out of the cell and two K⁺ ions into the cell.

6–14 *Three types of transport molecules. In the simplest, known as a uniport, one particular solute is moved directly across the membrane in one direction. In the type of cotransport system known as a symport, two different solutes are moved across the membrane, simultaneously and in the same direction. Often, a concentration gradient involving one of the transported solutes powers the transport of the other solute; for example, a concentration gradient of Na⁺ ions frequently powers cotransport of glucose molecules. In another type of cotransport system, known as an antiport, two different solutes are moved across the membrane, either simultaneously or sequentially, in opposite directions. The Na⁺–K⁺ pump is an example of a cotransport system involving an antiport.*

Types of Transport Molecules

Many ingenious models have been proposed to show how transport proteins, such as the sodium-potassium pump, might accept and eject their passengers. One of the first suggested that the protein rotated, like a revolving door. More recent evidence indicates that although membrane proteins can move laterally, they are not free to flip-flop through the membrane as that model required. A current model hypothesizes that transport proteins have hydrophilic cores, which the transported molecules are squeezed through, propelled by changes in the configuration of the protein. These changes in configuration may be triggered directly, by the binding of the molecule to be transported, or indirectly, by the interaction of a receptor on the protein surface with some other molecule or ion that is not actually transported through the membrane. The cell membrane of most nerve cells, for example, contains a complex protein molecule that is a receptor for a molecule known as acetylcholine. When acetylcholine binds to this receptor, a channel is opened in another protein molecule, closely associated with the acetylcholine receptor; this channel allows sodium ions (Na⁺) to flow into the cell, down the concentration gradient created and maintained by the action of the sodium-potassium pump.

Current evidence indicates that there are at least three general types of transport proteins (Figure 6–14). The simplest transfers one particular kind of molecule or ion directly across the membrane. More complex proteins function as cotransport systems, in which the transport of a particular molecule or ion depends upon the simultaneous or sequential transport of a different molecule or ion. In some cotransport systems, both solutes are transported in the same direction; in others, exemplified by the sodium-potassium pump, the two different solutes are transported in opposite directions.

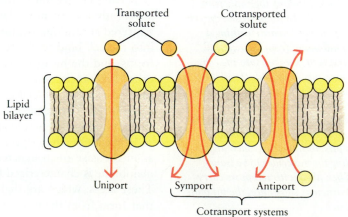

137

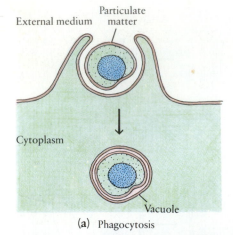

External medium / Particulate matter

Cytoplasm

Vacuole

(a) Phagocytosis

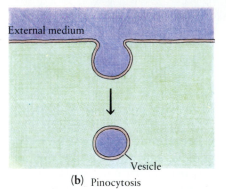

External medium

Vesicle

(b) Pinocytosis

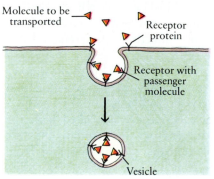

Molecule to be transported

Receptor protein

Receptor with passenger molecule

Vesicle

(c) Receptor-mediated endocytosis

6–15 Three types of endocytosis. (a) In phagocytosis, contact between the cell membrane and particulate matter causes the cell membrane to extend around the particle, engulfing it in a vacuole. If the particle is a food item, lysosomes fuse with the vacuole, spilling their digestive enzymes into it. (b) In pinocytosis, the cell membrane pouches inward, forming a vesicle around liquid from the external medium that is to be taken into the cell. (c) In receptor-mediated endocytosis, the substances to be transported into the cell must first bind to specific receptor molecules. The receptors are either localized in indented areas of the cell membrane, known as pits, or migrate to such areas after binding the molecules to be transported. When filled with receptors carrying their particular substance, the pit buds off as a vesicle.

VESICLE-MEDIATED TRANSPORT

Although crossing the cell membrane, with or without the assistance of transport proteins, is one of the principal ways substances get into and out of the cell, it is not the only way. Another type of transport process involves vesicles or vacuoles that are formed from or that fuse with the cell membrane. For example, many substances are exported from cells in vesicles formed by the Golgi complexes. As we saw in Figure 5–21 (page 118), vesicles move from the Golgi complexes to the surface of the cell. When a vesicle reaches the cell surface, its membrane fuses with the membrane of the cell, thus expelling its contents to the outside. This process is known as **exocytosis.** Transport by means of vesicles or vacuoles can also work in the opposite direction. In **endocytosis,** material to be taken into the cell induces the membrane to bulge inward, producing a vesicle enclosing the substance. This vesicle is released into the cytoplasm. Three different forms of endocytosis are known: **phagocytosis** ("cell-eating"), **pinocytosis** ("cell-drinking"), and **receptor-mediated endocytosis** (Figure 6–15).

When the substance to be taken into the cell in endocytosis is a solid, such as a bacterial cell, the process is usually called phagocytosis. Many heterotrophic protists, such as amoebas, feed in this way; similarly, macrophages (Figure 5–28, page 121) and other types of white blood cells in our own bloodstreams engulf bacteria and other invaders in phagocytic vacuoles. Often lysosomes fuse with these vacuoles, emptying their enzymes into them and so digesting or destroying their contents.

The taking in of liquids, as distinct from particulate matter, is given the special name of pinocytosis, although it is the same in principle as phagocytosis. Pinocytosis occurs not only in single-celled organisms but also in multicellular animals. One type of cell in which it has been frequently observed is the human egg cell. As the egg cell matures in the ovary of the female, it is surrounded by "nurse cells." These cells apparently transmit dissolved nutrients to the egg cell, which takes them in by pinocytosis.

In receptor-mediated endocytosis, currently the subject of a great deal of research, particular membrane proteins serve as receptors for specific molecules that are to be transported into the cell. Cholesterol, for example, is carried into animal cells by receptor-mediated endocytosis. As we noted earlier (page 71), cholesterol circulates in the bloodstream in the form of LDL particles, which interact with specific receptors on the cell surface. Binding of LDL particles to the receptor molecules triggers the formation of a vesicle that transports the cholesterol molecules into the cell. The protein forming the LDL receptor is a complex molecule that includes three different functional regions: a large segment that projects outside the cell, to which the LDL particles bind; a single, short segment that crosses the membrane; and a "tail" segment of about 50 amino acids that projects into the cytoplasm of the cell.

Membrane receptors for some substances, such as the hormone insulin, are apparently free to move laterally in the membrane and, when unoccupied, are scattered at random locations on its surface. As the molecules to be transported into the cell bind to the receptors, the receptors move close together. A vesicle forms, and the hormone-laden receptors are carried into the cell. Receptors for other substances, such as LDL particles, appear to be localized in groups in specific areas of the cell membrane even before binding of the substance to be transported. The segment of the LDL receptor that projects into the cytoplasm is thought to play a key role in this clustering of the receptors.

In the areas where specific receptors are localized—or to which they migrate, as in the case of insulin receptors—the inner, or cytoplasmic, face of the cell membrane is characterized by a peripheral membrane protein known as clathrin. These areas, which are slightly indented, are known as coated pits. The vesicles that form from them, containing receptor molecules and their passengers, thus

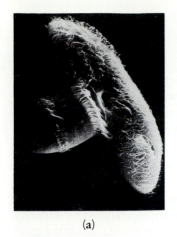
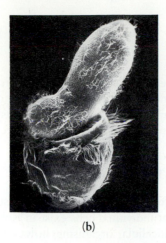
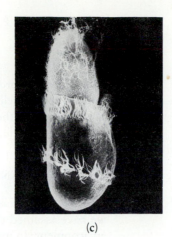
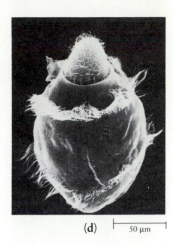

(a)   (b)   (c)   (d)   50 μm

6–16 *Phagocytosis of* Paramecium *by* Didinium. *(See Figure 5–30, on page 122, for the preamble to their encounter.)* (**a**) *Ingestion of the* Paramecium *has begun. The concave area just above the oral rim of the* Didinium *is the oral groove of the* Paramecium. Paramecium, *a heterotroph, feeds largely on bacteria.* (**b**) *Because the* Paramecium *is larger than the* Didinium, *folding helps.* (**c**) *The* Paramecium *is half*

"swallowed"; the part that is within the Didinium *is surrounded by a membrane composed of the cell membrane of the* Didinium. *The process of compression has begun, as you can see in the tip of the* Paramecium *protruding from the oral rim of the* Didinium. *This compression is largely a matter of squeezing out water.* (**d**) *Once the* Paramecium *is completely*

inside, the cell membrane of the Didinium *will fuse over it, forming a food vacuole. The* Paramecium, *however, must provide the means for its own demolition: the* Didinium *apparently lacks a crucial digestive enzyme that the* Paramecium *supplies. A* Didinium *can eat a dozen* Paramecium, *each larger than itself, in a single day.*

acquire an external, cagelike coating of clathrin. The formation of such a coated vesicle is illustrated in Figure 6–17.

As you can see by studying Figures 6–15 and 6–17, the surface of the membrane facing the interior of a vesicle or vacuole is equivalent to the surface facing the exterior of the cell; similarly, the surface of the vesicle or vacuole membrane facing the cytoplasm is equivalent to the cytoplasmic surface of the cell membrane. As we noted in the last chapter, new material needed for expansion of the cell membrane is transported, ready-made, from the Golgi complexes to the

6–17 *The formation of a coated vesicle in the developing egg cell of a hen. The vesicles are much larger than those of smaller cells and are thus easier to visualize.* (**a**) *A coated pit in the cell membrane is covered on the cytoplasmic face with a latticework of clathrin molecules. The large particles clustered in the shallow pit on the external face are lipoprotein molecules, gathered from the surrounding medium and bound to specific membrane receptors that are associated with the underlying clathrin layer.* (**b**) *The pit deepens, and then* (**c**) *the cell membrane closes around the pit to form the vesicle.* (**d**) *The completed vesicle with its outer coating of clathrin buds off and moves into the cell. The lipoproteins carried by this coated vesicle will be incorporated into the egg yolk.*

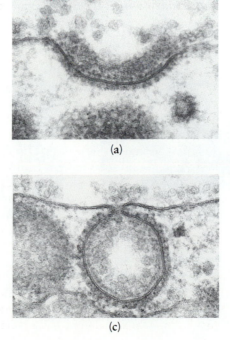
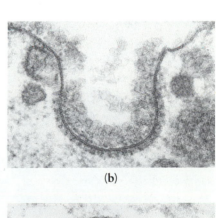
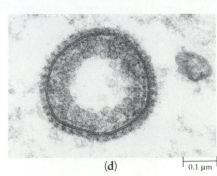

(a)   (b)

(c)   (d)   0.1 μm

membrane by a process similar to exocytosis. Current evidence indicates that the portions of the cell membrane used in forming endocytic vesicles or vacuoles are also returned to the membrane in exocytosis, thus recycling the membrane lipids and proteins, including the specific receptor molecules.

CELL-CELL JUNCTIONS

Thus far in our consideration of the transport of substances into and out of cells, we have assumed that individual cells exist in isolation, surrounded by a watery environment. In multicellular organisms, however, this is generally not the case. Cells are organized into **tissues,** groups of specialized cells with common functions. For example, in animals, the four principal types of tissues are muscle, nerve, connective, and epithelial (covering). Tissues are further organized in concert to form **organs,** such as the heart, brain, or kidney, each of which, like a subcellular organelle, has a design that suits it for a specific function.

As you might imagine, in multicellular organisms it is essential that individual cells communicate with one another so that they can collaborate to create a harmonious tissue or organ. Nerve impulses are transmitted from neuron to neuron, or from neuron to muscle or gland. Cells in the body of a plant or animal release hormones that travel over a distance and affect other cells in the same organism. In the course of development, embryonic cells influence the differentiation of neighboring cells into tissues and organs. These communications are all accomplished by means of chemical signals—that is, by substances that are transported out of one cell and travel to another cell. When they reach the cell membrane of the target cell, they may actually be transported into the cell, by any one of the processes we have considered, or they may bind to specific membrane receptors at the surface of the target cell, thereby triggering chemical reactions within the cell.

Often, however, the cells within a tissue or organ are tightly packed, allowing for direct and intimate contacts of various types between the cells. Among plant cells, which are separated from one another by cell walls, channels called **plasmodesmata** traverse the walls, directly connecting the cytoplasm of adjacent cells (Figure 6–18). Plasmodesmata, which are from 30 to 60 nanometers in diameter, appear to be lined by the cell membrane; in addition, they generally contain tubular extensions of the endoplasmic reticulum known as desmotubules.

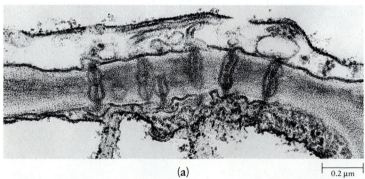

(a)

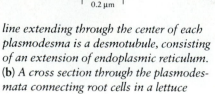

0.2 μm

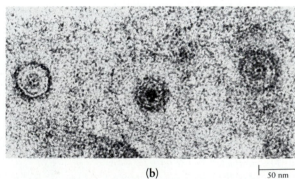

(b)

50 nm

6–18 (a) *Plasmodesmata connecting two leaf cells from a corn plant. The wide gray area running horizontally in this electron micrograph is the cell wall, which is traversed by the plasmodesmata. The dark line extending through the center of each plasmodesma is a desmotubule, consisting of an extension of endoplasmic reticulum. (b) A cross section through the plasmodesmata connecting root cells in a lettuce plant reveals their tubular structure. As you can see, each plasmodesma is encircled by an extension of the cell membrane and contains a desmotubule in its center.*

Communication in the Cellular Slime Mold

A cellular communication system of particular interest to biologists, because of the comparative ease with which it can be studied, is seen among a group of organisms known as the cellular slime molds. The slime mold *Dictyostelium discoideum* is an example. At one stage in its life cycle, it exists as a swarm of small individual amoebas (a), which divide and grow and feed, amoeba-fashion, until their food supply (mostly bacteria) gives out. At this point, the cells alter both their shape and behavior: they become sausage-shaped (b) and begin to migrate toward the center of the group. (The direction in which the stream is moving is indicated by the arrow.) Eventually, they pile up in a heap; the heap gradually takes on the form of a multicellular mass somewhat resembling a garden slug (c), slowly migrating and depositing a thick slime sheath that collapses behind it. The sluglike mass soon stops its migration, gathers itself into a mound, and sends up a long stalk at the tip of which a fruiting

body forms (d, e, f). The fruiting body matures (g) and eventually bursts open, releasing a new swarm of tiny amoebas, and the cycle begins again.

The chemical that spreads from cell to cell to initiate this remarkable sequence of events was first called acrasin, after Acrasia, the cruel witch in Spenser's *Faerie Queene* who attracted men and turned them into beasts. Acrasin was later identified as the chemical compound cyclic AMP (adenosine monophosphate). In recent years, it has become clear that many of the communications among cells in the human body also involve cyclic AMP. As we shall see in Chapter 40, the interaction of a number of vertebrate hormones (principally hormones that are proteins or amino acid derivatives) with their specific receptors in the cell membrane triggers a sequence of events within the cell in which cyclic AMP plays a central role.

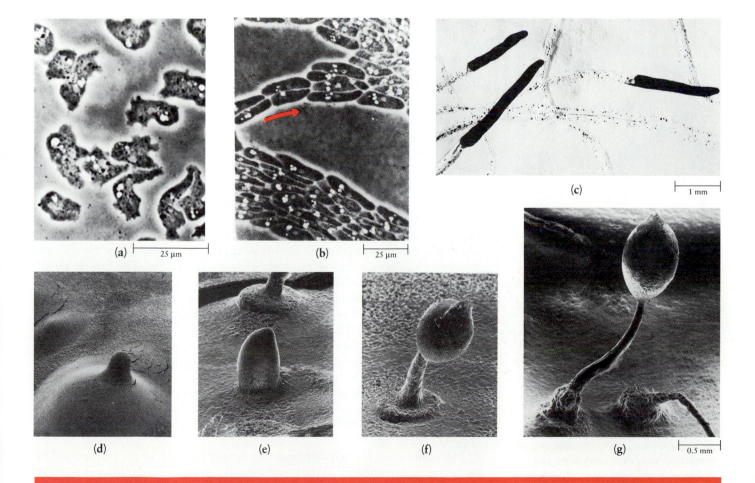

(a) 25 μm (b) 25 μm (c) 1 mm (d) (e) (f) (g) 0.5 mm

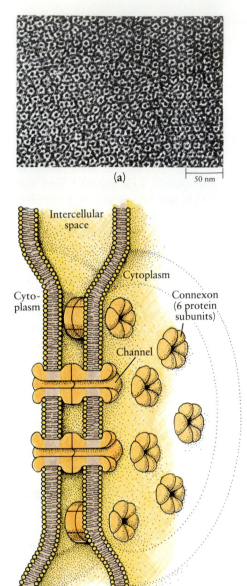

(a)

50 nm

Intercellular
space

Cytoplasm

Cyto-
plasm

Connexon
(6 protein
subunits)

Channel

(b)

6-19 (a) *A portion of a gap junction between two liver cells, as seen in cross section. In this electron micrograph, the narrow channels through which small molecules can pass between the cells are filled with an electron-opaque staining substance and appear as black dots.* (b) *A model of a gap junction. Embedded in the cell membranes are structures that have been appropriately named "connexons." Each connexon consists of six identical membrane protein subunits, arranged in a hexagonal pattern with a space through the center. Connexons in the adjacent cell membranes abut each other in perfect alignment, providing a channel connecting the cytoplasm of the two cells.*

In animal tissues, structures known as **gap junctions** permit the passage of materials between cells. These junctions appear as fixed clusters of very small channels (about 2 nanometers in diameter) surrounded by an ordered array of proteins (Figure 6–19). Experiments with radioactively labeled molecules have shown that small messenger molecules pass through these channels. Gap junctions also serve to transmit electrical signals in the form of ions. For example, the contractions of muscle cells in the heart are synchronized by the flow of sodium ions (Na^+) through gap junctions.

The transport of materials into and out of cells through the channels of plasmodesmata or gap junctions, through integral membrane proteins, and by means of endocytosis and exocytosis appear superficially to be three quite different processes. They are fundamentally similar, however, in that they all depend on the precise, three-dimensional structure of a great variety of specific protein molecules. These protein molecules not only form channels through which transport can occur but also endow the cell membrane with the capacity to "recognize" particular molecules. This capacity is the result of billions of years of an evolutionary process that began, as far as we are able to discern, with the formation of a fragile film around a few organic molecules. This film separated the molecules from their external environment and permitted them to maintain the particular kind of organization that we recognize as life. It is one of the many critical capacities transmitted from parent to offspring each time a cell divides, a process we shall examine in the next chapter.

SUMMARY

The cell membrane regulates the passage of materials into and out of the cell, a function that makes it possible for the cell to maintain its structural and functional integrity. This regulation depends on interactions between the membrane and the materials that pass through it.

One of the principal substances passing into and out of cells is water. Water potential determines the direction in which water moves; that is, water moves from where the water potential is higher to where it is lower. Water movement takes place by bulk flow and diffusion.

Bulk flow is the overall movement of water molecules and dissolved solutes as a group, as when water flows in response to gravity or pressure. The circulation of blood through the human body is an example of bulk flow.

Diffusion involves the random movement of individual molecules or ions and results in net movement down a concentration gradient. It is most efficient when the surface area is large in relation to volume, when the distance involved is short, and when the concentration gradient is steep. By their metabolic activities, cells maintain steep concentration gradients of many substances. The rate of movement of substances within cells is also increased by cytoplasmic streaming. In several anatomical systems of multicellular organisms, countercurrent exchange maintains a constant concentration gradient over a large surface area, thus maximizing the rate of diffusion.

Osmosis is the diffusion of water through a membrane that permits the passage of water but inhibits the movement of most solutes; such a membrane is said to be selectively permeable. In the absence of other forces, the net movement of water in osmosis is from a region of lower solute concentration (a hypotonic medium), and therefore of higher water potential, to one of higher solute concentration (a hypertonic medium), and so of lower water potential. Turgor in plant cells is a consequence of osmosis.

Molecules cross the cell membrane by simple diffusion or are transported by carrier proteins embedded in the membrane. If carrier-assisted transport is driven by the concentration gradient, the process is known as facilitated diffusion. If the transport requires the expenditure of energy by the cell, it is known as active transport. Active transport can move substances against their concentration gradients. One of the most important active-transport systems is the sodium-potassium pump, which maintains sodium ions at a relatively low concentration and potassium ions at a relatively high concentration in the cytoplasm.

Controlled movement into and out of a cell may also occur by endocytosis or exocytosis, in which the substances are transported in vacuoles or vesicles composed of portions of the cell membrane. Three forms of endocytosis are phagocytosis, in which solid particles are taken into the cell; pinocytosis, in which liquids are taken in; and receptor-mediated endocytosis, in which molecules or ions to be transported into the cell are bound to specific receptors in the cell membrane.

In multicellular organisms, communication among cells is essential for coordination of the different activities of the cells in the various tissues and organs. Much of this communication is accomplished by chemical agents that either pass through the cell membrane or interact with receptors in its surface. Communication may also occur directly, through the channels of plasmodesmata (in plant tissues) or gap junctions (in animal tissues).

QUESTIONS

1. Distinguish among the following: bulk flow/diffusion/osmosis; water potential/hydrostatic pressure/osmotic potential; hypotonic/hypertonic/isotonic; endocytosis/exocytosis; phagocytosis/pinocytosis/receptor-mediated endocytosis; coated pit/coated vesicle; plasmodesmata/gap junctions.

2. What is a concentration gradient? How does a concentration gradient affect diffusion? How does a concentration gradient affect osmosis?

3. When diffusion of dye molecules in a tank of water is complete, random movement of molecules continues (as long as the temperature remains the same). However, net movement stops. How do you reconcile these two facts?

4. Why is diffusion more rapid in gases than in liquids? Why is it more rapid at higher temperatures than at lower temperatures?

5. Three funnels have been placed in a beaker containing a solution (see the figure below). What is the concentration of the solution? Explain your answer.

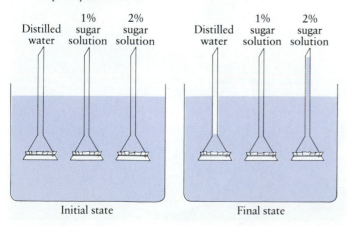

| Initial state | Final state |

6. How does countercurrent exchange affect diffusion?

7. Imagine a pouch with a selectively permeable membrane containing a saltwater solution. It is immersed in a dish of fresh water. Which way will the water move? If you add salt to the water in the dish, how will this affect water movement? What living systems exist under analogous conditions? How do you think they maintain water balance?

8. When you forget to water your house plants, they wilt and the leaves (and sometimes the stems) become very limp. What has happened to the plants to cause this change in appearance and texture? Within a few hours after you remember to water your plants, they resume their normal, healthy appearance. What has occurred within the plants to cause this restoration? Sometimes, if you wait too long to water your plants, they never revive. What do you suppose has happened?

9. What limits the passage of water and other polar molecules and ions through the cell membrane? How do such molecules get into and out of the cell? Describe four possible routes.

10. In what three ways does active transport differ from simple diffusion? How does it differ from facilitated diffusion?

11. Justify the conclusion that differences in ion concentration between cells and their surroundings (see Figure 6–1) indicate that cells regulate the passage of materials across membranes.

12. In Figure 6–16d, the *Paramecium* is sinking below the oral rim of *Didinium*. What will happen next? Complete the scenario, giving as many details as possible. (You might want to end your account with the fact that *Didinium* divides once for every two *Paramecium* consumed.)

CHAPTER 7

How Cells Divide

Cell division is the process by which cellular material is divided between two new daughter cells. In one-celled organisms, it increases the number of individuals in the population. In many-celled plants and animals, it is the means by which the organism grows, starting from one single cell, and also by which injured or worn-out tissues are replaced and repaired. An individual cell grows by assimilating materials from its environment and synthesizing these materials into new structural and functional molecules. When a cell reaches a certain critical size and metabolic state, it divides. The two daughter cells, each of which has received about half of the mass of the parent cell, then begin growing again. A bacterial cell may divide every six minutes. In a one-celled eukaryote such as *Paramecium,* cell division may occur every few hours.

The new cells produced are structurally and functionally similar both to the parent cell and to one another. They are similar, in part, because each new cell usually receives about half of the parent cell's cytoplasm and organelles. More important, in terms of structure and function, each new cell inherits an exact replica of the hereditary information of the parent cell.

7–1 *Single-celled eukaryotic organisms typically reproduce by simple cell division. In this scanning electron micrograph of the ciliated protist* Opisthonecta, *the separation of the two daughter cells is almost complete. Each cell has received not only an exact replica of the parent cell's hereditary information but also approximately half of its organelles and cytoplasm.*

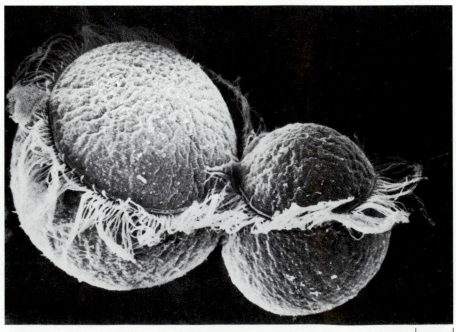

10 μm

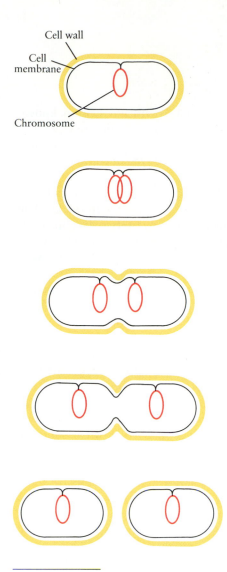

Cell wall

Cell membrane

Chromosome

7–2 *Schematic diagram of cell division in a bacterium. Attachment of the chromosome to an inward fold of the cell membrane ensures that one chromosome replicate is distributed to each daughter cell as the cell membrane elongates.*

CELL DIVISION IN PROKARYOTES

The distribution of exact replicas of the hereditary information is comparatively simple in prokaryotic cells. In such cells, most of the hereditary material is in the form of a single, long, circular molecule of DNA, with which a variety of proteins are associated. This molecule, the cell's chromosome, is replicated before cell division. According to present evidence, each of the two daughter chromosomes is attached to a different spot on the interior of the cell membrane. As the membrane elongates, the chromosomes move apart (Figure 7–2). When the cell has approximately doubled in size and the chromosomes are separated, the cell membrane pinches inward, and a new cell wall forms that separates the two new cells and their chromosome replicas.

The prokaryotic chromosome has been the subject of an enormous amount of research. In Section 3, we shall consider this marvelous structure, its replication and its functions, in more detail.

CELL DIVISION IN EUKARYOTES

In eukaryotic cells, the problem of exactly dividing the genetic material is much more complex. A typical eukaryotic cell contains about a thousand times more DNA than a prokaryotic cell, and this DNA is linear, forming a number of distinct chromosomes. For instance, human somatic ("body") cells have 46 chromosomes, each different from the others; when these cells divide, each daughter cell has to receive one copy, and only one copy, of each of the 46 chromosomes. Moreover, as we have seen, eukaryotic cells contain a variety of organelles, and these must also be apportioned between the daughter cells.

The solutions to these problems are, as you will see, ingenious and elaborate. In a series of steps called, collectively, **mitosis,** a complete set of chromosomes is allocated to each of two daughter nuclei. Mitosis is usually followed by **cytokinesis,** a process that divides the cell into two new cells, each of which contains not only a nucleus with a full chromosome complement but also approximately half of the cytoplasm and organelles of the parent cell.

Although mitosis and cytokinesis are the culminating events of cell division in eukaryotes, they represent only two stages of a larger process.

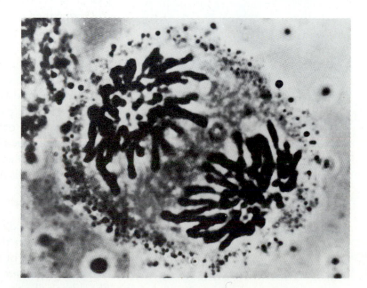

7–3 *A human cell divides. The long, dark bodies are chromosomes, the carriers of the hereditary information. The chromosomes have replicated and moved apart. Each set of chromosomes is an exact copy of the other. Thus the two new cells, the daughter cells, will contain the same hereditary material.*

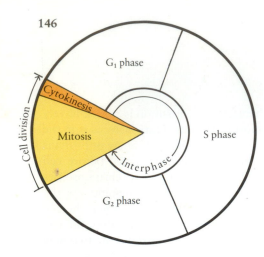

7–4 *The cell cycle. Cell division, which consists of mitosis (the division of the nucleus) and cytokinesis (the division of the cytoplasm), takes place after the completion of the three preparatory phases that constitute interphase. During the S (synthesis) phase, the chromosomal material is replicated. Separating cell division and the S phase are two G (gap) phases. The first of these (G_1) is a period of general growth and replication of cytoplasmic organelles. During the second (G_2), structures directly associated with mitosis and cytokinesis begin to assemble. After the G_2 phase comes mitosis, which is usually followed immediately by cytokinesis. In cells of different species or of different tissues within the same organism, the different phases occupy different proportions of the total cycle.*

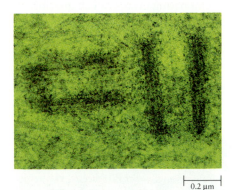

0.2 μm

7–5 *A centriole pair from a dividing cell of the fruit fly Drosophila, as seen in longitudinal section. The many thin fibers also visible in this micrograph are microtubules.*

Centrioles are formed either from preexisting centrioles, with the newly formed centriole appearing at right angles to the previously existing one, or from basal bodies. The structure of a centriole, as seen in cross section, is identical to that of a basal body (see Figure 5–32, page 123).

THE CELL CYCLE

Dividing cells pass through a regular sequence of cell growth and division, known as the **cell cycle** (Figure 7–4). The cycle consists of five major phases: G_1, S, G_2, mitosis, and cytokinesis. Completion of the cycle requires varying periods of time from a few hours to several days, depending on both the type of cell and external factors, such as temperature or available nutrients.

Before a cell can begin mitosis and actually divide, it must replicate its DNA, synthesize more of the histones and other proteins associated with the DNA in the chromosomes, produce a supply of organelles adequate for two daughter cells, and assemble the structures needed to carry out mitosis and cytokinesis. These preparatory processes occur during the G_1, S, and G_2 phases of the cell cycle, which are known collectively as **interphase.**

The key process of DNA replication occurs during the S (synthesis) phase of the cell cycle, a time in which many of the histones and other DNA-associated proteins are also synthesized. G (gap) phases precede and follow the S phase; during the G phases, no DNA synthesis can be detected in the nucleus of the cell.

The G_1 phase, which follows cytokinesis and precedes the S phase, is a period of intensive biochemical activity. The cell increases in size, and its enzymes, ribosomes, mitochondria, and other cytoplasmic molecules and structures also increase in number. Some of the cellular structures can be synthesized entirely *de novo* ("from scratch") by the cell; these include microtubules, actin filaments, and ribosomes, all of which are composed, at least in part, of protein subunits. Membranous structures, such as the Golgi complexes, lysosomes, vacuoles, and vesicles, are all apparently derived from the endoplasmic reticulum, which is renewed and enlarged by the synthesis of lipid and protein molecules. In those cells that contain centrioles (that is, virtually all eukaryotic cells except those of fungi, flowering plants, and nematodes), the two centrioles begin to separate from each other and to replicate. Each member of the original centriole pair gives rise to a smaller daughter centriole by a copying process that is not yet understood. Mitochondria and chloroplasts, which are produced only from previously existing mitochondria and chloroplasts or plastids (page 120), also replicate. Each of these organelles has its own chromosome, which is organized much like the single chromosome of the bacterial cell. (These are two of the reasons that many biologists hypothesize that mitochondria and chloroplasts originated as separate organisms and then took up a new way of life inside early eukaryotic cells, more than a billion years ago. This question will be discussed further in Chapter 22.)

During the G_2 phase, which follows the S phase and precedes mitosis, the final preparations for cell division occur. The newly replicated chromosomes, which are dispersed in the nucleus in the form of threadlike strands of chromatin, slowly begin to coil and condense into a compact form; this condensation appears to be necessary for the complex movements and separation of the chromosomes that will occur in mitosis. Replication of the centriole pair is completed, with the two mature centriole pairs lying just outside the nuclear envelope, somewhat separated from each other (Figure 7–5). Also during this period, the cell begins to assemble the special structures required for the allocation of a complete set of chromosomes to each daughter cell during mitosis and for the separation of the two daughter cells during cytokinesis.

Regulation of the Cell Cycle

Dividing cells of different species show characteristic variations in the pattern of the cell cycle. In the common bean, for example, the complete cycle requires about 19 hours, of which 7 hours are taken up by the S phase; G_1 and G_2 are of equal length (about 5 hours each), and mitosis lasts 2 hours. By contrast, in mouse fibroblast cells, the cell cycle is approximately 22 hours, of which mitosis is less than 1 hour, S is almost 10 hours, G_1 is 9 hours, and G_2 is a little more than 2 hours.

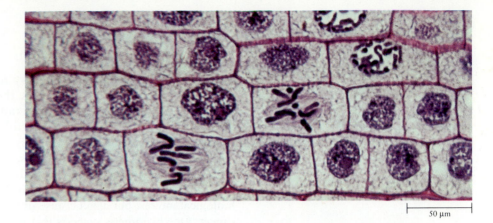

7–6 *Cross section of the tip of an onion root. The cells in this region undergo repeated divisions, providing new cells for the growth of the root. Different cells in this tissue section are in different phases of the cell cycle. The cells that are significantly larger than the others and that contain dark, rodlike structures are in mitosis.*

50 μm

Some cell types pass through successive cell cycles throughout the life of the organism. This group includes the one-celled organisms and certain cells in growth centers of both plants and animals (Figure 7–6). An example is the cells in the human bone marrow that give rise to red blood cells. The average red blood cell lives only about 120 days, and there are about 25 trillion (2.5×10^{13}) of them in an adult. To maintain this number, about 2.5 million new red blood cells must be produced by cell division each second. At the other extreme, some highly specialized cells, such as nerve cells, lose their capacity to replicate once they are mature. A third group of cells retains the capacity to divide but does so only under special circumstances. Cells in the human liver, for example, do not ordinarily divide, but if a portion of the liver is removed surgically, the remaining cells (even if as few as a third of the total remain) continue to replicate themselves until the liver reaches its former size. Then they stop. All told, about 2 trillion (2×10^{12}) cell divisions occur in an adult human every 24 hours, or about 25 million per second.

In a multicellular organism, it is of critical importance that cells of the various different types divide at a sufficient rate to produce as many cells as are needed for growth and replacement—and only that many. If any particular cell type divides more rapidly than is necessary, the normal organization and functions of the organism may be disrupted as specialized tissues are invaded and overwhelmed by the rapidly dividing cells. Such is the course of events in cancer. In both multicellular and unicellular organisms, it is also important that cells divide only when they have reached a size large enough to ensure that the resulting daughter cells will contain all of the metabolic machinery needed for survival.

A number of environmental factors, including the depletion of nutrients and changes in temperature or pH, can cause cells to stop growing and dividing. In multicellular organisms, contact with adjacent cells can have the same effect. If normal cells from vertebrates (including humans) are isolated from one another and grown in a nutrient medium on a smooth glass surface, they move, amoeba-like, ruffling their cell borders until they encounter another cell, at which time they stop. More significantly, they undergo repeated cell cycles until enough cells have been produced that each cell is touching another cell; then cell division ceases (Figure 7–7). This phenomenon, known as density-dependent inhibition, does not occur in cancer cells; they pile on top of one another, moving, multiplying, crowding each other until all of the nutrients are used up.

7–7 *An experimental demonstration of density-dependent inhibition, which is also known as contact inhibition. (a) When isolated cells are grown in a nutrient medium, they divide until they form a continuous layer, one cell thick, across the surface of the culture dish. (b) If several rows of cells are removed, for example, by scraping them off, the adjacent cells ruffle their borders and flatten out (c). These cells then begin dividing, stopping once more when the dish is completely covered by a single layer of cells (d).*

Culture dish containing nutrient medium

Individual cells

(a)

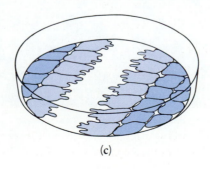

(b)   **(c)**

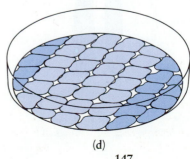

(d)

147

When normal cells stop growing, as a result of depletion of nutrients, density-dependent inhibition, or other factors, they stop at a point late in the G₁ phase. This point is known as the R ("restriction") point of the cell cycle. Once a cell passes the R point, it is committed to follow through the remaining parts of the cycle and then to divide. The G_1 phase is rapidly completed, synthesis of DNA and histone proteins in the S phase begins, and the cell moves steadily through the remaining phases of the cycle. The nature of the control or controls that act at the R point is currently the subject of intense research, not only because of its biological interest but also because of its potential importance in the control of cancer. One hypothesis suggests that passing the R point of the cycle requires a specific concentration of a particular protein synthesized in small quantities during the G_1 phase; only when this protein reaches the necessary concentration, thereby signaling that the cell has attained a suitable size and metabolic state, do the subsequent events of the cycle occur. Other hypotheses suggest that regulation involves a variety of stimulatory and inhibitory growth factors, some of which may be synthesized by the cell itself, and some of which may be synthesized and released into the surrounding medium by neighboring cells. A number of factors that influence the growth and division of cells in tissue culture have been discovered in the last few years; work is now under way characterizing their structure, the structure of the membrane receptors to which they bind, and the events triggered within the cell in response to that binding.

MITOSIS

The function of mitosis is to maneuver the replicated chromosomes so that each new cell gets a full complement—one of each. The capacity of the cell to accomplish this distribution depends on the condensed state of the chromosomes during mitosis and on an assembly of microtubules known as the **spindle.** Let us examine these structures before considering the "dance of the chromosomes."

The Condensed Chromosomes

As we noted previously, the threadlike chromosomes slowly begin to condense after their synthesis in the S phase of the cell cycle. By the beginning of mitosis, they are sufficiently condensed to become visible under the light microscope. Each chromosome can be seen to consist of two replicas, called **chromatids** (Figure 7–8a), joined together by a constricted area common to both chromatids.

7–8 (a) *A fully condensed chromosome. The chromosomal material was replicated during the S phase of the cell cycle, and each chromosome now consists of two identical parts called chromatids. The centromere, the constricted area at the center, is the site of attachment of the two chromatids. The kinetochores are protein-containing structures, one on each chromatid, associated with the centromere. Attached to the kinetochores are microtubules that form part of the spindle.*

(b) In this electron micrograph of a portion of a dividing green alga, spindle fibers can be seen extending from the kinetochores. The dark material is the centromere region of the chromosome, most of which is out of the plane of the thin section prepared for the micrograph. The kinetochores are the two disk-shaped areas at either side of the chromosome.

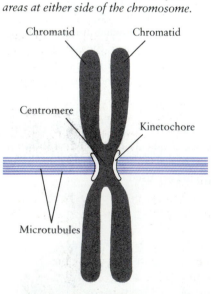

Chromatid   Chromatid

Centromere

Kinetochore

Microtubules

A replicated, condensed chromosome

(a)

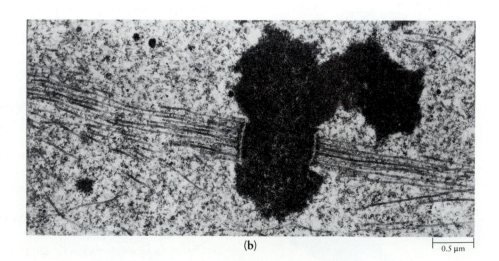

(b)   0.5 µm

This region of attachment is known as the **centromere.** Within the constricted region are disk-shaped protein-containing structures, the **kinetochores,** to which microtubules of the spindle are attached (Figure 7–8b).

The Spindle

When completely formed, the spindle (Figure 7–9) is a three-dimensional football-shaped structure, consisting of at least two groups of microtubules: (1) polar fibers, which reach from each pole of the spindle (analogous to the ends of the football) to a central region midway between the poles, and (2) kinetochore fibers, which are attached to the kinetochores of the replicated chromosomes. As we shall see, these two groups of spindle fibers are responsible for separating the sister chromatids during mitosis.

In those cells that contain centrioles, each pole of the spindle is marked by a pair of newly replicated centrioles. Such cells also contain a third group of shorter spindle fibers, extending outward from the centrioles. These additional fibers are known collectively as the **aster.** It has been hypothesized that the fibers of the aster may brace the poles of the spindle against the cell membrane during the movements of mitosis; in cells that lack centrioles and asters, the rigid cell wall may perform a similar function.

Most of the tubulin dimers from which the microtubules of the spindle are formed (see Figure 3–25, page 78) are apparently borrowed from the cytoskeleton. Immunofluorescence micrographs have shown that the network of cytoskeletal microtubules that radiate outward from the center of a nondividing cell (see Figure 5–14a, page 111) is disassembled at the beginning of mitosis. As a consequence, dividing cells take on a characteristic rounded appearance. Following cell division, the spindle is disassembled, the cytoskeletal network of microtubules is reassembled, and the cell assumes its nondividing shape.

Centrioles and the Microtubule Organizing Center

As we noted in Chapter 5, basal bodies and centrioles are the same structure used, perhaps, for different purposes. Basal bodies organize the microtubules of flagella and cilia, and centrioles have long been thought to play a role in organizing the microtubules of the spindle fibers—spinning them out somewhat as a spider spins out silk. A striking example of the interchangeability of basal bodies and centrioles is provided by the alga *Chlamydomonas.* At the beginning of mitosis, its two flagella are reabsorbed by the cell, and the basal bodies move near the nucleus—to the same location occupied in other cells by the centrioles. During mitosis, they behave exactly like centrioles, appearing to organize the spindle. When mitosis is complete, they migrate to the ends of the daughter cells, giving rise to new flagella. Despite this evidence for the role of the centrioles in spindle formation, cells that do not have centrioles or basal bodies also form spindles with microtubules. In some animal cells, moreover, it is possible to remove the centrioles from the cells—yet spindle formation proceeds normally.

The explanation for these seemingly contradictory observations may lie in a densely staining region seen around the centrioles in many electron micrographs. Such a densely staining region is also present in cells without centrioles and is the area from which both the spindle fibers and the microtubules of the cytoskeleton originate. The material in this region, rather than the centrioles themselves, is now thought to be the microtubule organizing center. It has also been suggested that the spindles, instead of forming from the centrioles, separate them, pushing them apart and so ensuring that each daughter cell receives an adequate supply of basal bodies from which to construct flagella or cilia.

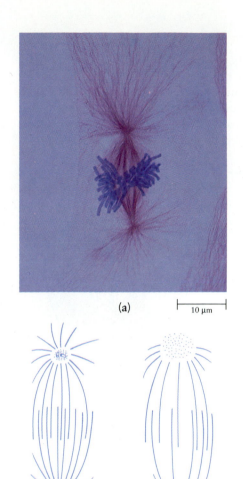

(a)

|← 10 µm →|

Typical animal cell   Typical plant cell
(b)   **(c)**

7–9 (a) *This micrograph of a dividing cell of the lung epithelium of an Oregon newt, an amphibian, illustrates the spindle's three-dimensional quality. The red fibers are the spindle microtubules. The large blue bodies near the equator of the spindle are the chromosomes.*

The basic framework of the spindle in (b) *an animal cell and* (c) *a plant cell. In the animal cell, a centriole pair is present at each pole; the polar fibers, which form the bulk of the spindle, are sharply focused on the centrioles, and additional fibers radiate outward from the centrioles, forming the aster. In plant cells, by contrast, centrioles are absent, the spindle is less sharply focused at the poles, and no aster is formed. Not shown in these diagrams are the replicated chromosomes and the spindle fibers attached to their kinetochores.*

The Phases of Mitosis

The process of mitosis is conventionally divided into four phases: prophase, metaphase, anaphase, and telophase. Of these, prophase is usually by far the longest. If a mitotic division takes 10 minutes (which is about the minimum time required), during about six of these minutes the cell will be in prophase. The schematic drawings that follow show mitosis as it takes place in an animal cell. (Similar schematic drawings of mitosis in a plant cell are shown in Figure 7–16 on page 154.)

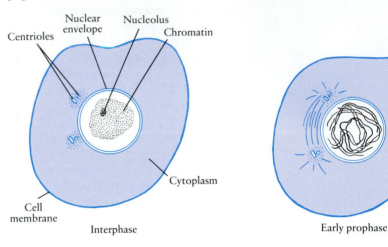

Interphase

Early prophase

During the interphase portions of the cell cycle, little can be seen in the nucleus. By early **prophase**, however, the chromatin has condensed sufficiently that the individual chromosomes become visible under the light microscope. Each chromosome consists of two duplicate chromatids pressed closely together longitudinally and connected at the centromere. In the cells of most organisms (fungi, flowering plants, and nematodes are the principal exceptions), two centriole pairs can be seen at one side of the nucleus, outside the nuclear envelope. As we noted previously, replication of the original centriole pair began during the G_1 phase of the cell cycle and was completed late in the G_2 phase. The cell becomes more spheroid and the cytoplasm more viscous at this stage, as the microtubules of the cytoskeleton are disassembled in preparation for the formation of the spindle.

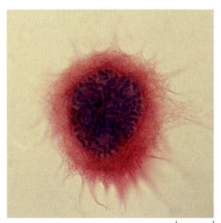

25 µm

7–10 *Early prophase in a cell from a seed of the African globe lily,* Haemanthus katherine. *The microtubules of the cytoskeleton are still in a meshwork surrounding the nucleus and have not yet become reorganized into a spindle. At this stage, the chromosomes are condensing. Their threadlike appearance when they first become visible under the microscope is the source of the name mitosis. Mitos is the Greek word for "thread."*

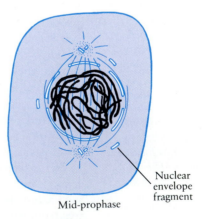

Mid-prophase

During prophase, the centriole pairs move apart. Between the centriole pairs, forming as they separate (or, more likely, separating them as they form), are the microtubules that become the polar fibers of the spindle. In those cells that have centrioles, the microtubules that form the aster radiate outward from the centrioles. By this time the nucleoli usually have disappeared from view. As the chromosomes continue to condense, the nuclear envelope breaks down, dispersing into membranous fragments similar to fragments of endoplasmic reticulum.

By the end of prophase, the chromosomes are fully condensed and are no longer separated from the cytoplasm. The centriole pairs have reached the poles of the cell, and the members of each pair are of equal size. The polar fibers of the spindle are fully formed, and the kinetochore fibers, attached to the kinetochores of the chromosomes, have also formed.

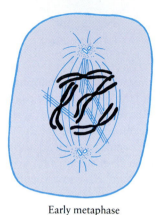
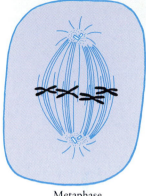

Early metaphase   Metaphase

During early **metaphase**, the chromatid pairs move back and forth within the spindle, apparently maneuvered by the spindle fibers, as if they were being tugged first toward one pole and then the other. Finally the chromatid pairs become arranged precisely at the midplane (equator) of the cell. This marks the end of metaphase.

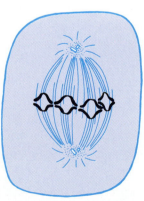
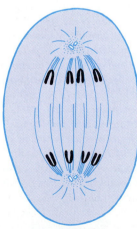

Early anaphase   Late anaphase

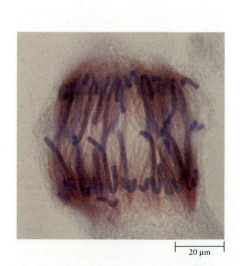

├─────────┤ 20 µm

7–11 *Middle anaphase in a cell from the seed of an African globe lily. The chromosomes have moved halfway toward the poles.*

At the beginning of **anaphase**, the most rapid stage of mitosis, the centromeres separate simultaneously in all the chromatid pairs. The chromatids of each pair then move apart, each chromatid becoming a separate chromosome, each apparently drawn toward the opposite pole by the kinetochore fibers. The centromeres move first, while the arms of the chromosomes seem to drag behind. In most cells, the spindle as a whole also elongates, with the poles appearing to be pushed farther apart. As anaphase continues, the two identical sets of newly separated chromosomes move rapidly toward the opposite poles of the spindle.

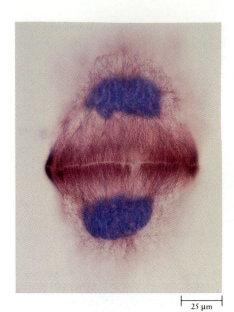

7–12 *Middle telophase in a cell from a seed of an African globe lily. Note the unstained region at the equator of the cell. It is here that the cell plate will form during cytokinesis.*

7–13 *Mitosis in embryonic cells of a whitefish. (a) Prophase. The chromosomes have become visible, the nuclear envelope is breaking down, and the spindle apparatus is forming. Note the prominent asters. (b) Metaphase. The chromatid pairs are lined up at the equator of the cell. (c) Anaphase. The two sets of chromosomes are moving apart. (d) Telophase. The chromosomes are completely separated, the spindle apparatus is disappearing, and a new cell membrane is forming that will complete the separation of the two daughter cells.*

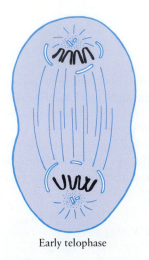

Early telophase

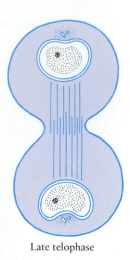

Late telophase

By the beginning of **telophase,** the chromosomes have reached the opposite poles and the spindle begins to disperse into tubulin dimers. During late telophase, nuclear envelopes re-form around the two sets of chromosomes, which once more become diffuse. In each nucleus, the nucleoli reappear. Often, a new centriole begins to form adjacent to each of the previous ones. As we saw earlier, replication of the centrioles continues during the subsequent cell cycle, so that each daughter cell has two centriole pairs by prophase of the next mitotic division.

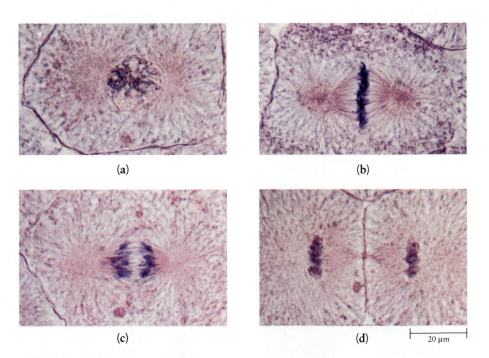

(a)

(b)

(c)

(d)

20 μm

The Mechanism of Chromosome Movement

Although there is little doubt that the movement of the chromosomes toward the poles and the separation of the poles from each other is the result of interactions among the kinetochore fibers and the polar fibers of the spindle, the exact mechanism or mechanisms remain unknown. One possibility is suggested by the fact that the kinetochore fibers lengthen during prophase and then shorten during anaphase—without getting either thinner or thicker. This indicates that the fibers probably do not contract but that material is added to or removed from them during different phases of mitosis, leading to the pushing apart or pulling together of the structures to which the fibers are attached. Another mechanism is suggested by the fact that the microtubules of the spindle fibers have little protein "arms," analogous to those seen in the microtubules of cilia and flagella (see Figure 5–31a, page 123). These "arms," like those of cilia and flagella, contain an enzyme involved in energy-releasing chemical reactions. It is thus hypothesized that the spindle microtubules may also "walk" along each other, tractor-fashion, powered by the energy released in the enzymatic reactions. Although microtubules constitute the bulk of the spindle fibers, there are other proteins associated with the spindle; it remains possible that these proteins, about which very little is presently known, also play key roles in the movements of mitosis.

CYTOKINESIS

Cytokinesis, the division of the cytoplasm, usually but not always accompanies mitosis, the division of the nucleus. The visible process of cytokinesis generally begins during telophase of mitosis, and it usually divides the cell into two nearly equal parts. Although the spindle does not seem to be directly involved in the division of the cytoplasm, the cleavage always occurs at the midline of the spindle, in the region where the polar fibers overlap. There is evidence that these spindle microtubules play a role in positioning the other structures that are responsible for the actual division of the cytoplasm. If, for example, the spindle is pushed to the side of the cell shortly after its formation is complete—so that its equator extends only halfway across the cell—only the cytoplasm of that half of the cell is ultimately divided in cytokinesis. The result is a two-lobed cell, with a daughter nucleus in each lobe.

Cytokinesis differs significantly in plant and animal cells. In animal cells, during early telophase, the cell membrane begins to constrict along the circumference of the cell in the plane of the equator of the spindle. At first, a furrow appears on the surface, and this gradually deepens into a groove (Figure 7–14). Eventually, the connection between the daughter cells dwindles to a slender thread, which soon parts. Actin filaments (page 110), seen in large numbers near the furrows, are thought to play a role in the constriction. They are believed to act as a sort of "purse string," gathering in the membrane of the parent cell at its midline, thus pinching apart the two daughter cells.

In plant cells, the cytoplasm is divided at the midline by a series of polysaccharide-containing vesicles produced from the Golgi complexes (Figure 7–15). The

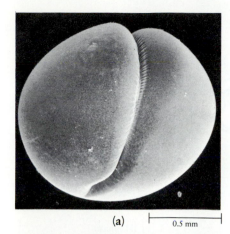

(a)

0.5 mm

(b)

50 µm

7–14 *Cytokinesis in an animal cell, a frog egg. (a) The egg is dividing in two; (b) close-up showing the constriction furrows.*

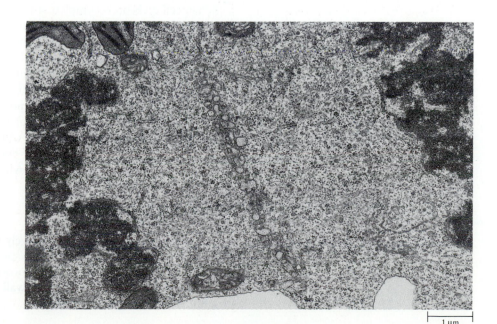

1 µm

7–15 *In plants, the separation of the two cells is effected by the formation of a structure known as the cell plate. Vesicles appear across the equatorial plane of the cell and gradually fuse, forming a flat, membrane-bound space, the cell plate, which extends outward until it reaches the wall of the dividing cell. The large, dark forms on either side of the micrograph are the chromosomes.*

7-16 *Mitosis in a plant cell with four chromosomes. Note that a spindle forms, although no centrioles are present and no aster is visible.*

The plane of cell division is established late in the G$_2$ phase of the cell cycle when the microtubules of the cytoskeleton are reorganized into a circular structure, known as the preprophase band, just inside the cell wall. Although this band disappears early in prophase, it determines the future location of the equator and the cell plate. The microtubules of the band are later reassembled into the spindle in a clear zone that develops around the nucleus in the course of prophase.

In cytokinesis, which begins during telophase, the cell plate gradually extends outward until it meets the exact region of the cell wall occupied earlier by the preprophase band. The vesicles that give rise to the cell plate are apparently guided into position by the spindle fibers remaining between the daughter nuclei.

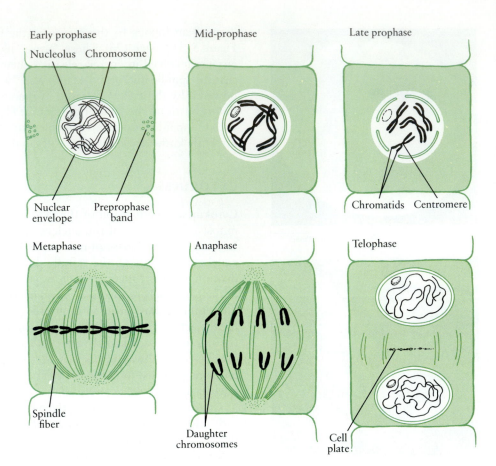

Early prophase — Nucleolus — Chromosome — Nuclear envelope — Preprophase band

Mid-prophase

Late prophase — Chromatids — Centromere

Metaphase — Spindle fiber

Anaphase — Daughter chromosomes

Telophase — Cell plate

vesicles eventually fuse to form a flat, membrane-bound space, the **cell plate.** As more vesicles fuse, the edges of the growing plate fuse with the membrane of the cell. In this way, a layer of polysaccharides is established between the two daughter cells, completing their separation. This layer becomes impregnated with pectins, ultimately forming the middle lamella (see page 107). Each new cell then constructs its own cell wall, laying down cellulose and other polysaccharides against the outer surface of its cell membrane.

When cell division is complete, two daughter cells are produced, smaller than the parent cell but otherwise indistinguishable from it and from each other.

Although we have focused in this chapter on the processes directly involved in cell division, it is important to realize that the other activities characteristic of the living cell are also occurring throughout the cell cycle. The cell is synthesizing the macromolecules necessary to maintain its structure, it is degrading other molecules, it is regulating both the internal movement of substances and the movement of substances between the cell interior and the exterior environment, and it is responding to a variety of stimuli. All of these activities—as well as cell division itself—require a steady expenditure of energy by the cell. In the next section, we shall see how cells provide themselves with this essential energy.

SUMMARY

Cell division in prokaryotes is a relatively simple process, in which two daughter chromosomes are attached to different spots on the interior of the cell membrane. As the membrane elongates, the chromosomes are separated. The cell membrane then pinches inward, and a new cell wall forms, completing the division of the daughter cells.

Cell division is a more complex process in eukaryotes, which contain a vast amount of genetic material, organized into a number of different chromosomes. Dividing cells pass through a regular sequence of cell growth and cell division known as the cell cycle. The cycle consists of a G_1 phase, during which the cytoplasmic molecules and structures increase; an S phase, during which the chromosomes are replicated; a G_2 phase, during which condensation of the chromosomes and assembly of the special structures required for mitosis and cytokinesis begin; mitosis, during which the replicated chromosomes are apportioned between two daughter nuclei; and cytokinesis, during which the cytoplasm is divided, separating the parent cell into two daughter cells. The first three phases of the cell cycle are known, collectively, as interphase. Regulation of the cell cycle occurs late in the G_1 phase and may involve a number of interacting factors.

When the cell is in the interphase portions of the cycle, the chromosomes are visible only as thin strands of threadlike material (chromatin) within the nucleus. As mitosis begins, the condensing chromosomes, previously replicated during the S phase, become visible under the light microscope. At these early stages of mitosis, the chromosomes consist of pairs of identical replicas, called chromatids, held together at the centromere. Simultaneously, the spindle is forming. In animal cells, it forms between the centrioles as they separate. In both animal and plant cells, the framework of the spindle is formed by fibers that extend from the poles to the equator of the cell; other fibers are attached to the chromatids at their kinetochores, protein-containing structures associated with the centromeres. Prophase ends with the breakdown of the nuclear envelope and disappearance of the nucleoli. During metaphase, the chromatid pairs, maneuvered by the spindle fibers, move toward the center of the cell. At the end of metaphase, they are arranged on the equatorial plane. During anaphase, the sister chromatids separate, and each chromatid, now an independent chromosome, moves to an opposite pole. During telophase, a nuclear envelope forms around each group of chromosomes. The spindle begins to break down, the chromosomes uncoil and once more become extended and diffuse, and the nucleoli reappear.

Cytokinesis in animal cells results from constrictions in the cell membrane between the two nuclei. In plant cells, the cytoplasm is divided by the coalescing of vesicles to form the cell plate, within which the cell wall is subsequently laid down. In both cases, the result is the production of two new, separate cells. As a result of mitosis, each has received an exact copy of the enormous skein of genetic material of the parent cell, and, as a result of cytokinesis, approximately half of the cytoplasm and organelles.

QUESTIONS

1. Distinguish among the following terms: cell cycle/cell division; mitosis/cytokinesis; chromatid/chromosome; centriole/centromere/kinetochore.

2. Describe the activities occurring during each phase of the cell cycle and the role of each phase in the overall process of cell division.

3. What is a chromosome? How is it related to chromatin?

4. Why do we often refer to chromatids as sister chromatids? When are sister chromatids formed? How? When do they first become visible under the microscope?

5. Describe the structure of the spindle in a typical animal cell. What are thought to be the functions of each of the different groups of spindle fibers?

6. In what ways does cell division in plant cells differ from that in animal cells?

7. What is the function of cell division in the life of an organism? Suppose you, as an organism, were made up of a large single cell rather than trillions of small ones. How would you differ from your present self?

SUGGESTIONS FOR FURTHER READING

Books

ALBERTS, BRUCE, DENNIS BRAY, JULIAN LEWIS, MARTIN RAFF, KEITH ROBERTS, and JAMES D. WATSON: *Molecular Biology of the Cell*, 2d ed., Garland Publishing, Inc., New York, 1989.

> *Progressing from the molecules of which cells are composed, through an examination of cellular structure and function, to the interactions of cells within tissues, this outstanding text describes not only our most current knowledge and how it was attained but also the many areas still to be explored. It is clearly written and filled with wonderful micrographs and explanatory diagrams. Highly recommended.*

DARNELL, JAMES, HARVEY F. LODISH, and DAVID BALTIMORE: *Molecular Cell Biology*, W. H. Freeman and Company, New York, 1986.

> *A comprehensive treatment of modern cell biology, richly illustrated with diagrams and micrographs. This text places particular emphasis on molecular genetics, membrane structure and function, cytoplasmic organelles, and the cytoskeleton, as well as on the techniques used in the contemporary study of cell biology.*

DE DUVE, CHRISTIAN: *A Guided Tour of the Living Cell*, Scientific American Library, W. H. Freeman and Company, New York, 1984.*

> *In this beautifully illustrated two-volume set, de Duve, one of the pioneers of modern cell biology, takes the reader—imagined to be a "cytonaut," a bacterium-sized tourist—on a journey through the eukaryotic cell. The first portion of the journey explores the cellular membranes; the second, the cytoplasm and its organelles; and the third, the nucleus. At the conclusion of the journey, de Duve considers such key questions of modern biology as the origin of life and the mechanisms of evolution.*

LEDBETTER, M. C., and KEITH R. PORTER: *Introduction to the Fine Structures of Plant Cells*, Springer-Verlag, New York, 1970.

> *An excellent atlas of electron micrographs of plant cells, with detailed explanations.*

LEHNINGER, ALBERT L.: *Principles of Biochemistry*, Worth Publishers, Inc., New York, 1982.

> *This introductory text is outstanding both for its clarity and for its consistent focus on the living cell.*

OPARIN, A. I.: *The Origin of Life*, Dover Publications, Inc., New York, 1938.*

> *Oparin, a Russian biochemist, was the first to argue that life arose spontaneously in the oceans of the primitive earth. Although his concepts have been somewhat modified in detail, they form the basis for the present scientific theories on the origin of living things.*

PORTER, KEITH R., and MARY A. BONNEVILLE: *An Introduction to the Fine Structures of Cells and Tissues*, 4th ed., Lea & Febiger, Philadelphia, 1973.

> *An atlas of electron micrographs of animal cells; detailed commentaries accompany each. These are magnificent micrographs,* and the commentaries describe not only what the pictures show but also the experimental foundations of our knowledge of cell ultrastructures.

PRESCOTT, DAVID M.: *Cells: Principles of Molecular Structure and Function*, Jones and Bartlett Publishers, Boston, 1988.

> *An up-to-date, yet concise, textbook of cell biology, written for a first course at the undergraduate level. It is a wonderful introduction for any reader wishing to gain an overview of the exciting developments in contemporary cell biology.*

SCIENTIFIC AMERICAN: *The Molecules of Life*, W. H. Freeman and Company, New York, 1986.*

> *This reprint of the October 1985 issue of* Scientific American *includes 11 articles on molecules that play key roles in the living cell. The articles on proteins, the molecules of the cell membrane, and the molecules of the cytoskeleton are of particular interest at this point in your study of biology. You will find the other articles in this collection useful at later points in the course.*

SCIENTIFIC AMERICAN: *Molecules to Living Cells*, W. H. Freeman and Company, New York, 1980.*

> *A collection of outstanding articles from* Scientific American. *Chapters 1, 2, 4, 10, and 11 cover the origin of life, evolution of the earliest cells, the cell cycle, and cell membranes and their assembly. Highly recommended.*

SILK, JOSEPH: *The Big Bang: The Creation and Evolution of the Universe*, 2d ed., W. H. Freeman and Company, New York, 1988.*

> *A discussion of modern evidence concerning the formation of the solar system and the planet earth. An excellent, well-written introduction to cosmology.*

STRYER, LUBERT: *Biochemistry*, 3d ed., W. H. Freeman and Company, New York, 1988.

> *An introductory text, with many examples of medical applications of biochemistry. Handsomely illustrated.*

THOMAS, LEWIS: *The Lives of a Cell: Notes of a Biology Watcher*, Viking Press, Inc., New York, 1974.*

THOMAS, LEWIS: *The Medusa and the Snail: More Notes of a Biology Watcher*, Viking Press, Inc., New York, 1979.*

> *Thomas, a physician and medical researcher, reveals the extent to which science can tune our intellectual antennae, broaden our perceptions, and extend our appreciation of ourselves and of the world around us. Anyone who wants to refute the contention that science destroys human values need look no further than these short, sensitive essays.*

WEINBERG, STEVEN: *The Discovery of Subatomic Particles*, Scientific American Library, W. H. Freeman and Company, New York, 1984.

> *In this handsome book, an introduction to the structure of the atom is combined with a lively history of twentieth-century physics. This revolution in physics profoundly influenced modern biology.*

---

* Available in paperback.

WEINBERG, STEVEN: *The First Three Minutes: A Modern View of the Origin of the Universe*, Basic Books, Inc., New York, 1977.*

A wonderful story, written for the intelligent nonscientist (characterized by the author as a smart old attorney who expects to hear some convincing arguments before he makes up his mind).

Articles

ALBERSHEIM, PETER: "The Walls of Growing Plant Cells," *Scientific American*, April 1975, pages 81–95.

ALLEN, ROBERT DAY: "The Microtubule as an Intracellular Engine," *Scientific American*, February 1987, pages 42–49.

BONNER, JOHN TYLER: "Chemical Signals of Social Amoebae," *Scientific American*, April 1983, pages 114–120.

BRETSCHER, MARK S.: "Endocytosis: Relation to Capping and Cell Locomotion," *Science*, vol. 224, pages 681–686, 1984.

BRETSCHER, MARK S.: "How Animal Cells Move," *Scientific American*, December 1987, pages 72–90.

BROWN, MICHAEL S., and JOSEPH L. GOLDSTEIN: "A Receptor-Mediated Pathway for Cholesterol Homeostasis," *Science*, vol. 232, pages 34–47, 1986.

DAUTRY-VARSAT, ALICE, and HARVEY F. LODISH: "How Receptors Bring Proteins and Particles into Cells," *Scientific American*, May 1984, pages 52–58.

DE DUVE, CHRISTIAN: "Microbodies in the Living Cell," *Scientific American*, May 1983, pages 74–84.

DUSTIN, PIERRE: "Microtubules," *Scientific American*, August 1980, pages 67–76.

ELGSAETER, ARNLJOT, BJORN T. STOKKE, ARNE MIKKELSEN, and DANIEL BRANTON: "The Molecular Basis of Erythrocyte Shape," *Science*, vol. 234, pages 1217–1223, 1986.

FAUL, HENRY: "A History of Geologic Time," *American Scientist*, vol. 66, pages 159–165, 1978.

GRIFFITHS, GARETH, and KAI SIMONS: "The *trans* Golgi Network: Sorting at the Exit Site of the Golgi Complex," *Science*, vol. 234, pages 438–443, 1986.

KASTING, JAMES F., OWEN B. TOON, and JAMES B. POLLACK: "How Climate Evolved on the Terrestrial Planets," *Scientific American*, February 1988, pages 90–97.

KELLY, REGIS B.: "Pathways of Protein Secretion in Eukaryotes," *Science*, vol. 230, pages 25–32, 1985.

KOSHLAND, DOUGLAS E., T. J. MITCHISON, and MARC W. KIRSCHNER: "Polewards Chromosome Movement Driven by Microtubule Depolymerization *in vitro*," *Nature*, vol. 331, pages 499–504, 1988.

LAZARIDES, ELIAS, and JEAN PAUL REVEL: "The Molecular Basis of Cell Movement," *Scientific American*, May 1979, pages 100–113.

MARX, JEAN L.: "A Potpourri of Membrane Receptors," *Science*, vol. 230, pages 649–651, 1985.

MERTZ, WALTER: "The Essential Trace Elements," *Science*, vol. 213, pages 1332–1338, 1981.

MILLER, JULIE ANN: "Cell Communication Equipment: Do-It-Yourself Kit," *Science News*, April 14, 1984, pages 236–237.

MOHNEN, VOLKER A.: "The Challenge of Acid Rain," *Scientific American*, August 1988, pages 30–38.

PETERSON, IVARS: "A Biological Antifreeze," *Science News*, November 22, 1986, pages 330–332.

RACKER, EFRAIM: "Structure, Function, and Assembly of Membrane Proteins," *Science*, vol. 235, pages 959–961, 1987.

ROTHMAN, JAMES E.: "The Compartmental Organization of the Golgi Apparatus," *Scientific American*, September 1985, pages 74–89.

SATIR, BIRGIT: "The Final Steps in Secretion," *Scientific American*, October 1975, pages 28–37.

SATIR, PETER: "How Cilia Move," *Scientific American*, October 1974, pages 44–52.

SCHINDLER, D. W.: "Effects of Acid Rain on Freshwater Ecosystems," *Science*, vol. 239, pages 149–157, 1988.

SHARON, NATHAN: "Carbohydrates," *Scientific American*, November 1980, pages 90–116.

SLOBODA, ROGER D.: "The Role of Microtubules in Cell Structure and Cell Division," *American Scientist*, vol. 68, pages 290–298, 1980.

SPORN, MICHAEL B., and ANITA B. ROBERTS: "Peptide Growth Factors Are Multifunctional," *Nature*, vol. 332, pages 217–219, 1988.

STAEHELIN, L. ANDREW, and BARBARA E. HULL: "Junctions between Living Cells," *Scientific American*, May 1978, pages 141–152.

STILLINGER, FRANK H.: "Water Revisited," *Science*, vol. 209, pages 451–457, 1980.

UNWIN, NIGEL, and RICHARD HENDERSON: "The Structure of Proteins in Biological Membranes," *Scientific American*, February 1984, pages 78–94.

WEISSKOPF, VICTOR F.: "The Origin of the Universe," *American Scientist*, vol. 71, pages 473–480, 1983.

WICKNER, WILLIAM T., and HARVEY F. LODISH: "Multiple Mechanisms of Protein Insertion Into and Across Membranes," *Science*, vol. 230, pages 400–407, 1985.

---

* Available in paperback.

SECTION 2

Energetics

The energy of the summer sun is stored in these wheat plants, ready for harvest. Of the radiant energy of the sun striking the wheat field, less than 10 percent is converted into stored chemical energy.

The Flow of Energy

Life here on earth depends on the flow of energy from the thermonuclear reactions taking place at the heart of the sun. The amount of energy delivered to the earth by the sun is about 13×10^{23} (the number 13 followed by 23 zeros) calories per year. It is a difficult quantity to imagine. For example, the amount of solar energy striking the earth every day is about 1.5 billion times greater than the amount of electricity generated in the United States each year.

About one-third of this solar energy is reflected back into space as light. Much of the remaining two-thirds is absorbed by the earth and converted to heat. Some of this absorbed heat energy serves to evaporate the waters of the oceans, producing the clouds that, in turn, produce rain and snow. Solar energy, in combination with other factors, is also responsible for the movements of air and of water that help set patterns of climate over the surface of the earth.

A small fraction—less than 1 percent—of the solar energy reaching the earth becomes, through a series of operations performed by the cells of plants and other photosynthetic organisms, the energy that drives all the processes of life. Living systems change energy from one form to another, transforming the radiant energy from the sun into the chemical and mechanical energy used by everything that is alive.

This flow of energy is the essence of life. As we noted in the Introduction, evolution may be viewed as a competition among organisms for the most efficient use of energy resources. A cell can be best understood as a complex of systems for transforming energy. At the other end of the biological scale, the structure of an ecosystem or of the biosphere itself is determined by the energy exchanges occurring among the groups of organisms within it.

In this chapter, we shall look first at the general principles governing all energy transformations and then at the characteristic ways in which cells regulate the energy transformations that take place within living systems. In the chapters that follow, the principal and complementary processes of energy flow through the biosphere will be examined—glycolysis and respiration in Chapter 9 and photosynthesis in Chapter 10.

8–1 *Harvest mice, feeding on wheat kernels. Of the chemical energy stored in the wheat kernels, less than 10 percent will be converted to chemical energy stored in the body tissues of the mice. Life runs downhill and is sustained only by a constant flow of radiant energy from the sun.*

8–2 *Another downhill step. Less than 10 percent of the chemical energy stored in the body tissues of the mouse will be converted to chemical energy stored in the body tissues of its predator, the red rat snake.*

THE LAWS OF THERMODYNAMICS

Energy is such a common term today that it is surprising to learn that the word was coined less than 200 years ago, at the time of the development of the steam engine. It was only then that scientists and engineers began to understand that heat, motion, light, electricity, and the forces holding the atoms together in molecules are all different forms of the same capacity to cause change, or, as it is often expressed, to do work. This new understanding led to the study of **thermodynamics**—the science of energy transformations—and to the formulation of its laws.

The First Law

The **first law of thermodynamics** states, quite simply: *Energy can be changed from one form to another, but it cannot be created or destroyed.* The total energy of any system plus its surroundings thus remains constant, despite any changes in form.

Electricity is a form of energy, as is light. Electrical energy can be changed to light energy (for example, by letting an electric current flow through the tungsten wire in a light bulb). Conversely, light energy can be changed to electrical energy, a transformation that is the essential first step of photosynthesis, as we shall see in Chapter 10.

Energy can be stored in various forms and then changed into other forms. In automobile engines, for example, the energy stored in the chemical bonds of gasoline is converted to heat (kinetic energy), which is then partially converted to mechanical movements of the engine parts. Some of the energy is converted back to heat by the friction of the moving engine parts, and some of it leaves the engine in the exhaust products. Similarly, when organisms oxidize carbohydrates, they convert the energy stored in chemical bonds to other forms. On a summer evening, for example, a firefly converts chemical energy to mechanical energy, to heat, to flashes of light, and to electrical impulses that travel along the nerves of its body. Birds and mammals convert chemical energy into the heat necessary to maintain their body temperature, as well as into mechanical energy, electrical energy, and other forms of chemical energy. According to the first law of thermodynamics, in these energy conversions, and in all others, energy is neither created nor destroyed.

In all energy conversions, however, some useful energy is converted to heat and dissipates. In an automobile engine, for example, the heat produced by friction and lost in the exhaust, unlike the heat confined in the engine itself, cannot produce work—that is, it cannot drive the pistons and turn the gears—because it is dissipated into the surroundings. But it is nevertheless part of the total equation. In a gasoline engine, about 75 percent of the energy originally present in the fuel is transferred to the surroundings in the form of heat—that is, it is converted to increased motion of atoms and molecules in the air. Similarly, the heat produced by the metabolic processes of animals is dissipated into the surrounding air or water.

It was in the course of studies of engine efficiency that the notion of potential energy was first developed. A barrel of gasoline or a ton of coal could be assigned a certain amount of potential energy, measured in terms of the amount of heat it would liberate when burned. The efficiency of the conversion of the potential energy to "useful" energy depended on the design of the system.

Although these concepts were formulated in terms of engines running on heat energy, they apply to other systems as well. As we saw in Chapter 1, a boulder pushed to the top of a hill gains energy—potential energy. Given a little push, it rolls down the hill again, converting that potential energy to motion and the heat produced by friction. Water, as we saw in Chapter 6, may also possess potential energy. As it moves by bulk flow from the top of a waterfall or over a dam, it can turn waterwheels that turn gears and, for example, grind corn. Thus the potential

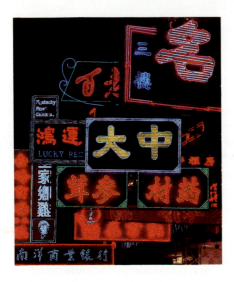

8–3 *Electrical energy can be converted to light energy, as in these Hong Kong signs, for example. The energy emitted as an electron falls from one energy level to another is a discrete amount, characteristic for each atom. When electricity is passed through a tube of gas, electrons in the atoms of the gas are boosted to higher energy levels. As they fall back, light energy is emitted, producing, for example, the red glow characteristic of neon and the yellow glow characteristic of sodium vapor.*

$E = mc^2$

Protons and neutrons, as we noted in Chapter 1, are arbitrarily assigned an atomic weight of 1. One would therefore expect that an element with, for example, twice as many protons and neutrons as another element would weigh twice as much. This assumption is true—almost. If the weights of nuclei are measured with great accuracy, as they can be by instruments developed by modern physics, small nuclei always have proportionately slightly greater weights than larger nuclei. For example, the most common isotope of carbon, as you know, has a combined total of 12 protons and neutrons, and carbon, by convention, is assigned an atomic weight of 12. The hydrogen atom, however, has an atomic weight not exactly of 1, as would be expected, but of 1.008. Helium has two protons and two neutrons. It does not have a weight of exactly 4, however, or of 4.032 (four times the weight of hydrogen); it weighs, in relation to carbon, 4.0026. Similarly, oxygen, with a combined total of 16 protons and neutrons, has an atomic weight—in relation to that of carbon—of 15.995. In short, when protons and neutrons are assembled into an atomic nucleus, there are slight changes in weight, which reflect changes in mass.

One of the oldest and most fundamental concepts of chemistry is the law of conservation of mass—that mass is never created or destroyed. We assume the law of conservation of mass every time we write a chemical equation. Yet under conditions of extremely high temperature, atomic nuclei fuse to make new elements and there is a measurable decrease in mass. What happens to the mass "lost" in the course of this fusion? This is the question answered by Einstein's fateful equation $E = mc^2$. E stands for energy, m stands for mass, and c is a constant equal to the speed of light. Einstein's equation means simply that under certain extreme and unusual conditions, mass is turned into energy.

The sun consists largely of hydrogen nuclei. At the extremely high temperatures at the core of the sun, hydrogen nuclei strike each other with enough velocity to fuse. In a series of steps, four hydrogen nuclei fuse to form one helium nucleus. In the course of these steps, energy is released, enough to keep the fusion reaction going and to emit tremendous amounts of radiant energy into space. Life on this planet depends on energy emitted by the sun in the course of this fusion reaction. This same reaction provides, of course—as Einstein foresaw—the energy of the hydrogen bomb.

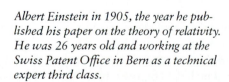

Albert Einstein in 1905, the year he published his paper on the theory of relativity. He was 26 years old and working at the Swiss Patent Office in Bern as a technical expert third class.

energy of water, in this system, is converted to the mechanical energy of the wheels and gears and to heat, produced by the movement of the water itself and also by the turning wheels and gears. Molecules also contain potential energy, stored in the chemical bonds between their constituent atoms. When these bonds are broken in chemical reactions, the energy they contain can be used to form other chemical bonds or can be released as heat.

The first law of thermodynamics states that in energy exchanges and conversions, wherever they take place and whatever they involve, the total energy of the system and its surroundings after the conversion is equal to the total energy before the conversion. In the case of chemical reactions, this means that the energy of the products of the reaction plus the energy released in the reaction is equal to the initial energy of the reactants.

The Second Law

The energy that is dissipated as heat as the result of an energy conversion has not been destroyed—it is still present in the random motion of atoms and molecules—but it has been "lost" for all practical purposes. It is no longer available to do useful work. This brings us to the **second law of thermodynamics,** which is the more interesting one, biologically speaking. It predicts the direction of all events involving energy exchanges; thus it has been called "time's arrow." The second law states that *in all energy exchanges and conversions, if no energy leaves or enters the system under study, the potential energy of the final state will always be less than the potential energy of the initial state.* The second law is entirely in keeping with everyday experience. A boulder will roll downhill but never uphill. Heat will flow from a hot object to a cold one and never the other way. A ball that is dropped will bounce—but not back to the height from which it was dropped.

A process in which the potential energy of the final state is less than that of the initial state is one that releases energy (otherwise it would be in violation of the first law). An energy-releasing process is called an **exergonic** ("energy-out") reaction. As the second law predicts, only exergonic reactions can take place spontaneously—that is, without an input of energy from outside the system. (Spontaneously, though the word has an explosive sound to it, says nothing about the rate of the reaction, just whether or not it can take place at all.) By contrast, a process in which the potential energy of the final state is greater than that of the initial state is one that requires energy. Such energy-requiring processes are known as **endergonic** ("energy-in") reactions, and in order for them to proceed, an input of energy is required that is greater than the difference in energy between the products and the reactants.

One important factor in determining whether or not a reaction is exergonic is already familiar to us: ΔH, the change in heat content of a system. As we noted in Chapter 3, the energy change that takes place when glucose, for instance, is oxidized can be measured in a calorimeter and expressed in terms of ΔH. The oxidation of a mole of glucose yields 673 kilocalories. Or,

$$C_6H_{12}O_6 + 6O_2 \longrightarrow 6CO_2 + 6H_2O$$
$$\Delta H = -673 \text{ kcal/mole}$$

Generally speaking, an exergonic chemical reaction is also an exothermic reaction—that is, it gives off heat and thus has a negative ΔH. However, there are exceptions. One of the most dramatic is found with a substance known as dinitrogen pentoxide, which decomposes spontaneously and with explosive force to nitrogen dioxide and oxygen, and in so doing, absorbs heat:

$$2N_2O_5 \longrightarrow 4NO_2 + O_2$$
$$\Delta H = +26.18 \text{ kcal/mole}$$

In short, another factor besides the gain or loss of heat affects the change in potential energy and thus the direction of the process. This factor is given the formal name of **entropy,** and it is a measurement of the disorder or randomness of a system.

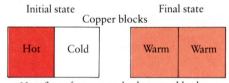

Initial state   Final state
Copper blocks

Heat flows from warm body to cool body

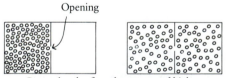

Opening

Gas molecules flow from zone of high pressure to zone of low pressure

Order becomes disorder

8–4 *Some illustrations of the second law of thermodynamics. In each case, a concentration of energy—in the hot copper block, in the gas molecules under pressure, and in the neatly organized books—is dissipated. In nature, processes tend toward randomness, or disorder. Only an input of energy can reverse this tendency and reconstruct the initial state from the final state. Ultimately, however, disorder will prevail, since the total amount of energy in the universe is finite.*

Before we examine more closely why the decomposition of dinitrogen pentoxide is exergonic despite its positive ΔH, let us return to the more familiar example of water. The change from ice to liquid water and the change from liquid water to water vapor are both endothermic processes—a considerable amount of heat is removed from the surrounding air as they take place. Yet, under the appropriate conditions, they proceed spontaneously. The key factor in all three of these examples is the increase in entropy. In the case of the dinitrogen pentoxide, a solid is being changed into two gases, and two molecules are being converted into five. In the case of ice and liquid water, a solid is being turned into a liquid, and some of the bonds that hold the water molecules together in a crystal (ice) are being broken. As the liquid water turns to vapor, the rest of the hydrogen bonds are ruptured as the individual water molecules dance off, one by one. In every case, the disorder of the system has increased.

The notion that there is more disorder associated with more numerous and smaller objects than with fewer, large ones is in keeping with our everyday experience. If I have 20 papers on my desk, the possibilities for disorder are greater than if I have 2 or even 10. If I cut each of the 20 in half, the entropy of the system—the capacity for randomness—increases. Also, the relationship between entropy and energy is a commonplace idea. If you were to find your room tidied up and your books in alphabetical order on the shelf, you would recognize that someone had been at work—that energy had been expended. For me to organize the papers on my desk similarly requires that I expend energy. Furthermore, it would be feasible to measure the energy expenditure in calories.

Now let us return to the question of the energy changes that determine the course of chemical reactions. Both the change in the heat content of the system (ΔH) and the change in entropy (which is symbolized as ΔS) contribute to the overall change in energy. This total change—the one that takes into account both heat and entropy—is called the **free energy change** and is symbolized as ΔG, after the American physicist Josiah Willard Gibbs (1839–1903), who was one of the first to put all of these ideas together.

The relationship between ΔG, ΔH, and entropy is given in the following equation:

$$\Delta G = \Delta H - T\Delta S$$

It states that the free energy change is equal to the change in heat (a negative value in exothermic reactions, remember) minus the change in entropy multiplied by the absolute temperature, T. In exergonic reactions, ΔH may be zero or may even be positive, but ΔG is always negative. As you can see in the equation, $T\Delta S$ is preceded by a minus sign. The greater the increase in entropy, the more negative ΔG will be; that is, the more exergonic the reaction will be.

With ΔG in mind, let us examine once more the combustion of glucose. The ΔH of that reaction is −673 kcal/mole. The free energy change, ΔG, is −686 kcal/mole. The increase in entropy has contributed 13 kcal/mole to the free energy change of the process. Thus both the change in heat and the change in entropy contribute to the lower energy state of the products of the reaction.

ΔG can also enable one to predict processes that occur when ΔH is zero or even positive, as with dinitrogen pentoxide. For instance, it confirms our earlier observations that heat will flow spontaneously from a hot object to a cold one, that dye molecules will diffuse spontaneously in a beaker of water, or that my desk will revert to disorder. In each of these processes, the final state has more entropy—and therefore less potential energy—than the initial state.

Previously we stated the second law in terms of the energy change between the initial and final states of a process. The second law can also be stated in another, simpler way: *All natural processes tend to proceed in such a direction that the disorder or randomness of the universe increases.*

(a)

(b)

(c)

(d)

8–5 Although energy is dissipated in every conversion from one form to another, considerable work can be accomplished in the process. For example, the transformation of stored chemical energy to mechanical energy can be used by an organism to move to a more satisfactory position. (a) A queen conch, a mollusk, resting on the surface of the sand. (b, c) Vigorous muscular movements of the conch's spade foot create a depression of just the right size to reorient the animal in relation to its world (d).

Living Systems and the Second Law

The universe, according to the present model, is a closed system—that is, neither matter nor energy either enters or leaves the system. The matter and energy present in the universe at the time of the primordial explosion (see page 23) are all the matter and energy it will ever have. Moreover, after each and every energy exchange and transformation, the universe as a whole has less potential energy and more entropy than it did before. In this view, of course, the universe is running down. The stars will flicker out, one by one; life—any form of life on any planet—will come to an end. Finally, even the motion of individual molecules will cease. However, even the most pessimistic among us do not believe this will occur for another 20 billion years or so.

In the meantime, life can exist *because* the universe is running down. Although the universe as a whole is a closed system, the earth is not. As we noted at the beginning of this chapter, it is receiving an energy input of 13×10^{23} calories per year from the sun. Photosynthetic organisms are specialists at capturing the light energy released by the sun as it slowly burns itself out. They use this energy to organize small, simple molecules (water and carbon dioxide) into larger, more complex molecules (sugars). In the process, the captured light energy is stored in the chemical bonds of sugars and other molecules. Living cells—including photosynthetic cells—can convert this stored energy into motion, electricity, light, and, by shifting the energy from one type of chemical bond to another, more convenient forms of chemical energy. At each transformation, energy is lost to the surroundings as heat. But before the energy captured from the sun is completely dissipated, organisms use it to create and maintain the complex organization of structures and activities that we know as life.

OXIDATION-REDUCTION

You will recall from Chapter 1 that electrons possess differing amounts of potential energy depending on their distance from the atomic nucleus and the attraction of the nucleus for electrons. An input of energy will boost an electron to a higher energy level, but without added energy an electron will remain at the lowest energy level available to it.

Chemical reactions are essentially energy transformations in which energy stored in chemical bonds is transferred to other, newly formed chemical bonds. In such transfers, electrons shift from one energy level to another. In many reactions, electrons pass from one atom or molecule to another. These reactions, which are of great importance in living systems, are known as oxidation-reduction (or redox) reactions. The *loss* of an electron is known as **oxidation,** and the atom or molecule that loses the electron is said to be oxidized. The reason electron loss is called oxidation is that oxygen, which attracts electrons very strongly, is most often the electron acceptor.

Reduction is, conversely, the *gain* of an electron. Oxidation and reduction always take place simultaneously because an electron that is lost by the oxidized atom is accepted by another atom, which is reduced in the process.

Redox reactions may involve only a solitary electron, as when sodium loses an electron and becomes oxidized to Na^+, and chlorine gains an electron and is reduced to Cl^-. Often, however, the electron travels with a proton, that is, as a hydrogen atom. In such cases, oxidation involves the removal of hydrogen atoms, and reduction the gain of hydrogen atoms. For example, when glucose is oxidized, hydrogen atoms are lost by the glucose molecule and gained by oxygen:

$$C_6H_{12}O_6 + 6O_2 \longrightarrow 6CO_2 + 6H_2O + Energy$$

The electrons are moving to a lower energy level, and energy is released.

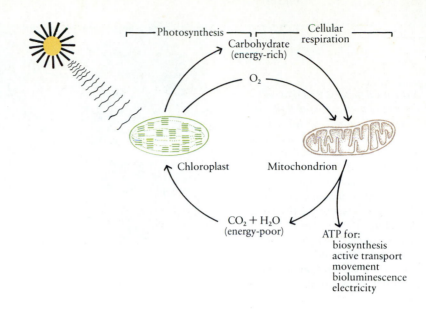

8–6 *The flow of biological energy. Chloroplasts, present in all photosynthetic eukaryotic cells, capture the radiant energy of sunlight and use it to convert water and carbon dioxide into carbohydrates, such as glucose, starch, and other foodstuff molecules. Oxygen is released as a product of the photosynthetic reactions.*

Mitochondria, present in all eukaryotic cells, carry out the final steps in the breakdown of these carbohydrates and capture their stored energy in ATP molecules. This process, cellular respiration, consumes oxygen and produces carbon dioxide and water, completing the cycling of the molecules.

With each transformation, some energy is dissipated to the environment in the form of heat. Thus the flow of biological energy is one-way and can continue only so long as there is an input of energy from the sun.

Conversely, in the process of photosynthesis, hydrogen atoms are transferred from water to carbon dioxide, thereby reducing the carbon dioxide to form glucose:

$$6CO_2 + 6H_2O + Energy \longrightarrow C_6H_{12}O_6 + 6O_2$$

In this case, the electrons are moving to a higher energy level, and an energy input is required to make the reaction occur.

In living systems, the energy-capturing reactions (photosynthesis) and energy-releasing reactions (glycolysis and respiration) are oxidation-reduction reactions. As we have seen, the complete oxidation of a mole of glucose releases 686 kilocalories of free energy (conversely, the reduction of carbon dioxide to form a mole of glucose stores 686 kilocalories of free energy in the chemical bonds of glucose). If this energy were to be released all at once, most of it would be dissipated as heat. Not only would it be of no use to the cell, but the resulting high temperature would be lethal. However, mechanisms have evolved in living systems that regulate these chemical reactions—and a multitude of others—in such a way that energy is stored in particular chemical bonds from which it can be released in small amounts as the cell needs it. These mechanisms generally involve sequences of reactions, some of which are oxidation-reduction reactions. Although each reaction in the sequence represents only a small change in the free energy, the overall free energy change for the sequence can be considerable.

METABOLISM

In any living system, energy exchanges occur through thousands of different chemical reactions, many of them taking place simultaneously. As we noted in Chapter 5, the sum of all these reactions is referred to as metabolism (from the Greek *metabole*, meaning "change"). If we were merely to list the individual chemical reactions, it would be difficult indeed to understand the flow of energy through a cell. Fortunately, there are some guiding principles that lead one through the maze of cell metabolism. First, virtually all the chemical reactions that take place in a cell involve enzymes—large protein molecules that play very specific roles. Second, biochemists are able to group these reactions in an ordered series of steps, commonly called a pathway; a pathway may have a dozen or more sequential reactions or steps. Each pathway serves a function in the overall life of the cell or organism. Furthermore, certain pathways have many steps in common —for instance, those that are concerned with the synthesis of the different amino acids or the various nitrogenous bases. Some pathways converge; for example, the pathway by which fats are broken down to yield energy leads to the pathway by which glucose is broken down to yield energy.

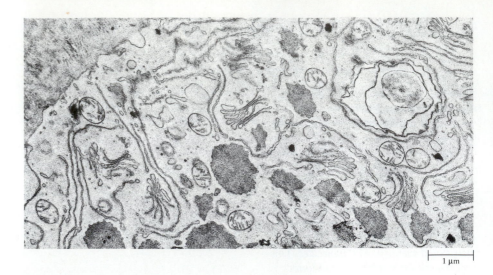

1 μm

8–7 *A chemical factory. Part of a cell from the root of a wheat plant. Cells such as this contain thousands of different organic molecules. These molecules form the many intricate structures of which the cell is composed and carry out the multitude of chemical reactions necessary for the life of the cell, its maintenance and growth, and its interactions with the cells around it. Many of the chemical activities of the cell are compartmentalized within the organelles, vesicles, and vacuoles, on the surfaces of the membranes, and in different regions of the cytoplasmic solution.*

Many types of living systems have pathways unique to them. Plant cells, for example, expend much of their energy building their cell walls, an activity not engaged in by animal cells. Red blood cells specialize in the synthesis of hemoglobin molecules, not made anywhere else in the animal body. It is not surprising that the distinctive differences in function among cells and organisms are correlated not only with their forms but also with their biochemistry. What *is* surprising, however, is that much of the metabolism of even the most diverse of organisms is exceedingly similar; the differences in many of the metabolic pathways of humans, oak trees, mushrooms, and jellyfish are very slight. Some pathways—for example, those of glycolysis and respiration—are virtually universal, found in almost all living systems.

The magnitude of the chemical work carried out by a cell—and its consequent energy expenditure—can be understood if one recognizes that, for the most part, the thousands of different molecules, large and small, found within a cell are synthesized there. The total of chemical reactions involved in synthesis is called **anabolism.** Cells also are constantly involved in the breakdown of larger molecules; these activities are known, collectively, as **catabolism.** Catabolism serves two purposes: (1) it releases the energy for anabolism and other work of the cell, and (2) it provides raw materials for anabolic processes.

Not only do living systems carry out this multitude of chemical activities, but they do so under what might seem, at first glance, extraordinarily difficult conditions. Most of their chemical reactions are carried out within individual living cells, and in cells, not two or three but thousands of different kinds of molecules are present. As we saw in Chapter 5, however, the cytoplasm of the living cell is highly structured, with the organelles, endoplasmic reticulum, vesicles, vacuoles, and the cytoskeleton itself effectively compartmentalizing the cell into different "work areas" (Figure 8–7). This segregates different reaction pathways from one another, enabling different chemical reactions to take place without mutual interference.

However, for any particular molecules to react with one another, it is not enough that they be in the same general region of the cell; they must be in extremely close proximity and, moreover, must collide with sufficient force to overcome the mutual repulsion of their electron clouds. The force required varies with the nature of the molecules; the more stable their initial state, the more forceful the collision must be. The force with which molecules collide depends on their kinetic energy, and the average kinetic energy of the molecules in a cell is quite moderate, as reflected in the moderate temperatures of living systems. In a group of molecules, it is likely that some proportion is moving with sufficient energy to cause a reaction to occur, but often this proportion is so small that the reaction, for all practical purposes, does not take place. How, then, can the complex chemical work of a cell be accomplished? The question can be answered in a single word: enzymes.

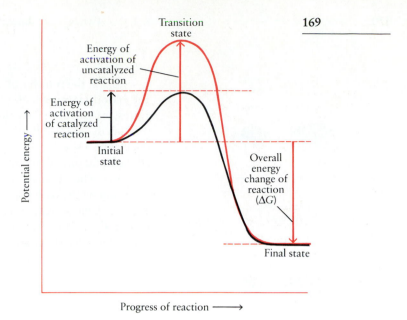

8-8 *In order to react, molecules must possess enough energy—the energy of activation—to collide with sufficient force to overcome their mutual repulsion and to weaken existing chemical bonds. An uncatalyzed reaction requires more activation energy than a catalyzed one, such as an enzymatic reaction. The lower activation energy in the presence of the catalyst is often within the range of energy possessed by the molecules, and so the reaction can occur at a rapid rate with little or no added energy. Note, however, that the overall energy change (ΔG) from the initial state to the final state is the same with and without the catalyst.*

ENZYMES

To proceed at a reasonable rate, most chemical reactions require an initial input of energy to get started. This is true even for exergonic reactions such as the oxidation of glucose or the burning of natural gas (methane). The added energy increases the kinetic energy of the molecules, enabling a greater number of them to collide with sufficient force not only to overcome their mutual repulsion but also to break existing chemical bonds within the molecules. The energy that must be possessed by the molecules in order to react is known as the **energy of activation.** Sometimes, as in the case of natural gas, a spark is all that is needed to supply enough energy. Once the reaction begins, it liberates energy that is transferred to the other methane molecules until all are moving rapidly enough to react almost simultaneously with explosive force.

In the laboratory, the energy of activation is usually supplied as heat. But, in a cell, many different reactions are going on at the same time, and heat would affect all of these reactions indiscriminately. Moreover, heat would break hydrogen bonds and would have other generally destructive effects on the cell. Cells get around this problem by the use of enzymes, globular proteins that are specialized to serve as catalysts. A **catalyst** is a substance that lowers the activation energy required for a reaction by forming a temporary association with the molecules that are reacting (Figure 8–8). This temporary association brings the reacting molecules close to one another and may also weaken the existing chemical bonds, making it easier for new ones to form. The lower activation energy in the presence of the catalyst is within the range of energy possessed by a greater proportion of the reacting molecules; as a result, the reaction goes more rapidly than it would in the absence of a catalyst. The catalyst itself is not permanently altered in the process, and so it can be used over and over again.

Because of enzymes, cells are able to carry out chemical reactions at great speed and at comparatively low temperatures. For instance, the combination of carbon dioxide with water,

$$CO_2 + H_2O \rightleftharpoons \underset{\substack{\text{Carbonic} \\ \text{acid}}}{H_2CO_3}$$

can take place spontaneously, as it does in the oceans. In the human body, however, this reaction is catalyzed by an enzyme, carbonic anhydrase. This is one of the fastest enzymes known, with each enzyme molecule catalyzing the production of 6×10^5 (600,000) molecules of carbonic acid per second. The catalyzed reaction is 10^7 times faster than the uncatalyzed one. In animals, this reaction is

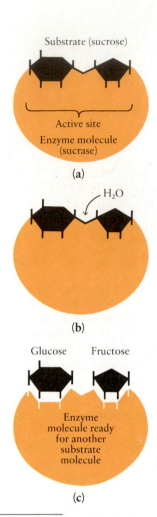

Substrate (sucrose)

Active site
Enzyme molecule
(sucrase)

(a)

H_2O

(b)

Glucose   Fructose

Enzyme
molecule ready
for another
substrate
molecule

(c)

8–9 *A model of enzyme action.* (**a**)
Sucrose, a disaccharide, (**b**) *is hydrolyzed
to yield* (**c**) *a molecule of glucose and a
molecule of fructose. The enzyme involved
in this reaction, sucrase, is specific for this
process; as you can see, the active site of
the enzyme fits the opposing surface of
the sucrose molecule. The fit is so exact
that a molecule composed, for example, of
two subunits of glucose would not be
affected by this enzyme.*

essential in the transfer of carbon dioxide from the cells, where it is produced, to
the bloodstream, which transports it to the lungs. As carbonic anhydrase illustrates, enzymes are typically effective in very small amounts.

Almost 2,000 different enzymes are now known, each of them capable of
catalyzing a specific chemical reaction. However, different types of cells are able
to manufacture different types of enzymes—no cell contains all the known
enzymes. The particular enzymes that a cell can manufacture are a major factor in
determining the biological activities and functions of that cell. A cell can carry out
a given chemical reaction at a reasonable rate only if it has a specific enzyme that
can catalyze that reaction. The molecule (or molecules) on which an enzyme acts
is known as its **substrate.** For example, in the reaction diagrammed in Figure 8–9,
sucrose is the substrate and sucrase is the enzyme. (Note that enzyme names often
end in "-ase.")

Enzyme Structure and Function

Enzymes are large to very large,* complex globular proteins consisting of one or
more polypeptide chains. They are folded so as to form a groove or pocket on the
surface into which the reacting molecule or molecules—the substrate—fit and
where the reactions take place. This portion of the enzyme is known as the **active
site.** Only a few amino acids of the enzyme are involved in any particular active
site. Some of these may be adjacent to one another in the primary structure, but
often the amino acids of the active site are brought close to one another by the
intricate folding of the amino acid chain that produces the tertiary structure. In an
enzyme with a quaternary structure, the amino acids of the active site may even be
on different polypeptide chains (Figure 8–10).

The active site not only has a three-dimensional shape complementary to that
of the substrate, but it also has a complementary array of charged or uncharged,
hydrophilic or hydrophobic, areas on the binding surface. If a particular portion
of the substrate has a negative charge, any corresponding feature on the active site
is likely to have a positive charge, and so on. Thus the active site not only
recognizes and confines the substrate molecule but also orients it in a particular
direction.

The Induced-Fit Hypothesis

When the existence of active sites was first postulated by Emil Fischer in 1894, he
compared the relationship between the active site and the substrate to that
between a lock and a key. Within the last few years, however, studies of enzyme
structure have suggested that the active site is considerably more flexible than a
keyhole. The binding between enzyme and substrate appears to alter the conformation of the enzyme, thus inducing a close fit between the active site and the
substrate (Figure 8–11). It is believed that this induced fit may put some strain on
the reacting molecules and so further facilitate the reaction.

Cofactors in Enzyme Action

The catalytic activity of some enzymes appears to depend only upon the physical
and chemical interactions between the amino acids of the active site and the
substrate. Many enzymes, however, require additional nonprotein low-molecu-
lar-weight substances in order to function. Such nonproteins that are essential for
enzyme function are known as **cofactors.**

---

* Different enzymes vary in their molecular weights from about 12,000 to more than 1 million. Amino
acids have an average molecular weight of about 120, and this figure is used to estimate the number of
amino acids in a polypeptide of known molecular weight.

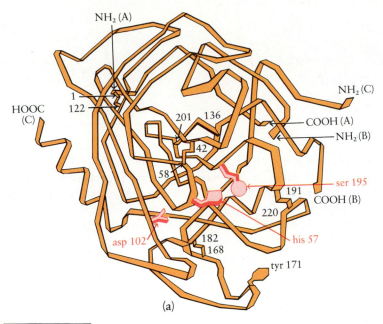

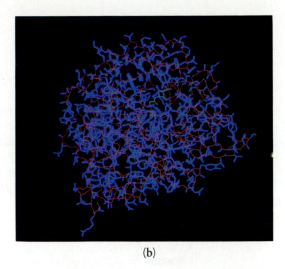

(a) | (b)

8–10 (a) *A model of the digestive enzyme chymotrypsin. This enzyme is composed of three polypeptide chains. The amino (NH₂) and carboxyl (COOH) ends of each are labeled. The numbers represent the positions of particular amino acids in the chains. Five disulfide bridges connect amino acids 1 and 122, 42 and 58, 136 and 201, 168 and 182, and 191 and 220. The three-dimensional shape of the molecule is a result of a combination of disul-* fide bridges and of interactions among the chains and between the chains and the surrounding water molecules. These interactions are based on the positive or negative charges or the polarity of the various amino acids. As a result of the bending and twisting of the polypeptide chains, particular amino acids come together in a highly specific configuration to form the active site of the enzyme. Three amino acids known to be part of the active site *are shown in red. (b) A computer-generated model of the digestive enzyme carboxypeptidase. The backbone of the polypeptide chain is shown in red, and the side (R) groups of the amino acids are in blue. The yellow-green structure represents the substrate, a dipeptide, nestled in the active site. The small white star burst just below the substrate is the zinc ion (Zn²⁺), an essential cofactor for this enzyme.*

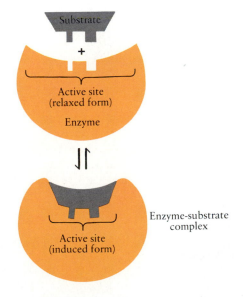

8–11 *The induced-fit hypothesis. The active site is believed to be flexible and to adjust its conformation to that of the substrate molecule. This induces a close fit between the active site and the substrate and may also put some strain on the substrate molecule.*

Ions as Cofactors

Certain ions are cofactors for particular enzymes. For example, the magnesium ion (Mg²⁺) is required in all enzymatic reactions involving the transfer of a phosphate group from one molecule to another. As you will recall (see Table 3–1, page 57), the phosphate group is usually negatively charged in solution; its two negative charges are attracted by the two positive charges of the magnesium ion, holding it in position at the active site of the enzyme. K⁺, Ca²⁺, and other ions play similar roles in other reactions. In some cases, bonds between ions and the R groups of particular amino acids help to maintain certain folds in the tertiary structure or to hold a quaternary structure together.

Coenzymes and Vitamins

Nonprotein organic molecules may also function as cofactors in enzyme-catalyzed reactions. Such molecules are called **coenzymes;** they are bound, either temporarily or permanently, to the enzyme, usually fairly close to the active site. Some coenzymes function as electron-acceptors in oxidation-reduction reactions, receiving electrons—often a pair of electrons accompanied by a hydrogen ion (that is, a proton)—and then passing them on to another molecule. There are several different kinds of electron-accepting coenzymes in any given cell, each capable of holding electrons at a slightly different energy level.

One of the most frequently encountered coenzymes is known as nicotinamide adenine dinucleotide (NAD). A nucleotide is, as we saw in Chapter 3 (page 80), a complex molecule composed of three subunits: a phosphate group, a five-carbon sugar, and a nitrogenous base. A dinucleotide, such as NAD, is a molecule that consists of two nucleotides. In NAD, the two nucleotides forming the molecule

Nicotinamide
(a)

Adenine
(b)

8–12 *The nitrogenous bases* (**a**) *nicotinamide and* (**b**) *adenine. Each of the nitrogen atoms in these molecules has an unshared pair of electrons that exerts a weak attraction for hydrogen ions* (H^+). *Thus nicotinamide and adenine are bases, combining with hydrogen ions and thereby increasing the relative number of hydroxide ions in a solution (see page 48).*

Ribose

8–13 *The five-carbon sugar ribose. Ribose is an essential component of many biologically important molecules. These include not only many coenzymes but also ATP and related compounds and, as we shall see in Chapter 15, several types of RNA molecules involved in protein synthesis.*

contain two different nitrogenous bases, nicotinamide and adenine (Figure 8–12). Adenine is a molecule we shall be mentioning frequently in subsequent chapters; in addition to its role as a subunit of NAD and other coenzymes, it is part of the ATP molecule (see page 180) and is one of the principal components of the nucleic acids DNA and RNA. (The use of adenine for three quite different purposes is an example of the economy with which the living cell operates.) Nicotinamide occurs as a component of the vitamin niacin. Niacin, like other vitamins, is a compound, required in small quantities, that we and other animals cannot synthesize ourselves and so must obtain in our diets (see Table 34–3). Thus we must eat foods containing niacin (which includes both nicotinamide and nicotinic acid but should not be confused with nicotine, found in tobacco). When nicotinamide is present, our cells can use it to make NAD. A number of other coenzymes are also vitamins or contain vitamins as subunits of the molecule.

The other components of NAD are two molecules of the five-carbon sugar ribose (Figure 8–13) and two phosphate groups. In the NAD molecule, the ribose subunits are linked together by the phosphate groups. One of the ribose units is linked, in turn, to the nicotinamide subunit, and the other to the adenine subunit (Figure 8–14).

The nicotinamide ring is the business end of NAD, the part that accepts—and subsequently releases—the electrons. In its oxidized, electron-accepting state, the molecule has a positive charge and is written as NAD^+. When it accepts two electrons and one proton, it is reduced to NADH. Like other coenzymes, this

NAD^+

$$H^+ + 2e^-$$

NADH

8–14 *Nicotinamide adenine dinucleotide (NAD) in its oxidized form, NAD^+, and its reduced form, NADH. Notice how the bonding within the nicotinamide ring shifts as the molecule changes from the oxidized to the reduced form, and vice versa.*

As indicated above the arrows, reduction of NAD^+ to NADH requires two electrons and one hydrogen ion (H^+). The two elec- trons, however, generally travel as components of two hydrogen atoms; thus, there is one hydrogen ion "left over" when NAD^+ is reduced. As we shall see in the next chapter, H^+ ions released into the surrounding solution when NAD^+ is reduced play a critical role in powering vital cellular processes.

Auxotrophs

Sometimes, as a result of a genetic mutation, an organism is unable to synthesize a particular enzyme in an active form. When this occurs, the reactions of the pathway in which that enzyme participates cannot proceed to completion. The end product, which may be of critical importance to the organism, is not formed, and there may be, moreover, an accumulation of the substrate of the defective or missing enzyme. As we shall see in Chapter 19, such accumulations can lead to serious human disease or even death.

Although mutations resulting in defective or missing enzymes are no less serious for microorganisms, such as bacteria and fungi, they do provide biologists with a valuable tool for elucidating enzymatic pathways. Cells with a defect in a biosynthetic pathway are known as auxotrophs. They grow normally only if they are supplied with either the end product of the entire pathway or the product of the specific reaction normally catalyzed by the defective or missing enzyme. Auxotrophs can be produced by irradiating cells with x-rays; they are identified and isolated by their inability to grow on chemical media that will support the growth of normal cells. Studies of the exact chemical requirements of different auxotrophs have revealed the details of many biosynthetic pathways, particularly those involved in the synthesis of amino acids. As we shall see in Chapter 15, auxotrophs have also made a major contribution to our understanding of the mechanism by which hereditary information is converted into the structures and processes of the living cell.

One group of auxotrophs of the red bread mold Neurospora crassa *have defective enzymes (indicated in red) at different points in the biosynthesis of the amino acid arginine (arg). These mutants, each deficient in one enzyme, retain the activity of the other enzymes of the arginine pathway. Four of the enzymes (E_1 through E_4) and four of the intermediate products (A through D) are shown in the diagram. The substance accumulated (indicated in pale blue) by each auxotroph and its requirements for growth reveal the sequence of the enzymes and the intermediates in the reaction pathway.*

$$A \xrightarrow{E_1} B \xrightarrow{E_2} C \xrightarrow{E_3} D \xrightarrow{E_4} arg \qquad \text{Normal cell}$$

$$A \xrightarrow{E_1} B \xrightarrow{E_2} C \xrightarrow{E_3} D \; \blacksquare^{E_4} \; arg \qquad \text{Auxotroph I: accumulates D and requires arg for growth}$$

$$A \xrightarrow{E_1} B \xrightarrow{E_2} C \; \blacksquare^{E_3} \; D \xrightarrow{E_4} arg \qquad \text{Auxotroph II: accumulates C and requires either D or arg for growth}$$

$$A \xrightarrow{E_1} B \; \blacksquare^{E_2} \; C \xrightarrow{E_3} D \xrightarrow{E_4} arg \qquad \text{Auxotroph III: accumulates B and requires C, D, or arg for growth}$$

molecule is recycled—that is, NAD^+ is regenerated when NADH passes its two electrons and one proton on to another electron acceptor. Thus, although this coenzyme is involved in many cellular reactions, the actual number of NAD^+/NADH molecules required is relatively small.

Enzymatic Pathways

Enzymes typically work in series—the pathways we referred to earlier.

$$\text{Product 1} \xrightarrow{\text{Enzyme 1}} \text{Product 2} \xrightarrow{\text{Enzyme 2}} \text{Product 3} \xrightarrow{\text{Enzyme 3}}$$
$$\text{Product 4} \xrightarrow{\text{Enzyme 4}} \text{Product 5} \xrightarrow{\text{Enzyme 5}} \text{End product}$$

Cells derive several advantages from this sort of arrangement. First, the groups of enzymes making up a common pathway can be segregated within the cell. Some are found in solution, as in the lysosomes, whereas others are embedded in the membranes of particular organelles. The enzymes located in membranes appear to be lined up in sequence, so the product of one reaction moves directly to the adjacent enzyme for the next reaction of the series. A second advantage is that there is little accumulation of intermediate products, since each product tends to

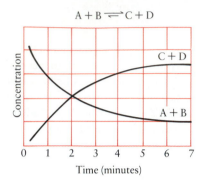

A + B ⇌ C + D

8–15 *The changes in concentration of products and reactants in a reversible reaction. At first, only molecules of A and B are present. The reaction begins when A and B start to yield products C and D. At the end of two minutes, the concentrations of A + B and C + D are equal. As the reaction proceeds, the concentration of C + D will continue to increase to the point of chemical equilibrium (at about the sixth minute) and will thereafter remain greater than the concentration of A + B. This is the proportion at which the rates of forward and reverse reactions are the same.*

be used up in the next reaction along the pathway. A third and most important advantage can be understood by considering the nature of chemical equilibrium.

As we have noted previously (page 47), chemical reactions can go in either direction. When net change ceases, the reaction is said to be at equilibrium. In the reaction

$$A + B \rightleftharpoons C + D$$

the point of equilibrium is reached when as many molecules of C and D are being converted to molecules of A and B as molecules of A and B are being converted to molecules of C and D.

The concentration of reactants does *not* have to equal the concentration of products in order for equilibrium to be established; only the *rates* of the forward and reverse reactions must be the same. Consider the reaction shown above. The different lengths of the arrows indicate that at equilibrium there is more C + D present than A + B. If only A and B molecules are present initially, the reaction occurs at first to the right, with A and B molecules converting into C and D molecules. Figure 8–15 shows the relative changes in concentration as the reaction continues. As C and D accumulate, the rate of the reverse reaction increases, and at the same time, the rate of the forward reaction decreases because of the decreasing concentrations of A and B. At about minute 6, the rates of the forward and reverse reactions equalize and no further changes in concentration take place. The proportions of A + B and C + D will remain the same. There will always be more C + D molecules in the system than A + B molecules, but the reaction will not go to completion—that is, not all of the A + B molecules will be converted to C + D molecules.

The relative proportions of A + B and C + D at equilibrium are determined by the free energy change (ΔG) of the reaction. Only if there were no net change in the free energy (that is, $\Delta G = 0$) would the concentrations of A + B and C + D be equal at equilibrium. The fact that the concentration of C + D molecules is greater than that of A + B molecules at equilibrium tells us that the potential energy of C + D is less than the potential energy of A + B. The reaction proceeding from A + B to C + D is exergonic (that is, it has a negative ΔG); conversely, the reaction from C + D to A + B is endergonic (positive ΔG). In any reversible reaction, the point of equilibrium will lie in the direction for which ΔG is negative. The larger the negative value of ΔG, the more strongly the reaction will be pulled in that direction.

This has important consequences in the sequential reactions that occur within living cells. If A + B and C + D molecules are in a closed system, equilibrium will be attained and there will be no subsequent change in their concentrations. If, however, the system is open and C + D molecules are continually removed from it, equilibrium will never be attained, and the conversion of A + B molecules into C + D molecules will continue. The sequential reactions of enzymatic pathways have the effect of removing the product of each reaction from the system, so that equilibrium is not attained. If, for example, product 2 in the enzymatic pathway shown on page 173 is used up (by being converted into product 3) almost as rapidly as it is formed, the reaction product 1 → product 2 can never reach equilibrium. If the eventual end product is also used up, the whole series of reactions will move toward completion. Moreover, if any of the reactions along the pathway are highly exergonic (large negative ΔG), they will rapidly use up the products of the preceding reactions, pulling those reactions forward; similarly, the accumulation of the products from the exergonic reactions will push the subsequent reactions forward by increasing the concentrations of the reactants. The linking of reactions in enzymatic pathways, with exergonic reactions moving the entire series forward, is a key factor in the remarkable efficiency with which living organisms carry out their chemical activities.

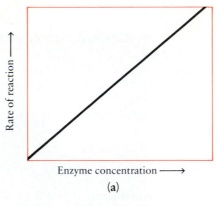

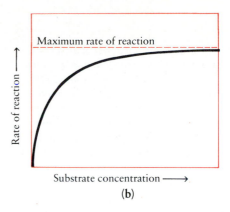

(a)   (b)

8-16 *The effect of enzyme and substrate concentrations on the rate of a typical enzymatic reaction for which all necessary cofactors are available in ample supply. (a) In the presence of excess substrate, the reaction rate increases in direct proportion to the concentration of enzyme. Regulation of enzyme concentrations is a principal means by which cells regulate the rate of chemical reactions. (b) If the enzyme concentration remains constant as the substrate concentration increases, the rate of reaction increases until it approaches a maximum. Further increases in the substrate concentration have no additional effect on the reaction rate. The maximum rate is attained when all of the enzyme molecules are occupied by substrate molecules.*

Regulation of Enzyme Activity

Another remarkable feature of the metabolic activity of cells is the extent to which each cell regulates the synthesis of the products necessary to its well-being in the amounts and at the rates required, while avoiding overproduction, which would waste both energy and raw materials. This regulation depends, in turn, on the regulation of enzyme activity.

The concentrations of enzyme and substrate molecules, as well as the availability of cofactors, are principal factors in limiting enzyme action (Figure 8–16). Most enzymes probably work at a rate well below their maximum because of these limitations. Moreover, many enzymes are broken down rapidly, characteristically by other enzymes that hydrolyze peptide bonds. A highly efficient means of regulation of these quickly degraded enzymes is for the cell to produce them only when they are needed. The way in which bacterial cells turn on and off their enzyme production at the source is described in some detail in Chapter 16.

Some enzymes are produced only in an inactive form and, just when they are needed, are activated, usually by another enzyme. Chymotrypsin, a digestive enzyme, is controlled in this way. It is synthesized by cells in the pancreas in the form of chymotrypsinogen, which consists of a single, very long polypeptide chain and is inactive. When this molecule is released into the small intestine, where it performs its digestive work, the enzyme trypsin snips out dipeptides at two points in the chain. The resulting three segments constitute the active chymotrypsin molecule (Figure 8–10a, page 171). In this way, chymotrypsin molecules (and other digestive enzymes) are prevented from digesting the proteins in the cells in which they are synthesized.

Living systems also have several other ways of turning enzyme activity on and off and of regulating its level.

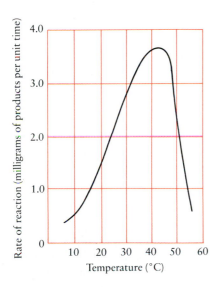

8-17 *The effect of temperature on the rate of an enzyme-controlled reaction. The concentrations of enzyme and substrate molecules were kept constant. As you can see, the rate of the reaction, as in most chemical reactions, approximately doubles for every 10°C rise in temperature. In the enzymatic reactions of humans (body temperature 37°C) and other mammals, the maximum rate of reaction is attained at about 40°C. Above this temperature, the rate decreases, and at about 60°C the reaction stops altogether, presumably because the enzyme is denatured. Although the shape of the curve is similar for all enzymatic reactions, the temperature range through which an enzyme is active varies with the type of organism and the particular enzyme.*

Effects of Temperature and pH

As we noted earlier, an increase in temperature increases the rate of uncatalyzed chemical reactions. This temperature effect also holds true for enzyme-catalyzed reactions—but only up to a point. As you can see in Figure 8–17, the rate of most enzymatic reactions approximately doubles for each 10°C rise in temperature and then drops off very quickly above 40°C. The increase in reaction rate occurs because, at higher temperatures, more of the substrate molecules possess sufficient energy to react; the decrease in the reaction rate occurs as the movement and vibration of the enzyme molecule itself increases, disrupting the hydrogen bonds and other relatively fragile forces that maintain its tertiary structure. A molecule that has lost its characteristic three-dimensional structure in this way is said to be **denatured.** Partially denatured enzymes (in which the structure is only slightly distorted) regain their activity on being cooled, indicating that their polypeptide chains have regained their necessary shape. If the denaturation is sufficiently severe, however, it is irreversible, leaving the polypeptide chains permanently tangled and inactivated.

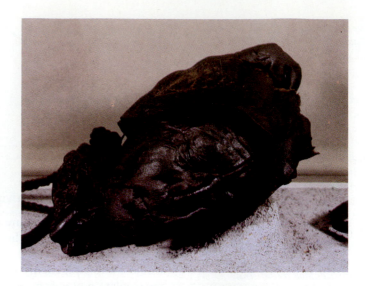

8-18 *The body of a man with a noose around his neck was found in a Danish peat bog in 1950. He died some 2,000 years ago. The remarkable preservation of the body is the result of the extremely acidic pH of the peat bog, which almost completely inhibited the enzymatic activities of the microorganisms that customarily decompose organic molecules.*

The pH of the surrounding solution also affects enzyme activity. The conformation of an enzyme depends, among other factors, on attractions and repulsions between negatively charged (acidic) and positively charged (basic) amino acids. As the pH changes, these charges change, and so the shape of the enzyme changes until it is so drastically altered that it is no longer functional. More important, probably, the charges of the active site and the substrate are changed so that the binding capacity is affected. The optimum pH of one enzyme is not the same as that of another. The digestive enzyme pepsin, for example, works at the very low (highly acidic) pH of the stomach (page 52), in an environment where most other proteins would be permanently denatured. Some enzymes are usually found at a pH that is not their optimum, suggesting that this discrepancy may not be an evolutionary oversight but a way of damping enzyme activity.

Allosteric Interactions

An ingenious mechanism by which an enzyme may be temporarily activated or inactivated is known as **allosteric interaction.** Allosteric interactions occur among enzymes that have at least two binding sites, one the active site and another, into which a second molecule, known as an allosteric effector, fits. The binding of the effector changes the shape of the enzyme molecule and either activates or inactivates it (Figure 8–19).

8-19 *An allosteric ("other shape") effector can bind to an enzyme and, by altering the bonds determining its tertiary structure, change the conformation of the active site. As a consequence, the enzyme may be altered so that it cannot interact with its substrate, as shown here, or it may be activated.*

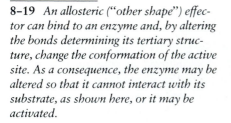
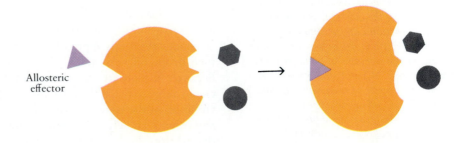

Allosteric effector

Allosteric interactions are frequently involved in **feedback inhibition,** which is a common means of biological control. A familiar nonbiological example of feedback inhibition is a thermostat that turns off the furnace when the room temperature reaches a desired level. In feedback inhibition of enzymatic reactions,

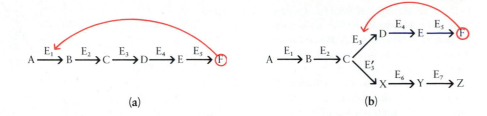

8-20 *Feedback inhibition.* **(a)** *In this series of reactions, each step (the black arrows) is catalyzed by a specific enzyme. Enzyme E_1, which converts A to B, is allosteric, and the allosteric inhibitor is the product F. Thus, enzyme E_1 will be more active when amounts of F are low.* **(b)** *A branched metabolic pathway, A being converted to B, B being converted to C, and C being converted to both D and X. In this case, the enzyme at the branch, E_3, is allosteric. As the quantity of the allosteric inhibitor F increases, the upper branch of the pathway will become less active, and the reaction series will be primarily shunted to the production of Z.*

one of the products, often the last in the series, acts as an allosteric effector, inhibiting the function of one of the enzymes, often the first in the series (Figure 8–20a). Or, in a reaction that may take one of two directions, the effector may act to shunt the reactions along another pathway (Figure 8–20b).

Competitive Inhibition

Some compounds inhibit enzyme activity by temporarily occupying the active site of the enzyme; regulation in this way is known as competitive inhibition, because the regulatory compound and the substrate compete with each other for binding to the active site. Competitive inhibition is completely reversible; the result of the competition at any particular time depends on how many of each kind of molecule are present. For example, in the reaction series

$$A \xrightarrow{E_1} B \xrightarrow{E_2} C \xrightarrow{E_3} D \xrightarrow{E_4} E \xrightarrow{E_5} F$$

the final product F might be rather similar in structure to product D. F could occupy the active site of enzyme E_4, preventing D, the normal substrate, from binding to the enzyme. As F was used up by the cell, the active site of enzyme E_4 would once more become available to D.

Competitive inhibition is the mechanism of action of some drugs used to treat bacterial infections in animals. For instance, bacteria make the vitamin folic acid, which animal cells do not make (animals obtain folic acid from their food). One of the compounds in the metabolic pathway leading to folic acid is *para*-aminobenzoic acid (PABA). As you can see in Figure 8–21, the drug sulfanilamide has a structure very similar to that of PABA. The two structures are so similar, in fact, that the enzyme involved in converting PABA to folic acid combines with the drug rather than with the PABA. Without folic acid, the bacterial cell dies, leaving the animal cell, which lacks this enzyme, unharmed.

para-Aminobenzoic acid
(PABA)

Sulfanilamide

8-21 *para-Aminobenzoic acid (PABA) is one of the compounds in the metabolic pathway to folic acid in bacterial cells. Sulfanilamide, a drug, has a similar structure. It can combine with the enzyme that converts PABA to folic acid, thereby blocking the synthesis of folic acid, without which the bacterial cell cannot live.*

Noncompetitive Inhibition

In noncompetitive inhibition, the inhibitory chemical, which need not resemble the substrate, binds with the enzyme at a site on the molecule other than the active site. Lead, for instance, forms covalent bonds with sulfhydryl (SH) groups. Many enzymes contain cysteine, which has a sulfhydryl group. The binding of lead to such enzymes disrupts their tertiary structure and deactivates them, producing the symptoms associated with lead poisoning. Like competitive inhibition, noncompetitive inhibition is often reversible, but such reversal is not accomplished by increased concentrations of substrate. In the case of lead, for example, the inhibition can be reversed by treatment with other sulfhydryl-containing compounds that bind the lead atoms more tightly than cysteine does.

Irreversible Inhibition

Some substances inhibit enzymes irreversibly, either binding permanently with key functional groups of the active site or so thoroughly denaturing the protein that

8–22 *The bacterial cell wall has polysac-charide backbones formed by alternating molecules of two sugars, NAG and NAM. The backbones (indicated by the thick black lines) are cross-linked by short pep-tide chains. One of the amino acids in these chains, lysine (shown in red), forms three peptide bonds—one with the amino acid above it, one with the amino acid below it, and another with the adjacent amino acid. These bonds are formed in an unusual way—by the enzymatic transfer of a peptide bond from one molecule to another. Penicillin blocks this transfer. Its structure mimics that of a dipeptide, and it binds to the active site of the enzyme, irreversibly inhibiting it. Without cross-links, the cell wall cannot hold the cell together. Thus penicillin acts specifically on growing bacterial cell walls.*

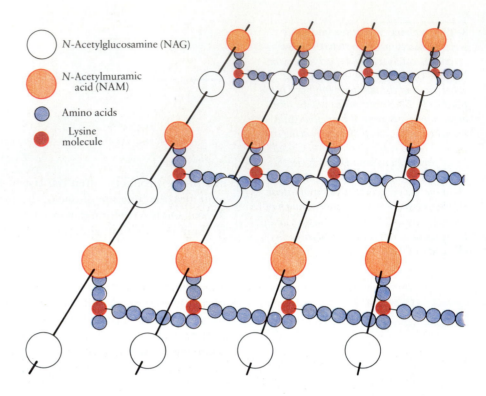

N-Acetylglucosamine (NAG)

N-Acetylmuramic acid (NAM)

Amino acids

Lysine molecule

its tertiary structure cannot be restored. The nerve gases, used widely during World War I and now banned, are among the most potent poisons known. They irreversibly inhibit enzymes involved in transmission of the nerve impulse, result-ing in paralysis and death. Many useful drugs, including the antibiotic penicillin (Figure 8–22), are also irreversible inhibitors of enzyme activity.

Membrane Transport Proteins and Receptors Revisited

In Chapter 6, we noted that the integral membrane proteins that transport molecules and ions across the lipid bilayer are sometimes called permeases because of their similarities to enzymes. We are now in a position to see just how similar they are. Both enzymes and membrane transport proteins are large, globular protein molecules with complex tertiary or quaternary structures. In both types of molecules, the tertiary or quaternary structure produces a precisely configured region on the surface—an active site or a binding site—into which another specific molecule or ion fits. In enzymes, association of a substrate molecule with the active site leads to a chemical change in the substrate; in transport proteins, the association of a molecule or ion with the binding site leads to the movement of that molecule or ion across the membrane. The changes in protein conformation that result from the binding play an important role in either catalyzing the chemical reaction or propelling the molecule or ion through the protein to the other side of the membrane.

These obvious similarities are, however, just the beginning. The rate at which a specific membrane protein transports its particular substance across the lipid bilayer—like the rate at which an enzyme catalyzes its particular reaction—is affected by the number of molecules of that specific protein in the membrane, the concentration of the molecules or ions to be transported, the temperature, and the pH. Moreover, the transport functions of integral membrane proteins are similarly regulated by allosteric interactions, reversible inhibitions (both competi-tive and noncompetitive), and irreversible inhibitions. An example of allosteric

Some Like It Cold

Enzyme action is exquisitely dependent on the tertiary and quaternary structure of the globular protein, particularly as it affects the active site. As a consequence, many enzymes are thermolabile—that is, they do not function at higher temperatures, even within the normal physiological range. One such enzyme is responsible for color in Siamese cats. It functions adequately in the cooler, peripheral areas of the body, such as the ears, nose, paws, and tip of the tail, but becomes inactive in the warmer areas of the body. For similar reasons, Himalayan rabbits are all black when raised at temperatures of about 5°C; white with black ears, forepaws, noses, and tails when raised at normal room temperatures; and all white when raised at temperatures above 35°C.

Thermolability has its advantages. In the northern seal, the newborns are white, as a result of their developing at a warm (internal) temperature. The newborns cannot swim and so are restricted to the ice floes, where their white coats provide them with color camouflage. By the time they are able to swim, their coats have turned brown and so blend with the dark Arctic waters. The thermolability of this same enzyme (tyrosinase) provides similar camouflage for the Arctic fox. During the summer it develops a white coat, which provides color protection during the winter months. During the winter, its dark coat develops, which is revealed when the white coat is shed in the spring. All of which illustrates that, as we shall see often, evolution is opportunistic.

(a)

(b)

(a) *The characteristic color pattern of the Siamese cat is a result of the thermolability of an enzyme affecting coat color. This enzyme, involved in controlling the synthesis of a dark pigment, is active only in the cooler, more peripheral areas of the* body. (b) *Similarly, the enzyme controlling dark pigmentation in northern seals is active only at low temperatures. The newborns, which, like other mammals, develop within the warm bodies of their mothers, are white. New fur, growing in* after the seals are exposed to the colder external environment, is brown. Although the white fur protects the newborn seals against most predators, it makes them highly desirable to human hunters, now their principal predators.

regulation of a transport protein is provided by the acetylcholine receptor. As we noted previously (page 137), when acetylcholine binds to this particular protein in the cell membrane of a nerve cell, a channel is opened in a closely associated protein through which sodium ions (Na^+) move down a concentration gradient into the cell.

Some of the protein receptors located on the surface of cell membranes are not involved in transport at all. They are instead the allosteric binding sites of enzyme molecules that are integral membrane proteins. In at least some cases, the allosteric effector binds to its receptor—located in the portion of the protein exposed on the external face of the membrane—and triggers a change in the

8-23 *Adenosine triphosphate (ATP) is the cell's chief energy currency. The bonds between the three phosphate groups in the molecule are important in ATP function. In order to represent more accurately the position of the adenine subunit in the ATP molecule, we have rotated it 180° (right to left) from the orientation depicted in Figures 8–12 and 8–14.*

Adenosine triphosphate (ATP)

conformation of the active site of the enzyme—located on the cytoplasmic face of the membrane. The result is a chemical reaction within the cell. In other cases, such as the cell membrane receptor for the hormone adrenaline, the receptor and the enzyme in which activity is triggered are apparently two distinct proteins. In Chapter 40, we shall examine the sequence of enzymatic reactions that result from the binding of adrenaline to its receptor.

Transport processes across cell membranes are, like chemical reactions, governed by the laws of thermodynamics. Substances can move spontaneously only from a state of higher potential energy to a state of lower potential energy. Such spontaneous movement in diffusion—either simple or facilitated—is driven by differences in potential energy that result from differences in concentration. Often, however, cells must transport substances against a concentration gradient —that is, from a state of lower potential energy to a state of higher potential energy. For such active transport processes, as for many chemical reactions, an input of energy is required.

THE CELL'S ENERGY CURRENCY: ATP

All of the biosynthetic activities of the cell, many of its transport processes, and a variety of other activities require energy. A large proportion of this energy is supplied by a single substance, **adenosine triphosphate,** or ATP. Glucose and other carbohydrates are storage forms of energy and also forms in which energy is transferred from cell to cell and organism to organism. In a sense, they are like money in the bank. ATP, however, is like the change in your pocket—it is the cell's immediately spendable energy currency.

At first glance, ATP, as shown in Figure 8–23, appears to be a complex, unfamiliar molecule. If you look at it more closely, however, you will find that you recognize all of its component parts. It is made up of the nitrogenous base adenine, the five-carbon sugar ribose, and three phosphate groups. These three phosphate groups with strong negative charges are covalently bonded to one another; this is an important feature in ATP function. Three linked phosphate groups are also characteristic of other molecules that play a role similar to that of ATP in certain cellular reactions; for example, guanosine triphosphate (GTP), another energy-carrying molecule, differs from ATP only in the substitution of the nitrogenous base guanine for adenine.

To understand the role of ATP and related triphosphate compounds, we must return briefly to the concept of the chemical bond. Because a chemical bond is a stable configuration of electrons, reacting molecules must possess a certain amount of energy in order to collide with sufficient force to overcome their mutual repulsion and to weaken existing chemical bonds, allowing the formation of new bonds. This energy is the energy of activation (Figure 8–8). Because of enzymes, which reduce the required energy of activation to a level already possessed by a significant proportion of the reacting molecules, the reactions essential to life are able to proceed at an adequate rate. As we have seen, however, the *direction* in which a reaction proceeds is determined by the free energy change, ΔG. Only if the reaction is exergonic (negative ΔG), will it proceed to any significant extent. Yet many cellular reactions, including synthetic reactions such as the formation of a disaccharide from two monosaccharide molecules, are endergonic (positive ΔG). In such a reaction, the electrons forming the chemical bonds of the product are at a higher energy level than the electrons in the bonds of the starting materials—that is, the potential energy of the product is greater than the potential energy of the reactants, an apparent violation of the second law of thermodynamics. Cells circumvent this difficulty by **coupled reactions** in which endergonic reactions (or transport processes, such as the active transport of a substance against a concentration gradient) are linked to exergonic reactions that provide a surplus of energy, making the entire process exergonic and thus able to proceed spontaneously. The molecule that most frequently supplies energy in such coupled reactions is ATP.

The internal structure of the ATP molecule makes it unusually suited to this role in living systems. In the laboratory, energy is released from the ATP molecule when the third phosphate is removed by hydrolysis, leaving ADP (adenosine diphosphate) and a phosphate:

$$ATP + H_2O \longrightarrow ADP + Phosphate$$

In the course of this reaction, about 7 kilocalories of energy are released per mole of ATP. Removal of the second phosphate produces AMP (adenosine monophosphate) and releases an equivalent amount of energy.

$$ADP + H_2O \longrightarrow AMP + Phosphate$$

The covalent bonds linking these two phosphates to the rest of the molecule are symbolized by a squiggle, $\sim$, and were, for many years, called "high-energy" bonds—an incorrect and confusing term. These bonds are not strong bonds, like the covalent bonds between carbon and hydrogen, which have a bond energy of 98.8 kilocalories per mole. They are instead bonds that are easily broken, releasing an amount of energy—about 7 kilocalories per mole—adequate to drive many of the essential endergonic reactions of the cell. Moreover, the energy released does not arise entirely from the movement of the bonding electrons to lower energy levels. It is also a result of a rearrangement of the electrons in other orbitals of the ATP or ADP molecules. The phosphate groups each carry negative charges and so tend to repel each other. When a phosphate group is removed, the molecule undergoes a change in electron configuration that results in a structure with less energy.

ATP in Action

In living cells, ATP is sometimes directly hydrolyzed to ADP plus phosphate, releasing energy for a variety of activities. ATP hydrolysis provides, for example, a means for producing heat, as in those animals, such as birds and mammals, that generally maintain a high and constant body temperature. Enzymes catalyzing the hydrolysis of ATP are known as ATPases; a variety of different ATPases have been identified. The protein "arms" in cilia and flagella (page 123) and on microtubules of the mitotic spindle (page 153), for example, are ATPase molecules, catalyzing the energy release that causes the microtubules to move past one another. Many of the proteins that move molecules and ions through cell membranes against a concentration gradient are not only transport proteins but also ATPases, releasing energy to power the transport process.

(a)

(b)

8–24 *Living organisms use the energy stored in the phosphate bonds of ATP for a variety of purposes. (a) In the skunk cabbage, a common plant in bogs and marshy areas of the northeastern United States, ATP is hydrolyzed to produce heat. The plant produces enough heat to melt surrounding snow or ice, while maintaining a nearly constant internal temperature of about 22°C (72°F). (b) Certain organisms, such as these bioluminescent mushrooms, transform ATP energy into light energy, thereby glowing in the dark.*

Usually, however, the terminal phosphate group of ATP is not simply removed but is transferred to another molecule. This addition of a phosphate group is known as **phosphorylation;** enzymes that catalyze such transfers are known as kinases. Phosphorylation reactions transfer some of the energy of the phosphate group in the ATP molecule to the phosphorylated compound, which, thus energized, participates in a subsequent reaction.

For example, in the reaction

$$W + X \longrightarrow Y + Z$$

if the potential energy of W plus the potential energy of X were less than that of Y plus Z, the reaction would not take place to any significant extent. Chemists could drive the reaction forward by supplying outside energy, probably in the form of heat. The cell might handle it in a two-step process. First:

$$W + ATP \longrightarrow W\text{–}P + ADP$$

The potential energy of the products is less than that of the reactants, so this reaction will take place. However, much of the energy made available when the phosphate group was removed from ATP is conserved in the new compound W–phosphate, or W–P. The next step in the process becomes

$$W\text{–}P + X \longrightarrow Y + Z + P$$

With the release of the phosphate from W, this second reaction also becomes one in which the potential energy of the products is less than the potential energy of the reactants and which, therefore, can take place.

Take, for instance, the formation of sucrose in sugarcane.

$$\text{Glucose} + \text{Fructose} \longrightarrow \text{Sucrose} + H_2O$$

In this reaction, the potential energy of the products is 5.5 kilocalories per mole greater than the potential energy of the reactants (that is, $\Delta G = +5.5$ kcal/mole). However, the sugarcane plant carries out this synthesis through a series of reactions coupled to the breakdown of ATP and the accompanying phosphorylation of the glucose and fructose molecules. The overall reaction is

$$\text{Glucose} + \text{Fructose} + 2\text{ATP} \longrightarrow \text{Sucrose} + 2\text{ADP} + 2\text{P}$$

Since the potential energy of 2 ADPs is about 14 kilocalories per mole less than the potential energy of 2 ATPs, the overall difference in products and reactants becomes 8.5 kilocalories per mole (that is, $\Delta G = -8.5$ kcal/mole). The coupling of reactions permits sugarcane to form sucrose.

Where does the ATP come from? As we shall see in the next chapter, energy released in the cell's catabolic reactions, such as the breakdown of glucose, is used to "recharge" the ADP molecule to ATP. Thus the ATP/ADP system serves as a universal energy-exchange system, shuttling between energy-releasing reactions and energy-requiring ones.

One marvels at the process of evolution as exemplified by the intricate flower of an orchid, the shell of a chambered nautilus, or the opposable thumb and forefinger of the human hand. Remember that ATP, NAD, and indeed the place of each amino acid in the polypeptide chain of an enzyme are also the products of evolution and also, you must admit, quite marvelous. Perhaps even beautiful.

SUMMARY

Living systems convert energy from one form to another as they carry out essential functions of maintenance, growth, and reproduction. In these energy conversions, as in all others, some useful energy is lost to the surroundings at each step.

The laws of thermodynamics govern transformations of energy. The first law states that energy can be converted from one form to another but cannot be created or destroyed. The potential energy of the initial state (or reactants) is equal to the potential energy of the final state (or products) plus the energy released in the process or reaction. The second law of thermodynamics states that in the course of energy conversions, the potential energy of the final state will always be less than the potential energy of the initial state. The difference in potential energy between the initial and final states is known as the free energy change and is symbolized as ΔG. Exergonic (energy-releasing) reactions have a negative ΔG, and endergonic (energy-requiring) reactions have a positive ΔG. Factors that determine ΔG include ΔH, the change in heat content, ΔS, the change in entropy, which, multiplied by the absolute temperature (T), is a measure of randomness or disorder:

$$\Delta G = \Delta H - T\Delta S$$

Another way of stating the second law of thermodynamics is that all natural processes tend to proceed in such a direction that the entropy of the universe increases. To maintain the organization on which life depends, living systems must have a constant supply of energy to overcome the tendency toward increasing disorder. The sun is the original source of this energy.

The energy transformations in living cells involve the movement of electrons from one energy level to another and, often, from one atom or molecule to another. Reactions in which electrons move from one atom to another are known as oxidation-reduction reactions. An atom or molecule that loses electrons is oxidized; one that gains electrons is reduced.

Metabolism is the total of all the chemical reactions that take place in cells. Reactions resulting in the breakdown or degradation of molecules are known, collectively, as catabolism; biosynthetic reactions are known, collectively, as anabolism. Metabolic reactions take place in series, called pathways, each of which serves a particular function in a cell. Each step in the pathway is controlled by a specific enzyme. The stepwise reactions of enzymatic pathways enable cells to carry out their chemical activities with remarkable efficiency in terms of both energy and materials.

Enzymes serve as catalysts, lowering the energy of activation and thus enormously increasing the rate at which reactions take place. They are large globular protein molecules folded in such a way that particular groups of amino acids form an active site. The reacting molecules, known as the substrate, fit precisely into this active site. Although the conformation of an enzyme may change temporarily in the course of a reaction, it is not permanently altered. Many enzymes require cofactors, which may be simple ions, such as Mg^{2+} or Ca^{2+}, or nonprotein organic molecules known as coenzymes. Many coenzymes, such as NAD, function as electron carriers, with different coenzymes holding electrons at slightly different energy levels. Many vitamins are parts of coenzymes.

Enzyme-catalyzed reactions are under tight cellular control. Principal factors in the rate of enzymatic reactions are the concentrations of enzyme and substrate and the availability of required cofactors. Many enzymes are synthesized by the cell or activated only when they are needed. The rate of enzymatic reactions is also affected by temperature and pH, which affect the attractions among the amino acids of the protein molecule and also between the active site and substrate.

A precise means of enzyme control is allosteric interaction. Allosteric interaction occurs when a molecule other than the substrate combines with an enzyme at a site other than the active site and in so doing alters the shape of the active site to render it either functional or nonfunctional. Feedback inhibition occurs when the product of an enzymatic reaction at the end or at a branch of a particular pathway acts as an allosteric effector, temporarily inhibiting the activity of an enzyme earlier in the pathway and thus temporarily stopping the series of chemical reactions.

Enzymes may also be regulated by competitive inhibition, in which another molecule, similar to the normal substrate, competes for the active site. Competitive inhibition can be reversed by increased concentrations of the substrate. Noncompetitive inhibitors bind elsewhere on the molecule, altering the tertiary structure so that the enzyme cannot function. Noncompetitive inhibition is usually reversible, but not by the substrate. Irreversible inhibitors bind permanently to the active site or irreparably disrupt the tertiary structure.

The transport proteins of cell membranes resemble enzymes in their complex protein structures, their specificity, and in the variety of ways in which their activity is regulated. The protein receptors found on the surface of the cell membrane are often allosteric binding sites, regulating the conformation—and thus the activity—of either transport proteins or membrane-bound enzymes.

ATP participates as an energy carrier in most series of reactions that take place in living systems. The ATP molecule consists of the nitrogenous base adenine, the five-carbon sugar ribose, and three phosphate groups. The three phosphate groups are linked by two covalent bonds that are easily broken, each yielding about 7 kilocalories of energy per mole. Cells are able to carry out endergonic reactions and processes (such as biosynthetic reactions, active transport, or the movement of microtubules) by coupling them with exergonic reactions that provide a surplus of energy. Such coupled reactions usually involve ATP or related triphosphate compounds.

QUESTIONS

1. Distinguish among the following terms: the first law of thermodynamics/the second law of thermodynamics; $\Delta H/\Delta S/\Delta G$; exergonic/endergonic; oxidation/reduction; metabolism/catabolism/anabolism; active site/substrate; competitive inhibition/noncompetitive inhibition/irreversible inhibition; ATP/ADP/AMP.

2. At present, at least four types of energy conversions are going on in your body. Name them.

3. All natural processes proceed with an increase in entropy. How then do you explain the freezing of water?

4. The laws of thermodynamics apply only to closed systems, that is, to systems into which no energy is entering. Is an aquarium ordinarily a closed system? Could you convert it to one? A space-ship may or may not be a closed system, depending on certain features of its design. What would these features be? Is the earth a closed system?

5. Explain why it is that living systems, despite appearances, are not in violation of the second law of thermodynamics.

6. What is there about the orderliness of a living organism that most significantly distinguishes it from the orderliness of a machine, such as a computer or the telephone system?

7. What is the basis for the specificity of enzyme action? What is the advantage to the cell of such specificity? What might be its disadvantages to the cell?

8. Turn back to Figure 3–18, in which all the amino acids are shown, and try to make some educated guesses about which amino acids can substitute for one another in the structure of an enzyme, and which substitutions would have drastic effects.

9. When a plant does not have an adequate supply of an essential mineral, such as magnesium, it is likely to become sickly and may die. When an animal is deprived of a particular vitamin in its diet, it too is likely to become ill and may die. What is a reasonable explanation of such phenomena?

10. Most organisms cannot live at high temperatures. Explain at least one way in which high temperatures are harmful to organisms. Some bacteria and algae, however, live in hot springs, at temperatures far above those that can be tolerated by most organisms. How might such bacteria and algae differ from most other organisms?

11. In enzyme regulation by allosteric interaction, the inhibitor often works on the first enzyme of the series. In regulation by competitive inhibition, it often works on the last. Why this difference?

12. In a series of experiments with an enzyme that catalyzes a reaction involving substrate A, it was found that a particular substance X inhibited the enzyme. When the concentration of A was high and the concentration of X was low, the reaction proceeded rapidly; as the concentration of X was increased and that of A was decreased, the reaction slowed down; when the concentration of X was high and that of A was low, the reaction stopped. If the concentration of A was again increased, the reaction resumed. How can you explain these results?

13. When a sulfa drug, such as sulfanilamide, is prescribed for a bacterial infection, it is very important to remember to take the drug at the prescribed times and in the prescribed quantity. Why is this essential? Suppose you were instructed to take two tablets every three hours, and instead you took only one tablet every five hours. What do you think would happen?

14. Some human societies use the barter system for exchange of goods and services. However, all complex societies have some form of monetary exchange. What are the advantages of a monetary exchange? Relate your answer to the ADP/ATP system.

15. Why, in the accompanying photograph, are there more plants than zebras and more zebras than lions? (Explain in terms of thermodynamics.)

How Cells Make ATP: Glycolysis and Respiration

ATP is the principal energy carrier in living systems. It participates in a great variety of cellular events, from chemical biosyntheses, to the flick of a cilium, the twitch of a muscle, or the active transport of a molecule across a cell membrane. It is involved in the propagation of an electric impulse along a nerve or, in some remarkable organisms, the electrocution of prey (Figure 9–1). In the following pages, we shall show in some detail how a cell breaks down carbohydrates and captures and stores a portion of their potential energy in the terminal phosphate bonds of ATP. The oxidation of glucose (or other carbohydrates) is complicated in detail—so go slowly—but simple in its overall design.

AN OVERVIEW OF GLUCOSE OXIDATION

Oxidation, as you will recall (page 166), is the loss of an electron. Reduction is the gain of an electron. Since, in spontaneous oxidation-reduction reactions, electrons go from higher to lower energy levels, a molecule usually releases energy as it is oxidized. In the oxidation of glucose, carbon-carbon bonds, carbon-hydrogen bonds, and oxygen-oxygen bonds are exchanged for carbon-oxygen and hydrogen-oxygen bonds, as oxygen atoms attract and hoard electrons. The summary equation for this process is

$$\text{Glucose} + \text{Oxygen} \longrightarrow \text{Carbon dioxide} + \text{Water} + \text{Energy}$$

Or,

$$C_6H_{12}O_6 + 6O_2 \longrightarrow 6CO_2 + 6H_2O$$

$$\Delta G = -686 \text{ kcal/mole}$$

9–1 (a) *The electric ray converts the chemical energy of ATP—provided by the oxidation of glucose and other organic compounds—to electrical energy, stunning and immobilizing its prey with electric discharges.* (b) *The prey shown here, a small reef fish, is then moved to the mouth by the pectoral fins and* (c) *is swallowed. The reef fish will be converted to chemical energy, which will, in turn, be converted to mechanical and electrical energy for the capture of additional prey. Only a fraction of the potential energy is transferred at each passage.*

Some other types of electric fish produce discharges of lower voltage that are used in establishing territories, in locating objects (including both potential prey and potential predators), and, perhaps, in communicating with members of the same species.

(a)

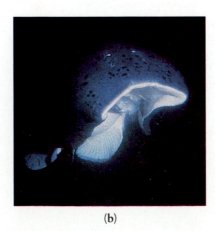

(b)

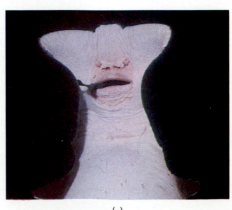

(c)

Living systems are experts at energy conversions. They are organized to trap this free energy so that it will not be dissipated randomly but can be used to do the work of the cell. About 40 percent of the free energy released by the oxidation of glucose is conserved in the conversion of ADP to ATP. As you will recall, about 75 percent of the energy in gasoline is "lost" as heat in an automobile engine, and only 25 percent is converted to useful forms of energy. The living cell is significantly more efficient.

In living systems, the oxidation of glucose takes place in two major stages. The first is known as **glycolysis.** The second is **respiration,** which, in turn, consists of two stages: the **Krebs cycle** and terminal **electron transport.** Glycolysis occurs in the cytoplasm of the cell, and the two stages of respiration take place within the mitochondrion.

In glycolysis, the six-carbon glucose molecule is split into two molecules of a three-carbon compound, pyruvic acid (Figure 9–2). Four hydrogen atoms (that is, four electrons and four protons) are removed from the glucose molecule in this process. The electrons and two of the protons are accepted by NAD^+ molecules (page 172), while the two other protons remain in solution as hydrogen ions (H^+). The free energy change, ΔG, of this stage is -143 kcal/mole; this represents a relatively small proportion of the potential energy stored in the glucose molecule.

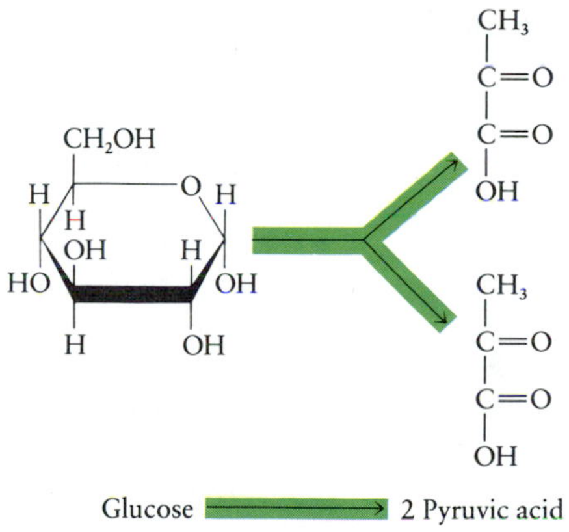

9–2 *In glycolysis, the six-carbon glucose molecule is split into two molecules of a three-carbon compound known as pyruvic acid.*

In respiration, the remaining hydrogen atoms are removed from the pyruvic acid molecules, and the carbon atoms are oxidized to carbon dioxide. The hydrogen atoms, in the form of electrons and protons, are initially accepted by NAD^+ and a related electron acceptor. Ultimately, all of the electrons and protons removed from the carbon atoms of the original glucose molecule are transferred to oxygen, forming water. The free energy change, ΔG, of respiration is -543 kcal/mole, a comparatively large energy yield.

In the course of glycolysis and respiration, about 38 molecules of ATP are regenerated from ADP in the breakdown of each molecule of glucose. As we shall see, the exact number of ATP molecules produced depends on at least two variables.

GLYCOLYSIS

Glycolysis—the lysis (splitting) of glucose—exemplifies the way the biochemical processes of a living cell proceed in small sequential steps. It takes place in a series of nine reactions, each catalyzed by a specific enzyme.

As we examine the details of glycolysis, notice how the carbon skeleton is dismembered and its atoms rearranged step by step. Note especially the formation of ATP from ADP and of NADH and H+ from NAD+. ATP and NADH represent the cell's net energy harvest from this reaction pathway.

Step 1. The first steps in glycolysis require an input of energy, which is supplied by coupling these steps to the ATP/ADP system. The terminal phosphate group is transferred from an ATP molecule to the carbon in the sixth position of the glucose molecule, to make glucose 6-phosphate. (The symbol Ⓟ represents a phosphate group.) The reaction of ATP with glucose to yield glucose 6-phosphate and ADP is an exergonic reaction. Some of the free energy is conserved in the chemical bond linking the phosphate to the glucose molecule, which then becomes energized. This reaction is catalyzed by a specific enzyme (hexokinase), and each of the reactions that follows is also catalyzed by a specific enzyme.

Step 2. The molecule is reorganized, again with the help of a particular enzyme. The six-sided ring characteristic of glucose becomes the five-sided fructose ring. (As you know, glucose and fructose both have the same number of atoms—$C_6H_{12}O_6$—and differ only in the arrangement of these atoms.) This reaction can proceed approximately equally well in either direction; it is pushed forward by the accumulation of glucose 6-phosphate and the removal of fructose 6-phosphate as the latter enters Step 3.

Step 3. In this step, which is similar to Step 1, fructose 6-phosphate gains a second phosphate by the investment of another ATP. The added phosphate is bonded to the first carbon, producing fructose 1,6-diphosphate, that is, fructose with phosphates in the 1 and 6 positions. Note that in the course of the reactions thus far, two molecules of ATP have been converted to ADP and no energy has been recovered.

The enzyme catalyzing this step, phosphofructokinase, is an allosteric enzyme, and ATP is an allosteric effector inhibiting its activity. The allosteric interaction between them is the chief regulatory mechanism of glycolysis. If the ATP concentration in the cell is high—that is, if ATP is present in quantities more than adequate to meet the various needs of the cell—ATP inhibits the activity of phosphofructokinase. Glycolysis, and thus ATP production, cease, and glucose is conserved. As the cell uses up its supply of ATP and the concentration drops, the enzyme is released from inhibition, and the breakdown of glucose resumes. This is one of the major control points of ATP production.

Step 4. The six-carbon sugar molecule is split into two three-carbon molecules, dihydroxyacetone phosphate and glyceraldehyde phosphate. The two molecules

9–3 *The steps of glycolysis.*

Glucose

Step 1   Hexokinase   ATP → ADP

Glucose
6-phosphate

Step 2   Phosphogluco-
isomerase

Fructose
6-phosphate

Step 3   Phospho-
fructokinase   ATP → ADP

Fructose
1,6-diphosphate

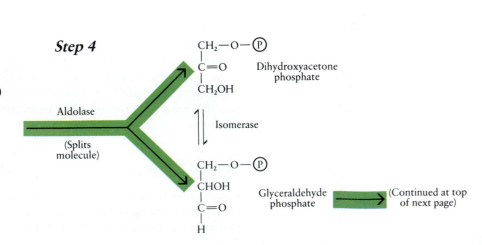

Step 4

Aldolase

(Splits
molecule)

Dihydroxyacetone
phosphate

Isomerase

Glyceraldehyde
phosphate

(Continued at top
of next page)

are interconvertible by the enzyme isomerase. However, because the glyceraldehyde phosphate is used up in subsequent reactions, all of the dihydroxyacetone phosphate is eventually converted to glyceraldehyde phosphate. Thus, *the products of all subsequent steps must be counted twice to account for the fate of one glucose molecule.* With the completion of Step 4, the preparatory reactions are complete.

Step 5. Glyceraldehyde phosphate molecules are oxidized—that is, hydrogen atoms with their electrons are removed—and NAD^+ is reduced to NADH and H^+ (a total of two molecules of NADH and two H^+ ions per molecule of glucose). This is the first reaction from which the cell harvests energy. Some of the energy from this oxidation reaction is also conserved in the attachment of a phosphate group to what is now the 1 position of the glyceraldehyde phosphate molecule. (The designation P_i represents inorganic phosphate available as a phosphate ion in solution in the cytoplasm.) The properties of this bond are similar to those of the phosphate bonds of ATP, as indicated by the squiggle.

Step 6. This phosphate is released from the diphosphoglycerate molecule and used to recharge a molecule of ADP (a total of two molecules of ATP per molecule of glucose). This is a highly exergonic reaction (ΔG is negative and large), and so it pulls all the preceding reactions forward.

Step 7. The remaining phosphate group is enzymatically transferred from the 3 position to the 2 position.

Step 8. In this step, a molecule of water is removed from the three-carbon compound. This internal rearrangement of the molecule concentrates energy in the vicinity of the phosphate group.

Step 9. The phosphate is transferred to a molecule of ADP, forming another molecule of ATP (again, a total of two molecules of ATP per molecule of glucose). This is also a highly exergonic reaction and thus pulls forward the preceding two reactions (Steps 7 and 8).

Summary of Glycolysis

The complete sequence begins with one molecule of glucose. Energy is invested at Steps 1 and 3 by the transfer of a phosphate group from an ATP molecule—one at each step—to the sugar molecule. The six-carbon molecule splits at Step 4, and from this point onward, the sequence yields energy. At Step 5, a molecule of NAD^+ is reduced to NADH and H^+, storing some of the energy from the oxidation of glyceraldehyde phosphate. At Steps 6 and 9, molecules of ADP take energy from the system, becoming phosphorylated to ATP.

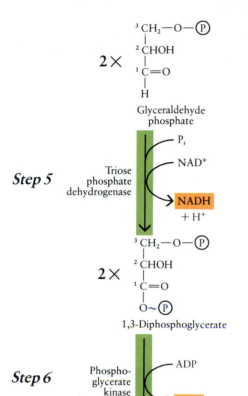

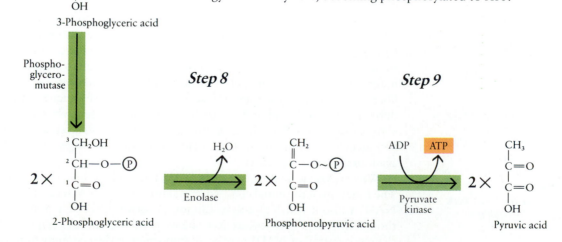

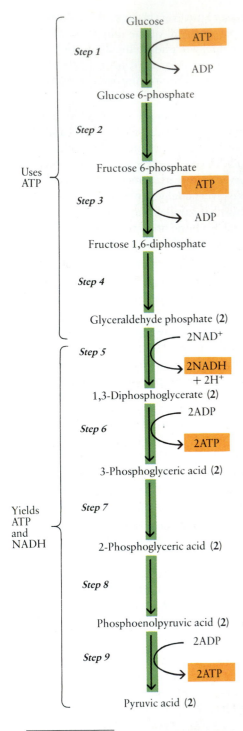

Glucose

Step 1   ATP
              ADP

Glucose 6-phosphate

Step 2

Fructose 6-phosphate

Step 3   ATP
              ADP

Fructose 1,6-diphosphate

Step 4

Glyceraldehyde phosphate (2)

Step 5   2NAD$^+$
              2NADH
              + 2H$^+$

1,3-Diphosphoglycerate (2)

Step 6   2ADP
              2ATP

3-Phosphoglyceric acid (2)

Step 7

2-Phosphoglyceric acid (2)

Step 8

Phosphoenolpyruvic acid (2)

Step 9   2ADP
              2ATP

Pyruvic acid (2)

Uses ATP

Yields ATP and NADH

9–4 *Summary of the two stages of glycolysis. The first stage utilizes 2ATP; the second stage yields 4ATP and 2NADH. Compounds other than glucose, such as the hexose galactose and a number of pentoses, as well as glycogen and starch, can undergo glycolysis once they have been converted to glucose 6-phosphate.*

To sum up: The energy from the phosphate bonds of two ATP molecules is needed to initiate the glycolytic sequence. Subsequently, two NADH molecules are produced from two NAD$^+$ and four ATP molecules from four ADP:

$$\text{Glucose} + 2\text{ATP} + 4\text{ADP} + 2\text{P}_i + 2\text{NAD}^+ \longrightarrow$$
$$2 \text{ Pyruvic acid} + 2\text{ADP} + 4\text{ATP} + 2\text{NADH} + 2\text{H}^+ + 2\text{H}_2\text{O}$$

Thus one glucose molecule has been converted to two molecules of pyruvic acid. The net harvest—the energy recovered—is two molecules of ATP and two molecules of NADH per molecule of glucose. The two molecules of pyruvic acid still contain a large amount of the potential energy that was stored in the original glucose molecule. This series of reactions is carried out by virtually all living cells—from prokaryotes to the eukaryotic cells of our own bodies.

ANAEROBIC PATHWAYS

Pyruvic acid can follow one of several pathways. One pathway is aerobic (with oxygen), and the others are anaerobic (without oxygen). We shall briefly discuss two of the most interesting anaerobic pathways and then follow the aerobic one, which is the principal pathway of energy metabolism for most cells in the presence of oxygen.

In the absence of oxygen, pyruvic acid can be converted to ethanol (ethyl alcohol) or to one of several different organic acids, of which lactic acid is the most common. The reaction product depends on the type of cell. For example, yeast cells, present as a "bloom" on the skin of grapes, can grow either with or without oxygen. When the sugar-filled juices of grapes and other fruits are extracted and stored under anaerobic conditions, the yeast cells turn the fruit juice to wine by converting glucose into ethanol (Figure 9–5). When the sugar is exhausted, the yeast cells cease to function; at this point, the alcohol concentration is between 12 and 17 percent, depending on the variety of the grapes and the season at which they were harvested.

The formation of alcohol from sugar is called **fermentation.** Because of the economic importance of the wine industry, fermentation was the first enzymatic process to be intensively studied. In fact, before their effects were known to be so diverse, enzymes were commonly referred to as "ferments."

Lactic acid is formed from pyruvic acid by a variety of microorganisms and also by some animal cells when O_2 is scarce or absent (Figure 9–6). It is produced, for example, in vertebrate muscle cells during strenuous exercise, as by an athlete during a sprint. We breathe hard when we run fast, thereby increasing the supply of oxygen, but even this increase may not be enough to meet the immediate needs of the muscle cells. These cells, however, can continue to work by accumulating what is known as an oxygen debt. Glycolysis continues, using glucose released from glycogen stored in the muscle, but the resulting pyruvic acid does not enter the aerobic pathway of respiration. Instead, it is converted to lactic acid, which, as it accumulates, lowers the pH of muscle and reduces the capacity of the muscle fibers to contract, producing the sensations of muscle fatigue. The lactic acid diffuses into the blood and is carried to the liver. Later, when oxygen is more abundant (as a result of the deep breathing that follows strenuous exercise) and ATP demand is reduced, the lactic acid is resynthesized to pyruvic acid and back again to glucose or glycogen.

Why is pyruvic acid converted to lactic acid, only to be converted back again? The function of the initial conversion is simple: it uses NADH and regenerates the NAD$^+$ without which glycolysis cannot go forward (see Step 5, page 189). Even though the overall process seems to be wasteful in terms of energy consumption, the regeneration of NAD$^+$ may be all-important in the economy of the organism, spelling the difference between life and death when an animal "out of breath" needs one last burst of ATP to escape from a predator or catch a prey.

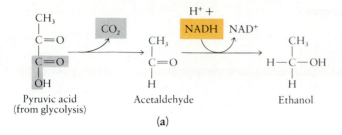

(a)

9–5 (a) *The steps by which pyruvic acid, formed by glycolysis, is converted anaerobically to ethanol (ethyl alcohol). In the first step, carbon dioxide is released. In the second, NADH is oxidized, and acetaldehyde is reduced. Most of the potential energy of the glucose remains in the alcohol, which is the end product of the sequence. However, by regenerating NAD^+, these steps allow glycolysis to continue, with its small but sometimes vitally necessary yield of ATP.*

(b) Yeast cells on the skins of these grapes give them their dust-like "bloom." When the grapes are crushed, the yeast cells mix with the juice. Storing the mixture under anaerobic conditions causes the yeast to break down the glucose in the grape juice to alcohol.

(b)

9–6 *The enzymatic reaction that produces lactic acid from pyruvic acid anaerobically in muscle cells. In the course of this reaction, NADH is oxidized and pyruvic acid is reduced. The NAD^+ molecules produced in this reaction and the one shown in Figure 9–5 are recycled in the glycolytic sequence. Without this recycling, glycolysis cannot proceed. Lactic acid accumulation results in muscle soreness and fatigue.*

An oxygen debt is usually accumulated during a short burst of intensive exercise—for example, a rapid sprint. During sustained moderate exercise, the intake of oxygen by the lungs and the circulation of the blood supplying oxygen to the muscle tissues often catch up with the oxygen consumption in the muscle, thus producing the "second wind" phenomenon well known to runners.

The fact that glycolysis does not require oxygen suggests that the glycolytic sequence evolved early, before free oxygen was present in the atmosphere. Presumably, primitive one-celled organisms used glycolysis (or something very much like it) to extract energy from the organic compounds they absorbed from their watery surroundings. Although anaerobic glycolysis generates only two molecules of ATP for each glucose molecule processed (a small fraction—about 5 percent—of the ATP that can be generated through aerobic processes), it was and is adequate for the needs of many organisms.

RESPIRATION

In the presence of oxygen, the next stage in the breakdown of glucose involves the stepwise oxidation of pyruvic acid to carbon dioxide and water—the process known as respiration. Respiration has two meanings in biology. One is the breathing in of oxygen and breathing out of carbon dioxide; this is also the ordinary, nontechnical meaning of the word. The second meaning of respiration is the oxidation of food molecules by cells. This process, sometimes qualified as cellular respiration, is what we are concerned with here.

As we noted previously, cellular respiration takes place in two stages: the Krebs cycle and terminal electron transport. In eukaryotic cells, these reactions take place within the mitochondria. Mitochondria, as we saw in Chapter 5, are surrounded by two membranes. The outer one is smooth, and the inner one folds inward. The folds are called cristae. Within the inner compartment of the mitochondrion, surrounding the cristae, is a dense solution known as the **matrix.** The matrix contains enzymes, coenzymes, water, phosphates, and other molecules involved in respiration. The outer membrane is permeable to most small molecules, but the inner one permits the passage of only certain molecules, such as pyruvic acid and ATP, and restrains the passage of others. As we shall see, this selective permeability of the inner membrane is critical to the ability of the mitochondria to harness the power of respiration to the production of ATP.

Dissecting the Cell

In every living cell, many hundreds of chemical reactions proceed simultaneously. Molecules are continuously synthesized via certain enzymatic pathways, and molecules are broken down by other enzymatic pathways. Many of these reactions are mutually incompatible, as can be demonstrated by destroying the structure of cells and mixing their enzymes in a test tube. Chemical chaos results, and the enzymes are soon inactivated. In the cell, however, anabolic and catabolic pathways operate in harmony because biochemical reactions are spatially localized and compartmentalized within specific subcellular organelles. A living cell is the most intensely concentrated set of chemical reactions known. A cell carries out many more chemical reactions than any apparatus devised by chemical engineers, and all within the space of a few cubic micrometers. The cell's unique chemical versatility results from the compartmentalization of biochemical pathways within organelles.

In order to study the specific functions of any organelle type, the organelle must be dissected free from all other cell structures and collected in large quantities. Mitochondria, lysosomes, and other organelles are, of course, far too small for hand dissection, but cell biologists can prepare pure samples of any organelle type by the technique of preparative centrifugation.

Small particles, ranging in size from cells to macromolecules, can be separated by centrifugation if the particle types differ in size and density. Particles suspended in fluid and then subjected to strong gravitational force will move through the fluid at varying rates, the largest, densest particles settling most rapidly. Forces up to 400,000 times the force of gravity (400,000 g) can be generated in a test tube of suspended particles by rotating the tube at very high speeds in an ultracentrifuge. Thus, subcellular structures, such as mitochondria, nuclei, and intracellular membranes, can be separated into purified fractions by spinning fragmented cells at appropriate centrifugal forces.

For instance, in order to determine which enzymatic pathways are present in mitochondria, a tissue, such as rat liver, is minced into small pieces and homogenized—that is, the cells are gently broken up by grinding the tissue, for example, in a glass tube fitted with a Teflon pestle. The tube contains a sucrose solution that is isotonic with intracellular fluid. The resulting suspension of cell organelles is then placed in an unbreakable test tube and spun in the centrifuge at low speed (700 g) for 10 minutes, so as to drive the bulkiest structures, the nuclei, to the bottom of the tube. All the lighter organelles remain suspended in the fluid, which is called the supernatant. The supernatant is transferred to another centrifuge tube and spun at a higher speed (10,000 g for 20

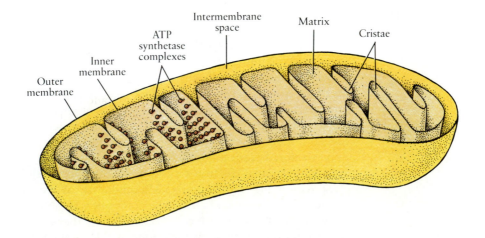

9–7   A mitochondrion is surrounded by two membranes. The inner membrane folds inward to make a series of shelves, or cristae. Many of the enzymes and electron carriers involved in cellular respiration are built into these internal membranes. Among the enzymes are the ATP synthetase complexes, which, as we shall see, play a critical role in the formation of ATP in the final stage of cellular respiration. The matrix is a dense solution containing enzymes involved in earlier stages of cellular respiration, plus coenzymes, phosphates, and other solutes.

minutes), which sediments particles such as mitochondria and lysosomes into a pellet at the bottom of the tube. The supernatant, containing ribosomes and various membranes, is discarded, and the pellet is retained.

At this point, the mitochondria have been partially purified by means of differential centrifugation. This type of centrifugation separates particles of quite different size and density by "spinning down" the large particles to form a pellet. To separate particles of rather similar size, such as mitochondria and lysosomes, the more subtle technique of zonal centrifugation is used. To return to the experiment we have outlined, the pellet containing the mitochondria and lysosomes is resuspended and gently layered atop a sucrose density gradient in a centrifuge tube. A sucrose density gradient is prepared by layering sucrose solutions of differing densities one above the other so that the densest solution is at the bottom of the tube and the least dense solution is at the top of the tube. When organelles are centrifuged in a density gradient (120,000 g for 8 hours), each organelle type moves through the gradient at a different rate, depending on the organelle's density. Following centrifugation, mitochondria will occupy one zone in the gradient, lysosomes another, and other organelles will be found in other zones. By puncturing the bottom of the tube and removing the contents drop by drop, we

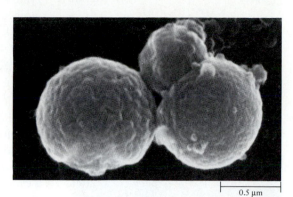

0.5 μm

Scanning electron micrograph of isolated, intact liver mitochondria.

can collect a pure sample of mitochondria. The purity can be verified by the electron microscope. If the procedure has been performed correctly, the isolated mitochondria will emerge with membranes intact and all enzymatic pathways functioning. It is then possible to test for the activities of specific enzymes, thus determining which biochemical functions are compartmentalized within mitochondria. Similarly, other cell constituents can be isolated and their biochemical activities determined.

Some of the enzymes of the Krebs cycle are in solution in the matrix. Other Krebs cycle enzymes and the enzymes and other components of the electron transport chains are built into the membrane of the cristae. These inner membranes of the mitochondria are about 80 percent protein and 20 percent lipid. In the mitochondria, pyruvic acid from glycolysis is oxidized to carbon dioxide and water, completing the breakdown of the glucose molecule. Ninety-five percent of the ATP generated by heterotrophic cells is produced in the mitochondria.

A Preliminary Step: The Oxidation of Pyruvic Acid

Pyruvic acid passes from the cytoplasm, where it is produced by glycolysis, and crosses the outer and inner membranes of the mitochondria. Before entering the Krebs cycle, the three-carbon pyruvic acid molecule is oxidized (Figure 9–8). The carbon and oxygen atoms of the carboxyl group are removed in the form of carbon dioxide, and a two-carbon acetyl group (CH_3CO) remains. In the course of this exergonic reaction, the hydrogen of the carboxyl group reduces a molecule of NAD^+ to NADH. The original glucose molecule has now been oxidized to two

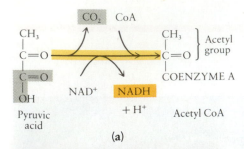

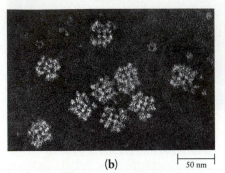

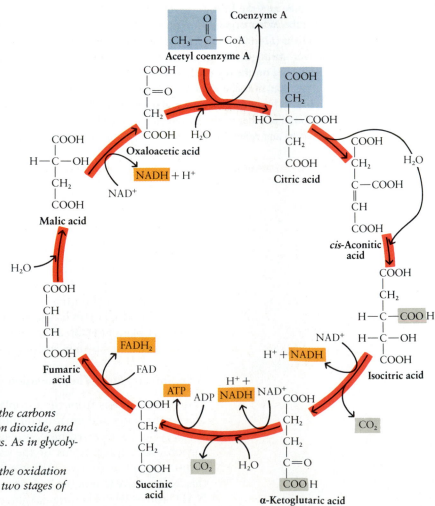

(a)

(b)   50 nm

9–8 (a) *The three-carbon pyruvic acid molecule is oxidized to the two-carbon acetyl group, which is combined with coenzyme A to form acetyl CoA. The oxidation of the pyruvic acid molecule is coupled to the reduction of NAD⁺. Acetyl CoA enters the Krebs cycle.* **(b)** *Electron micrograph showing the enzymes involved in the oxidation of pyruvic acid to acetyl CoA. Each of the complexes visible here represents multiple copies of the three different enzymes required for this reaction sequence.*

CO_2 molecules and two acetyl groups, and, in addition, four NADH molecules have been formed (two in glycolysis and two in the oxidation of pyruvic acid).

Each acetyl group is momentarily accepted by a compound known as coenzyme A. Like the coenzymes we have examined previously, coenzyme A is a large molecule, a portion of which is a nucleotide and a portion of which is a vitamin (pantothenic acid, one of the B complex vitamins). The combination of the acetyl group and CoA is abbreviated acetyl CoA. Its formation is the link between glycolysis and the Krebs cycle.

The Krebs Cycle

Upon entering the Krebs cycle (Figure 9–9), the two-carbon acetyl group is combined with a four-carbon compound (oxaloacetic acid) to produce a six-carbon compound (citric acid). In the course of the cycle, two of the six carbons are oxidized to CO_2, and oxaloacetic acid is regenerated—making this series literally a cycle. Each turn around the cycle uses up one acetyl group and regenerates a molecule of oxaloacetic acid, which is then ready to begin the sequence again.

In the course of these steps, some of the energy released by the oxidation of the carbon-hydrogen and carbon-carbon bonds is used to convert ADP to ATP (one molecule per cycle), and some is used to produce NADH and H⁺ from NAD⁺

9–9 *The Krebs cycle. In the course of the cycle, the carbons donated by the acetyl group are oxidized to carbon dioxide, and the hydrogen atoms are passed to electron carriers. As in glycolysis, a specific enzyme is involved at each step.*

Coenzyme A shuttles back and forth between the oxidation of pyruvic acid and the Krebs cycle, linking these two stages of respiration.

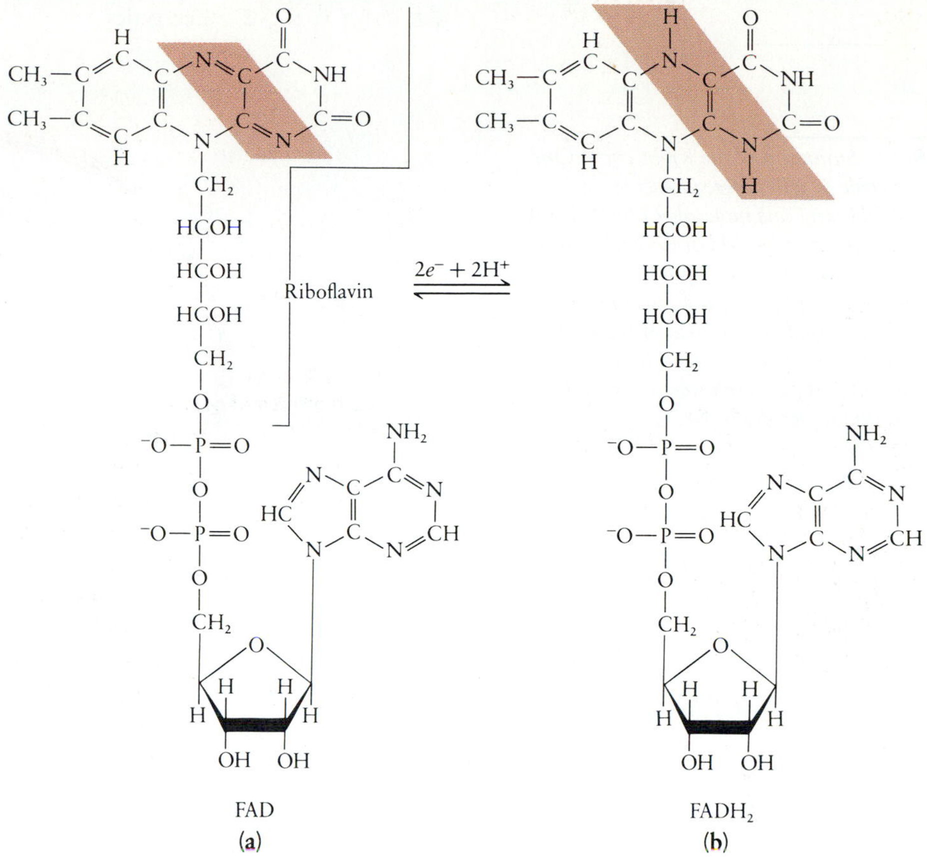

9–10 *Flavin adenine dinucleotide, an electron acceptor, in (**a**) its oxidized form (FAD) and (**b**) its reduced form (FADH$_2$). Riboflavin is a vitamin made by all plants and many microorganisms. It is also known as vitamin B$_2$. It is a pigment; in its oxidized form it is a bright yellow.*

A related electron acceptor, flavin mononucleotide (FMN), consists of riboflavin and the first phosphate group shown here. It accepts electrons from NADH in the electron transport chain.

In the living cell, both flavin adenine dinucleotide and flavin mononucleotide are bound to specific proteins, forming macromolecules known as flavoproteins.

(three molecules per cycle). In addition, some energy is used to reduce a second electron carrier, flavin adenine dinucleotide, abbreviated FAD (Figure 9–10). One molecule of FADH$_2$ is formed from FAD per turn of the cycle. No O$_2$ is required for the Krebs cycle; the electrons and protons removed in the oxidation of carbon are accepted by NAD$^+$ and FAD.

Oxaloacetic acid + Acetyl CoA + ADP + P$_i$ + 3NAD$^+$ + FAD $\longrightarrow$
    Oxaloacetic acid + 2CO$_2$ + CoA + ATP + 3NADH + FADH$_2$ +
                                                                3H$^+$ + H$_2$O

Note that the oxaloacetic acid molecule with which the cycle ends is not the same molecule with which the cycle began. If one begins with a glucose molecule in which the carbon atoms are radioactive, radioactive carbon atoms will appear among the four carbons of the oxaloacetic acid.

Electron Transport

The carbon atoms of the glucose molecule are now completely oxidized. Some of the potential energy of the glucose molecule has been used to produce ATP from ADP. Most of the energy, however, remains in electrons removed from the C—C and C—H bonds and passed to the electron carriers NAD$^+$ and FAD. These electrons are still at a high energy level.

In the final stage of respiration, these high-energy-level electrons are passed step-by-step to the low energy level of oxygen. The energy they yield in the course of this passage is ultimately used to regenerate ATP from ADP. This step-by-step passage is made possible by a series of electron carriers, each of which holds the electrons at a slightly lower level.

These carriers make up what is known as an **electron transport chain.** At the top of the energy hill the electrons are held by NADH and FADH$_2$. Most of the potential energy of the glucose molecule now resides in these electron acceptors.

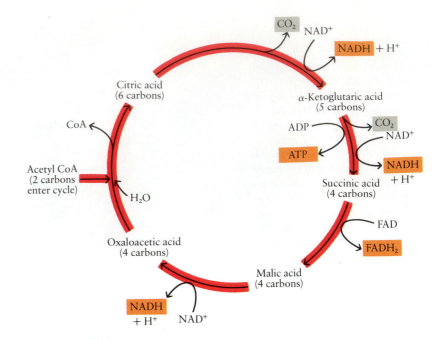

9–11 *Summary of the Krebs cycle. One molecule of ATP, three molecules of NADH, and one molecule of FADH$_2$ represent the energy yield of the cycle. Two turns of the cycle are required to complete the oxidation of one molecule of glucose. Thus the total energy yield of the Krebs cycle for one glucose molecule is two molecules of ATP, six molecules of NADH, and two molecules of FADH$_2$.*

The Krebs cycle yielded two molecules of FADH$_2$ and six molecules of NADH for each molecule of glucose. The oxidation of pyruvic acid to acetyl CoA yielded two molecules of NADH. Also, you will recall, two molecules of NADH were produced in glycolysis. In the presence of oxygen, the electrons held by these two NADH molecules are also transported into the mitochondrion where they are fed into the electron transport chain. In the process, NAD$^+$ is regenerated in the cytoplasm, allowing glycolysis to continue.

Among the principal components of the electron transport chain are molecules known as cytochromes. These molecules consist of a protein and a heme group, analogous to that of hemoglobin, in which an atom of iron is enclosed in a porphyrin ring (Figure 9–12). Although similar, the protein structures of the individual cytochromes differ enough to enable them to hold electrons at different energy levels. The iron atom of each cytochrome alternately accepts and releases an electron, passing it along to the next cytochrome at a slightly lower energy level until the electrons, their energy spent, are accepted by oxygen (Figure

9–12 *Cytochromes are molecules that participate in electron transfer in the mitochondria. A cytochrome molecule consists of a heme group held in an intricate protein structure. (a) The heme group of cytochrome c. The iron (Fe) atom, enclosed in a porphyrin ring (a nitrogen-containing ring), combines with and then releases electrons. (b) The overall structure of the cytochrome c molecule, showing the position of the heme group (color) within the globular protein.*

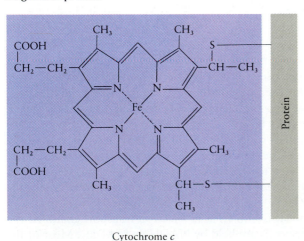

Cytochrome *c*

(a)

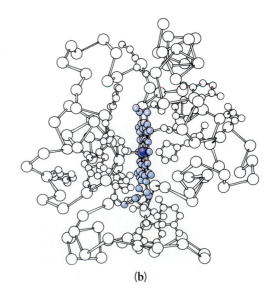

(b)

9-13 *The principal electron-carrier molecules of the electron transport chain. At least nine other carrier molecules function as intermediates between the carriers shown here.*

Flavin mononucleotide (FMN) and coenzyme Q (CoQ) transfer electrons and protons. The cytochromes transfer only electrons. The electrons carried by NADH enter the chain when they are transferred to FMN; those carried by FADH$_2$ enter the chain farther down the line at CoQ. The electrons are ultimately accepted by oxygen, which combines with protons (hydrogen ions) in the solution to form water.

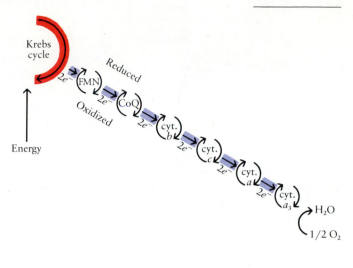

9-13). The energy released in this downhill passage of electrons is harnessed, as we shall see, to form ATP molecules from ADP. Such ATP formation is known as **oxidative phosphorylation.** At the end of the chain, the electrons are accepted by oxygen, which then combines with protons (H$^+$ ions) from the solution to produce water.

Quantitative measurements show that for every two electrons that pass from NADH to oxygen, three molecules of ATP are formed from ADP and phosphate. For every pair of electrons that passes from FADH$_2$, which holds them at a slightly lower energy level than NADH, two molecules of ATP are formed. In oxidative phosphorylation, the electron transfer potential of NADH and FADH$_2$ is converted to the phosphate transfer potential of ATP.

The Mechanism of Oxidative Phosphorylation: Chemiosmotic Coupling

For many years, the mechanism of oxidative phosphorylation—that is, the way in which ATP is formed from ADP and phosphate as electrons pass down the electron transport chain—was a puzzle. A major breakthrough occurred when a British biochemist, Peter Mitchell, proposed that the process is powered by a gradient of protons (H$^+$ ions) established across the inner mitochondrial membrane. Continuing studies have revealed many details of this mechanism, known as **chemiosmotic coupling,** although much remains to be learned.

The term "chemiosmotic" reflects the fact that the production of ATP in oxidative phosphorylation includes both chemical processes and transport processes across a selectively permeable membrane. Two distinct events take place in chemiosmotic coupling: (1) a proton gradient is established across the inner membrane of the mitochondrion, and (2) potential energy stored in the gradient is released and captured in the formation of ATP from ADP and phosphate.

The proton gradient is established as electrons move down the electron transport chain. At three transition points in this chain, significant drops occur in the amount of potential energy held by the electrons. As a consequence, a relatively large amount of free energy is released at each of these steps—as the electrons move from FMN to coenzyme Q, as they move from cytochrome *b* to cytochrome *c*, and as they move from cytochrome *a* to cytochrome *a$_3$*. This energy powers the pumping of protons from the mitochondrial matrix through the inner membrane to the intermembrane space—that is, the space between the inner and outer membranes of the mitochondrion. From the intermembrane space, some of the protons pass through the outer membrane into the cytoplasm (the outer membrane, you will recall, is freely permeable to most ions and small molecules).

Exactly how the pumping of protons is accomplished is a matter of current investigation. One hypothesis proposes that the electron carriers in the chain are

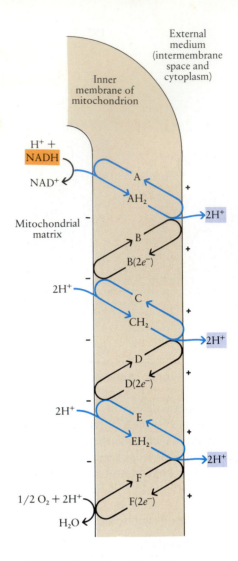

External medium (intermembrane space and cytoplasm)

Inner membrane of mitochondrion

H$^+$ + NADH

NAD$^+$

Mitochondrial matrix

A
AH$_2$
2H$^+$
B
B(2e$^-$)
2H$^+$
C
CH$_2$
2H$^+$
D
D(2e$^-$)
2H$^+$
E
EH$_2$
2H$^+$
F
F(2e$^-$)
1/2 O$_2$ + 2H$^+$
H$_2$O

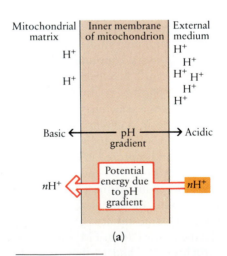

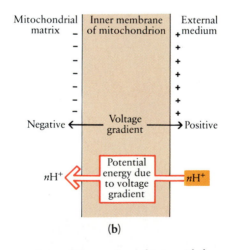

positioned so that the electrons travel a zigzag course from the inner to the outer surface of the inner membrane, traversing it three times in the course of their passage from NADH to oxygen. According to this hypothesis, each time two electrons travel from the inside of the membrane to the outside, they pick up two protons from the matrix and release them to the outside (Figure 9–14). Another hypothesis proposes that the energy released by the electrons triggers conformational changes in specific membrane transport proteins that enable them to carry protons against the concentration gradient from the matrix into the intermembrane space. The precise number of protons transported as each electron pair moves down the chain is uncertain; it is thought to be at least six, but it may be more.

Like the boulder at the top of the hill, the water at the top of the falls, or the chemical energy in a stick of dynamite, the difference in the concentration of protons between the matrix and the outside represents potential energy. This potential energy results not only from the difference in pH (more H$^+$ ions outside than inside) but also from the difference in electric charge (Figure 9–15). Because the inner membrane is impermeable to virtually all charged particles, other positive ions cannot move into the matrix to neutralize the negative charge created when the protons are pumped out. The movement of negative ions out of the matrix, which would also neutralize the charge difference, is similarly blocked. The result is potential energy, available to power any process that provides a channel allowing the protons to flow down the electrochemical gradient back into the matrix.

Such a channel is provided by a large enzyme complex known as **ATP synthetase**. This complex consists of two major portions, or factors, known as F$_O$ and F$_1$ (Figure 9–16a). F$_O$ is embedded in the inner mitochondrial membrane, traversing the membrane from outside to inside. It is thought to have an inner channel through which protons can pass. F$_1$ is a large globular structure, consisting of nine polypeptide subunits, that is attached to F$_O$ on the matrix side of the membrane.

9–14 *According to one hypothesis, the chemiosmotic proton gradient is established as the electrons contributed by NADH follow a looping pattern across the inner mitochondrial membrane during their transit down the electron transport chain. With each passage of two electrons from the matrix side of the membrane to its outer side, two protons are transported by carrier molecules (loops indicated in blue) and then discharged into the external medium. After discharge of the protons, the two electrons are transported back to the matrix side of the membrane by other carrier molecules (black loops). This hypothesis assumes that the carrier molecules (here designated as A through F) are fixed in the membrane in the necessary positions.*

Mitochondrial matrix | Inner membrane of mitochondrion | External medium

H$^+$

H$^+$

Basic ← pH gradient → Acidic

Potential energy due to pH gradient

nH$^+$

(a)

Mitochondrial matrix | Inner membrane of mitochondrion | External medium

Negative ← Voltage gradient → Positive

Potential energy due to voltage gradient

nH$^+$

(b)

9–15 *The potential energy stored by the pumping of protons from the mitochondrial matrix has two components: (a) a chemical gradient resulting from the difference in pH between the matrix and the external medium, and (b) a voltage gradient, resulting from the difference in electric charge on the inside and outside of the* membrane. *When a channel is provided that allows protons to reenter the matrix, they descend both gradients simultaneously. The contribution of the voltage gradient to the total potential energy is, however, greater than that of the chemical gradient.*

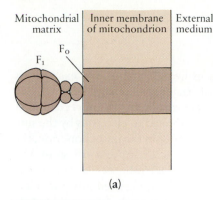

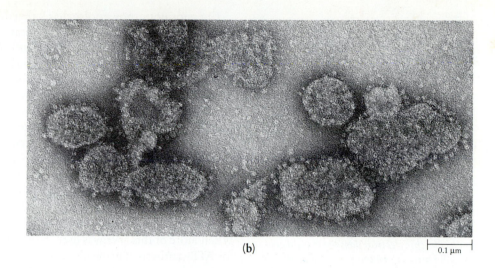

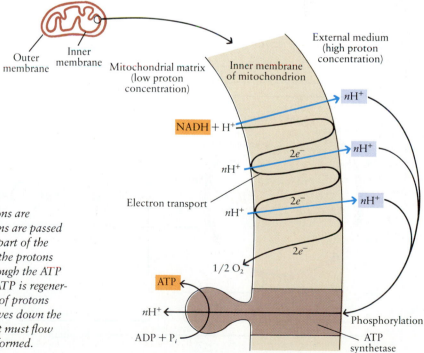

(a)

9–16 (a) *Diagram of the ATP synthetase complex. The F_O portion is contained within the inner membrane of the mitochondrion, and the F_1 portion, which consists of nine subunits, extends into the mitochondrial matrix.*

(b) The knobs protruding from the membrane of these vesicles are the F_1 portions of ATP synthetase complexes. The F_O portions to which they are attached are embedded in the membrane and are not visible in this electron micrograph. These vesicles were prepared by disrupting the inner mitochondrial membrane with ultrasonic waves. When the membrane is disrupted in this fashion, the membrane fragments immediately reseal, forming closed vesicles. These vesicles are, however, inside-out; the outer surface here is the surface that faces the matrix in the intact mitochondrion. Such inside-out vesicles are an important tool in the continuing study of oxidative phosphorylation.

(b)

0.1 µm

In electron micrographs of the inner mitochondrial membrane, the F_1 units appear as protruding knobs (Figure 9–16b). With proper chemical treatment, the F_1 unit can be removed from the mitochondrial membrane and subjected to detailed study. It has been shown to have binding sites for ATP and ADP, and, in solution, it catalyzes the hydrolysis of ATP to ADP, thus functioning as an ATPase. Its usual function when attached to the F_O unit in the intact mitochondrion is, however, the reverse. As protons flow down the electrochemical gradient from the outside into the matrix, passing through the F_O unit and then the F_1 unit, the free energy released powers the synthesis of ATP from ADP and phosphate. It is not certain whether the phosphorylation of one molecule of ADP to ATP requires the passage of two, three, or four protons through the ATP synthetase complex. Figure 9–17 summarizes the chemiosmotic coupling of oxidative phosphorylation.

9–17 *According to the chemiosmotic theory, protons are pumped out of the mitochondrial matrix as electrons are passed down the electron transport chain, which forms a part of the inner mitochondrial membrane. The movement of the protons down the electrochemical gradient as they pass through the ATP synthetase complex provides the energy by which ATP is regenerated from ADP and phosphate. The exact number of protons pumped out of the matrix as each electron pair moves down the chain is still to be determined, as is the number that must flow through ATP synthetase for each molecule of ATP formed.*

Chemiosmotic power also has other uses in living systems. For example, it provides the power that drives the rotation of bacterial flagella, which we shall discuss in Chapter 21. In photosynthetic cells, as we shall see in the next chapter, it is involved in the formation of ATP using energy supplied to electrons by the sun. And, it can be used to power other transport processes. In the mitochondrion, the energy stored in the proton gradient is used not only to drive the synthesis of ATP but also to carry other substances through the inner membrane via cotransport systems (page 137). Both phosphate and pyruvic acid are carried into the mitochondrion by integral membrane proteins that simultaneously transport protons down the gradient.

You will recall that we mentioned earlier that "about" 38 molecules of ATP are formed for each molecule of glucose oxidized to carbon dioxide and water. One of the reasons for this vagueness is that the exact amount of ATP formed depends on how the cell apportions the energy made available by the proton gradient. When more of this energy is used in other transport processes, less of it is available for ATP synthesis. The needs of the cell vary according to the circumstances, and so does the amount of ATP synthesized.

Control of Oxidative Phosphorylation

Electrons continue to flow along the electron transport chain, providing energy to create and maintain the proton gradient, only if ADP is available to be converted to ATP. Thus oxidative phosphorylation is regulated by supply and demand. When the energy requirements of the cell decrease, fewer molecules of ATP are used, fewer molecules of ADP become available, and electron flow is decreased.

OVERALL ENERGY HARVEST

We are now in a position to see how—and how much of—the potential energy originally present in the glucose molecule has been recovered in the form of ATP. Bear in mind that because the proton gradient in the mitochondrion can be used for purposes other than ATP synthesis, the figures we give represent the maximum energy harvest.

9–18 *Summary of glycolysis and respiration. Glucose is first broken down to pyruvic acid, with a yield of two ATP molecules and the reduction (dashed arrows) of two NAD^+ molecules to NADH. Pyruvic acid is oxidized to acetyl CoA, and one molecule of NAD^+ is reduced. (Note that this and subsequent reactions occur twice for each glucose molecule; this electron passage is indicated by solid arrows.) In the Krebs cycle, the acetyl group is oxidized and the electron acceptors NAD^+ and FAD are reduced. NADH and $FADH_2$ then transfer their electrons to the series of cytochromes and other electron carriers that make up the electron transport chain. As the electrons are passed "downhill," relatively large amounts of free energy are released during the passage from FMN to CoQ, from cytochrome b to cytochrome c, and from cytochrome a to cytochrome a_3. These bursts of free energy transport protons through the inner mitochondrial membrane, establishing the proton gradient that powers the synthesis of ATP from ADP.*

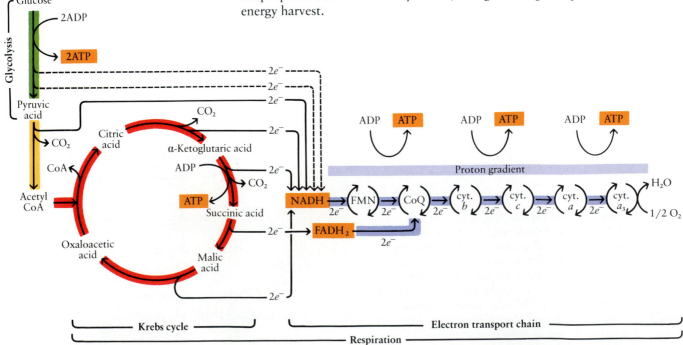

Glycolysis, in the presence of oxygen, yields two molecules of ATP directly and two molecules of NADH. These NADH molecules, however, cannot cross the inner membrane of the mitochondrion, and the electrons they carry must be "shuttled across" the membrane. In most cells, the energy cost of this process is quite low, and each NADH formed in glycolysis ultimately results in the synthesis of three molecules of ATP. In these cells, the total gain from glycolysis is 8 ATP. In other cells, including those of brain, skeletal muscle, and insect flight muscle, the energy cost of the shuttle is higher; the electrons are at a lower energy level when they reach the electron transport chain, and they enter the chain at coenzyme Q, rather than at FMN. Thus, like the electrons carried by $FADH_2$ from the Krebs cycle, they yield only two molecules of ATP per electron pair. In such cells, the total gain from glycolysis is only 6 ATP. (This was the second factor in our earlier hedging about the number of ATPs formed.)

The conversion of pyruvic acid to acetyl CoA, which occurs inside the mitochondrion, yields two molecules of NADH for each molecule of glucose and so produces six molecules of ATP.

The Krebs cycle, which also occurs inside the mitochondrion, yields two molecules of ATP, six of NADH, and two of $FADH_2$, or a total of 24 ATP, for each molecule of glucose.

As a balance sheet (Table 9–1) shows, the complete yield from a single molecule of glucose is a maximum of 38 molecules of ATP. Note that all but 2 of the 38 molecules of ATP have come from reactions taking place in the mitochondrion, and all but 4 result from the passage down the electron transport chain of electrons carried by NADH or $FADH_2$.

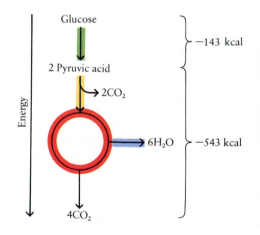

9–19 *Energy changes in the oxidation of glucose. The complete respiratory sequence (glucose + $6O_2$ ⟶ $6CO_2$ + $6H_2O$) proceeds with an energy drop of 686 kcal/mole. Of this, almost 40 percent (266 kilocalories) is conserved in 38 ATP molecules. In anaerobic glycolysis (glucose ⟶ lactic acid), by contrast, only 2 ATP molecules are produced, representing only about 2 percent of the available energy of glucose.*

TABLE 9–1   Summary of Maximum Energy Yield from the Oxidation of One Molecule of Glucose

In the Cytoplasm				
Glycolysis:		2 ATP	⟶	2 ATP
In the Mitochondria				
From glycolysis:	2 NADH ⟶	6 ATP	⟶	6 ATP*
From respiration:				
Pyruvic acid ⟶ Acetyl CoA:	1 NADH ⟶	3 ATP	(× 2) ⟶	6 ATP
Krebs cycle:		1 ATP ⎫		
	3 NADH ⟶	9 ATP ⎬ (× 2)	⟶	24 ATP
	1 $FADH_2$ ⟶	2 ATP ⎭		
Total:				38 ATP

* In some cells, the energy cost of transporting the electrons from the NADH molecules formed in glycolysis across the inner mitochondrial membrane lowers the net yield from these 2 NADH to 4 ATP; thus the total maximum yield in these cells is 36 ATP.

The free energy change (ΔG) that occurs during glycolysis and respiration is -686 kilocalories per mole. About 266 kilocalories per mole (7 kilocalories per mole of ATP × 38 moles of ATP) have been captured in the phosphate bonds of the ATP molecules, an efficiency of almost 40 percent.

The ATP molecules, once formed, are exported across the membrane of the mitochondrion by a shuttle system that simultaneously brings in one molecule of ADP for each ATP exported.

Ethanol, NADH, and the Liver

The human body can dispose fairly readily of most toxic products of its own manufacture, such as carbon dioxide and nitrogenous wastes. In contrast, most ingested toxic substances, such as ethanol (beverage alcohol), must first be broken down by the liver, which possesses special enzymes not present in other tissues.

It has been known for many years that heavy drinkers are at great risk for severe, and often fatal, liver disease. Studies conducted by Charles S. Lieber and his colleagues at the Bronx Veterans Administration Hospital and the Mount Sinai School of Medicine in New York City, have demonstrated that the origin of the problem lies in the simple chemical steps involved in the breakdown of ethanol. Enzymes in the liver first oxidize ethanol (CH_3CH_2OH) to acetaldehyde (CH_3CHO), removing two hydrogen atoms and reducing a molecule of NAD^+; this is the reverse of the second reaction shown in Figure 9–5a on page 191. The acetaldehyde is then oxidized to acetic acid, which is, in turn, oxidized to carbon dioxide and water and eliminated from the body.

Although the intoxicating effects of alcohol are due mostly to the acetaldehyde, which stimulates the release of adrenalinelike agents, the chief culprits in the development of liver disease are the hydrogen atoms (electrons and protons) removed from ethanol. These "extra" hydrogens—carried by NADH—follow two principal pathways within the cell. Most are fed directly into the electron transport chain, producing water and ATP. Because of the high levels of NADH present in the cell from the oxidation of ethanol, the production of NADH by glycolysis and the Krebs cycle is reduced. As a result, sugars, amino acids, and fatty acids are not broken down but are instead converted to fats. The fats accumulate in the liver. The mitochondria also swell, presumably as a result of the distortion of their normal function—the electron transport chain is doing very heavy duty, while the Krebs cycle is effectively shut down.

Other hydrogen atoms are used in the synthesis of fatty acids from the carbohydrate skeletons that are not being processed in glycolysis and the Krebs cycle. More fats accumulate. It does not take long. In human volunteers fed a good high-protein, low-fat diet, six drinks (about 10 ounces) a day of 86 proof alcohol produced an eightfold increase in fat deposits in the liver in only 18 days. Fortunately, these early effects are completely reversible.

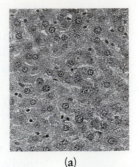
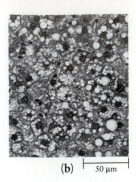

(a)   (b) ⊢ 50 μm

(a) *Normal liver tissue from a rat fed a balanced liquid diet for 24 days.* (b) *In this liver tissue from another rat fed a liquid diet in which ethanol provided 36 percent of the total calories, many globular fat droplets have accumulated. This rat was also maintained on its special diet for 24 days.*

The liver cells work hard to get rid of the excess fats, which are not soluble in blood plasma. Before being released into the bloodstream, the fats are coated with a thin layer of protein in a process carried out on the membranes of the endoplasmic reticulum. The liver cells of heavy drinkers show enormous proliferation of the endoplasmic reticulum.

After a few years—depending on how much alcohol is consumed—liver cells, engorged with fat, begin to die, triggering the inflammatory process known as alcoholic hepatitis. Liver function becomes impaired. Cirrhosis is the next step; it is the formation of scar tissue, which interferes with the function of the individual cells and also with the supply of blood to the liver. This leads to the death of more cells. The liver can no longer carry out its normal activities—such as breaking down nitrogenous wastes—which is why cirrhosis is a cause of death. In fact, cirrhosis of the liver is the ninth leading cause of death in the United States.

Not so long ago, it was commonly believed that a good diet was all that was required to protect even a heavy drinker from the deleterious effects of alcohol. In fact, if one were just to add a few vitamins to the alcohol itself, some sophisticates maintained, most of the long-term physical damage of alcohol would disappear. This new evidence refutes these comforting notions, and it comes at a time when alcohol is enjoying a resurgence of popularity among persons of high school and college age. (In populations of postgraduate age, as in other human societies the world over, it never lost its status as the drug of choice for abuse.)

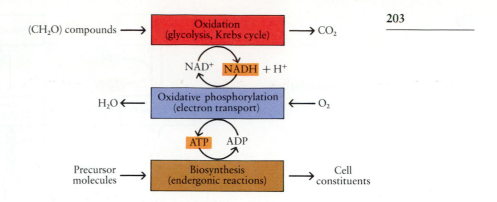

9–20 *The strategy of energy metabolism. Organisms extract energy from compounds by oxidizing them to carbon dioxide and water. NAD$^+$ is the major oxidizing agent. The NADH gives up its high-energy electrons and associated protons to electron acceptors in the electron transport chain. The electrons are passed down the chain, ultimately to oxygen. The energy released in this process is used to phosphorylate ADP, converting it to ATP, which is used to drive the endergonic reactions of the organism. Note that both NADH and ATP are used in a cyclic fashion. Although they are critically important compounds, they are present in very small amounts, accomplishing their work as a result of constant and rapid turnover. The need for a constant supply of these molecules explains why a deprivation of oxygen brings about death in a very few minutes.*

OTHER CATABOLIC PATHWAYS

Most organisms do not feed directly on glucose. How do they extract energy from, for example, fats or proteins? The answer lies in the fact that the Krebs cycle is a Grand Central Station for energy metabolism. Other foodstuffs are broken down and converted into molecules that can feed into this central pathway.

Polysaccharides, such as starch, are broken down to their constituent monosaccharides and phosphorylated to glucose 6-phosphate; in this form, they enter the glycolytic pathway. Fats are first split into their glycerol and fatty acid components. The fatty acids are then chopped up into two-carbon fragments and slipped into the Krebs cycle as acetyl CoA. Proteins are broken down into their constituent amino acids. The amino acids are deaminated (the amino groups removed), and the residual carbon skeleton is either converted to an acetyl group or to one of the larger carbon compounds of the glycolytic pathway or the Krebs cycle so that it can be processed at this stage of the central pathway. The amino groups, if not reutilized, are eventually excreted as urea or other nitrogen-containing wastes. These various degradative pathways are, collectively, catabolism.

BIOSYNTHESIS

The pathways of glucose breakdown, central to catabolism, are also central to the biosynthetic, or anabolic, processes of life. These processes are the pathways of synthesis of the various molecules and macromolecules that make up an organism.

Since many of these substances, such as proteins and lipids, can be broken down and fed into the central pathway, you might guess that the reverse process can occur—namely, that the various intermediates of glycolysis and the Krebs cycle can serve as precursors for biosynthesis. This is in fact the case. However, the biosynthetic pathways, while similar to the catabolic ones, are distinctive. Different enzymes control the steps, and various critical steps of anabolism differ from those of the catabolic processes. These general pathways, which are followed in the cells of virtually all living organisms, are outlined in Figure 9–21.

In order for the reactions of the catabolic and anabolic pathways to occur, there must be a steady supply of organic molecules that can be broken down to yield energy and building block molecules. Without a supply of such molecules, the metabolic pathways cease to function and the life of the organism ends. Heterotrophic cells (including the heterotrophic cells of plants, such as the cells of the roots) are dependent on external sources—specifically, autotrophic cells—for the organic molecules that are essential to life. Autotrophic cells, however, are able to synthesize monosaccharides from simple inorganic molecules and an external energy source. These monosaccharides are then used not only to supply energy but also as building blocks for the variety of organic molecules synthesized in the anabolic pathways. By far the most important autotrophic cells are the photosynthetic cells of algae and plants. In the next chapter, we shall examine how these cells capture the energy of sunlight and use it to synthesize the monosaccharide molecules on which life on this planet depends.

9–21 *Major pathways of catabolism and anabolism in the cell.*

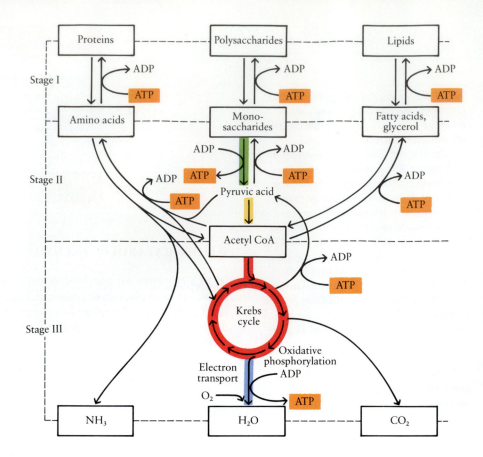

SUMMARY

The oxidation of glucose is a chief source of energy in most cells. As the glucose is broken down in a series of small enzymatic steps, a significant proportion of the potential energy of the molecule is repackaged in the phosphate bonds of ATP molecules.

The first phase in the breakdown of glucose is glycolysis, in which the six-carbon glucose molecule is split into two three-carbon molecules of pyruvic acid. A net yield of two molecules of ATP (from ADP) and two of NADH (from NAD⁺) results from the process. Glycolysis takes place in the cytoplasm of the cell.

The second phase in the breakdown of glucose and other fuel molecules is respiration. It requires oxygen and, in eukaryotic cells, takes place in the mitochondria. It occurs in two stages: the Krebs cycle and terminal electron transport. (In the absence of oxygen, the pyruvic acid produced by glycolysis is converted to either ethanol or lactic acid by the process of fermentation. NAD⁺ is regenerated, allowing glycolysis to continue, producing a small but vital supply of ATP for the organism.)

In the course of respiration, the three-carbon pyruvic acid molecules from glycolysis are broken down to two-carbon acetyl groups, which then enter the Krebs cycle. In a series of reactions in the Krebs cycle, the two-carbon acetyl group is oxidized completely to carbon dioxide. In the course of the oxidation of each acetyl group, four electron acceptors (three NAD⁺ and one FAD) are reduced, and another molecule of ATP is formed.

The final stage of respiration is terminal electron transport, which involves a chain of electron carriers and enzymes embedded in the inner membrane of the mitochondrion. Along this series of electron carriers, the high-energy electrons carried by NADH from glycolysis and by NADH and FADH₂ from the Krebs cycle pass downhill to oxygen. At three points in their passage down the complete electron transport chain, large quantities of free energy are released that power the pumping of protons (H^+ ions) out of the mitochondrial matrix. This creates an electrochemical gradient of potential energy across the inner membrane of the mitochondrion. When protons pass through the ATP synthetase complex as they flow down the electrochemical gradient back into the matrix, the energy released is used to form ATP molecules from ADP and phosphate. This mechanism, by which oxidative phosphorylation is accomplished, is known as chemiosmotic coupling.

In the course of the breakdown of the glucose molecule, a maximum of 38 molecules of ATP can be formed. The exact number of ATP molecules formed depends on how much of the energy of the proton gradient is used to power other mitochondrial transport processes and on the shuttle mechanism by which the electrons from the NADH molecules formed in glycolysis are brought into the mitochondrion. Generally, almost 40 percent of the free energy released in glucose oxidation is retained in the form of newly synthesized ATP molecules.

Other food molecules, including fats, polysaccharides, and proteins, are utilized by being degraded to compounds that can enter these central pathways at various steps. The biosynthesis of these substances also originates with precursor compounds derived from intermediates in the respiratory sequence and is driven by the energy derived from those processes.

QUESTIONS

1. Distinguish among the following: oxidation of glucose/glycolysis/respiration/fermentation; aerobic pathways/anaerobic pathways; $FAD/FADH_2$; Krebs cycle/electron transport.

2. Describe the process of fermentation. What conditions are essential if it is to occur? With some strains of yeast, fermentation stops before the sugar is exhausted, usually at an alcohol concentration in excess of 12 percent. What is a plausible explanation?

3. If aerobic (oxygen-utilizing) organisms are so much more efficient than anaerobes in converting energy, why are there any anaerobes left on this planet? Why didn't they all become extinct long ago?

4. Sketch the structure of a mitochondrion. Describe where the various stages in the breakdown of glucose take place in relation to mitochondrial structure. What molecules and ions cross the mitochondrial membranes during these processes?

5. Each $FADH_2$ molecule produced in the Krebs cycle results in the formation of only two molecules of ATP when its electrons pass down the electron transport chain. It was long thought that because these electrons enter the chain at coenzyme Q rather than at FMN they "missed" one of the sites of phosphorylation. What is a more accurate explanation?

6. Cyanide can combine with—and deactivate—cytochrome a and cytochrome a_3. In our bodies, however, cyanide tends to react first with hemoglobin and to make it impossible for oxygen to bind to the hemoglobin. Either way, cyanide poisoning has the same effect: it inhibits the synthesis of ATP. Explain how this is so.

7. When the F_1 portion of the ATP synthetase complex is removed from the mitochondrial membrane and studied in solution, it functions as an ATPase. Why does it not function as an ATP synthetase?

8. Certain chemicals function as "uncoupling" agents when they are added to respiring mitochondria. The passage of electrons down the chain to oxygen continues, but no ATP is formed. One of these agents, the antibiotic valinomycin, is known to transport K^+ ions through the inner membrane into the matrix. Another, 2,4-dinitrophenol, transports H^+ ions through the membrane. How do these substances prevent the formation of ATP? Which would you expect to have the most profound effect on ATP formation? Why?

9. In the cells of a specialized tissue known as brown fat, the inner membrane of the mitochondrion is permeable to H^+ ions. These cells contain large stores of fat molecules, which are gradually broken down and the resulting acetyl groups are fed into the Krebs cycle. The electrons captured by NADH and $FADH_2$ are, in turn, fed into the electron transport chain and ultimately accepted by oxygen. No ATP is synthesized, however. Why not? Brown fat tissue is found in some hibernating animals and in mammalian infants that are born hairless, including human infants. What do you suppose the function of brown fat tissue is?

10. (a) As we have seen, a cell can obtain a maximum of 38 molecules of ATP from each molecule of glucose that is completely oxidized. Account for the production of each molecule of ATP. (b) In the course of glycolysis, the Krebs cycle, and electron transport, 40 molecules of ATP are actually formed. Why is the net yield for the cell only 38 molecules? (c) What other factors can reduce the yield of ATP?

11. Describe how the processes of the cell are adapted to the efficient use of a variety of foodstuffs, and to the efficient production of the variety of materials that the cell needs to manufacture for its own use.

12. In terms of the cell's economy, what do anabolic processes provide for the cell? What do catabolic processes provide? How are they dependent on each other?

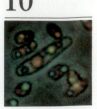

C H A P T E R 10

Photosynthesis, Light, and Life

The first photosynthetic organism probably appeared 3 to 3.5 billion years ago. Before the evolution of photosynthesis, the physical characteristics of earth and its atmosphere were the most powerful forces in shaping the course of natural selection. With the evolution of photosynthesis, however, organisms began to change the face of our planet and, as a consequence, to exert strong influences on each other. Organisms have continued to change the environment, at an ever-increasing rate, up to the present day.

As we saw earlier, the atmosphere in which the first cells evolved lacked free oxygen. These earliest organisms were, of course, adapted to living in an environment without free oxygen; in fact, oxygen, with its powerful electron-attracting capacities, would have been poisonous to them (as it is to many modern anaerobes). Their energy came from anaerobic processes, most likely glycolysis and fermentation, which would have resulted in a gradual accumulation of carbon dioxide in the atmosphere. The organic molecules they used as fuel may have been formed by nonbiological processes (page 87), or they may have been produced by chemosynthetic autotrophs (page 89) or by primitive photosynthetic cells that, like some modern photosynthetic bacteria, did not release oxygen to the environment.

Then, it is hypothesized, there slowly evolved photosynthetic organisms that used carbon dioxide as their carbon source and released oxygen, as do most modern photosynthetic forms. As these photosynthetic organisms multiplied, they provided a new supply of organic molecules, and free oxygen began to accumulate. In response to these changing conditions, cell species arose for which oxygen was not a poison but rather a requirement for existence. (It has been proposed that the original function of the electron transport chain found in the cell membrane of aerobic bacteria—thought to be the forerunner of the mito-chondrial electron transport chain—was to protect the cell from oxygen.)

As we saw in the last chapter, oxygen-utilizing organisms have an advantage over those that do not use oxygen. A higher yield of energy can be extracted per molecule from the aerobic breakdown of carbon-containing compounds than from anaerobic processes, in which fuel molecules are not completely oxidized. Energy released in cells by reactions using oxygen made possible the development of increasingly active, increasingly complex organisms. Without oxygen, the complex forms of life that now exist on earth could not have evolved.

Life on earth continues to be dependent on photosynthesis both for its oxygen and for its carbon-containing fuel molecules. Photosynthetic organisms capture light energy and use it to form carbohydrates and free oxygen from carbon dioxide and water, in a complex series of reactions. The overall equation for photosynthesis can be summarized as:

$$CO_2 + H_2O + \text{Light energy} \longrightarrow \underset{\text{Carbohydrate}}{(CH_2O)} + O_2$$

To understand how organisms are able to capture light energy and convert it into stored chemical energy, we must first look at the characteristics of light itself.

10–1 *A freshwater green alga of the genus* Micrasterias. *In this alga and other unicellular algae, each cell is an individual, self-sufficient photosynthetic organism. The green color is due to chlorophyll, in which the radiant energy of sunlight is converted to chemical energy.*

20 μm

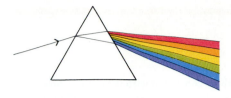

10–2 *White light is actually a mixture of different colors, ranging from violet at one end of the spectrum to red at the other. It is separated into its component colors when it passes through a prism—"the celebrated phaenomena of colors," as Newton referred to it.*

THE NATURE OF LIGHT

Over 300 years ago, the English physicist Sir Isaac Newton (1642–1727) separated visible light into a spectrum of colors by passing it through a prism (Figure 10–2). Then by passing the light through a second prism, he recombined the colors, producing white light once again. By this experiment, Newton showed that white light is actually made up of a number of different colors, ranging from violet at one end of the spectrum to red at the other. Their separation is possible because light of different colors is bent at different angles in passing through the prism. Newton believed that light was a stream of particles (or, as he termed them, "corpuscles"), in part, because of its tendency to travel in a straight line.

In the nineteenth century, through the genius of James Clerk Maxwell (1831–1879), it became known that what we experience as light is in truth a very small part of a vast continuous spectrum of radiation, the electromagnetic spectrum (Figure 10–3). As Maxwell showed, all the radiations in this spectrum act as if they travel in waves. The wavelengths—that is, the distances from one wave peak to the next—range from those of gamma rays, which are measured in nanometers (1 nanometer = 10^{-9} meter), to those of low-frequency radio waves, which are measured in kilometers (1 kilometer = 10^3 meters). Within the spectrum of visible light, red light has the longest wavelength, violet the shortest. Another feature that these radiations have in common is that, in a vacuum, they all travel at the same speed—300,000 kilometers per second.

By 1900, it had become clear, however, that the wave model of light was not adequate. The key observation, a very simple one, was made in 1888: when a zinc plate is exposed to ultraviolet light, it acquires a positive charge. The metal, it was soon deduced, becomes positively charged because the light energy dislodges electrons, forcing them out of the metal atoms. Subsequently, it was discovered that this photoelectric effect, as it is called, can be produced in all metals. Every metal has a critical wavelength for the effect; the light (visible or invisible) must be of that wavelength or a shorter wavelength for the effect to occur.

With some metals, such as sodium, potassium, and selenium, the critical wavelength is within the spectrum of visible light, and as a consequence, visible light striking the metal can set up a moving stream of electrons (such a stream is an electric current). Burglar alarms, exposure meters, television cameras, and the electric eyes that open doors for you at supermarkets or airline terminals all operate on this principle of turning light energy into electrical energy.

10–3 *Visible light is only a small portion of the vast electromagnetic spectrum. For the human eye, the visible spectrum ranges from violet light, which is made up of comparatively short rays, to red light, the longest visible rays.*

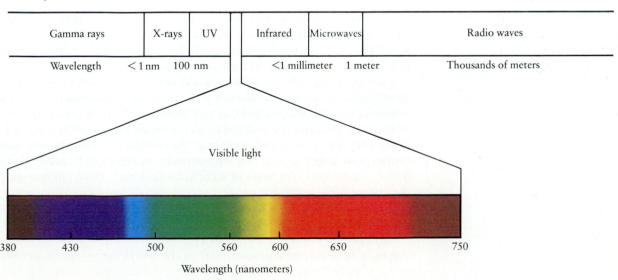

| Gamma rays | X-rays | UV | Infrared | Microwaves | Radio waves |

Wavelength   < 1 nm   100 nm   <1 millimeter   1 meter   Thousands of meters

Visible light

| 380 | 430 | 500 | 560 | 600 | 650 | 750 |

Wavelength (nanometers)

No Vegetable Grows in Vain

Until about 350 years ago, observers of the biological world, noting that the life processes of animals were dependent on the food they ate, thought that plants derived their food from the soil in a similar way. This concept was widely accepted until the Belgian physician Jan Baptista van Helmont (1577–1644) offered the first experimental evidence to the contrary.

Van Helmont grew a small willow tree in an earthenware pot for five years, adding only water to the pot. At the end of five years, the willow had increased in weight by 74 kilograms, while the earth in the pot had decreased in weight by only 57 grams. On the basis of these results, van Helmont concluded that all the substance of the plant was produced from the water and none from the soil. (This experiment is of general interest to those concerned with tracing the history of science, because it is one of the first carefully designed biological experiments ever reported. Van Helmont's conclusions, however, were too broad.)

The next stage in our understanding of plant nutrition resulted from studies of combustion, a topic that intrigued not only the medieval alchemists but also their successors who laid the foundations of modern chemistry. One of the fascinating problems about combustion was that it in some way "injured" air. For example, if a candle was burned in a closed container, it would soon go out; if a mouse was then put in the container, it would die.

One of those concerned with the changes produced in air by burning was Joseph Priestley (1733–1804), an English clergyman and chemist. On August 17, 1771, Priestley "put a sprig of mint into air in which a wax candle had burned out and found that, on the 27th of the same month, another candle could be burned in this same air." Priestley believed, as he reported, that he had "accidentally hit upon a method of restoring air that has been injured by the burning of candles." The "restorative which nature employs for this purpose," he stated, is "vegetation." Priestley extended his observations and soon showed that air "restored" by vegetation was not "at all inconvenient to a mouse." These experiments offered the first logical explanation of how the air remained "pure" and able to support life despite the burning of countless fires and the breathing of many animals. When Priestley was presented with a medal for his discovery, the citation read in part: "For these discoveries we are assured that no vegetable grows in vain ... but cleanses and purifies our atmosphere."

Priestley's reports that plants purify the air were of great interest to his fellow chemists, but they soon attracted criticism because the experiments could not be confirmed. In fact, when Priestley tried to do the experiments again himself, he did not get the same results. (We think now that he may have moved his equipment to a dark corner of his laboratory.) It was a Dutch physician, Jan Ingenhousz (1730–1799), who

The wave model of light would lead you to predict that the brighter the light—that is, the stronger, or more intense, the beam—the greater the force with which the electrons would be dislodged. But as we have already seen, whether or not light can eject the electrons of a particular metal depends not on the brightness of the light but on its wavelength. A very weak beam of the critical wavelength or a shorter wavelength is effective, while a stronger beam of a longer wavelength is not. Furthermore, as was shown in 1902, increasing the brightness of the light increases the number of electrons dislodged but not the velocity at which they are ejected from the metal. To increase the velocity, one must use a shorter wavelength of light. Nor is it necessary for energy to be accumulated in the metal. With even a dim beam of a critical wavelength, electrons may be emitted the instant the light hits the metal.

To explain such phenomena, the particle model of light was resurrected by Albert Einstein in 1905. According to this model, light is composed of particles of energy called **photons.** The energy of a photon is not the same for all kinds of light but is, in fact, inversely proportional to the wavelength—the longer the wave-

was finally able to confirm Priestley's work with an important addition. He found that the purification takes place only in sunlight. Plants at night or in the shade, he reported, "contaminate the air which surrounds them, throwing out an air hurtful to animals." He also observed that only the green parts of plants restore the air and, on the basis of control experiments, that "the sun by itself has no power to mend air without the concurrence of plants."

While Ingenhousz was performing his experiments on plants, Antoine Lavoisier (1743–1794) was carrying out the experiments that put chemistry on a modern basis. Among Lavoisier's many discoveries, those that had the most impact on studies of plant processes concerned the exchanges of gases that take place when animals breathe. Working with the mathematician P. S. Laplace (1749–1827), Lavoisier confined a guinea pig for about 10 hours in a jar containing oxygen and measured the carbon dioxide produced. He also measured the amount of oxygen used by a man active and at rest. In these experiments, he was able to show that the combustion of carbon compounds with oxygen is the true source of animal heat and that oxygen consumption increases during physical work. "Respiration is merely a slow combustion of carbon and hydrogen, which is similar in every respect to that which occurs in a lighted lamp or candle, and, from this point of view, animals that breathe are really combustible bodies which burn and are consumed."

The work of Ingenhousz spanned the prematurely terminated career of Lavoisier, who was guillotined on May 8, 1794, during the French Revolution. (The judge presiding over the case is reported to have said, "The Republic has no need of savants.") Quick to adopt Lavoisier's ideas about gases, Ingenhousz hypothesized that the plant was not just exchanging "good air" for "bad" and so making the world habitable for animal life. In the sunshine, he suggested, a plant absorbs the carbon from carbon dioxide, "throwing out at that time the oxygen alone, and keeping the carbon to itself as nourishment."

Nicholas Theodore de Saussure (1767–1845) later showed that equal volumes of CO_2 and O_2 are exchanged during photosynthesis and that the plant does indeed retain the carbon. He also showed that more weight was gained by the plant during photosynthesis than could be accounted for by the carbon taken in as carbon dioxide. In other words, the carbon in the dry matter of plants comes from carbon dioxide, but equally important, the rest of the dry matter, with the exception of minerals from the soil, comes from water. Thus, all the components—carbon dioxide, water, and light—were identified, and it became possible to write the overall photosynthetic equation, as shown on page 206.

length, the lower the energy. Photons of violet light, for example, have almost twice the energy of photons of red light, the longest visible wavelength.

The wave model of light permits physicists to describe certain aspects of its behavior mathematically, and the photon model permits another set of mathematical calculations and predictions. These two models are no longer regarded as opposed to one another; rather, they are complementary, in the sense that both—or a totally new model—are required for a complete description of the phenomenon we know as light.

The Fitness of Light

Light, as Maxwell showed, is only a tiny band in a continuous spectrum. From the physicist's point of view, the difference between radiations we can see and radiations we cannot see—so dramatic to the human eye—is only a few nanometers of wavelength, or, expressed differently, a small amount of energy. Why does this particular group of radiations, rather than some other, make the leaves

(a)

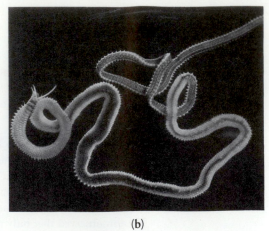

(b)

10-4 (a) *Gathering palolo worms at dawn in Western Samoa. Once a year, following the first full moon after the autumnal equinox, Pacific palolo worms leave their undersea burrows and swarm. The rear portion of each worm, filled with eggs or sperm, breaks off and, attracted to the* moonlight, *rises to the water's surface. There, in the early morning hours, before daylight has stimulated the release of eggs and sperm, people gather the wriggling worms for a communal feast. The sea is said to resemble vermicelli soup. Because* it occurs so accurately, the day of the "big rising" marked the beginning of the Samoan New Year in premissionary days. (b) A palolo worm. Its narrower posterior region is filled with sperm and will break off when the worm swarms.

grow and the flowers burst forth, cause the mating of fireflies and the spawning of palolo worms (Figure 10–4), and, when reflecting off the surface of the moon, excite the imagination of poets and lovers? Why is it that this tiny portion of the electromagnetic spectrum is responsible for vision, for the rhythmic day-night regulation of many biological activities, for the bending of plants toward the light, and also for photosynthesis, on which life depends? Is it an amazing coincidence that all these biological activities are dependent on these same wavelengths?

George Wald of Harvard, an expert on the subject of light and life, says no. He thinks that if life exists elsewhere in the universe, it is probably dependent on the same fragment of the vast spectrum. Wald bases this conjecture on two points. First, living things, as we have seen, are composed of large, complicated molecules held in special configurations and relationships to one another by hydrogen bonds and other weak bonds. Radiation of even slightly higher energies than the energy of violet light breaks these bonds and so disrupts the structure and function of the molecules. Radiations with wavelengths less than 200 nanometers—that is, with still higher energies—drive electrons out of atoms. On the other hand, light of wavelengths longer than those of the visible band—that is, with less energy than red light—is absorbed by water, which makes up the great bulk of all living things on earth. When this light is absorbed by molecules, its lower energy causes them to increase their motion (increasing heat) but does not trigger changes in their electron configurations. Only those radiations within the range of visible light have the property of exciting molecules—that is, of moving electrons into higher energy levels—and so of producing chemical and, ultimately, biological changes.

The second reason that the visible band of the electromagnetic spectrum has been "chosen" by living things is that it, above all, is what is available. Most of the radiation reaching the surface of the earth from the sun is within this range.

White light

Prism

V B G Y O R

Movable slit

Test sample (chlorophyll *a*)

Mirror

Half-silvered mirror

Reference (solvent only)

Photo-electric cell

Mirror

Recording device displaying the absorption spectrum of chlorophyll *a*

V B G Y O R

10-5 *The absorption spectrum of a pigment is measured with a spectrophotometer. This device directs a beam of light of each wavelength at the substance to be analyzed and records what percentage of light of each wavelength is absorbed by the pigment sample as compared to a ref-* erence sample. Because the mirror is lightly (half) silvered, half of the light is reflected and half is transmitted. The photoelectric cell is connected to an electronic device that automatically records the percentage absorption at each wavelength.

Higher-energy wavelengths are screened out by the oxygen and ozone high in the modern atmosphere. Much infrared radiation is screened out by water vapor and carbon dioxide before it reaches the earth's surface.

This is an example of what has been termed "the fitness of the environment"; the suitability of the environment for life and that of life for the physical world are exquisitely interrelated. If they were not, life could not exist.

CHLOROPHYLL AND OTHER PIGMENTS

In order for light energy to be used by living systems, it must first be absorbed. A pigment is any substance that absorbs light. Some pigments absorb all wavelengths of light and so appear black. Some absorb only certain wavelengths, transmitting or reflecting the wavelengths they do not absorb. Chlorophyll, the pigment that makes leaves green, absorbs light in the violet and blue wavelengths and also in the red; because it reflects green light, it appears green. Different pigments absorb light energy at different wavelengths. The absorption pattern of a pigment is known as the **absorption spectrum** of that substance (Figure 10–5).

Different groups of plants and algae use various pigments in photosynthesis. There are several different kinds of chlorophyll that vary slightly in their molecular structure (Figure 10–6). In plants, chlorophyll *a* is the pigment directly involved in the transformation of light energy to chemical energy. Most photosynthetic cells also contain a second type of chlorophyll—in plants, it is chlorophyll *b*—and a representative of another group of pigments called the carotenoids. One of the carotenoids found in plants is beta-carotene. The carotenoids are red, orange, or yellow pigments. In the green leaf, their color is masked by the chlorophylls, which are more abundant. In some tissues, however, such as those of a ripe tomato, the carotenoid colors predominate, as they do also when leaf cells stop synthesizing chlorophyll in the fall.

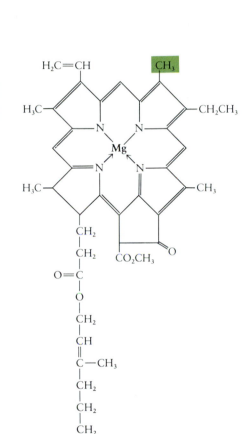

(a)

10–6 (a) *Chlorophyll* a *is a large molecule with a central atom of magnesium held in a porphyrin ring. Attached to the ring is a long, hydrophobic carbon-hydrogen chain that may help to anchor the molecule in the internal membranes of the chloroplast. Chlorophyll b differs from chlorophyll a in having an aldehyde (CHO) group in* place of the CH₃ *group indicated in green. Alternating single and double bonds, such as those in the chlorophylls, are common in pigments.* (b) *The estimated absorption spectra of chlorophyll a and chlorophyll b within the chloroplast. (Prepared by Govindjee.)*

(b)

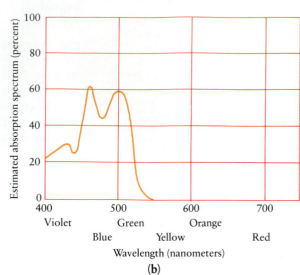

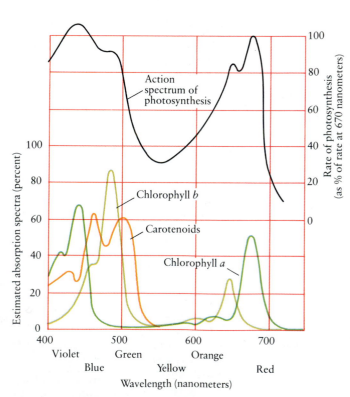

10-7 (a) *Related carotenoids. Cleavage of the beta-carotene molecule at the point indicated by the arrow yields two molecules of vitamin A. Oxidation of vitamin A yields retinal, the pigment involved in vision. Absorption of light energy changes the electron configuration of retinal and triggers a nerve impulse in the retina. All of this explains why you were told to eat your carrots.* (b) *The estimated absorption spectrum of carotenoids in the chloroplast. (Prepared by Govindjee.)*

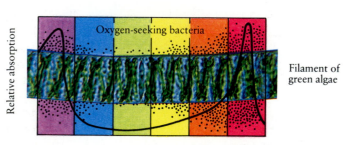

10-8 *Results of an experiment performed in 1882 by T. W. Englemann revealing the action spectrum of photosynthesis in a filamentous alga. Like more recent investigators, Englemann used the rate of oxygen production to measure the rate of photosynthesis. Unlike his successors, however, he lacked sensitive devices for detecting oxygen. As his oxygen indicator, he chose motile bacteria that are attracted by oxygen. In place of the mirror and diaphragm usually used to illuminate objects under view in his microscope, he substituted a "microspectral apparatus," which, as its name implies, produced a tiny spectrum of colors that it projected upon the slide under the microscope. Then he arranged a filament of algal cells parallel to the spread of the spectrum. The oxygen-seeking bacteria congregated mostly in the areas where the violet and red wavelengths fell upon the algal filament. As you can see, the action spectrum for photosynthesis Englemann revealed in this elegant experiment paralleled the absorption spectrum of chlorophyll. He therefore concluded that photosynthesis depends on the light absorbed by chlorophyll.*

10-9 *The upper curve shows the action spectrum for photosynthesis and the lower curves, absorption spectra for chlorophyll a, chlorophyll b, and carotenoids in the chloroplast. Note that the action spectrum of photosynthesis indicates that chlorophyll a, chlorophyll b, and carotenoids all absorb light used in photosynthesis. (Prepared by Govindjee.)*

The other chlorophylls and the carotenoids are able to absorb light at wavelengths different from those absorbed by chlorophyll *a*. They apparently can pass the energy on to chlorophyll *a*, thus extending the range of light available for photosynthesis (Figure 10–7).

An **action spectrum** defines the relative effectiveness (per number of incident photons) of different wavelengths of light for light-requiring processes, such as photosynthesis, flowering, phototropism (the bending of a plant toward light), and vision. Similarity between the absorption spectrum of a pigment and the action spectrum of a process is considered evidence that that particular pigment is responsible for that particular process (Figures 10–8 and 10–9).

When pigments absorb light, electrons within the pigment molecules are boosted to a higher energy level. Three possible consequences are: (1) the energy may be dissipated as heat; (2) it may be re-emitted immediately as light energy of a longer wavelength, a phenomenon known as fluorescence; (3) the energy may trigger a chemical reaction, as happens in photosynthesis. Whether or not a chemical reaction occurs depends not only on the structure of the particular pigment but also on its relationship with neighboring molecules. For example, if chlorophyll molecules are isolated in a test tube and light is permitted to strike them, they fluoresce. In other words, the molecules absorb light energy, and the electrons are momentarily raised to a higher energy level and then fall back again to a lower one. As they fall to a lower energy level, they release much of this energy as light. None of the light absorbed by isolated chlorophyll molecules is converted to any form of energy useful to living systems. Chlorophyll can convert light energy to chemical energy only when it is associated with certain proteins and embedded in a specialized membrane.

PHOTOSYNTHETIC MEMBRANES: THE THYLAKOID

The structural unit of photosynthesis is the **thylakoid,** which usually takes the form of a flattened sac, or vesicle. In the photosynthetic prokaryotes, thylakoids may form a part of the cell membrane, or they may occur singly in the cytoplasm, or, as in the cyanobacteria, they may be part of an elaborate internal membrane structure (see Figure 4–9, page 92). In eukaryotes, the thylakoids form a part of the internal membrane structure of specialized organelles, the chloroplasts (Figure 10–10). The alga *Chlamydomonas*, for instance, has a single very large chloroplast; the cell of a leaf characteristically has 40 to 50 chloroplasts, and there are often 500,000 chloroplasts per square millimeter of leaf surface.

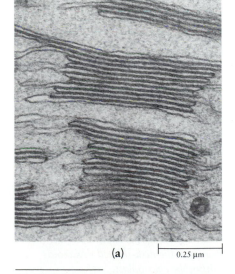

(a) 0.25 μm

10–10 *The unit of photosynthesis is the thylakoid, a flattened sac, whose membranes contain chlorophyll and other pigments. In plants and algae, thylakoids are part of an elaborate membrane system enclosed in a special organelle, the chloroplast. (a) Stacks of thylakoids (grana) from a plant cell. The inner compartments of the thylakoids are interconnected, forming the thylakoid space, which contains a solution whose composition differs from that of the stroma and the cytoplasm. (b) A chloroplast, showing the elaborate system of internal membranes comprising interconnected stacks of thylakoids.*

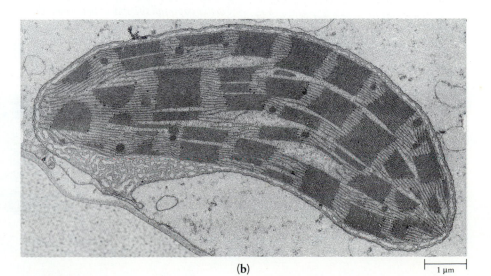

(b) 1 μm

The Structure of the Chloroplast

Chloroplasts, like mitochondria, are surrounded by two membranes that are separated by an intermembrane space. The inner membrane, unlike that of the mitochondrion, is smooth. The thylakoids, in the interior of the chloroplast, constitute a third membrane system. Surrounding the thylakoids, and filling the interior of the chloroplast, is a dense solution, the **stroma,** which (like the matrix of the mitochondrion) is different in composition from the cytoplasm. The thylakoids enclose an additional compartment, known as the thylakoid space, which contains a solution of still different composition. Thus, whereas the mitochondrion has two membrane systems (the outer and the inner) and two compartments (the intermembrane space and the matrix), the chloroplast has three membrane systems (outer, inner, and thylakoid) and three compartments (intermembrane space, stroma, and thylakoid space).

With the light microscope under high power, it is possible to see little spots of green within the chloroplasts of leaves. The early microscopists called these green specks **grana** ("grains"), and this term is still in use. Under the electron microscope, it can be seen that the grana are stacks of thylakoids. Some of the thylakoid membranes have extensions that interconnect the grana through the stroma that separates them.

All the thylakoids in a chloroplast are oriented parallel to each other. Thus, by swinging toward the light, the chloroplast can simultaneously aim all of its millions of pigment molecules for optimum reception, as if they were miniature electromagnetic antennae (which, of course, they are).

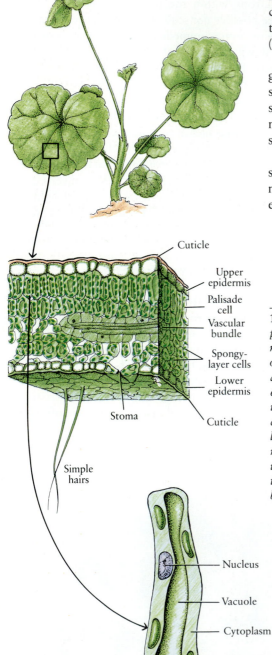

Cuticle

Upper epidermis

Palisade cell

Vascular bundle

Spongy-layer cells

Lower epidermis

Stoma

Cuticle

Simple hairs

Nucleus

Vacuole

Cytoplasm

Chloroplast

10–11 *Journey into a chloroplast. The plant shown is a geranium, which you may recognize by the characteristic shape of its leaves. The inner tissues of the leaf are completely enclosed by transparent epidermal cells that are coated with a waxy layer, the cuticle. Oxygen, carbon dioxide, and other gases enter the leaf largely through special openings, the stomata (singular, stoma). These gases and water vapor fill the spaces between cells in the spongy layer, leaving and entering cells by diffusion. Water, taken up by the roots,* enters the leaf by way of the vascular bundle, and sugars, the products of photosynthesis, leave the leaf by this route, traveling to nonphotosynthetic parts of the plant. Much of the photosynthesis takes place in the palisade cells, elongated cells directly beneath the upper epidermis. They have a large central vacuole and numerous chloroplasts that move within the cell, orienting themselves with respect to the light. Light is captured in the membranes of the disk-shaped thylakoids within the chloroplast.

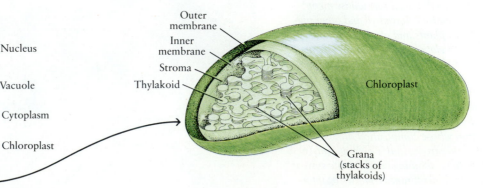

Outer membrane

Inner membrane

Stroma

Thylakoid

Chloroplast

Grana (stacks of thylakoids)

THE STAGES OF PHOTOSYNTHESIS

As we noted earlier (page 208), it was demonstrated about 200 years ago that light is required for the process we know as photosynthesis. It is now known that photosynthesis actually takes place in two stages, only one of which requires light. Evidence for this two-stage mechanism was first presented in 1905 by the English plant physiologist F. F. Blackman, as the result of experiments in which he measured the rate of photosynthesis under varying conditions.

Blackman first plotted the rate of photosynthesis at various light intensities. In dim to moderate light, increasing the light intensity increased the rate of photosynthesis, but at higher intensities, a further increase in light intensity had no effect. He then studied the combined effects of light and temperature on photosynthesis. In dim light, an increase in temperature had no effect. However, Blackman found that if he increased the light and also increased the temperature, the rate of photosynthesis was greatly accelerated (Figure 10–12). As the temperature increased above 30°C, the rate of photosynthesis slowed and finally the process ceased.

On the basis of these experiments, Blackman concluded that more than one set of reactions was involved in photosynthesis. First, there was a group of light-dependent reactions that were temperature-independent. The rate of these reactions could be accelerated in the dim-to-moderate light range by increasing the amount of light, but it was not accelerated by increases in temperature. Second, there was a group of reactions that were dependent not on light but rather on temperature. Both sets of reactions seemed to be required for the process of photosynthesis. Increasing the rate of only one set of reactions increased the rate of the entire process only to the point at which the second set of reactions began to hold back the first (that is, it became rate-limiting). Then it was necessary to increase the rate of the second set of reactions in order for the first to proceed unimpeded.

In Blackman's experiments, the temperature-dependent reactions increased in rate as the temperature was increased, but only up to about 30°C, after which the rate began to decrease. From this evidence it was concluded that these reactions were controlled by enzymes, since this is the way enzymes are expected to respond to temperature (see Figure 8–17, page 175). This conclusion has since proved to be correct.

10–12 (a) *An increase in light intensity beyond about 1,200 candelas does not produce a corresponding increase in the rate of photosynthesis. A curve such as the one shown here indicates that some other factor—known as a rate-limiting factor—is involved in the process under study. Under field conditions, CO_2 concentration is commonly the rate-limiting factor.* (b) *At a low intensity of light, an increase in temperature does not increase the rate of photosynthesis. At a high intensity, however, an increase in temperature has a very marked effect. From these data, Blackman concluded that photosynthesis includes both light-dependent and light-independent reactions.*

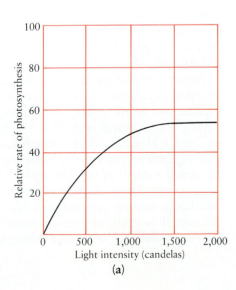

(a)

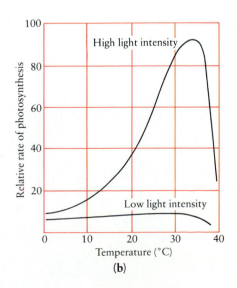

(b)

10–13 *Although NAD+ (Figure 8–14) and NADP+ resemble one another very closely, their biological roles are distinctly different. NADH generally transfers its electrons to other electron carriers, which continue to pass them on down to successively lower energy levels in discrete steps. In the course of this electron transfer, ATP molecules are formed. NADPH, by contrast, provides energy directly to biosynthetic processes of the cell that require large energy inputs.*

Photosynthesis was thus shown to have both a light-dependent stage, the so-called "light" reactions, and an enzymatic, light-independent stage, the "dark" reactions. The terms "light" and "dark" reactions have created much confusion, for although the "dark" reactions do not require light as such—only the chemical products of the "light" reactions—they can occur in either light or darkness. Moreover, recent work has shown that the enzyme controlling one of the key "dark" reactions is indirectly stimulated by light. As a result, these terms are now falling into disfavor. They are being replaced by terms that more accurately describe the processes occurring during each stage of photosynthesis.

In the first stage of photosynthesis—the energy-capturing reactions—light strikes chlorophyll *a* molecules that are packed in a special way in the thylakoid membranes. Electrons from the chlorophyll *a* molecules are boosted to higher energy levels, and, in a series of reactions, their added energy is used to form ATP from ADP and to reduce an electron-carrier molecule known as NADP+ (Figure 10–13). NADP+ closely resembles NAD+, and it too is reduced by the addition of two electrons and a proton, forming NADPH. Water molecules are also broken apart in this stage of photosynthesis, supplying electrons that replace those boosted from the chlorophyll *a* molecules.

In the second stage of photosynthesis, the ATP and NADPH formed in the first stage are used to reduce the carbon in carbon dioxide to a simple sugar. Thus the chemical energy temporarily stored in ATP and NADPH molecules is transferred to molecules suitable for transport and storage in the algal cell or plant body. At the same time, a carbon skeleton is formed from which other organic molecules can be built. This incorporation of CO_2 into organic compounds is known as the **fixation of carbon.** The steps by which it is accomplished, called the carbon-fixing reactions, occur in the stroma of the chloroplast.

THE ENERGY-CAPTURING REACTIONS

The Photosystems

In the thylakoids, chlorophyll and other molecules are, according to the present model, packed into units called photosystems. Each unit contains from 250 to 400 molecules of pigment, which serve as light-trapping antennae. Once a photon of light energy is absorbed by one of the antenna pigments, it is bounced around (like a hot potato) among the other pigment molecules of the photosystem until it reaches a special form of chlorophyll *a*, which is the reaction center. When this particular chlorophyll molecule absorbs the energy, an electron is boosted to a higher energy level from which it is transferred to another molecule, a primary electron acceptor. The chlorophyll molecule is thus oxidized (minus an electron) and positively charged.

Present evidence indicates that there are two different photosystems. In Photosystem I, the reactive chlorophyll *a* molecule is known as P_{700} (P is for pigment) because one of the peaks of its absorption spectrum is at 700 nanometers, a slightly longer wavelength than the usual chlorophyll *a* peak. When P_{700} is oxidized, it bleaches, which is how it was detected. No one has managed to isolate pure P_{700}. Recent evidence indicates that P_{700} is not a different kind of chlorophyll but rather a dimer of two chlorophyll *a* molecules; its unusual properties result from its association with special proteins in the thylakoid membrane and its position in relation to other molecules. Photosystem II also contains a reactive chlorophyll *a* molecule, which passes its electron on to a different primary electron acceptor. The reactive chlorophyll *a* molecule of Photosystem II is P_{680}.

Van Niel's Hypothesis

For more than 100 years after the completion of the work of Ingenhousz (page 208), it was generally assumed that in the equation

$$CO_2 + H_2O + Light \longrightarrow (CH_2O) + O_2$$

the carbohydrate (CH_2O) resulted from the combination of carbon atoms with water molecules and that the oxygen was released from the carbon dioxide molecule. This entirely reasonable hypothesis was widely accepted. But, as it turned out, it was wrong.

The investigator who upset this long-held assumption was the late C. B. van Niel of Stanford University. Van Niel, then a graduate student, was investigating photosynthesis in different types of photosynthetic bacteria. In their photosynthetic reactions, bacteria reduce carbon to carbohydrates, but they do not release oxygen. Among the types of bacteria van Niel was studying were the purple sulfur bacteria, which require hydrogen sulfide for photosynthesis. In the course of photosynthesis, globules of sulfur (S) are excreted or accumulated inside the bacterial cells. In these bacteria, van Niel found that this reaction takes place during photosynthesis:

$$CO_2 + 2H_2S \xrightarrow{Light} (CH_2O) + H_2O + 2S$$

This finding was simple enough and did not attract much attention until van Niel made a bold extrapolation. He proposed that the generalized equation for photosynthesis is

$$CO_2 + 2H_2A \xrightarrow{Light} (CH_2O) + H_2O + 2A$$

In this equation, H_2A stands for some oxidizable substance such as hydrogen sulfide, free hydrogen, or any one of several other compounds used by photosynthetic bacteria—or water. In the cyanobacteria, the algae, and the green plants, H_2A is water. In short, van Niel proposed that it was the water that was the source of oxygen in photosynthesis, *not* the carbon dioxide.

This brilliant speculation, first proposed in the early 1930s, was not proved until many years later. Eventually, investigators, using a heavy isotope of oxygen (^{18}O), traced the oxygen from water to oxygen gas:

$$CO_2 + 2H_2{}^{18}O \xrightarrow{Light} (CH_2O) + H_2O + {}^{18}O_2$$

This confirmed van Niel's hypothesis. The overall concept of photosynthesis has remained unchanged from the time of van Niel's proposal. However, many of its details have subsequently been worked out, and more are still under active investigation.

Purple sulfur bacteria. In these cells, hydrogen sulfide plays the same role as water does in the photosynthetic process of plants. The hydrogen sulfide (H_2S) is split, and the sulfur accumulates as globules, visible within the cells.

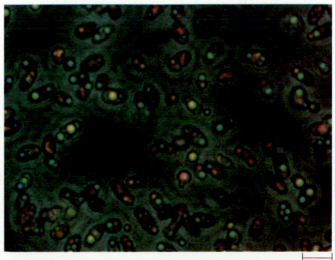

5 μm

10–14 *Light energy trapped in the reactive chlorophyll* a *molecule of Photosystem II boosts electrons to a higher energy level. These electrons are replaced by electrons pulled away from water molecules, releasing protons (H+ ions) and oxygen gas. The electrons are passed from the primary electron acceptor along an electron transport chain to a lower energy level, the reaction center of Photosystem I. As they pass along this electron transport chain, some of their energy is packaged in the form of ATP. Light energy absorbed by Photosystem I boosts electrons to another primary electron acceptor. From this acceptor, they are passed via other electron carriers to NADP+ to form NADPH. The electrons removed from Photosystem I are replaced by those from Photosystem II.*

ATP and NADPH represent the net gain from the energy-capturing reactions. To generate one molecule of NADPH, two electrons must be boosted from Photosystem II and two from Photosystem I. Two molecules of water are split into protons and oxygen gas, making available the two replacement electrons needed by Photosystem II. One molecule of water is regenerated in the formation of ATP.

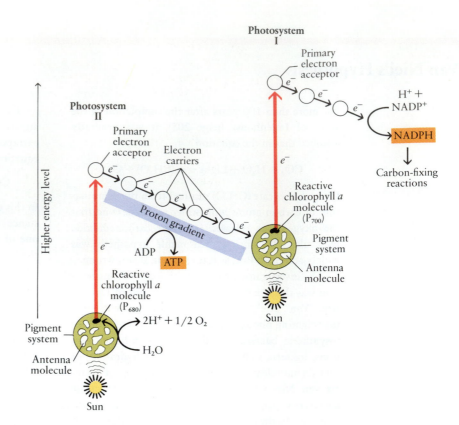

The Light-Trapping Reactions

The two photosystems probably evolved separately, with Photosystem I coming first. As we shall see, Photosystem I can operate independently. In general, however, the two systems work together simultaneously and continuously, as shown in Figure 10–14. According to the current model, light energy enters Photosystem II, where it is trapped by the reactive chlorophyll *a* molecule P_{680}. An electron from the P_{680} molecule is boosted to a higher energy level from which it is transferred to a primary electron-acceptor molecule. The electron then passes downhill along an electron transport chain to Photosystem I. As electrons pass along this transport chain, a proton gradient is established across the thylakoid membrane; the potential energy of this electrochemical gradient is used to form ATP from ADP in a chemiosmotic process similar to that in the mitochondrion. This process is known as **photophosphorylation.**

Three other events are taking place simultaneously:

1. The P_{680} chlorophyll molecule, having lost its electron, is avidly seeking a replacement. It finds it in the water molecule, which, while bound to a manganese-containing protein, is stripped of an electron and then broken into protons and oxygen gas.

2. Additional light energy is trapped in the reactive chlorophyll molecule (P_{700}) of Photosystem I. The molecule is oxidized, and an electron is boosted to a primary electron acceptor from which it goes downhill to NADP+.

3. The electron removed from the P_{700} molecule of Photosystem I is replaced by the electron that moved downhill from the primary electron acceptor of Photosystem II.

Thus in the light there is a continuous flow of electrons from water to Photosystem II to Photosystem I to NADP+. In the words of the late Nobel laureate Albert Szent-Györgi: "What drives life is . . . a little electric current, kept up by the sunshine."

The energy harvest from these steps is represented by an ATP molecule (whose formation releases a water molecule) and NADPH, which then become the chief sources of energy for the reduction of carbon dioxide. To generate one molecule of NADPH, four photons must be absorbed, two by Photosystem II and two by Photosystem I.

Cyclic Electron Flow

As we mentioned previously, there is also evidence that Photosystem I can work independently. When this occurs, no NADPH is formed. In this process, called **cyclic electron flow,** electrons are boosted from P_{700} to the primary electron acceptor of Photosystem I. They do not, however, pass down the series of electron carriers leading to NADP$^+$. Instead, they are shunted to the electron transport chain that connects Photosystems I and II and pass downhill through that chain back into the reactive P_{700} molecule (Figure 10–15). ATP is produced in the course of this passage. In the absence of NADP$^+$ (as, for example, when all of the available NADP$^+$ has been reduced to NADPH but has not been reoxidized in the carbon-fixing reactions), or when the cell needs additional supplies of ATP but not of NADPH, photosynthetic eukaryotic cells are able to synthesize ATP using the energy of sunlight to power cyclic electron flow. However, no oxygen is released and no carbon dioxide is reduced.

It is believed that the most primitive photosynthetic mechanisms worked by cyclic electron flow. This is also apparently the way in which some photosynthetic bacteria carry out photosynthesis. What is an alternative shunt pathway in eukaryotes is, in these bacteria, the principal pathway of electron flow in photosynthesis.

Photosynthetic Phosphorylation

The photophosphorylation of ADP to ATP as electrons pass down the electron transport chain from Photosystem II to Photosystem I (or, as they pass down a portion of that chain during cyclic electron flow) is a chemiosmotic process, similar in many ways to oxidative phosphorylation in the mitochondria. In both the mitochondria and the chloroplasts, the electron transport chains contain cytochromes, and the electron carriers and enzymes of these chains are embedded

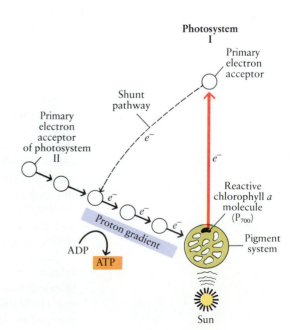

10–15 *When cyclic electron flow occurs in eukaryotic photosynthetic cells, Photosystem II is bypassed. Only Photosystem I and a portion of the electron transport chain between the two photosystems are utilized. ATP is produced from ADP, but oxygen is not released and NADP$^+$ is not reduced. In some photosynthetic bacteria, which have only Photosystem I, cyclic electron flow is the principal photosynthetic mechanism.*

10–16 *The chemiosmotic mechanism of photophosphorylation. In this process, electrons from chlorophyll a—boosted to a high energy level by sunlight—flow down an electron transport chain in the thylakoid membrane. The energy they release as they move to a lower energy level is used to pump protons from the stroma into the thylakoid space, creating an electrochemical gradient of potential energy. As the protons flow down the gradient from the thylakoid space back into the stroma, ADP is phosphorylated to ATP by ATP synthetase. The chemical structures of the electron carriers and enzymes of the thylakoid membrane (including ATP synthetase) are only slightly different from those of the mitochondrial membrane.*

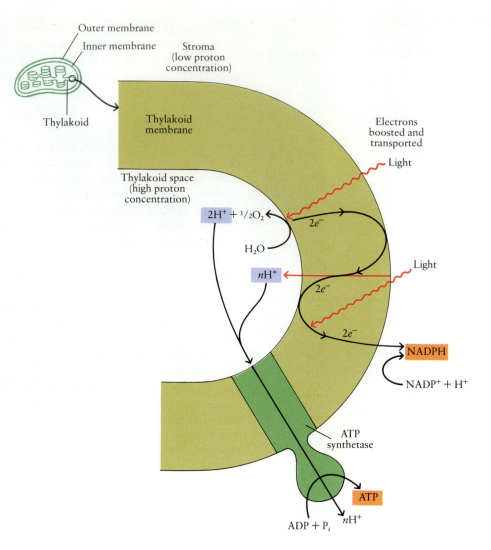

in membranes (the inner membrane of the mitochondrion and the thylakoid membrane of the chloroplast) that are impermeable to protons (H+ ions). In both organelles, an electrochemical gradient of potential energy is established as protons are pumped through the membrane using energy released as electrons pass down the chain. Moreover, ADP is phosphorylated to ATP as protons flow down the potential energy gradient through ATP synthetase complexes consisting of two factors.

The topography of this chemiosmotic process is, however, slightly different in chloroplasts. In mitochondria, as we saw on page 197, protons are pumped out of the matrix into the external medium; they flow down the gradient from the exterior back into the matrix. Similarly, in chloroplasts the protons are pumped out of the stroma (analogous to the mitochondrial matrix); they are not, however, pumped out of the organelle itself but rather *into* the thylakoid space (Figure 10–16). The potential energy gradient is between this internal, third compartment and the stroma. When protons flow down the gradient, they move from the thylakoid space back into the stroma, where the ATP is synthesized.

Many of the details of photophosphorylation, like those of oxidative phosphorylation, remain to be worked out. For example, it is clear that protons are released into the thylakoid space when water is split into protons, free oxygen, and electrons at the reaction center of Photosystem II. Additional protons are pumped into the thylakoid space as electrons flow down the transport chain, but

Photosynthesis without Chlorophyll

Halobacteria are rod-shaped cells, quite similar in appearance to *Escherichia coli.* They grow best in very salty water, about seven times as salty as sea water. If the salt concentration is reduced to only about three times that of sea water, the cell wall falls apart, and as the concentration is reduced still further, the cell membrane begins to break up. Walther Stoeckenius, then at Rockefeller University, separated the membrane fragments by centrifugation. One of the fractions, it turned out, was purple, and this, though the investigators did not know it at the time, was the first clue to a major energy source of the salt-loving bacteria.

Halobacteria are aerobes, and when suitable substrates and adequate oxygen are available, they oxidize organic molecules, producing ATP by oxidative phosphorylation. Oxygen, however, is often unavailable in the salty waters in which the halobacteria live. The secret of their success, it has now been shown, lies in the purple patches of the cell membrane, which provide an alternative, photosynthetic mechanism for producing ATP. The photosynthetic pigment of the halobacteria is not a form of chlorophyll, as in all other photosynthetic organisms, but retinal (Figure 10–7, page 212), which is also the visual pigment of the vertebrate eye. The membrane of the halobacteria contains molecules of retinal plus protein—the complex is called bacteriorhodopsin. When bacteriorhodopsin is excited by light and then returns to its original energy level, the energy released pumps protons across the membrane, out of the cell. This pumping establishes a proton gradient that drives the phosphorylation of ADP to ATP, thus providing additional support that the chemiosmotic mechanism is indeed a universal one for the regeneration of ATP.

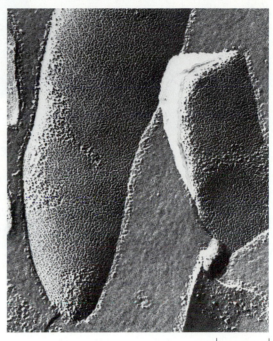

0.25 μm

A freeze-fracture preparation of the membrane of a halobacterium. The fine-grained regions with a hexagonal pattern are patches of purple membrane.

Darwin himself confessed a certain uneasiness when called upon to explain how an organ as complex as the eye might have arisen by the slow, cumulative steps of evolution. Was retinal "invented" twice? Or do these purple fragments of membrane hold clues both to the mechanisms of human vision and also to its origins?

the exact number pumped and the mechanism are not yet certain. A similar uncertainty concerns the number of protons that must flow through the ATP synthetase complex to synthesize one molecule of ATP. The best estimate at the present time is three protons for each ATP synthesized.

The proton gradient that drives photophosphorylation is established by the release of energy that originally entered the system from sunlight. Phosphorylation by isolated chloroplasts can also be driven by a proton gradient established artificially across the thylakoid membrane. In such experimental systems, phosphorylation proceeds in the dark, providing impressive evidence that the key factor in photophosphorylation, as in oxidative phosphorylation, is the establishment of a proton gradient.

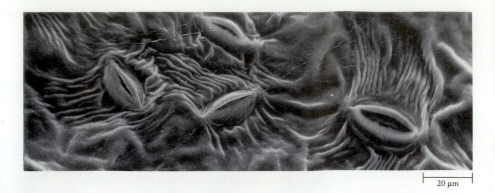

10–17 *Scanning electron micrograph of open stomata on the lower surface of a cottonwood leaf. The carbon dioxide used in photosynthesis reaches the photosynthetic cells through these openings.*

20 µm

THE CARBON-FIXING REACTIONS

The reactions that we have just described are the energy-capturing reactions of photosynthesis. In the course of these reactions, as we saw, light energy is converted to electrical energy—the flow of electrons—and the electrical energy is converted to chemical energy stored in the bonds of NADPH and ATP. In the second stage of photosynthesis, this energy is used to reduce carbon. Carbon is available to photosynthetic cells in the form of carbon dioxide. Algae, such as the cell shown in Figure 10–1, obtain dissolved carbon dioxide directly from the surrounding water. In plants, carbon dioxide reaches the photosynthetic cells through specialized openings in leaves and green stems, called **stomata** (Figure 10–17).

The Calvin Cycle: The Three-Carbon Pathway

The reduction of carbon takes place in the stroma in a cycle named after its discoverer, Melvin Calvin. The Calvin cycle is analogous to the Krebs cycle (page 194) in that, in each turn of the cycle, the starting compound is regenerated. The starting (and ending) compound is a five-carbon sugar with two phosphates attached, ribulose bisphosphate (RuBP).

The cycle begins when carbon dioxide is bound to RuBP, which then splits to form two molecules of phosphoglycerate, or PGA (Figure 10–18). (Each PGA molecule contains three carbon atoms, hence the name, the three-carbon pathway.) The enzyme catalyzing this crucial reaction, RuBP carboxylase, is very abundant in chloroplasts, making up more than 15 percent of the total chloroplast protein. RuBP carboxylase, said to be the most abundant protein in the world, is located on the surface of the thylakoid membranes.

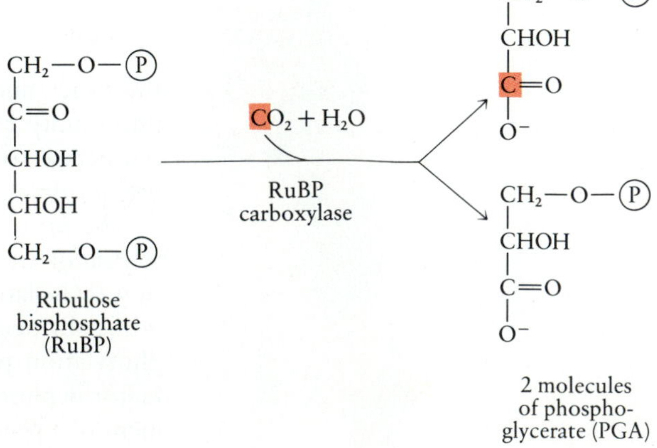

10–18 *Calvin and his collaborators briefly exposed photosynthesizing algae to radioactive carbon dioxide ($^{14}CO_2$). They found that the radioactive carbon is first bound to ribulose bisphosphate (RuBP), which then immediately splits to form two molecules of phosphoglycerate (PGA). The radioactive carbon atom, indicated in color, appears in one of the two molecules of PGA. This is the first step of the Calvin cycle.*

$CO_2 + H_2O$

RuBP carboxylase

Ribulose bisphosphate (RuBP)

2 molecules of phosphoglycerate (PGA)

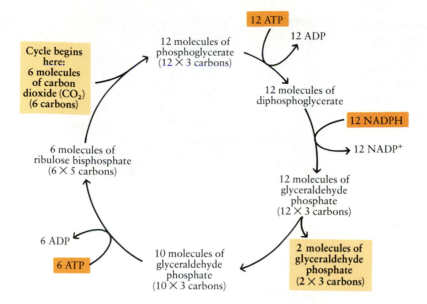

12 ATP

12 ADP

12 molecules of
phosphoglycerate
(12 × 3 carbons)

**Cycle begins
here:
6 molecules
of carbon
dioxide (CO₂)
(6 carbons)**

12 molecules of
diphosphoglycerate

12 NADPH

12 NADP⁺

6 molecules of
ribulose bisphosphate
(6 × 5 carbons)

12 molecules of
glyceraldehyde
phosphate
(12 × 3 carbons)

6 ADP

6 ATP

10 molecules of
glyceraldehyde
phosphate
(10 × 3 carbons)

**2 molecules of
glyceraldehyde
phosphate
(2 × 3 carbons)**

10–19 *Summary of the Calvin cycle. At each full "turn" of the cycle, one molecule of carbon dioxide enters the cycle. Six turns are summarized here—the number required to make two molecules of glyceraldehyde phosphate, the equivalent of one molecule of a six-carbon sugar. Six molecules of ribulose bisphosphate (RuBP), a five-carbon compound, are combined with six molecules of carbon dioxide, yielding twelve molecules of phosphoglycerate, a three-carbon compound. These are reduced to twelve molecules of glyceraldehyde phosphate. Ten of these three-carbon molecules are combined and rearranged to form six five-carbon molecules of RuBP. The two "extra" molecules of glyceraldehyde phosphate represent the net gain from the Calvin cycle. The energy that drives the Calvin cycle is in the form of ATP and NADPH, produced by the energy-capturing reactions in the first stage of photosynthesis.*

The complete cycle is diagrammed in Figure 10–19. As in the Krebs cycle, each step is catalyzed by a specific enzyme. At each full turn of the cycle, a molecule of carbon dioxide enters the cycle, is reduced, and a molecule of RuBP is regenerated. Three turns of the cycle introduce three molecules of carbon dioxide, the equivalent of one three-carbon sugar, and produce one molecule of glyceraldehyde phosphate, which is the immediate product of the Calvin cycle. This same three-carbon sugar-phosphate molecule is formed when the fructose diphosphate molecule is split at the fourth step in glycolysis (page 188).

Six revolutions of the cycle, with the introduction of six molecules of carbon dioxide, are necessary to produce the equivalent of a six-carbon sugar, such as glucose. These six revolutions of the cycle produce two molecules of glyceraldehyde phosphate, which can subsequently react to produce one molecule of a six-carbon sugar. The overall equation for the series of reactions required for the synthesis of glucose is

$$6RuBP + 6CO_2 + 18ATP + 12NADPH + 12H^+ + 12H_2O \longrightarrow$$
$$6RuBP + Glucose + 18P_i + 18ADP + 12NADP^+$$

The Four-Carbon Pathway

In most plants, the first step in the fixation of carbon is the binding of carbon dioxide to RuBP and its entrance into the Calvin cycle. Some plants, however, first bind carbon dioxide to a compound known as phosphoenolpyruvate (PEP) to form the four-carbon compound oxaloacetic acid. (Oxaloacetic acid, you may recall, is also an intermediate in the Krebs cycle.) The carbon dioxide incorporated into oxaloacetic acid is ultimately transferred to RuBP and enters the Calvin cycle,

10–20 *Carbon dioxide fixation by the C_4 pathway. Carbon dioxide is bound to phosphoenolpyruvate (PEP) by the enzyme PEP carboxylase. The resulting oxaloacetic acid is converted either to malic acid or aspartic acid. These steps later will be reversed, releasing carbon dioxide for use in the Calvin cycle.*

but not until it has passed through a series of reactions that transport it more deeply into the leaf. Plants that utilize this pathway, also known as the Hatch-Slack pathway, are commonly called C_4, or four-carbon, plants, as distinct from the C_3 plants in which carbon is bound first to RuBP to form the three-carbon compound phosphoglycerate (PGA).

In C_4 plants, the binding of carbon dioxide to PEP is catalyzed by the enzyme PEP carboxylase (Figure 10–20). The resulting oxaloacetic acid is then reduced to malic acid or converted (with the addition of an amino group) to aspartic acid. These steps take place in mesophyll cells, whose chloroplasts are characterized by an extensive network of thylakoids, organized into well-developed grana. The next step is a surprise: the malic acid (or aspartic acid, depending on the species) is transported to bundle-sheath cells. The chloroplasts of these cells, which form tight sheaths around the vascular bundles of the leaf, have poorly developed grana and often contain large grains of starch (Figure 10–21). In the bundle-sheath cells, the malic (or aspartic) acid is decarboxylated to yield CO_2 and pyruvic acid. The CO_2 then enters the Calvin cycle. This process, summarized in Figure 10–22, physically separates the capture of CO_2 by the plant from the reactions of the Calvin cycle.

One might well ask why C_4 plants should have evolved such an energetically expensive and seemingly clumsy method of providing carbon dioxide to the Calvin cycle. To answer this question, we must consider both the function of the leaf as a whole and the properties of PEP carboxylase and RuBP carboxylase, the enzymes that catalyze the first step of the carbon-fixing reactions in C_4 and C_3 plants, respectively.

10–21 *In C_4 plants, the chloroplasts of mesophyll cells differ from those of bundle-sheath cells. As shown in this micrograph of portions of chloroplasts in two adjacent cells of a corn leaf, the mesophyll chloroplast (top) contains well-developed grana, while the grana of the bundle-sheath chloroplast are poorly developed. Notice the plasmodesmata that connect the two cells, providing channels through which substances can flow from one cell to the other.*

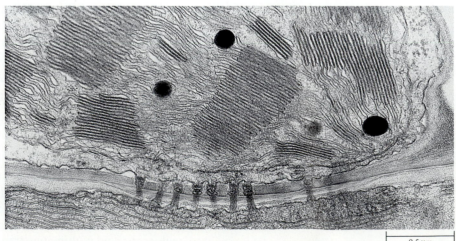

0.5 μm

10–22 *A pathway for carbon fixation in C₄ plants. CO₂ is first fixed in mesophyll cells as oxaloacetic acid. It is then transported to bundle-sheath cells, where the carbon dioxide is released. The CO₂ thus formed enters the Calvin cycle. Pyruvic acid returns to the mesophyll cell, where it is phosphorylated to PEP.*

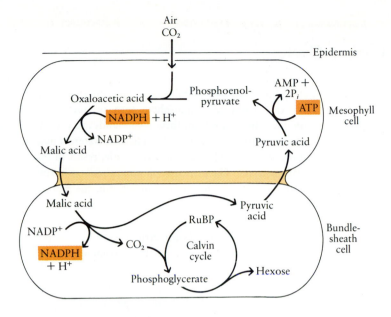

Carbon dioxide is not continuously available to the photosynthesizing cells. It enters the leaf by way of the stomata, specialized pores that open and close depending on, among other factors, water stress. Moreover, when plants are growing in close proximity to one another, the air surrounding the leaves may be quite still, with little gas exchange between the immediate environment and the atmosphere as a whole. Under such conditions, the concentration of carbon dioxide in the air closest to the leaves may be rapidly reduced to low levels by the photosynthetic activity of the plants. PEP carboxylase, the enzyme that catalyzes the formation of oxaloacetic acid in C₄ plants, has a higher affinity for carbon dioxide than does RuBP carboxylase. Even at low concentrations of carbon dioxide, the enzyme works rapidly to bind it to PEP. Compared with RuBP carboxylase, PEP carboxylase fixes carbon dioxide faster, at lower levels, keeping the CO_2 concentration lower within the cells near the surface of the leaf. This maximizes the gradient of carbon dioxide between these cells and the outside air. Thus, when the stomata are open, carbon dioxide readily diffuses down the concentration gradient into the leaf. If the stomata must be closed much of the time—as they must be to conserve water in a hot, dry climate—the plant with C₄ metabolism will take up more carbon dioxide with each gasp (so to speak) than the plant that has only C₃ metabolism. Hence, the C₄ plant is at a distinct advantage in drought-ridden areas.

In the presence of ample carbon dioxide, RuBP carboxylase fixes carbon dioxide efficiently, feeding it into the Calvin cycle. However, when the carbon dioxide concentration in the leaf is low in relation to the oxygen concentration, this same enzyme catalyzes a reaction of RuBP with oxygen, rather than with carbon dioxide. This reaction leads to the formation of glycolic acid, the substrate for a process known as photorespiration. Photorespiration, which occurs in the peroxisomes (page 118) of photosynthetic cells, is the oxidation of carbohydrates in the presence of light and oxygen. Unlike mitochondrial respiration, however, it yields neither ATP nor NADH. Under normal atmospheric conditions, as much as 50 percent of the carbon fixed in photosynthesis by a C₃ plant may be reoxidized to CO_2 during photorespiration. Thus photorespiration greatly reduces the photosynthetic efficiency of C₃ plants.

The Carbon Cycle

By photosynthesis, living systems incorporate carbon dioxide from the atmosphere into organic compounds. In respiration, these compounds are broken down again into carbon dioxide and water. These processes, viewed on a worldwide scale, result in the carbon cycle. The principal photosynthesizers in this cycle are plants and the phytoplankton, the marine algae. They synthesize carbohydrates from carbon dioxide and water and release oxygen into the atmosphere. About 100 billion metric tons of carbon per year are bound into carbon compounds by photosynthesis.

Some of the carbohydrates are used by the photosynthesizers themselves. Plants release carbon dioxide from their roots and leaves, and marine algae release it into the water where it maintains an equilibrium with the carbon dioxide of the air. Some 500 billion metric tons of carbon are "stored" as dissolved carbon dioxide in the seas, and some 700 billion metric tons in the atmosphere. Some of the carbohydrates are used by animals that feed on the living plants, on algae, and on one another, releasing carbon dioxide. An enormous amount of carbon is contained in the dead bodies of plants and other organisms plus discarded leaves and shells, feces, and other waste materials that settle into the soil or sink to the ocean floors where they are consumed by small invertebrates, bacteria, and fungi. Carbon dioxide is also released by these processes into the reservoir of the air and oceans.

Another, even larger store of carbon lies below the surface of the earth in the form of coal and oil, deposited there some 300 million years ago.

The natural processes of photosynthesis and respiration generally balance one another out. Over the long span of geologic time, the carbon dioxide concentration of the atmosphere has varied, but for the last 10,000 years it has remained relatively constant. By volume, it is a very small proportion of the atmosphere, only about 0.03 percent. It is important, however, because carbon dioxide, unlike most other components of the atmosphere, absorbs heat from the sun's rays. Since 1850, carbon dioxide concentrations in the atmosphere have been increasing, owing in large part to our use of fossil fuels, to our plowing of the soil, and to our destruction of forest land, particularly in the tropics. A recent study by the Environmental Protection Agency predicts that this increase in the carbon dioxide "blanket" will significantly increase the average temperatures here on earth, beginning within the next 20 years. Average increases of about 2°C by the year 2040 and 5°C by the year 2100 are expected. The consequences of these temperature increases cannot be known with certainty. In some parts of the world, there may be lengthened growing seasons, increased precipitation, and, in conjunction with the increased levels of carbon dioxide available to plants, greater agricultural productivity. In other parts of the world, however, it is thought that

High CO_2 and low O_2 concentrations limit photorespiration. Consequently, C_4 plants have another distinct advantage over C_3 plants. First, the RuBP carboxylase is sequestered in the bundle-sheath cells, in the interior of the leaf, where it is to some extent protected from atmospheric oxygen. Second, because the CO_2 fixed by the C_4 pathway is essentially "pumped" from the mesophyll cells into the bundle-sheath cells, it is delivered to RuBP carboxylase in a concentrated form. The ratio of CO_2 to O_2 is high enough that the enzyme catalyzes the reaction involving CO_2 and not the reaction leading to photorespiration. Any carbon dioxide that is released by photorespiration is immediately recaptured by PEP carboxylase in the mesophyll cells and fed back to RuBP and into the Calvin cycle. As a result of all of these factors, the net rates of photosynthesis in C_4 grasses, such as corn, sugarcane, and sorghum, can be two to three times the rates in C_3 grasses, such as wheat, oats, and rice.

C_4 plants evolved primarily in the tropics and are especially well adapted to high

precipitation will be significantly reduced, lowering crop yields and, in already arid areas, accelerating the spread of the great deserts of the world. Most biologists are greatly concerned that although we do not know the consequences—either for ourselves or for other organisms—of what we are doing, we keep right on doing it.

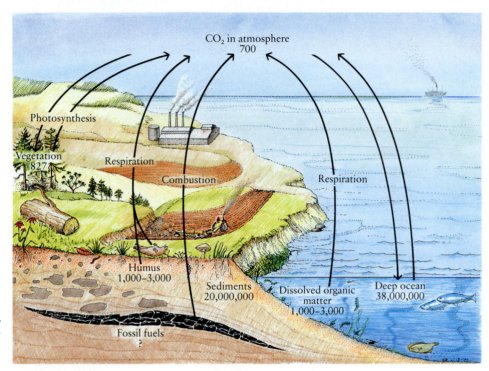

The carbon cycle. The arrows indicate the movement of carbon atoms. The numbers are all estimates of the amount of carbon stored, expressed in billions of metric tons. The amount of carbon released by respiration and combustion has begun to exceed the amount fixed by photosynthesis.

light intensities, high temperatures, and dryness. The optimal temperature range for C_4 photosynthesis is much higher than that for C_3 photosynthesis, and C_4 plants flourish even at temperatures that would eventually be lethal to many C_3 species. Because of their more efficient use of carbon dioxide, C_4 plants can attain the same photosynthetic rate as C_3 plants but with smaller stomatal openings and, hence, with considerably less water loss.

Perhaps the most familiar example of the competitive capacity of C_4 plants is seen in lawns in the summertime. In most parts of the United States, lawns consist mainly of C_3 grasses such as Kentucky bluegrass. As the summer days become hotter and drier, these dark green, fine-leaved grasses are often overwhelmed by rapidly growing crabgrass, which disfigures the lawn as its yellowish-green, broader-leaved plants slowly take over. Crabgrass, you will not be surprised to hear, is a C_4 plant.

The list of plants known to utilize the four-carbon pathway has grown to over 100 genera, at least a dozen of which include both C_3 and C_4 species. This pathway has undoubtedly arisen independently many times in the course of evolution and is another example of the exquisite adaptation of living systems to their environment.

THE PRODUCTS OF PHOTOSYNTHESIS

Glyceraldehyde phosphate, the three-carbon sugar produced by the Calvin cycle, may seem an insignificant reward, both for all the enzymatic activity on the part of the cell and for our own intellectual stress. However, this molecule and those derived from it provide (1) the energy source for virtually all living systems, and (2) the basic carbon skeleton from which the great diversity of organic molecules can be synthesized. Carbon has been fixed—that is, it has been brought from the inorganic world into the organic one.

Molecules of glyceraldehyde phosphate may flow into a variety of different metabolic pathways, depending on the activities and requirements of the cell. Often they are built up to glucose or fructose, following a sequence that is in many of its steps the reverse of the glycolysis sequence described in the previous chapter. (At some steps, the reactions are simply reversed and the enzymes are the same. Other steps—the highly exergonic ones of the downhill sequence—are bypassed.) Plant cells use these six-carbon sugars to make starch and cellulose for their own purposes and sucrose for export to other parts of the plant body. Animal cells store them as glycogen. All cells use sugars, including glyceraldehyde phosphate and glucose, as the starting point for the manufacture of other carbohydrates, fats and other lipids, and, with the addition of nitrogen, amino acids and nitrogenous bases. Finally, as we saw in the preceding chapter, oxidation of the carbon fixed in photosynthesis is the source of ATP energy for heterotrophic organisms and for the heterotrophic cells of plants.

SUMMARY

In photosynthesis, light energy is converted to chemical energy, and carbon is fixed into organic compounds. The generalized equation for this reaction is

$$CO_2 + 2H_2A + \text{Light energy} \longrightarrow (CH_2O) + H_2O + 2A$$

in which H_2A stands for water or some other substance from which electrons can be removed.

Light energy is captured by the living world by means of pigments. The pigments involved in photosynthesis in eukaryotes include the chlorophylls and the carotenoids. Light absorbed by the pigments boosts their electrons to higher energy levels. Because of the way the pigments are packed into membranes, they are able to transfer this energy to reactive molecules, probably chlorophyll *a* packed in a particular way.

Photosynthesis takes place within organelles known as chloroplasts, which are surrounded by two membranes. Contained within the membranes of the chloroplast are a solution of organic compounds and ions known as the stroma and a complex internal membrane system consisting of fused membranes that form sacs called thylakoids. The pigments and other molecules responsible for capturing light are located in and on the thylakoid membranes.

Photosynthesis takes place in two stages, as summarized in Table 10–1. In the currently accepted model of the energy-capturing reactions, light energy strikes antennae pigments of Photosystem II, which contains several hundred molecules of chlorophyll *a* and chlorophyll *b*. Electrons are boosted uphill from the reactive chlorophyll *a* molecule P_{680} to a primary electron acceptor. As the electrons are

TABLE 10-1 **Summary of the Stages of Photosynthesis**

	CONDITIONS	WHERE	WHAT APPEARS TO HAPPEN	RESULTS
Energy-capturing reactions	Light	Thylakoids	Light striking Photosystem II boosts electrons uphill. (These electrons are replaced by electrons from water molecules, which release O.) Electrons then pass downhill along an electron transport chain to Photosystem I; in the process, ATP is formed by a chemiosmotic process. Light hits Photosystem I, boosts electrons uphill to an electron acceptor from which they are passed to $NADP^+$ to form NADPH. Electrons are replaced in Photosystem I by electrons from Photosystem II.	Energy of light is converted to chemical energy stored in bonds of ATP and NADPH.
Carbon-fixing reactions	Do not require light, although some enzymes stimulated by light	Stroma	Calvin cycle. NADPH and ATP formed in energy-capturing reactions are used to reduce carbon dioxide. The cycle yields glyceraldehyde phosphate from which glucose or other organic compounds can be formed.	Chemical energy of ATP and NADPH is used to incorporate carbon into organic molecules.

removed, they are replaced by electrons from water molecules, with the simultaneous production of free O_2 and protons (H^+ ions). The electrons then pass downhill to Photosystem I along an electron transport chain; this passage generates a proton gradient that drives the synthesis of ATP from ADP (photophosphorylation). Light energy absorbed in antennae pigments of Photosystem I and passed to chlorophyll P_{700} results in the boosting of electrons to another primary electron acceptor. The electrons removed from P_{700} are replaced by the electrons from Photosystem II. The electrons are ultimately accepted by the electron carrier $NADP^+$. The energy yield from this sequence of reactions is contained in the molecules of NADPH and in the ATP formed by photophosphorylation.

Photophosphorylation also occurs as a result of cyclic electron flow, a process that bypasses Photosystem II. In cyclic electron flow, the electrons boosted from P_{700} in Photosystem I do not pass to $NADP^+$ but are instead shunted to the electron transport chain that links Photosystem II to Photosystem I. As they flow down this chain, back into P_{700}, ADP is phosphorylated to ATP.

Like oxidative phosphorylation in the mitochondria, photophosphorylation in the chloroplasts is a chemiosmotic process. As electrons flow down the electron transport chain from Photosystem II to Photosystem I, protons are pumped from the stroma into the thylakoid space, creating an electrochemical gradient of potential energy. As protons flow down this gradient from the thylakoid space back into the stroma, passing through ATP synthetase complexes, ATP is formed.

In the carbon-fixing reactions, which take place in the stroma, NADPH and ATP produced in the energy-capturing reactions are used to reduce carbon dioxide to organic carbon. This is accomplished by means of the Calvin cycle. In the Calvin cycle, a molecule of carbon dioxide is combined with the starting material, a five-carbon sugar called ribulose bisphosphate. At each turn of the cycle, one carbon atom enters the cycle. Three turns of the cycle produce a three-carbon molecule, glyceraldehyde phosphate. Two molecules of glyceraldehyde phosphate (six turns of the cycle) can combine to form a glucose molecule. At each turn of the cycle, RuBP is regenerated. The glyceraldehyde phosphate can also be used as a starting material for other organic compounds needed by the cell.

In C_4 plants, carbon dioxide is initially accepted by a compound known as PEP (phosphoenolpyruvate) to yield the four-carbon oxaloacetic acid. Following a series of reactions that transport it to cells more deeply within the leaf, the carbon dioxide is released, binding with RuBP and entering the Calvin cycle. Although C_4 plants use more energy to fix carbon, their net photosynthetic efficiency can be higher than that of C_3 plants. The greater affinity of PEP carboxylase for carbon dioxide enables C_4 plants to capture ample CO_2 with minimal water loss. Also, photorespiration, a process in which fixed carbon is reoxidized to carbon dioxide and lost to the plant, is limited in C_4 plants. Under conditions of intense sunlight, high temperatures, or drought, C_4 plants are more efficient than C_3 plants.

QUESTIONS

1. Distinguish between the following: absorption spectrum/action spectrum; grana/thylakoid; stroma/thylakoid space; NAD^+/$NADP^+$; energy-capturing reactions/carbon-fixing reactions; Photosystem I/Photosystem II; C_3 photosynthesis/C_4 photosynthesis.

2. Why is it plausible to argue, as the Nobel laureate George Wald does, that wherever in the universe we find living organisms, we will find them (or at least some of them) to be colored?

3. Predict what colors of light might be most effective at stimulating plant growth. (Such is the principle of the special light bulbs used for plants.)

4. Sketch a chloroplast and label all membranes and compartments. In what ways does the structure of a chloroplast resemble that of a mitochondrion? In what ways is it different?

5. Describe in general terms the events of photosynthesis. Compare your description with Table 10–1.

6. In what ways are photophosphorylation and oxidative phosphorylation alike? In what ways are they different?

7. The experiment shown in Figure 4–3 on page 87 was attempted in Stockton, California, in 1973, but instead of making amino acids, the electric sparks caused the apparatus to explode. (No one was hurt, fortunately.) What is present in today's atmosphere that was not present in the primitive atmosphere and that would account for the explosion?

8. Given the scarcity of high-salt environments and the difficulties of surviving in them, explain in evolutionary terms why the halobacteria (page 221) are found there.

9. Return to Figure 10–21 and describe the biochemical events taking place in each of the chloroplasts.

10. Consider the anatomy of the leaf of a C_4 plant. In which type of cell—mesophyll or bundle-sheath—would you expect the oxygen-releasing reaction of photosynthesis to occur? Why? Does this localization of oxygen release increase or decrease photorespiration?

11. Over 100 genera of plants have acquired C_4 photosynthesis in the course of evolution. How is this adaptation advantageous to these plants? Many plants, however, have not evolved C_4 photosynthesis. Why is it advantageous to such plants *not* to have C_4 photosynthesis?

12. Trace a carbon atom through a series of biological events, such as those illustrated on pages 158, 160, and 161.

SUGGESTIONS FOR FURTHER READING

Books

ALBERTS, BRUCE, DENNIS BRAY, JULIAN LEWIS, MARTIN RAFF, KEITH ROBERTS, and JAMES D. WATSON: *Molecular Biology of the Cell*, 2d ed., Garland Publishing, Inc., New York, 1989.

This outstanding cell biology text includes a clear, up-to-date discussion of our current understanding of the processes that occur in mitochondria and chloroplasts. The authors' discussion of the experimental procedures used to study these processes is especially helpful.

CONANT, JAMES BRYANT (ed.): *Harvard Case Histories in Experimental Science*, vol. 2, Harvard University Press, Cambridge, Mass., 1964.

Case #5, Plants and the Atmosphere, edited by Leonard K. Nash, describes the early work on photosynthesis, often presented in the words of the investigators themselves. The narrative illuminates the historical context in which the discoveries were made.

DARNELL, JAMES, HARVEY F. LODISH, and DAVID BALTIMORE: *Molecular Cell Biology*, W. H. Freeman and Company, New York, 1986.

A comprehensive treatment of modern cell biology, richly illustrated with diagrams and micrographs. Chapter 20 of this outstanding text is devoted to a thorough explication of our current understanding of the processes that occur in the mitochondria and chloroplasts.

LEHNINGER, ALBERT L.: *Principles of Biochemistry*, Worth Publishers, Inc., New York, 1982.

This introductory text is outstanding both for its clarity and for its consistent focus on the living cell. Lehninger was one of the foremost experts on cellular energetics, and this text is enriched by his vast experience and thorough understanding of the processes by which cells provide themselves with energy.

NEWSHOLME, ERIC, and TONY LEECH: *The Runner: Energy and Endurance*, Fitness Books, Roosevelt, N.J., 1984.*

A lively introduction to the energetics of the working human body. Equally useful as a primer for runners who want to know more about their own physiology and as an introduction to the biochemistry of carbohydrate and fat metabolism.

PRESCOTT, DAVID M.: *Cells: Principles of Molecular Structure and Function*, Jones and Bartlett Publishers, Boston, 1988.

An up-to-date, yet concise, textbook of cell biology, written for a first course at the undergraduate level. Chapter 4 is devoted to energy flow and metabolism.

RABINOWITCH, EUGENE, and GOVINDJEE: *Photosynthesis*, John Wiley & Sons, Inc., New York, 1969.*

Although now out of date in many respects, this book remains a lucid introduction, suitable for undergraduate students, to the processes of photosynthesis and to related physical and chemical concepts, such as entropy and free energy.

SCIENTIFIC AMERICAN: *Molecules to Living Cells*, W. H. Freeman and Company, New York, 1980.*

A collection of articles from Scientific American. *Chapters 5 and 6 are concerned with the structure and function of enzymes, Chapter 12 is an excellent presentation of the chemiosmotic synthesis of ATP in both mitochondria and chloroplasts, and Chapter 13 examines the photosynthetic membrane.*

STRYER, LUBERT: *Biochemistry*, 3d ed., W. H. Freeman and Company, New York, 1988.

A good introduction, handsomely illustrated, to cellular energetics.

Articles

AHERN, TIM J., and ALEXANDER M. KLIBANOV: "The Mechanism of Irreversible Enzyme Inactivation at 100°C," *Science*, vol. 228, pages 1280–1284, 1985.

BARBER, JIM: "Signals from the Reaction Centre," *Nature*, vol. 332, pages 111–112, 1988.

BJORKMAN, O., and J. BERRY: "High-Efficiency Photosynthesis," *Scientific American*, October 1973, pages 80–93.

CLOUD, PRESTON: "The Biosphere," *Scientific American*, September 1983, pages 176–189.

DETWILER, R. P., and C. A. S. HALL: "Tropical Forests and the Global Carbon Cycle," *Science*, vol. 239, pages 42–47, 1988.

DICKERSON, RICHARD E.: "Cytochrome *c* and the Evolution of Energy Metabolism," *Scientific American*, March 1980, pages 136–153.

DICKINSON, ROBERT E., and RALPH J. CICERONE: "Future Global Warming from Atmospheric Trace Gases," *Nature*, vol. 319, pages 109–115, 1986.

KARPLUS, MARTIN, and J. ANDREW McCAMMON: "The Dynamics of Proteins," *Scientific American*, April 1986, pages 42–51.

KERR, RICHARD A.: "Is the Greenhouse Here?" *Science*, vol. 239, pages 559–561, 1988.

KLIBANOV, ALEXANDER M.: "Immobilized Enzymes and Cells as Practical Catalysts," *Science*, vol. 219, pages 722–727, 1983.

KNOWLES, JEREMY R.: "Tinkering with Enzymes: What Are We Learning?" *Science*, vol. 236, pages 1252–1258, 1987.

KOLATA, GINA: "How Do Proteins Find Mitochondria?" *Science*, vol. 228, pages 1517–1518, 1985.

LANE, M. DANIEL, PETER L. PEDERSEN, and ALBERT S. MILDVAN: "The Mitochondrion Updated," *Science*, vol. 234, pages 526–527, 1986.

LEWIN, ROGER: "A Downward Slope to Greater Diversity," *Science*, vol. 217, pages 1239–1240, 1982.

NASSAU, KURT: "The Causes of Color," *Scientific American*, October 1980, pages 124–154.

NEURATH, HANS: "Evolution of Proteolytic Enzymes," *Science*, vol. 224, pages 350–357, 1984.

REVELLE, ROGER: "Carbon Dioxide and World Climate," *Scientific American*, August 1982, pages 35–43.

SCHOPF, J. WILLIAM: "The Evolution of the Earliest Cells," *Scientific American*, September 1978, pages 110–138.

SRIVASTAVA, D. K., and SIDNEY A. BERNHARD: "Metabolite Transfer via Enzyme-Enzyme Complexes," *Science*, vol. 234, pages 1081–1086, 1986.

STOECKENIUS, WALTHER: "The Purple Membrane of Salt-loving Bacteria," *Scientific American*, June 1976, pages 38–46.

WAGGONER, PAUL E.: "Agriculture and Carbon Dioxide," *American Scientist*, vol. 72, pages 179–184, 1984.

WEBB, A. DINSMOOR: "The Science of Making Wine," *American Scientist*, vol. 72, pages 360–367, 1984.

WESTHEIMER, F. H.: "Why Nature Chose Phosphates," *Science*, vol. 235, pages 1173–1178, 1987.

WOODWELL, G. M., et al.: "Global Deforestation: Contribution to Atmospheric Carbon Dioxide," *Science*, vol. 222, pages 1081–1086, 1983.

YOUVAN, DOUGLAS C., AND BARRY L. MARRS: "Molecular Mechanisms of Photosynthesis," *Scientific American*, June 1987, pages 42–48.

---

* Available in paperback.

SECTION 3

Genetics

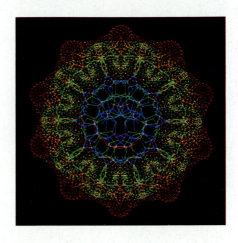

"*Of everything that creepeth on the earth, there went in two and two . . .*" *Knowledge that like begets like is at least as old as recorded history. The way in which this continuity of life is preserved from generation to generation, and the way that male and female each contribute to the characteristics of their offspring were fundamental questions of biological science. These ancient questions have now been answered in a few extraordinary decades of modern genetic research.*

233

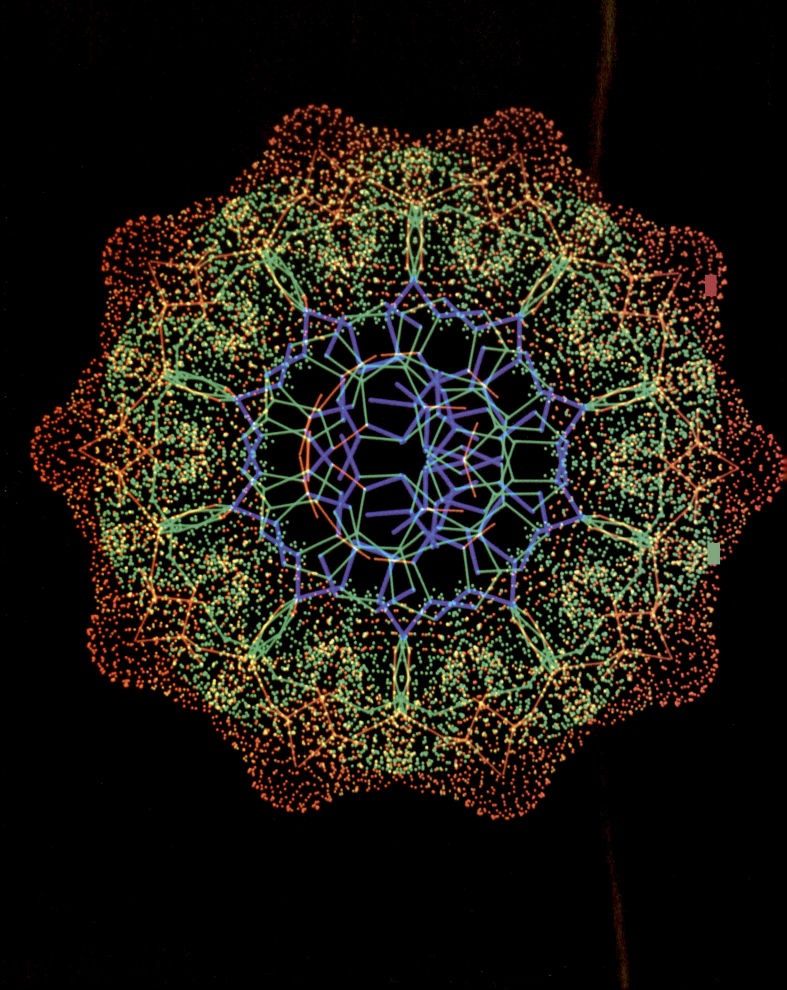

CHAPTER 11

From an Abbey Garden: The Beginning of Genetics

11–1 *The answers to the age-old questions of the nature of heredity were found in the structure of a remarkable molecule, deoxyribonucleic acid, or DNA. This molecular "rose window" is a computer-generated image of one complete turn of the DNA helix, with red representing oxygen, green for carbon, yellow for phosphorus, and blue for nitrogen. In this image, the helix has been collapsed into a single plane.*

Among all the symbols in biology, perhaps the most widely used and most ancient are the hand mirror and comb of Venus (♀) and the shield and spear of Mars (♂), the scientific shorthand for female and male. Ideas about the biological roles of male and female are even older than these familiar symbols. Very early, it must have been noticed that male and female were both necessary to produce children and that both transmitted characteristics—hair color, for example, a large nose, or a small chin—to their children. And throughout history, biological inheritance has been an important factor in human social organization, often determining the distribution of wealth, power, land, and royal privileges.

Sometimes a family characteristic is so distinctive that it can be traced through many generations. A famous example of such a characteristic is the Hapsburg lip (Figure 11–2), which has appeared in Hapsburg after Hapsburg, over and over again since at least the thirteenth century. Examples such as this have made it easy to accept the importance of inheritance, but it is only comparatively recently that we have begun to understand something about how this process works. In fact, heredity—the transmission of characteristics from parent to offspring—was not really studied as a science until the second half of the nineteenth century. Yet the questions addressed in this study are among the most fundamental in biology, since self-replication is the essence of the hereditary process and one of the principal properties of living systems.

(a)

(b)

(c)

11–2 *The protruding lip of the Hapsburgs is a famous example of an inherited characteristic. These portraits of members of the Hapsburg family encompass a period of about 200 years: (a) Ferdinand I (1503–1564), Holy Roman Emperor, (b)* Rudolph II (1552–1612), Holy Roman Emperor, and (c) Charles II (1661–1700), King of Spain.

11-3 *Only fairly recently has it been realized that living things come only from other living things of the same species. This picture from an old Turkish history of India shows a wakwak tree, which bears human fruit. According to the account, the tree is to be found on an island in the South Pacific.*

EARLY IDEAS ABOUT HEREDITY

Far back in human history, people learned to improve domestic animals and crops by selective breeding of individuals with desirable characteristics. The ancient Egyptians and Babylonians, for example, knew how to produce fruits by artificial fertilization, crossing male flowers borne on one date palm tree with female flowers from another tree. The nature of the difference between the male and female flowers was understood by the Greek philosopher and naturalist Theophrastus (371–287 B.C.). "The males should be brought to the females," he wrote, "for the male makes them ripen and persist." In the days of Homer, breeding a male donkey to a mare was known to produce a mule, although little explanation could be given for the manner in which the beast came by its unusual appearance.

Many legends were based on bizarre possibilities of matings between individuals of different species. The wife of Minos, according to Greek mythology, mated with a bull and produced the Minotaur. Folk heroes of Russia and Scandinavia were traditionally the sons of women who had been captured by bears, from which these men derived their great strength and so enriched the national stock. The camel and the leopard also mated from time to time, according to the early naturalists, who were otherwise unable—and it is hard to blame them—to explain an animal as improbable as the giraffe (the common giraffe still bears the scientific name of *Giraffa camelopardalis*). Thus folklore reflected early and imperfect glimpses into the nature of hereditary relationships.

The first scientist known to have pondered the mechanism of heredity was Hippocrates (460?-377? B.C.). He hypothesized that specific particles, or "seeds," are produced by all parts of the body and are transmitted to offspring at the time of conception, causing certain parts of the offspring to resemble those parts of the parents. A century later, Aristotle rejected the ideas of Hippocrates. Children often seem to inherit characteristics of their grandparents or even their great-grandparents rather than their parents, observed Aristotle. How could these distant relatives have contributed the "seeds" of flesh and blood that were transmitted from parent to offspring? To resolve the conflict, Aristotle postulated that the male semen was made up of imperfectly blended ingredients, some of which were inherited from past generations. At fertilization, he proposed, the male semen mixed with the "female semen," the menstrual fluid, giving form and power *(dynamis)* to the amorphous substance. From this material, flesh and blood formed as the offspring developed.

For two thousand years no one had a better idea. Indeed, there were not many new ideas at all. Seventeenth-century medical texts continued to show various stages in the coagulation of the embryo from the mixture of maternal and paternal semens. In fact, many scientists as well as laymen did not believe that such mixtures were even always necessary; they held that life, at least the "simpler" forms of life, could arise by spontaneous generation. Worms, flies, and various crawling things, it was commonly believed, took shape from putrid substances, ooze, or mud, and a lady's hair dropped in a rain barrel could turn into a snake. Jan Baptista van Helmont, a seventeenth-century physician known for his experiments on the growth of plants (see page 208), actually published his personal recipe for the production of mice: One need only place a dirty shirt in a pot containing a few grains of wheat, and in 21 days mice would appear. He had performed the experiment himself, he said. The mice would be adults, both male and female, he added, and would be able to produce more mice by mating. As we saw in Chapter 4 (page 86), spontaneous generation did not lose its grip on the imagination until Pasteur's decisive disproof in 1864.

THE FIRST OBSERVATIONS

In 1677, the Dutch lens maker Anton van Leeuwenhoek discovered living sperm —"animalcules," as he called them—in the seminal fluid of various animals, including man. Enthusiastic followers peered through Leeuwenhoek's "magic looking glass" (his homemade microscope) and believed they saw within each human sperm a tiny creature—the homunculus, or "little man" (Figure 11–4). This little creature was thought to be the future human being in miniature. Once implanted in the womb of the female, the future human being would be nurtured there, but the mother's only contribution would be to serve as an incubator for the growing fetus. Any resemblance a child might have to its mother, these theorists held, was because of the "prenatal influences" of the womb.

During the very same decade (the 1670s), another Dutchman, Régnier de Graaf, described for the first time the ovarian follicle, the structure in which the human egg cell (the ovum) forms. Although the actual human egg was not seen for another 150 years, the existence of a human egg was rapidly accepted. In fact, de Graaf attracted a school of followers, the ovists, who were as convinced of their opinions as the animalculists, or spermists, were of theirs and who soon contended openly with them. It was the female egg, the ovists said, that contained the future human being in miniature; the animalcules in the male seminal fluid merely stimulated the egg to grow. Ovists and spermists alike carried the argument one logical step further. Each homunculus was thought to have within it another perfectly formed but smaller being, and in that was still another one, and so on—children, grandchildren, and great-grandchildren, all stored away for future use. Some ovists even went so far as to say that Eve had contained within her body all the unborn generations yet to come, each egg fitting closely inside another like a child's hollow blocks. Each female generation since Eve had contained one fewer than the previous generation, they explained, and after 200 million generations, all the eggs would be spent and human life would come to an end.

BLENDING INHERITANCE

By the middle of the nineteenth century, the concepts of the ovists and spermists began to yield to new data. The facts that challenged these earlier hypotheses came not so much from scientific experiments as from practical attempts by master gardeners to produce new ornamental plants. Artificial crossings of such plants showed that, in general, regardless of which plant supplied the pollen— which contains the sperm cells—and which plant contributed the egg cells, both contributed to the characteristics of the new variety. But this conclusion raised even more puzzling questions: What exactly did each parent plant contribute? How did all the hundreds of characteristics of each plant get combined and packed into a single seed?

The most widely held hypothesis of the nineteenth century was that of blending inheritance. According to this concept, when the sperm and egg cells, or **gametes** (from the Greek word *gamos*, meaning "marriage"), combine, there is a mixing of hereditary material that results in a blend, analogous to a blend of two different-colored inks. On the basis of such a hypothesis, one would predict that the offspring of a black animal and a white animal would be gray, and their offspring would also be gray because the black and white hereditary material, once blended, could never be separated again.

You can see why this concept was unsatisfactory. It ignored the phenomenon of characteristics skipping a generation, or even several generations, and then reap-

11–4 *What the animalculists, or spermists, of the seventeenth and eighteenth centuries believed they saw when they looked through a microscope at sperm cells. This is a homunculus ("little man"), a future human being in miniature, in a sperm cell.*

pearing. To Charles Darwin and other proponents of the theory of evolution, it presented particular difficulties. As we noted in the Introduction (page 7), evolution, according to Darwin, takes place as natural selection acts on existing hereditary variations—that is, variations that can be inherited. If the hypothesis of blending inheritance were valid, the hereditary variations would disappear, like a single drop of ink in the many-colored mixture. Sexual reproduction would eventually result in complete uniformity, natural selection would have no raw material on which to act, and evolution would not occur.

THE CONTRIBUTIONS OF MENDEL

At about the same time that Darwin was writing *The Origin of Species,* an Austrian monk, Gregor Mendel, was beginning a series of experiments that would lead to a new understanding of the mechanism of inheritance. Mendel, who was born into a peasant family in 1822, entered a monastery in Brünn (now Brno, Czechoslovakia), where he was able to receive an education. He attended the University of Vienna for two years, pursuing studies in both mathematics and science. He failed his tests for the teaching certificate he was seeking and so retired to the monastery, of which he eventually became abbot. Mendel's work, carried on in a quiet monastery garden and ignored until after his death, marks the beginning of modern genetics.

Mendel's great contribution was to demonstrate that inherited characteristics are carried by discrete units that are parceled out separately (reassorted) in each generation. These discrete units, which Mendel called *Elemente,* eventually came to be known as **genes**.

Mendel's Experimental Method

For his experiments in heredity, Mendel chose the common garden pea. It was a good choice. The plants were commercially available, easy to cultivate, and grew rapidly. Different varieties had clearly different characteristics that "bred true," appearing unchanged from one crop to the next. For instance, a variety with tall plants always produced tall offspring, and one with yellow seeds always produced yellow seeds, generation after generation. Moreover, the reproductive structures of the pea flower are entirely enclosed by petals, even when they are mature (Figure 11–5). Consequently, the flower normally self-pollinates; that is, sperm cells from the flower's own pollen fertilize its egg cells. Although the plants could be crossbred experimentally, accidental crossbreeding could not occur to confuse the experimental results. As Mendel said in his original paper, "The value and utility of any experiment are determined by the fitness of the material to the purpose for which it is used."

Mendel's choice of the pea plant for his experiments was not original. However, he was successful in formulating the fundamental principles of heredity— where others had failed—because of his approach to the problem. First, he tested a very specific hypothesis in a series of logical experiments. He planned his experiments carefully and imaginatively, choosing for study only clear-cut, measurable hereditary differences. Second, he studied the offspring of not only the first generation but also the second and subsequent generations. Third, and most important, he counted the offspring and then analyzed the results mathematically. Even though his mathematics was simple, the idea that a biological problem could be studied quantitatively was startlingly new. Finally, he organized his data in such a way that his results could be evaluated simply and objectively. The experiments themselves were described so clearly that they could be repeated and checked by other scientists, as eventually they were.

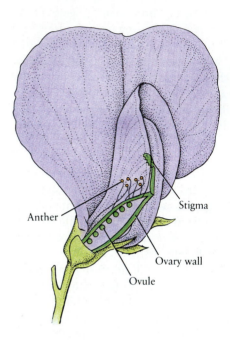

11–5 *In a flower, pollen develops in the anthers and the egg cells in the ovules. Pollination occurs when pollen grains, trapped on the stigma, germinate and grow down to the ovules, where they release sperm. Egg and sperm nuclei unite, and the fertilized eggs develop within the ovules, which are attached to the ovary wall. In the garden pea (Pisum sativum), the ovules with their enclosed embryos form the peas (the seeds), while the ovary wall becomes the pod.*

Pollination in most species of flowering plants involves the pollen from one plant (often carried by an insect) being caught on the stigma of another plant. This is called cross-pollination.

In the pea flower, however, the stigma and anthers are completely enclosed by petals, and the flower, unlike most, does not open until after fertilization has taken place. Thus the plant normally self-pollinates. In his crossbreeding experiments, Mendel pried open the bud before the pollen matured and removed the anthers with tweezers, preventing self-pollination. Then he artificially pollinated the flower by dusting the stigma with pollen collected from another plant.

Labels on figure: Anther, Stigma, Ovary wall, Ovule

Parental generation
(**P**)

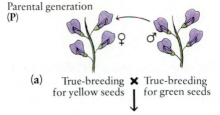

(a) True-breeding ✕ True-breeding
for yellow seeds | for green seeds

First filial generation
(**F₁**)

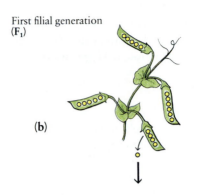

(b)

Second filial generation
(**F₂**)

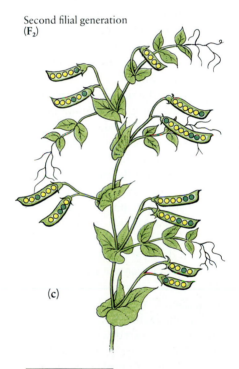

(c)

11–6 *An outline of one of Mendel's experiments.* **(a)** *A true-breeding yellow-seeded pea plant was crossed with a true-breeding green-seeded plant by removing pollen from the anthers of flowers on one plant and transferring it to the stigmas of flowers on the other plant.* **(b)** *Pea pods containing only yellow seeds developed from the fertilized flowers. These peas (seeds) were planted, and the resulting plants were allowed to self-pollinate.* **(c)** *Pea pods developing from the self-pollinated flowers contained both yellow and green peas in an approximate ratio of 3:1, that is, about 3/4 were yellow and 1/4 were green.*

The Principle of Segregation

Mendel began with 32 different types of pea plants, which he studied for several years before he began his quantitative experiments. As he said later in his report on this work, he did not want to experiment with traits in which the difference could be "of a 'more or less' nature, which is often difficult to define." As a result of his preliminary observations, Mendel selected for detailed study seven traits that appeared in two conspicuously different forms in different varieties of plants. One variety of plant, for example, always produced yellow peas (seeds), while another always produced green ones. In one variety, the seeds, when dried, had a wrinkled appearance; in another variety they were smooth. The complete list of traits is given in Table 11–1.

Mendel performed experimental crosses, removing the pollen-containing anthers from flowers and dusting their stigmas with pollen from a flower of another variety. He found that in every case in the first generation (now known in biological shorthand as the F_1, for "first filial generation"), all of the offspring showed only one of the two alternative characteristics; the other characteristic disappeared completely. For example, all of the plants produced as a result of a cross between true-breeding yellow-seeded plants and true-breeding green-seeded plants were as yellow-seeded as the yellow-seeded parent. Similarly, all of the flowers produced by plants resulting from a cross between a true-breeding purple-flowered plant and a true-breeding white-flowered plant were purple. Characteristics that appeared in the F_1 generation, such as yellow seeds and purple flowers, Mendel called **dominant.**

The interesting question was: What had happened to the alternative characteristic—the greenness of the seed or the whiteness of the flower—that had been passed on so faithfully for generations by the parent stock? Mendel let the pea plant itself carry out the next stage of the experiment by permitting the F_1 plants to self-pollinate (Figure 11–6). The characteristics that had disappeared in the first generation reappeared in the second, or F_2, generation. In Table 11–1 are the results of Mendel's actual counts. These characteristics, which were present in the parent generation and reappeared in the F_2 generation, must also have been present somehow in the F_1 generation, although they were not apparent there. Mendel called these characteristics **recessive.**

Looking at the results in Table 11–1, you will notice, as Mendel did, that the dominant and recessive characteristics appear in the second, or F_2, generation in ratios of about 3:1. How do the recessives disappear so completely and then reappear again, and always in such constant proportions? It was in answering this question that Mendel made his greatest contribution. He saw that the appearance and disappearance of alternative characteristics, as well as their constant proportions in the F_2 generation, could be explained if hereditary characteristics are

TABLE 11–1 Results of Mendel's Experiments with Pea Plants

	ORIGINAL CROSSES		SECOND FILIAL GENERATION (F_2)			
TRAIT	DOMINANT	✕ RECESSIVE	DOMINANT	RECESSIVE	TOTAL	RATIO
Seed form	Round	✕ Wrinkled	5,474	1,850	7,324	2.96:1
Seed color	Yellow	✕ Green	6,022	2,001	8,023	3.01:1
Flower position	Axial	✕ Terminal	651	207	858	3.14:1
Flower color	Purple	✕ White	705	224	929	3.15:1
Pod form	Inflated	✕ Constricted	882	299	1,181	2.95:1
Pod color	Green	✕ Yellow	428	152	580	2.82:1
Stem length	Tall	✕ Dwarf	787	277	1,064	2.84:1

11–7 *A pea plant homozygous for purple flowers is represented as* WW *in genetic shorthand. The allele for purple flowers is designated* W *because of a convention by which geneticists use the first letter of the less common form (white) of the gene. The capital indicates the dominant allele, the lowercase the recessive. A* WW *plant can produce only gametes with a purple-flower (*W*) allele. The female symbol* ♀ *indicates that this flower contributed the egg cells, or female gametes.*

♀   WW

*A white-flowered pea plant (*ww*) can produce only gametes with a white-flower (*w*) allele. The male symbol* ♂ *indicates that this flower contributed the sperm cells, or male gametes.*

♂   ww

When a w *sperm cell fertilizes a* W *egg cell, the result is a* Ww*, which, since the* W *allele is dominant, will produce purple flowers. However, this* Ww *plant can produce gametes with either a* W *or a* w *allele.*

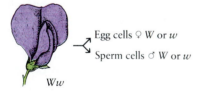

Egg cells ♀ W or w
Sperm cells ♂ W or w

Ww

And so, if the plant self-pollinates, four possible combinations can occur:
♀ W × ♂ W ⟶ *purple flowers*
♀ W × ♂ w ⟶ *purple flowers*
♀ w × ♂ W ⟶ *purple flowers*
♀ w × ♂ w ⟶ *white flowers*

determined by discrete (separable) factors. These factors, Mendel realized, must have occurred in the F_1 plants in pairs, with one member of each pair inherited from the maternal parent, the other from the paternal parent. The paired factors separated again when the mature F_1 plants produced sex cells, resulting in two kinds of gametes, with one member of the pair in each.

The hypothesis that every individual carries pairs of factors for each trait and that the members of the pair segregate (separate) during the formation of gametes has come to be known as Mendel's first law, or the **principle of segregation.**

Consequences of Segregation

We now recognize that any given gene, for instance, the gene for seed color, can exist in different forms. These different forms of a gene are known as **alleles.** For example, yellow seededness and green seededness are determined by different alleles—that is, different forms—of the gene for seed color. The alleles are represented in biological shorthand by letters; the allele for yellow seededness is represented by Y, and the allele for green seededness by y.

How a given trait is expressed in an organism is determined by the particular combination of two alleles for that trait carried by the organism. If the two alleles are the same (for example, YY or yy), then the organism is said to be **homozygous** for that particular trait. If the two alleles are different from one another (for example, Yy), then the organism is **heterozygous** for the trait.

When gametes are formed, alleles are passed on to them, but each gamete contains only one allele for any given gene. When two gametes combine in the fertilized egg, the alleles occur in matched pairs again. If the two alleles in a given pair are the same (a homozygous state), the characteristic they determine will be expressed. If they are different (a heterozygous state), one may be dominant over the other; a dominant allele is one that produces its particular characteristic in the heterozygous as well as in the homozygous state. The outward appearance and other observable characteristics of an organism constitute its **phenotype,** a term that is also used to describe a single characteristic, such as yellow or green seeds. Even though a recessive allele may not be expressed in the phenotype, each allele of a matched pair still exists independently and as a discrete unit in the genetic makeup, or **genotype,** of the organism. The two alleles of a pair will separate from each other when gametes are again formed. Only if two recessive alleles come together in the fertilized egg—one from the female gamete and one from the male gamete—will the phenotype then show the recessive characteristic.

When pea plants homozygous for purple flowers are crossed with pea plants having white flowers, only pea plants with purple flowers are produced. Each plant in this F_1 generation, however, carries both an allele for purple and an allele for white. Figure 11–7 shows what happens in the F_2 generation if the F_1 generation self-pollinates. One of the simplest ways to predict the types of offspring that might be produced from such a cross is to diagram it using a Punnett square, as shown in Figure 11–8. Notice that the result would be the same if an F_1 individual were cross-fertilized with another F_1 individual, which is how these experiments are performed with animals and with plants that are not self-pollinating.

In order to test the hypothesis that alleles occur in pairs and that the two alleles of a pair segregate during gamete formation, it is necessary to perform an additional experiment: cross purple-flowering F_1 plants (the result of a cross between purple- and white-flowering plants) with white-flowering plants. To the casual observer, it would appear as if this were simply a repeat of Mendel's first experiment, crossing plants having purple flowers with plants having white flowers. But if Mendel's hypothesis is correct, the results will be different from those of his first experiment. Can you predict the results of such a cross? Stop a moment and think about it.

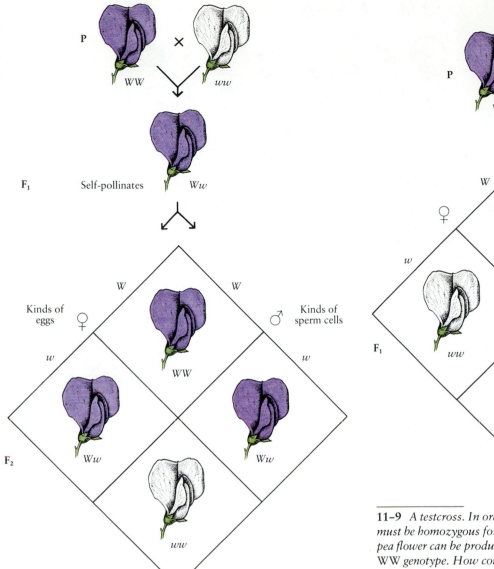

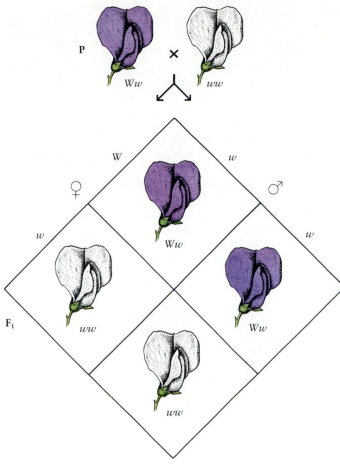

11–8 *Mendel's hypothesis, as exemplified in the F₁ and F₂ generations following a cross between a parent (P) pea plant with two dominant alleles for purple flowers (WW) and one with two recessive alleles for white flowers (ww). The phenotype of the offspring in the F₁ generation is purple, but note that the genotype is Ww. The F₁ heterozygote produces four kinds of gametes, ♀ W, ♂ W, ♀ w, ♂ w, in equal proportions. When this plant self-pollinates, the W and w sperm cells and eggs combine randomly to form, on the average, 1/4 WW, 2/4 (or 1/2) Ww (purple), and 1/4 ww offspring. It is this underlying 1:2:1 genotypic ratio that accounts for the phenotypic ratio of 3 dominants (purple) to 1 recessive (white). The distribution of characteristics in the F₂ is shown by a Punnett square, named after the English geneticist who first used this sort of checkerboard diagram for the analysis of genetically determined traits.*

11–9 *A testcross. In order for a pea flower to be white, the plant must be homozygous for the recessive allele (ww). But a purple pea flower can be produced by a plant with either a Ww or a WW genotype. How could you tell such plants apart? Geneticists solve this problem by breeding such plants with homozygous recessives. This sort of experiment is known as a testcross. As shown here, a phenotypic ratio in the F₁ generation of one purple to one white indicates that the purple-flowering parent used in the testcross must have been heterozygous.*

The easiest way to analyze the possible result of this cross is again to use a Punnett square, as in Figure 11–9. This type of experiment, which reveals the genotype of the parent with the dominant phenotype, is known as a **testcross**. A testcross is an experimental cross between an individual having the dominant phenotype (genotype unknown) for a given trait with another individual that is known to be homozygous for the recessive allele. Whether or not two different phenotypes are produced indicates whether the individual with the dominant phenotype is heterozygous or homozygous for the trait being studied. The testcross shown in Figure 11–9 reveals that the genotype of the plant being tested is *Ww* rather than *WW*. What would have been the results of the testcross if the plant tested had been homozygous for the purple-flower allele?

The Principle of Independent Assortment

In a second series of experiments, Mendel studied crosses between pea plants that differed in two characteristics; for example, one parent plant produced peas that were round and yellow, and the other had peas that were wrinkled and green. The

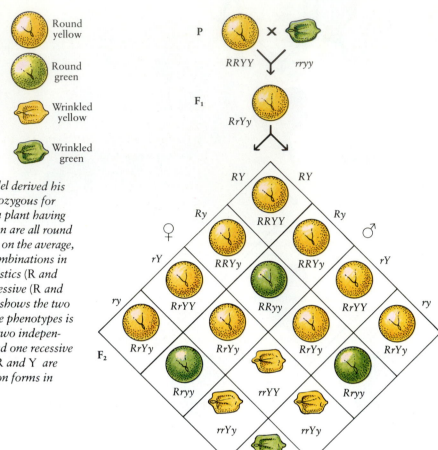

11–10 *One of the experiments from which Mendel derived his principle of independent assortment. A plant homozygous for round (RR) and yellow (YY) peas is crossed with a plant having wrinkled (rr) and green (yy) peas. The F₁ generation are all round and yellow, but notice how the characteristics will, on the average, appear in the F₂ generation. Of the 16 possible combinations in the offspring, 9 show the two dominant characteristics (R and Y), 3 show one combination of dominant and recessive (R and y), 3 show the other combination (r and Y), and 1 shows the two recessives (r and y). This 9:3:3:1 distribution of the phenotypes is always the expected result from a cross involving two independently assorting genes, each with one dominant and one recessive allele in each of the parents. (Note that the letters R and Y are used because round and yellow are the less common forms in nature.)*

round and yellow characteristics, you will recall (see Table 11–1), are dominant, and the wrinkled and green are recessive. As you would expect, all the seeds produced by a cross between the true-breeding parental types were round and yellow. When these F₁ seeds were planted and the resulting flowers allowed to self-pollinate, 556 seeds were produced. Of these, 315 showed the two dominant characteristics, round and yellow, but only 32 combined the recessive characteristics, green and wrinkled. All the rest of the seeds were unlike either parent; 101 were wrinkled and yellow, and 108 were round and green. Totally new combinations of characteristics had appeared.

This experiment did not contradict Mendel's previous results. If the two traits, seed color and seed shape, are considered independently, round and wrinkled still appeared in a 3:1 ratio (423 round to 133 wrinkled), and so did yellow and green (416 yellow to 140 green). But the seed shape and the seed color characteristics, which had originally been combined in a certain way in one plant (round only with yellow and wrinkled only with green), behaved as if color factors and shape factors were entirely independent of one another (yellow could now be found with wrinkled and green with round). From this, Mendel formulated his second law, the **principle of independent assortment.** This principle states that when gametes are formed, the alleles of a gene for one trait segregate independently of the alleles of a gene for another trait.

Figure 11–10 diagrams Mendel's interpretation of these results. It shows why, in a cross involving two independently assorting genes, with each gene having one dominant and one recessive allele, the phenotypes in the offspring will, on the

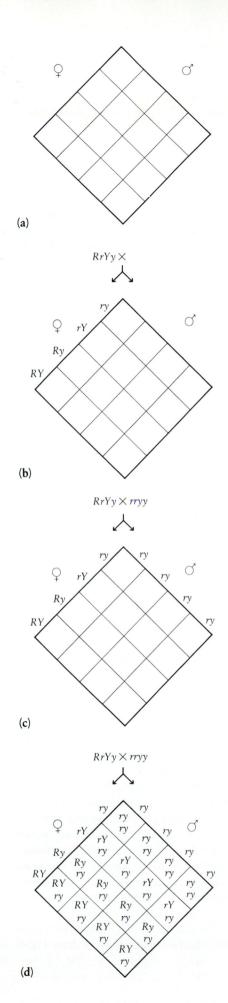

(a)

(b)

(c)

(d)

11-11 *A testcross involving two gene pairs.*

average, be in the ratio of 9:3:3:1. Nine out of the total 16, or 9/16, represents the proportion of F_2 offspring that will show the two dominant characteristics, 1/16 the proportion that will show the two recessive characteristics, and 3/16 and 3/16 the proportions that will show the two alternative combinations of dominants and recessives. In terms of probability (see essay), a seed has a 3/4 chance of being yellow and a 3/4 chance of being round and thus a 9/16 chance (3/4 × 3/4) of being both yellow and round. It has only a 1/4 chance of being wrinkled, however, so its chance of being yellow and wrinkled is 3/4 × 1/4, or 3/16.

The 9:3:3:1 ratio holds true when one of the original parents is homozygous for both recessive characteristics and the other homozygous for both dominant ones, as in the experiment just described *(RRYY × rryy)*, as well as when each original parent is homozygous for one recessive and also homozygous for one dominant characteristic *(rrYY × RRyy)*. The F_1 progeny from either of these crosses will always be heterozygous for both traits *(RrYy)*; crossing these heterozygotes produces the F_2 generation with the expected 9:3:3:1 ratio of phenotypes.

A Testcross

Can you predict the outcome of a testcross between an individual homozygous for the recessive alleles for each of two traits and an individual who is heterozygous for both traits? Such a cross is similar to the one analyzed in Figure 11–9 but involves alleles of two independently assorting genes instead of one. For simplicity, let us again study the distribution of the alleles for seed shape, round versus wrinkled *(R versus r)*, and the alleles for seed color, yellow versus green *(Y versus y)*, in a cross between a heterozygote and a homozygous recessive. Draw a Punnett square with 16 squares (Figure 11–11a). Put the female symbol on one side and the male on the other. Assume that the heterozygote contributes the female gametes. With a genotype of *RrYy*, the heterozygote can produce four kinds of gametes: *RY, Ry, rY,* and *ry*. At the head of each column at the left, where the female symbol is, put one of these possible combinations (Figure 11–11b). Notice that in this step we are assuming, as Mendel did, that each of the possible kinds of gametes is produced in equal numbers.

The homozygous recessive can produce only one type of gamete in terms of the traits being studied: *ry*. Put *ry* at the head of each column on the right (Figure 11–11c). The advantage of using a Punnett square is that it makes it impossible to overlook any combination of gametes.

Now, starting with the column on the far left, begin to fill in the squares. You are less likely to make a mistake if you fill in all the female gametes first (or all male gametes), a column at a time, than if you try to work with both male and female gametes at the same time.

Next, fill in the symbols for the alleles carried in the male gametes. Your square will then look like the one in Figure 11–11d.

Now count the phenotypes that this square predicts. Every capital *R* indicates a round seed (since *R* is dominant); there are eight capital *R*'s and, hence, eight round seeds. Conversely, you can count eight wrinkled *(rr)* seeds. Similarly, there are eight yellow *(Y)* seeds and eight green *(yy)* seeds.

Notice that each of the possible combinations of characteristics, round green, round yellow, wrinkled green, and wrinkled yellow, appears in equal proportions. This ratio of 1:1:1:1 is the typical result of a testcross involving an individual that is heterozygous for both traits.

Mendel and the Laws of Probability

In applying mathematics to the study of heredity, Mendel was asserting that the laws of probability apply to biology as they do to the physical sciences. Toss a coin. The probability that it will turn up heads is fifty-fifty, that is, one chance in two, or 1/2. The probability (or chance) that it will turn up tails is also fifty-fifty, or 1/2. The chance that it will turn up one or the other is certain, or one chance in one. Now toss two coins. The chance that one will turn up heads is again 1/2. The chance that the second will turn up heads is also 1/2. The chance that both will turn up heads is $1/2 \times 1/2$, or 1/4. The probability of two independent events occurring simultaneously is simply the probability of one occurring alone multiplied by the probability of the other occurring alone. This is known as the product rule of probability. The probability that both coins will turn up tails is similarly $1/2 \times 1/2$. The probability of the first coin turning up tails and the second coin turning up heads is $1/2 \times 1/2$, and the probability of the second turning up tails and the first heads is also $1/2 \times 1/2$.

We can diagram this in a Punnett square (see figure), which indicates that the combination in each square has an equal chance of occurring. Similarly, in Mendel's experiment diagrammed in Figure 11–8, the probability that a gamete produced by an F_1 plant of Ww genotype will carry the W allele is 1/2, and the probability that it will carry the w allele is 1/2. The probability, therefore, of any specific combination of the two alleles in the offspring—that is, WW, Ww, wW, or ww—is $1/2 \times 1/2$, or 1/4. It was undoubtedly the observation that one-fourth of the offspring in the F_2 generation showed the recessive phenotype that indicated to Mendel that he was dealing with a simple case of the laws of probability.

Returning to our coin toss, if there were three coins involved, the probability of any given combination would be simply the product of all three individual possibilities: $1/2 \times 1/2 \times 1/2$, or 1/8. Similarly, with four coins, the probability of any specific combination is $1/2 \times 1/2 \times 1/2 \times 1/2$, or 1/16. The Punnett square on page 245 expresses the probability of each of any one of four possible phenotype combinations.

When there is more than one possible arrangement of the events producing the specified outcome, however, the individual probabilities are added. For instance, what is the probability of throwing a head and a tail, in either order? There are two ways you could do this: by throwing a head first and then a tail (HT), or a tail first and then a head (TH). The probability of throwing a tail and a head or a head and a tail is the sum of their individual probabilities: $(1/2 \times 1/2) + (1/2 \times 1/2) = 1/4 + 1/4 = 1/2$. This is

11–12 *A flower of one species of evening primrose.*

MUTATIONS

In 1902, a Dutch botanist, Hugo de Vries, reported results of his studies on Mendelian inheritance in the evening primrose. Heredity in the primrose, he found, was generally orderly and predictable, as in the garden pea. Occasionally, however, a characteristic appeared that was not present in either parent or indeed anywhere in the lineage of that particular plant. De Vries hypothesized that such characteristics came about as the result of abrupt changes in genes and that the characteristic produced by a changed gene was then passed along like any other hereditary characteristic. De Vries called these abrupt hereditary changes **mutations,** and organisms exhibiting such changes came to be known as **mutants.** Different alleles of a gene, de Vries proposed, arose as a result of mutations. For example, the allele for wrinkled peas is thought to have arisen as a mutation of the gene for round peas.

As it turned out, only about 2 of some 2,000 changes in the evening primrose observed by de Vries were actually mutations. The vast majority were due to new

known as the sum rule of probability. In the cross diagrammed in Figure 11–8, a heterozygote is produced by either *Ww* or *wW*. The probability of a heterozygote in the F_2 generation is the sum of the probability of each of the two possible combinations: $1/4 + 1/4 = 1/2$.

The sum rule of probability, like the product rule, applies in more complex cases as well. For example, if you were asked the probability of throwing two heads and a tail, the answer would be 3/8. Three combinations are possible: HHT, HTH, and THH. For each of these combinations, the probability is $1/2 \times 1/2 \times 1/2 = 1/8$, that is, the product of three independent throws. Thus, the probability of throwing two heads and a tail is $1/8 + 1/8 + 1/8 = 3/8$.

Notice that in planning his experiments, Mendel made several assumptions: (1) of the male gametes produced, one-half contain one paternal allele and one-half contain the other paternal allele for each gene; (2) of the female gametes produced, one-half contain one maternal allele and one-half contain the other maternal allele for each gene; (3) the male and female gametes combine at random. Thus, the laws of probability could be employed—an elegant marriage of biology and mathematics.

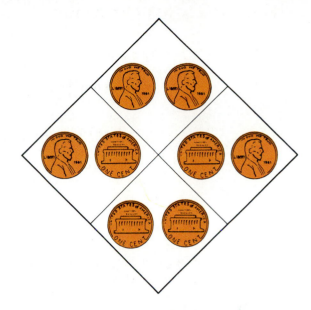

If you toss two coins 4 times, it is unlikely that you will get the precise results diagrammed above. However, if you toss two coins 100 times, you will come close to the proportions predicted in the Punnett square, and if you toss two coins 1,000 times, you will be very close indeed. As Mendel knew, the ratio of dominants to recessives in the F_2 generation might well not have been so clearly visible if he had been dealing with a small sample. The larger the sample, however, the more closely it will conform to results predicted by the laws of probability.

combinations of alleles, about which we shall have more to say in subsequent chapters, rather than to actual changes in any particular gene. However, de Vries's concept of mutation as the source of genetic variation proved of great importance, even though most of his examples were not valid.

Mutations and Evolutionary Theory

An important gap in Darwin's theory of evolution, first published in 1859, was the lack of explanation as to how variations can persist in populations. Mendel's work filled this gap. Segregation of alleles explained how variation is maintained from generation to generation. Independent assortment explained how individuals could have characteristics in combinations not present in either parent and so perhaps be better adapted, in evolutionary terms, than either parent. However, Mendelian principles presented new problems to the early evolutionists. If all hereditary variations were to be explained by the reshuffling process proposed by Mendel, there would be little or no opportunity for the kind of change in organisms envisioned by Darwin. As a result of mutations, however, there is a wide range of variability in natural populations. In a complex or shifting environment, a particular variation may give an individual or its offspring a slight edge. Although mutations seldom, if ever, determine the direction of evolutionary change, they are now recognized as the ultimate—and continual—source of the hereditary variations that make evolution possible.

THE INFLUENCE OF MENDEL

Mendel's experiments were first reported in 1865 before a small group of people at a meeting of the Brünn Natural History Society. None of them, apparently, understood what Mendel was talking about. But his paper was published the following year in the *Proceedings* of the Society, a journal that was circulated to libraries all over Europe. In spite of this, his work was ignored for 35 years, during most of which he devoted himself to the duties of an abbot, and he received no scientific recognition until after his death. He was, to use DuPraw's phrase (page 16), an odd traveler whose tale could not be made to fit.

(a)

(b)

11–13 **(a)** *Gregor Mendel, holding a fuchsia, is third from the right in this photograph of members of the Augustinian monastery in Brünn in 1862. In his experiments carried out in the monastery garden, Mendel showed that hereditary determinants are carried as separate units* *from generation to generation. His discoveries explained how inherited variations can persist for generation after generation.*

Although Mendel published only one other scientific paper during his lifetime, he continued his breeding experiments with a variety of plants until his election *as abbot of the monastery in 1871. Unfortunately, almost all of Mendel's papers relating to his scientific work were destroyed shortly before or after his death in 1884.* **(b)** *A facsimile of a page from one of Mendel's remaining handwritten manuscripts.*

It was not until 1900 that biologists were finally prepared to accept Mendel's findings. Within a single year, his paper was independently rediscovered by three scientists, one of whom was de Vries, each working in a different European country. Each of these scientists had done similar experiments and was searching the scientific literature for confirmation of his results. And each found, in Mendel's brilliant analysis, that much of his own work had been anticipated.

During the 35 years that Mendel's work remained in obscurity, great improvements were made in microscopy and, as a consequence, in the study of the structure of the cell (cytology). It was during this period that chromosomes were discovered and their movements during mitosis, which we described in Chapter 7, were first observed and recorded. Also discovered during this period was the process by which the gametes—the bearers of the hereditary information from one generation to the next—are formed; we shall examine this process in the next chapter.

SUMMARY

We have begun our consideration of genetics with the earliest ideas about biological inheritance, tracing the gradual development of these ideas into a science. The first question with which this new science was concerned was the mechanism of inheritance. How are hereditary characteristics passed from generation to generation?

By the middle of the nineteenth century, it was recognized that ova and sperm are specialized cells and that the ovum and sperm both contribute to the hereditary characteristics of the new individual. But how are these special cells, called gametes, able to pass on the many hundreds of characteristics involved in inheritance? Blending inheritance, which held that the characteristics of the parents blended in the offspring, like a mixture of two fluids, was one hypothesis. This explanation, however, did not allow for the inheritance of variations, which clearly occurred.

The revolution in genetics came when the blending concept was replaced by a unit concept. According to Mendel's principle of segregation, hereditary characteristics are determined by discrete factors (now called genes) that occur in pairs—one of each pair inherited from each parent. The members of the pair may be the same, in which case the individual is homozygous for the trait determined by the gene, or they may be different, in which case the individual is heterozygous for that trait. Different forms of the same gene are known as alleles.

The genetic makeup of an organism is known as its genotype. Its outward, observable characteristics are known as its phenotype. An allele that is expressed in the phenotype of a heterozygous individual to the exclusion of the other allele is a dominant allele; one whose effects are concealed in the phenotype is a recessive allele. In crosses involving two individuals heterozygous for the same gene, the ratio of dominant to recessive in the phenotypes of the offspring is 3:1.

A testcross, in which an individual with a dominant phenotypic characteristic but an unknown genotype is crossed with an individual homozygous for the recessive allele, reveals the genotype of the individual. If, in a testcross involving one gene, the two possible phenotypes appear in the offspring, the tested individual is heterozygous; if only the dominant phenotype appears in the offspring, the individual is homozygous for the dominant allele.

Mendel's other great principle, the principle of independent assortment, applies to the behavior of two or more different genes. This principle states that the alleles of one gene segregate independently of the alleles of another gene. When organisms heterozygous for each of two independently assorting genes are crossed, the expected phenotypic ratio in the offspring is 9:3:3:1.

Mutations are abrupt changes in the genotype—the ultimate source of the genetic variations studied by Mendel. Different mutations of a single gene increase the diversity of alleles of that gene in a population. As a consequence, mutation provides the variability among organisms that is the raw material for evolution.

QUESTIONS

1. Distinguish between the following terms: gene/allele; dominant/recessive; homozygous/heterozygous; genotype/phenotype; the F_1/the F_2; mutation/mutant.

2. In the experiments summarized in Table 11–1, which of the alternative characteristics appeared in the F_1 generation?

3. Why is a homozygous recessive always used in a testcross?

4. (a) What is the genotype of a pea plant that breeds true for tall? (Use the symbol T for tall, t for dwarf.) What possible gametes can be produced by such a plant? (b) What is the genotype of a pea plant that breeds true for dwarf? What possible gametes can be produced by such a plant? (c) What will be the genotype of the F_1 generation produced by a cross between a true-breeding tall pea plant and a true-breeding dwarf pea plant? (d) What will be the phenotype of this F_1 generation? (e) What will be the probable distribution of characteristics in the F_2 generation? Illustrate with a Punnett square.

5. The ability to taste a bitter chemical, phenylthiocarbamide (PTC), is due to a dominant allele. In terms of tasting ability, what are the possible phenotypes of a man both of whose parents are tasters? What are his possible genotypes?

6. If the man in Question 5 marries a woman who is a nontaster, what proportion of their children could be tasters? Suppose one of the children is a nontaster. What would you know about the father's genotype? Explain your results by drawing Punnett squares.

7. A taster and a nontaster have four children, all of whom can taste PTC. What is the probable genotype of the parent who is a taster? Is there another possibility?

8. What is the probability of drawing two aces out of a deck of 52 playing cards, one the ace of hearts and the other the ace of spades?

9. You have just flipped a coin five times and it has turned up heads every time. What is the chance that the next time you flip it, it will turn up tails?

10. (a) Suppose you would like to have a family consisting of two girls and a boy. What are your chances, assuming you have no children now? (b) If you already have one boy, what are your chances of completing your family as planned? (c) If you have two girls, what is the probability that the next child will be a boy?

11. PKU, phenylketonuria, is a disease caused by the presence of two recessive alleles of a particular gene. Individuals who are homozygous for the dominant allele of that gene or who are heterozygous show no signs of the disease. If two healthy parents have a child with PKU, what are their genotypes with respect to PKU? What are their chances of having another child with the same disease?

12. A pea plant that breeds true for round, green seeds *(RRyy)* is crossed with a plant that breeds true for wrinkled, yellow seeds *(rrYY)*. Each parent is homozygous for one dominant characteristic and for one recessive characteristic. (a) What is the genotype of the F_1 generation? (b) What is the phenotype? (c) The F_1 seeds are planted and their flowers are allowed to self-pollinate. Draw a Punnett square to determine the ratios of the phenotypes in the F_2 generation. How do the results compare with those of the experiment shown in Figure 11–10?

Meiosis and Sexual Reproduction

Mendelian genetics is concerned with the way in which hereditary traits are passed from parents to offspring. Most eukaryotic organisms—including, for example, pea plants, sea urchins, and human beings—reproduce sexually. Sexual reproduction generally requires two parents, and it always involves two events: **fertilization** and **meiosis**. Fertilization is the means by which the different genetic contributions of the two parents are brought together to form the new genetic identity of the offspring. Meiosis is a special kind of nuclear division that is believed to have evolved from mitosis and uses much of the same cellular machinery. However, as you will see, meiosis differs from mitosis in some important respects.

HAPLOID AND DIPLOID

To understand meiosis, we must look again at the chromosomes. Every organism has a chromosome number characteristic of its particular species. A mosquito has 6 chromosomes per somatic (body) cell; a cabbage, 18; corn, 20; a sunflower, 34; a cat, 38; a human being, 46; a potato, 46; a plum, 48; a dog, 78; and a goldfish, 94. However, in these organisms and most other familiar plants and animals, the sex cells—or gametes—have exactly half the number of chromosomes that is characteristic of the somatic cells of the organism. The number of chromosomes in the gametes is referred to as the **haploid** ("single set") number, and the number in the somatic cells as the **diploid** ("double set") number. Cells that have more than two sets of chromosomes are known as **polyploid** ("many sets").

12–1 (a) *Wake-robin* (Trillium erectum) *flowers in the early spring.* (b) *Meiosis leading to the formation of pollen grains occurs in the anthers. The clearly visible chromosomes shown here are almost completely separated. Each diploid nucleus has divided to form four haploid sets of chromosomes.*

(a)

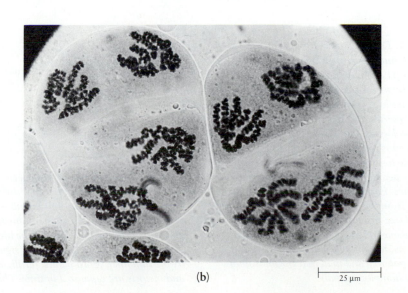

(b)

25 μm

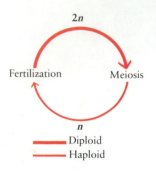

12–2 *Sexual reproduction is character-ized by two events: the coming together of the gametes (fertilization) and meiosis. Following meiosis, there is a single set of chromosomes, that is, the haploid number (n). Following fertilization, there is a dou-ble set of chromosomes, that is, the diploid number (2n).*

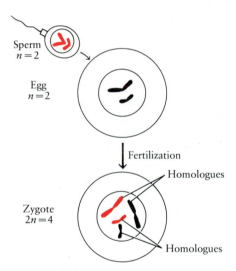

12–3 *During gamete formation, individ-ual homologues (members of a homolo-gous pair of chromosomes) are parceled out by meiosis so that a haploid (n) gam-ete, which is produced from a diploid (2n) cell, carries only one member of each homologous pair. At fertilization, the chromosomes in the sperm and egg nuclei come together in the zygote, producing, once again, pairs of homologous chromo-somes. Each pair consists of one homolo-gue from the father (paternal chromosome) and one from the mother (maternal chro-mosome). Here, and in subsequent dia-grams, red and black are used to indicate the paternal and maternal chromosomes of a homologous pair.*

For brevity, the haploid number is designated n and the diploid number as $2n$. In humans, for example, $n = 23$ and therefore $2n = 46$. When a sperm fertilizes an egg, the two haploid nuclei fuse, $n + n = 2n$, and the diploid number is restored (Figure 12–2). A diploid cell produced by the fusion of two gametes is known as a **zygote.**

In every diploid cell, each chromosome has a partner. These pairs of chromo-somes are known as homologous pairs, or **homologues.** The two resemble each other in size and shape and also, as we shall see, in the kinds of hereditary information (genes) each contains. One homologue comes from the gamete of one parent, and its partner is from the gamete of the other parent. After fertilization both homologues are present in the zygote (Figure 12–3).

In the special kind of nuclear division called meiosis, the diploid set of chromosomes, which contains the two homologues of each pair, is reduced to a haploid set, which contains only one homologue of each pair. Meiosis thus counterbalances the effects of fertilization. Cytologists predicted the existence of this so-called "reduction division" before it was actually observed; as they realized, without such a division, fertilization would double the chromosome number with each succeeding generation. In addition to maintaining a constant number of chromosomes from generation to generation, meiosis is, as we shall see shortly, a source of new combinations within the chromosomes themselves.

MEIOSIS AND THE LIFE CYCLE

Meiosis occurs at different times during the life cycle of different organisms (Figure 12–4). In many protists and fungi, such as the alga *Chlamydomonas* and the mold *Neurospora*, it occurs immediately after fusion of the mating cells (Figure 12–5). The cells are ordinarily haploid, and meiosis after fertilization restores the haploid number.

In plants, such as ferns, a haploid phase typically alternates with a diploid phase (Figure 12–6). The common and conspicuous form of a fern is the sporophyte, the diploid organism. By meiosis, fern sporophytes produce spores, usually on the undersides of their fronds (leaves). These spores have the haploid number of chromosomes. They germinate to form much smaller plants (gametophytes), typically only a few cell layers thick. In these plants, all the cells are haploid. The

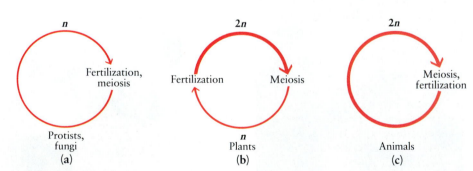

12–4 *Fertilization and meiosis occur at different points in the life cycle of different organisms. (a) In many—but not all—pro-tists and fungi, meiosis occurs immedi-ately after fertilization. Most of the life cycle is spent in the haploid state (signified by the thin line). (b) In plants, fertilization and meiosis are separated in time. The life* cycle of the organism consists of both a diploid phase and a haploid phase. (c) In animals, completion of meiosis is immedi-ately followed by fertilization. As a conse-quence, during most of the life cycle the organism is diploid (signified by the thick line).

12-5 *The life cycle of Chlamydomonas is of the type shown in Figure 12–4a. The organism is haploid for most of its life cycle (thin arrows). Fertilization, the fusion of cells of different mating strains (indicated here by + and −), temporarily produces the diploid zygote (thick arrows). The zygote produces a thick coat that allows it to remain dormant during harsh conditions. Following dormancy, the diploid zygote divides meiotically, forming four new haploid cells. Each haploid cell can reproduce asexually (by mitosis and cytokinesis) to form either more haploid cells or, in periods of stress, haploid cells of a particular mating type. These cells can fuse with cells of the opposite mating type, and another sexual cycle is underway.*

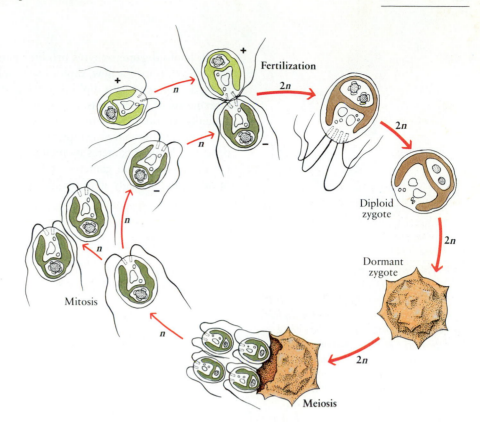

12-6 *The life cycle of a fern is of the type shown in Figure 12–4b. Following meiosis, spores, which are haploid, are produced in the sporangia and then are shed (extreme right). The spores develop into haploid gametophytes. In many species, the gametophytes are only one layer of cells thick and are somewhat heart-shaped, as shown here (bottom). From the lower surface of the gametophyte, filaments, the rhizoids, extend downward into the soil.*

On the lower surface of the gameto-phyte are borne the flask-shaped arche-gonia, which enclose the egg cells, and the antheridia, which enclose the sperm. When the sperm are mature and there is an adequate supply of water, the antheri-dia burst, and the sperm cells, which have numerous flagella, swim to the archegonia and fertilize the eggs. From the zygote, the diploid (2n) sporophyte develops, grow-ing out of the archegonium within the gametophyte. After the young sporophyte becomes rooted in the soil, the gameto-phyte disintegrates. The sporophyte matures, develops sporangia, in which meiosis occurs, and the cycle begins again.

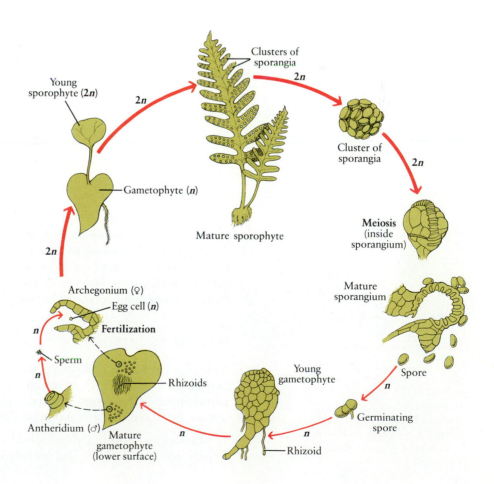

small, haploid gametophytes produce gametes by mitosis. The gametes fuse and then develop into a new, diploid sporophyte. This process, in which a haploid phase is followed by a diploid phase and again by a haploid phase, is known as **alternation of generations.** As we shall see, alternation of generations occurs in all sexually reproducing plants, although not always in the same form.

Human beings have the typical animal life cycle, in which the diploid individual produces haploid gametes by meiosis immediately preceding fertilization. Fusion of male and female gametes at fertilization restores the diploid chromosome number, and virtually all of the life cycle is spent in the diploid state (Figure 12–7).

12–7 *The life cycle of* Homo sapiens. *Gametes—egg cells and sperm cells—are produced by meiosis. At fertilization, the haploid gametes fuse, restoring the diploid number in the fertilized egg. The zygote develops into a mature man or woman, who again produces haploid gametes. As is the case with most other animals, the cells are diploid during almost the entire life cycle, the only exception being the gametes. This is the type of life cycle diagrammed in Figure 12–4c.*

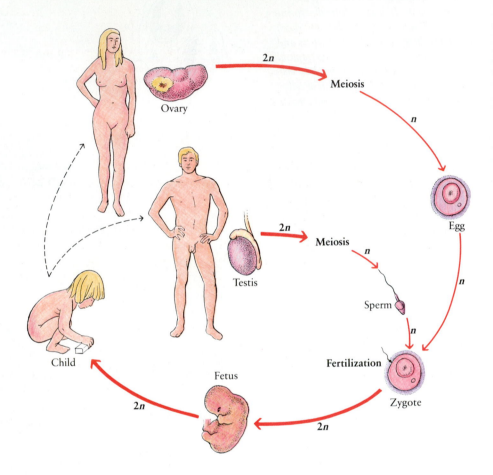

Note that although meiosis in animals produces gametes, meiosis in plants produces **spores.** A spore is a haploid reproductive cell that, unlike a gamete, can develop into a haploid organism without first fusing with another cell. With the formation of either gametes and spores, however, meiosis has the same result: at some point during the life cycle of a sexually reproducing organism, it reduces the diploid chromosome set to the haploid chromosome set.

MEIOSIS VS. MITOSIS

As we saw in Chapter 7, the events that occur in mitosis result in the formation of two daughter nuclei, each of which receives an exact copy of the parent cell's chromosomes. The events that take place during meiosis resemble those of mitosis, but there are some important differences:

1. Mitosis can occur in either haploid or diploid cells, whereas meiosis occurs only in cells with the diploid (or polyploid) number of chromosomes.

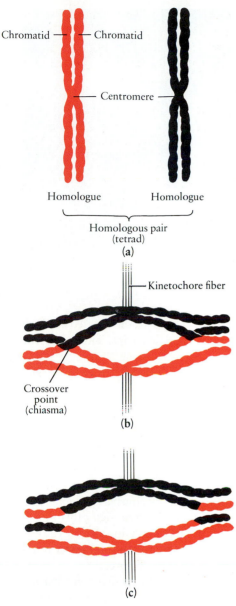

12–8 (a) *During prophase I of meiosis, chromosomes become arranged in homologous pairs. Each homologous pair consists of four chromatids and is therefore also known as a tetrad (from the Greek tetra meaning "four"). (b) Crossing over. During meiosis I, homologues, paired together as tetrads, connect at crossover points, where exchanges of segments of the chromosomes—crossing over—take place. Homologues remain associated at the crossover points, or chiasmata, until the end of prophase I. (c) Then, as the chromosomes separate slightly, the chiasmata appear to slip off. As you can see, crossing over results in a recombination of the genetic material of the two homologues.*

2. During meiosis, each diploid nucleus divides twice, producing a total of four nuclei. The chromosomes, however, replicate only once—prior to the first nuclear division.

3. Thus, each of the four nuclei produced contains half the number of chromosomes present in the original nucleus.

4. The haploid nuclei produced by meiosis contain new combinations of chromosomes. That is, the homologous chromosomes, originally derived from the organism's parents, are assorted randomly among the four new haploid nuclei. (For example, whether a given chromosome in a particular gamete produced by your body is the homologue derived originally from your mother or from your father is purely a matter of chance.)

Before looking at meiosis, you might wish to review the events of mitosis (pages 150–153).

THE PHASES OF MEIOSIS

Meiosis consists of two successive nuclear divisions, conventionally designated meiosis I and meiosis II. In meiosis I, homologous chromosomes pair and then separate from one another; in meiosis II, the chromatids of each homologue separate. In the following discussion we shall describe meiosis in a plant cell in which the diploid number is 6 *(n = 3)*. Three of the six chromosomes were originally derived from one parent and three from the other parent. For each chromosome from one parent there is a homologous chromosome, or homologue, from the other parent.

During interphase preceding meiosis, the chromosomes are replicated, so that by the beginning of meiosis each chromosome consists of two identical sister chromatids held together at the centromere region (Figure 12–8a). The first of the two nuclear divisions in meiosis then proceeds through the stages of prophase, metaphase, anaphase, and telophase (all of these are given the designation I to indicate that they are substages of meiosis I).

At the beginning of meiosis, **prophase I,** the chromatin condenses and the chromosomes come into view. By this time, an event has occurred for which the mechanism is completely unknown—the homologous chromosomes have come together in pairs. Once contact is made at any point between the two homologues, pairing extends, zipperlike, along the length of the chromatids in a process called **synapsis.** Since each chromosome consists of two identical chromatids, the pairing of the homologous chromosomes actually involves four chromatids; this complex of paired homologous chromosomes is known as a **tetrad.**

At this point, a crucial process occurs that can alter the genetic makeup of the chromosomes. This process, known as **crossing over,** involves the exchange of segments of one chromosome with corresponding segments from its homologous chromosome (Figure 12–8b). At the sites of crossing over, portions of the chromatids of one homologue are broken and exchanged with the corresponding portions of one or the other of the chromatids of the second homologue. The breaks are resealed, and the result is that the sister chromatids of a single homologue no longer contain identical genetic material (Figure 12–8c). The maternal homologue now contains portions of the paternal homologue, and vice versa. Thus crossing over is an important mechanism for recombining the genetic material from the two parents.

As prophase progresses, the homologues begin to pull away from each other—except in the areas of crossing over. Here, at the crossover points, or **chiasmata** (singular, chiasma), the homologues remain in close association until the end of prophase. Then the chiasmata seem to slip off the ends of the chromosomes. (If

12–9 (a) *Early prophase I in the formation of a sperm cell in a grasshopper. The homologous chromosomes are now paired; the individual chromatids are not visible, however, so each chromosome appears as a single structure, and the tetrads appear double-stranded (rather than four-stranded). The dark area in the upper right is a sex chromosome, which is very prominent in the grasshopper.* **(b)** *Late prophase I. All four chromatids can be seen in some of the tetrads, and chiasmata are evident.*

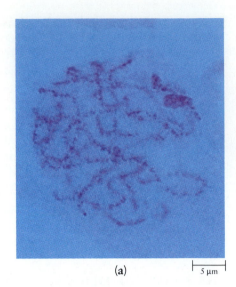

(a) ⊢ 5 µm ⊣

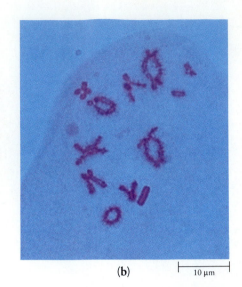

(b) ⊢ 10 µm ⊣

you cross your arms in front of you and then slide them apart, the crossed area will eventually reach your hands and then slide off your fingertips, much like the chiasmata.)

Although the homologous chromosomes have moved slightly apart by the end of prophase I, the homologues are still paired. Spindle microtubules can be seen radiating out from the two poles of the cell. The nucleoli and nuclear envelope disappear toward the end of this stage.

If you remember one important point—that the homologous chromosomes are arranged in pairs at this stage of meiosis—you will be able to remember all of the subsequent events with little difficulty.

In **metaphase I,** the homologous pairs (as you will recall, three pairs in this example) line up along the equatorial plane of the cell. (By contrast, in metaphase of mitosis, replicated chromosomes line up in single file with no sign of pairing of homologues.) The centromere region of each homologue has doubled by the end of metaphase, and spindle fibers have become associated with the kinetochores (see page 148). In an animal cell, centrioles and asters are also present.

During **anaphase I,** the homologues, each consisting of two sister chromatids, separate, as if pulled apart by the spindle fibers attached to the kinetochores. However, the two sister chromatids of each homologue do not separate as they did in mitosis.

Late prophase I

Metaphase I

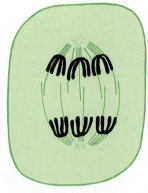

Anaphase I

By the end of the first meiotic division, **telophase I,** the homologues have moved to the poles. Each chromosome group now contains only half the number of chromosomes as the original nucleus.* Moreover, these chromosomes may be different from any of those present in the original cell because of exchanges that took place during crossing over. Depending on the species, new nuclear envelopes may or may not form, and cytokinesis may or may not take place. In some animal cells, but not all, the centrioles also divide at this stage.

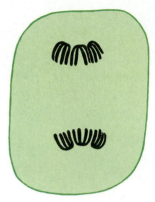

Telophase I

Meiosis, however, does not end here. Although two haploid nuclei have been formed, each nucleus contains double the haploid amount of hereditary material. Why? Because each chromosome consists of two chromatids.

Meiosis II resembles mitosis except that it is not preceded by replication of the chromosomal material. A short interphase may occur, during which the chromosomes partially unfold, but meiosis in many species proceeds from telophase directly to prophase II.

At the beginning of the second meiotic division, the chromosomes, if dispersed, condense fully again. Remember, in this example there are three chromosomes in each nucleus (the haploid number), and each is still in the form of two chromatids held together at the centromere region.

During **prophase II,** the nuclear envelopes, if present, disintegrate, and new spindle fibers begin to appear.

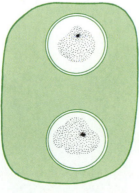

Interphase II

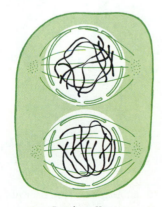

Prophase II

---

* In counting, it is often difficult to know whether to count a chromosome that has replicated but has not divided as 1 or 2 chromosomes. It is customary to count such a chromosome as 1. The trick is to count centromeres. If a chromatid has its own centromere, not associated with the centromere of its sister chromatid, then it can be called a chromosome.

During **metaphase II,** the three chromatid pairs in each nucleus line up on the equatorial plane. Spindle fibers are once again associated with the kinetochores, and other spindle fibers extend from the poles.

At **anaphase II,** as in anaphase of mitosis, the sister chromatids separate from one another. Each chromatid, which can now be called a chromosome, moves toward one of the poles.

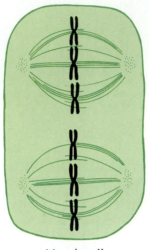

Metaphase II

Anaphase II

During **telophase II,** the spindle microtubules disappear and a nuclear envelope forms around each set of chromosomes. There are now four nuclei in all, each containing the haploid number of chromosomes. Cytoplasmic division (cytokinesis) proceeds as it does following mitosis. Cell walls form, dividing the cytoplasm, and these haploid plant cells begin to differentiate into spores.

Thus, beginning with one cell containing six chromosomes (three homologous pairs), we end with four cells, each with three chromosomes (no homologous pairs). The chromosome number has been reduced from the diploid to the haploid number.

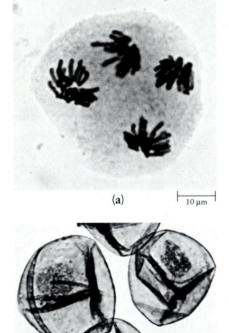

(a)   10 μm

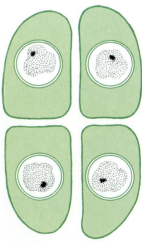

(b)   10 μm

12–10 (a) *Anaphase II in the royal fern* Osmunda regalis. *The chromatids have separated, and the daughter chromosomes are moving to the opposite poles of the spindles.* (b) *The end of spore formation in* Osmunda regalis. *Each of these haploid cells can germinate to produce a gametophyte, the haploid phase in the life cycle.*

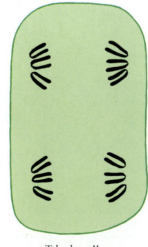

Telophase II

Four haploid cells

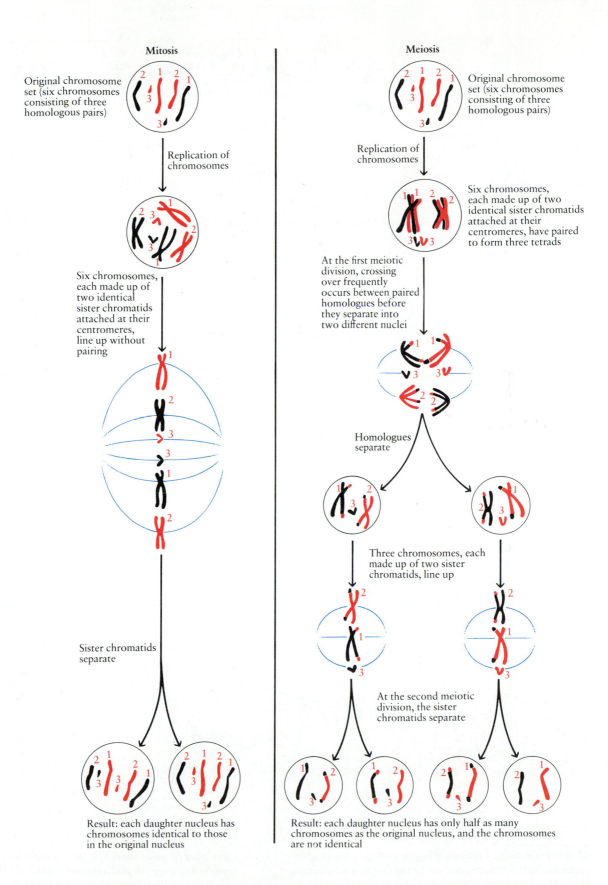

Mitosis

Original chromosome set (six chromosomes consisting of three homologous pairs)

Replication of chromosomes

Six chromosomes, each made up of two identical sister chromatids attached at their centromeres, line up without pairing

Sister chromatids separate

Result: each daughter nucleus has chromosomes identical to those in the original nucleus

Meiosis

Original chromosome set (six chromosomes consisting of three homologous pairs)

Replication of chromosomes

Six chromosomes, each made up of two identical sister chromatids attached at their centromeres, have paired to form three tetrads

At the first meiotic division, crossing over frequently occurs between paired homologues before they separate into two different nuclei

Homologues separate

Three chromosomes, each made up of two sister chromatids, line up

At the second meiotic division, the sister chromatids separate

Result: each daughter nucleus has only half as many chromosomes as the original nucleus, and the chromosomes are not identical

12–11 *A comparison of mitosis and meiosis. In these examples, each diploid cell has six chromosomes (2n = 6).*

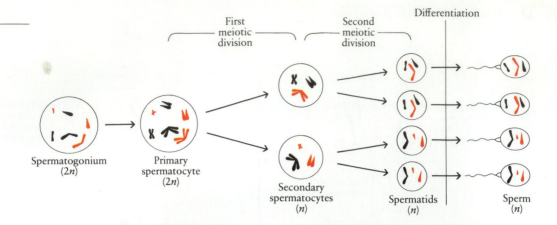

12-12 *The series of changes resulting in the formation of sperm cells begins with the growth of spermatogonia into primary spermatocytes. For simplicity, only six (n = 3) chromosomes are shown. At the first meiotic division, each primary spermatocyte divides into two haploid secondary spermatocytes. The second meiotic division results in the formation of four haploid spermatids, which differentiate into functional sperm. This process occurs continuously; the normal ejaculate of an adult human male contains between 300 and 400 million sperm cells.*

Meiosis in the Human Species

In all vertebrates, including humans, meiosis takes place in the reproductive organs, the testes of the male and the ovaries of the female. In the male, a cell known as a primary spermatocyte undergoes the two divisions of meiosis to produce four haploid spermatids, each of which then differentiates into a sperm cell (Figure 12–12).

In females, the meiotic divisions also produce haploid nuclei but the cytoplasm is apportioned unequally during cytokinesis in both meiosis I and II. One egg cell (the ovum) is produced, along with two or three polar bodies (Figure 12–13). The polar bodies contain the other post-meiotic nuclei and usually disintegrate. As a result of this unequal division, the ovum is well supplied with cytoplasmic materials, such as ribosomes, mitochondria, enzymes, and stored nutrients, important for the development of the embryo.

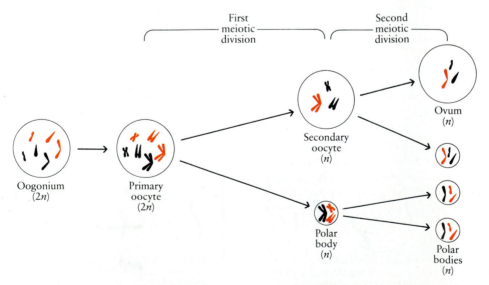

12-13 *Formation of the ovum begins with the growth of an oogonium into a primary oocyte. In the first meiotic division, this cell divides into a secondary oocyte and a polar body. The first meiotic division begins in the human female during the third month of fetal development and ends at ovulation, which may take place 50 years later. The second meiotic division, which produces the egg cell and a second polar body, does not take place until after the fertilizing sperm cell has penetrated the secondary oocyte. The first polar body may also divide.*

The Consequences of Sexual Reproduction

Many organisms can reproduce both asexually (by mitosis and cytokinesis) and sexually. Most unicellular eukaryotes have a life cycle similar to that of *Chlamydomonas* (Figure 12–5), in which either sexual or asexual reproduction may take place, often depending on environmental circumstances. Some unicellular eukaryotes, such as amoebas, reproduce only asexually. Many plants can also reproduce asexually; many grasses, for instance, spread by means of horizontal stems (rhizomes), growing either just above or just below the surface of the soil. In animals, asexual reproduction can take place by budding, as in *Hydra,* or by the breaking off of a fragment of the parent animal, as occurs in sponges, sea anemones, and certain types of worms. Because of the careful copying process of mitosis, asexually produced individuals are genetically identical to their parents.

By contrast, the potential for genetic variability in sexually produced individuals is enormous. The diagram below shows the possible distributions of chromosomes at meiosis in organisms with relatively few chromosomes. The red chromosomes were originally of paternal origin, and the black chromosomes of maternal origin. In the course of meiosis, these chromosomes are distributed among the haploid cells. As you can see, chromosomes of maternal or paternal origin do not stay together but are assorted independently. (At metaphase I of meiosis, the orientation of the homologous pairs is random, with no "rule" as to how many paternal or maternal homologues should be on either side of the equator.)

The number of possible chromosome combinations in the gametes is 2^n; 2 is the number of homologues in a pair, and n equals the haploid chromosome number. For example, (a) if the original number of chromosomes is 4 ($n = 2$), the number of possible combinations of chromosomes is 2^2, or 4. (b) If the original number is 6 ($n = 3$), the number of possible combinations is 2^3, or 8. (c) If there are 8 chromosomes ($n = 4$), 16 different combinations (2^4) are possible.

A human male with his 46 chromosomes is capable of producing 2^{23} kinds of sperm cells—8,388,608 different combinations of chromosomes, about equal in number to the population of New York City. Similarly, a human female is capable of producing 2^{23} kinds of egg cells—8,388,608 different combinations of chromosomes. And this does not take into account the additional variations that may be introduced by crossing over.

Sexual reproduction, however, is expensive and inefficient, since a sexual population wastes half of its reproductive capacity on producing males and can therefore reproduce only half as fast as an asexual population. This raises one of the most intriguing questions of modern evolutionary biology: Given that sexual reproduction is expensive and inefficient, why is it so widely practiced throughout the multicellular world? The answer to this question is not known, but, as we shall see in Chapter 47, several different possibilities are the subject of current discussion and debate.

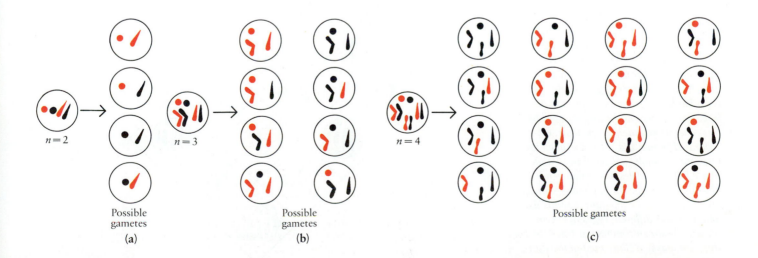

$n=2$

Possible gametes

(a)

$n=3$

Possible gametes

(b)

$n=4$

Possible gametes

(c)

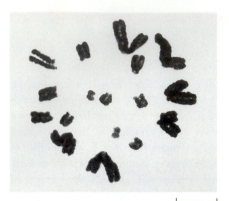

|———————————————|
| 10 μm |

12-14 *Chromosomes from a diploid cell of a grasshopper, during metaphase of a mitotic division. Each chromosome consists of two closely aligned chromatids. Note that even though these chromosomes are not paired, it is possible to pick out some of the homologues. The observation that chromosomes come in homologous pairs was one of Sutton's clues to the meaning of meiosis.*

CYTOLOGY AND GENETICS MEET: SUTTON'S HYPOTHESIS

In 1902, shortly after the rediscovery of Mendel's work, Walter S. Sutton, a graduate student at Columbia University, was studying the formation of sperm cells in male grasshoppers. Observing the process of meiosis, Sutton noticed that the chromosomes were paired at the beginning of the first meiotic division. He also noticed that the two chromosomes of any one pair had physical resemblances to one another. In diploid cells, he noted, chromosomes apparently come in pairs. The pairing was obvious only at meiosis, although the discerning eye might also find the matching, but unpaired, homologues during metaphase of mitosis (Figure 12–14).

Sutton was struck by the parallels between what he was seeing and the first principle of Mendel—the principle of segregation. Suddenly the facts fell into place. Suppose chromosomes carried genes, the *Elemente* described by Mendel. This idea may not seem very startling to you now, but remember that at the turn of

12-15 *The chromosome distributions in Mendel's cross of round yellow and wrinkled green peas, according to Sutton's hypothesis. Although the pea has 14 chromosomes (n = 7), only 4 are shown here, the two carrying the alleles for round or wrinkled and the two carrying the alleles for yellow or green. (This selection of specific chromosomes is analogous to Mendel's selection of specific traits to study.) As you can see, one parent is homozygous for the recessives, one for the dominants. Therefore, the only gametes they can produce are RY and ry. (Remember, R now stands not just for the allele but also for the chromosome carrying the allele, as do the other letters.) The F₁ generation, therefore, must be Rr and Yy. When a cell of this generation undergoes meiosis, R is separated from r and Y from y when the respective homologues separate at anaphase I. These alleles assort independently. Four different types of haploid egg nuclei are possible, as the diagram reminds us, and also four different types of sperm nuclei. These can combine in 4 × 4, or 16 different ways, as illustrated in the Punnett square.*

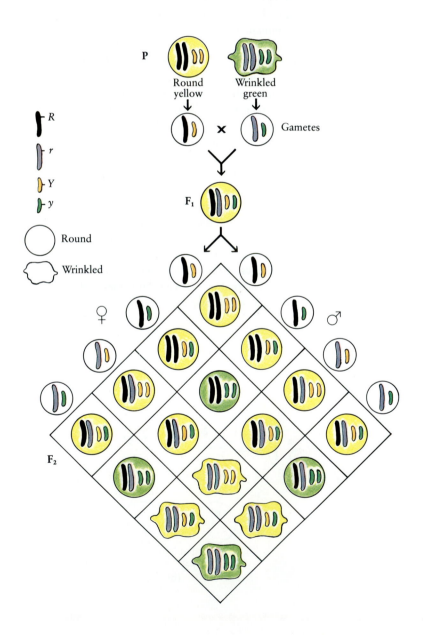

the century the gene was just an abstract idea or mathematical unit to the geneticist, and to the cytologist the chromosome was just an easily stained body with unknown function. Suppose, Sutton reasoned, alleles occurred on homologous chromosomes. Then the alleles would always remain independent and so would be separated at meiosis I as homologous chromosomes separated. New combinations of alleles would be formed as gametes fused at fertilization. Mendel's principle of the segregation of alleles could thus be explained by the segregation of the homologous chromosomes at meiosis.

What about Mendel's second principle in relation to the movement of the chromosomes at meiosis? This principle, as you will recall, states that the alleles of different genes assort independently. How one pair of alleles is segregated has no effect on the segregation of another pair of alleles, provided—and this is an important point—the two different pairs of alleles are on different pairs of chromosomes. The fact that Mendel chose traits for which the genes were found on different pairs of homologous chromosomes was essential to the success of his work (Figure 12–15).

As occurs often in the history of science, two other biologists recognized, at about the same time, the correlation between the behavior of Mendel's *Elemente* and the observed movement of the chromosomes. Young Sutton's paper appeared first, however, and his presentation was by far the most convincing. Nonetheless, much more evidence was required before, more than a decade later, most biologists were ready to concede that the little "colored bodies" performing their stereotyped, repetitive dance within the cell's nucleus actually held the secrets of the most ancient mysteries of heredity.

SUMMARY

Sexual reproduction involves a special kind of nuclear division called meiosis. Meiosis is the process by which the chromosomes are reassorted and cells are produced that have the haploid chromosome number (*n*). The other principal component of sexual reproduction is fertilization, the coming together of haploid cells to form the zygote; fertilization restores the diploid number (*2n*). There are characteristic differences among major groups of organisms as to when in the life cycle these events take place.

At the start of meiosis, the chromosomes arrange themselves in pairs. The members of the pairs are known as homologues. One homologue of each pair is of maternal origin, and one is of paternal origin. Each homologue consists of two identical sister chromatids, held together at the centromere. Early in meiosis, crossing over occurs between homologues, resulting in exchanges of chromosomal material.

In the first stage of meiosis, meiosis I, the homologues are separated. Two nuclei are produced, each with a haploid number of chromosomes, which, in turn, consist of two chromatids each. The nuclei enter interphase, but the chromosomal material is not replicated. In the second stage of meiosis, meiosis II, the sister chromatids of each chromosome separate as in mitosis. When the two nuclei divide, four haploid cells result.

Each of the haploid cells produced by meiosis contains a unique assortment of chromosomes due to crossing over and random assortment of homologues. Thus meiosis is a source of variation in the offspring.

Sutton was among the first to notice the analogy between the behavior of the chromosomes at meiosis and the segregation and assortment of the factors described by Mendel. On the basis of this observation, Sutton proposed that genes are carried on chromosomes.

QUESTIONS

1. Distinguish among the following: haploid/diploid/polyploid; sporophyte/gametophyte; gamete/zygote; meiosis I/meiosis II; homologues/tetrad; crossing over/chiasmata.

2. Dogs have a diploid chromosome number of 78. How many chromosomes would you expect to find in a gamete? In a liver cell? Plums have a haploid chromosome number of 24. How many chromosomes would you expect to find in a stem cell? In the nucleus of a pollen grain?

3. Draw a diagram of a cell with six chromosomes $(n = 3)$ at meiotic prophase I. Label each pair of chromosomes differently (for example, label one pair A^1 and A^2, and another B^1 and B^2, etc.).

4. Diagram the eight possible gametes resulting from meiosis in a plant cell with six chromosomes $(n = 3)$. Label each chromosome differently, as in Question 3. Assume that crossing over does not occur.

5. Identify the stages of meiosis in *Lilium* shown in the micrographs below. What stage of meiosis is visible in Figure 12–1?

6. (a) Compare metaphase of mitosis and metaphase II of meiosis. (b) Compare anaphase of mitosis with anaphase I and anaphase II of meiosis. In your answers, consider both the positions and composition of the chromosomes, as well as the consequences.

7. Compare and contrast the overall processes and the genetic consequences of meiosis and mitosis.

8. In our bodies and in those of most other animals, both mitosis and meiosis occur. What are the end products of these two processes? Where in our bodies do these processes occur?

9. Is sexual reproduction—that is, fertilization and meiosis—possible with only one parent? Explain.

10. Mendel did not know of the existence of chromosomes. Had he known, what change might he have made in his second principle?

11. You do not look exactly like your mother or your father. Why is this so? Explain how you might have inherited some of your maternal grandfather's characteristics. (Start with a gamete produced by your grandfather, and end with one of your somatic cells.)

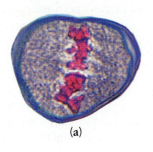

(a)

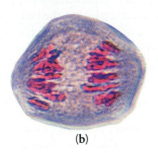

(b)

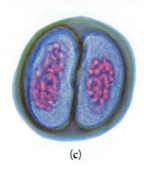

(c)

(d)

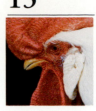

Genes and Gene Interactions

13–1 *Thomas Hunt Morgan at work in Columbia University's "Fly Room" in 1917. The camera-shy Morgan was photographed surreptitiously by a colleague who concealed a camera under a pile of milk bottles on his own desk. The half-pint milk bottles were used to house experimental* Drosophila.

When Mendel's work was rediscovered in Europe in 1900 by Hugo de Vries and others, it attracted wide attention throughout the world and stimulated many studies by investigators seeking to confirm and extend Mendel's observations. Prominent among these were the English scientists Reginald Punnett, the geneticist immortalized in the Punnett square, and William Bateson, a zoologist.

In 1909, Thomas Hunt Morgan, a biologist from the United States who had visited de Vries's laboratory in Holland and been impressed by his work, abandoned his previous work in embryology and began a study of genetics. At Columbia University, he founded what was to be the most important laboratory in the field for several decades. The wealth of data that emerged from these studies was so impressive that this period in genetics research, which lasted until World War II, has been characterized as the "golden age" of genetics (though some would argue that the golden age is now).

By a remarkable combination of insight and good fortune, Morgan selected the fruit fly *Drosophila melanogaster* as his experimental organism. Biologists have often used for their experiments "insignificant" plants and animals—such as Mendel's pea plants, for instance, or Hertwig's sea urchins (page 109). Underlying this approach is the assumption that basic biological principles are universal, applying equally to all living things. As it turned out, the little fruit fly proved to be a "fit material" for a wide variety of genetic investigations. In the decades that followed, *Drosophila* was to become famous as the biologist's principal tool in studying animal genetics.

Drosophila means "lover of dew," although actually this useful animal is not attracted by dew but feeds on the fermenting yeast that it finds in rotting fruit. The fruit fly was an excellent choice for genetic studies since it is easy to breed and maintain. Only 3 millimeters long, these tiny flies can produce a new generation every two weeks. Each female lays hundreds of eggs during her adult life, and very large numbers of flies can be kept in a half-pint bottle, as they were in Morgan's laboratory, familiarly known as the "Fly Room."

THE REALITY OF THE GENE

Perhaps the most important of the principles established by Morgan and his colleagues was that Mendel's factors—genes—are located on chromosomes. Early in the twentieth century, at the beginning of the "golden age," this idea, so commonplace to us now, evoked raging controversy. At this stage of genetics research, the gene still had no physical reality. It was a pure abstraction. The work of Sutton and other cytologists was known, but it seemed irrelevant to studies of inheritance. As late as 1916, Bateson wrote, "The supposition that particles of chromatin, indistinguishable from each other and indeed almost homogeneous under any known test, can by their material nature confer all the properties of life, surpasses the range of even the most convinced materialism."

Sex Determination

One line of cytological observations, however, was providing further evidence linking "particles of chromatin" with heredity. As early as the 1890s, cytologists noticed that male and female organisms often show chromosomal differences, and they began to speculate that these differences are related to sex determination.

As Sutton had observed, the chromosomes of a diploid organism come in pairs. In all of the pairs except one, the chromosomes in both males and females appear to be the same; these are called **autosomes.** The structure of one pair, however, may differ between males and females. The chromosomes of this pair are known as the **sex chromosomes.** In many species, the two sex chromosomes are identical in the female but are dissimilar in the male, with one male sex chromosome the same as the female sex chromosomes and the other usually smaller. The sex chromosome that is the same in the cells of both males and females is called the *X* chromosome, and the unlike chromosome characteristic of the cells of males is called the *Y* chromosome. Thus we can characterize the two sexes as *XX* (female) and *XY* (male). In some insects, such as the grasshopper, which Sutton studied, there is no *Y* chromosome. In such cases, the females are characterized as *XX* and the males as *XO* (the O does not represent a chromosome but indicates the absence of one). In species in which the male is *XY* or *XO*, the male is said to be **heterogametic,** since he can produce two types of gametes, and the female is said to be **homogametic.**

Not all organisms, however, have heterogametic males and homogametic females. In birds, moths, and butterflies (and in occasional species in other groups), the sex chromosomes are reversed; the male has the two *X* chromosomes, and the female only one. The *Y* chromosome may or may not be present. In these organisms, it is the female that is heterogametic.

Human beings have 22 pairs of autosomes, which are structurally the same in both sexes. Females have a twenty-third matching pair, the sex chromosomes, *XX*. Human males, as their twenty-third pair, have one *X* and one *Y* (see Figure 19–2, page 383). During meiosis, as each diploid spermatocyte undergoes meiotic division into four haploid sperm cells, two of the sperm cells receive *X* chromosomes and two receive *Y* chromosomes. The ovum always contains an *X* chromosome, since a human female does not normally possess the *Y* in any of her cells. Thus the zygote will be *XX* or *XY*, depending on whether an *X*-bearing sperm or a *Y*-bearing sperm fertilizes the egg (Figure 13–2). It is in this way that the sperm cell contributed by the male determines the sex of the offspring, and it is the process of meiosis that governs the almost equal production of male and female offspring.

The correlation of the appearance of chromosomes with a particular trait—sex—gave strength to the gene-chromosome hypothesis. As we shall see, however, a large body of evidence, from a variety of different observations and experiments, was required before biologists became convinced that genes are on the chromosomes.

Sex Linkage

One of the advantages of *Drosophila melanogaster* for genetic studies is that it has only four pairs of chromosomes. Three of these pairs are autosomes, and the fourth is a pair of sex chromosomes, an *XX* pair in the female and an *XY* pair in the male (Figure 13–3). This feature of the fruit fly turned out to be particularly useful, although Morgan could not have foreseen that when he first selected *Drosophila* as his experimental organism.

When he began his investigations in 1909, Morgan intended to use *Drosophila* for breeding experiments similar to those Mendel had carried out with the pea

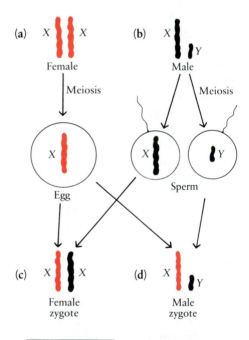

13–2 *Sex determination in an organism (such as humans) in which the male is heterogametic.* (**a**) *At meiosis, every egg cell receives an X chromosome from the mother.* (**b**) *A sperm cell may receive either an X chromosome or a Y chromosome.* (**c**) *If a sperm cell carrying an X chromosome fertilizes the egg, the offspring will be female (XX).* (**d**) *If a sperm cell carrying a Y chromosome fertilizes the egg, the offspring will be male (XY).*

13–3 *The fruit fly* (Drosophila melanogaster) *and its chromosomes. Fruit flies have only four pairs of chromosomes (2n = 8), a fact that simplified Morgan's experiments. Six of the chromosomes (three pairs) are autosomes (including the two dot-like chromosomes in the center), and two are sex chromosomes.*

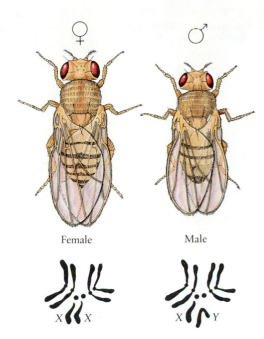

Female   Male

plant. He was looking, as Mendel had, for patterns of inheritance. Such experiments involved examining under a magnifying lens hundreds—eventually tens of thousands—of individual fruit flies.

Initially, the investigators in Morgan's laboratory were looking for genetic differences among individual flies that they could then study in the breeding experiments. A year after Morgan established his colony, such a difference appeared. One of the prominent and readily visible characteristics of fruit flies is their brilliant red eyes. One day, a white-eyed fly, a mutant, was observed in the colony (Figure 13–4). This fly, a male, was mated with a red-eyed female, and all of the F_1 offspring had red eyes.

Morgan then crossbred the F_1 offspring, just as Mendel had done in his pea experiments. However, instead of the expected 3:1 ratio of dominant to recessive phenotypes (that is, of red-eyed to white-eyed individuals), the ratio was closer to 4:1, and, moreover, all of the white-eyed individuals were males:

Red-eyed females	2,459
White-eyed females	0
Red-eyed males	1,011
White-eyed males	782

Why were there no white-eyed females? To explore the situation further, Morgan crossed the original white-eyed male with one of the F_1 females. The following results were obtained from this testcross:

Red-eyed females	129
White-eyed females	88
Red-eyed males	132
White-eyed males	86

In other words, females can be white-eyed. The characteristic behaves pretty much like a typical recessive (the expected ratio is 1:1:1:1). So why were there no white-eyed females in the F_2 generation? Stop here a moment and see if you can answer this question. Morgan was able to do so.

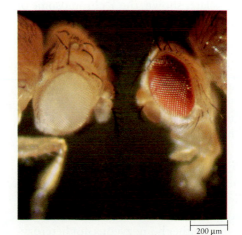

200 µm

13–4 *A mutant white-eyed fruit fly (left) and a normal, or wild-type, red-eyed fruit fly (right). While looking for genetic differences among* Drosophila, *Morgan discovered a single white-eyed fruit fly in his population of thousands. This chance occurrence launched an avalanche of studies and established a new concept of the gene.*

P $X^{w+}X^{w+}$ × X^wY
Red-eyed   White-eyed
female   male

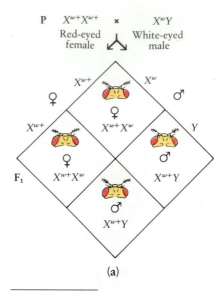

$X^{w+}X^w$ × $X^{w+}Y$
Red-eyed   Red-eyed
female from F_1 male from F_1

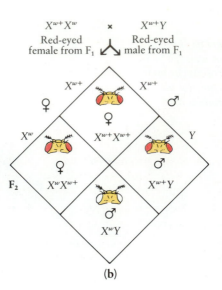

$X^{w+}X^w$ × X^wY
Red-eyed   White-eyed
female from F_1 male

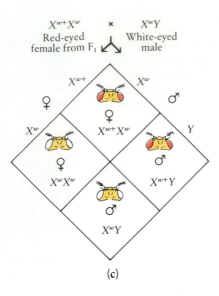

(a)   (b)   (c)

13–5 *Punnett square diagrams of the experiments Morgan performed after discovery of the white-eyed male Drosophila. The least common characteristic, white eyes, is represented by w; w$^+$ symbolizes the wild-type allele for red eyes. Alleles located on the sex chromosomes are commonly designated by superscripts to X and Y. (a) Morgan first mated a true-* *breeding red-eyed female to the white-eyed male. All of the offspring had red eyes. (b) Next, he mated an F_1 red-eyed female to an F_1 red-eyed male. Although both red-eyed and white-eyed males were produced in the F_2 generation, all F_2 females had red eyes, suggesting a relationship between the inheritance of eye color and the behavior of the sex chromosomes.* *(c) A testcross between a red-eyed F_1 female and the original white-eyed male produced both red-eyed and white-eyed flies of both sexes. This led to the conclusion that the gene for eye color must be carried on the X chromosome. The allele for red eyes (w$^+$) is dominant, and the allele for white eyes (w) is recessive.*

On the basis of these experiments, outlined in Figure 13–5, Morgan and his coworkers formulated the following hypothesis: The gene for eye color is carried only on the X chromosome. (In fact, as it was later shown, the Y chromosome of *Drosophila* carries very little genetic information.) The allele for white eyes must indeed be recessive, since all of the F_1 flies had red eyes. Thus a heterozygous female would have red eyes—which is why there were no white-eyed females in the F_2 generation. However, a male that received an X chromosome carrying the allele for white eyes would always be white-eyed since no other allele would be present.

Further experimental crosses, such as the one shown in Figure 13–6, confirmed Morgan's hypothesis. They also revealed that white-eyed fruit flies are more likely to die before they reach adulthood than are red-eyed fruit flies, which explains their lower-than-expected number in the F_2 generation and the testcross.

These experiments introduced the concept of **sex-linked traits,** which are, as we shall see in Chapter 19, important in the genetics of human beings as well as of fruit flies. With regard to the alleles for sex-linked traits, members of the heterogametic sex—most often, the male—are neither homozygous nor heterozygous; instead, they are said to be **hemizygous.** Moreover, with sex-linked traits, the terms dominant and recessive apply, strictly speaking, only in the case of the homogametic sex—most often, the female. In the heterogametic male, any allele carried on the X chromosome will be expressed in the phenotype. As a consequence, sex-linked "recessives" appear much more frequently in males.

The results of the breeding experiments with white-eyed and red-eyed fruit flies convinced Morgan, and many other geneticists as well, that Sutton's hypothesis was correct: Genes *are* on chromosomes. Conclusive demonstration of the physical location of the gene, however, depended on subsequent experiments, to which we shall return later in this chapter.

Tortoiseshell Cats, Barr Bodies, and the Lyon Hypothesis

A dark spot of chromatin—called a Barr body—can be seen at the outer edge of the nucleus of female mammalian somatic cells in interphase. This dark spot is an inactivated X chromosome. According to the Lyon hypothesis—named after the British geneticist Mary Lyon, who proposed it—early in the development of the embryo of the female mammal, one or the other X chromosome is inactivated in each somatic cell already formed. This inactivation occurs randomly, with the result that the embryo becomes a mosaic of cells, some with one X chromosome inactivated, and others with the other X chromosome inactivated. Thus, all the somatic cells of a female mammal are not identical but are one of two types, depending on which of the X chromosomes is active and which is inactive. Once an X chromosome is inactivated, all the daughter cells of that cell will have the same X chromosome inactivated. In the germ cells from which the egg cells will ultimately be produced by meiosis, one X chromosome appears to be inactivated early in development, but it is reactivated prior to meiosis.

In human females heterozygous for certain sex-linked traits, it has been found that the recessive is expressed to varying degrees, with some populations of cells expressing the recessive phenotype and other populations not expressing it. Whether a sex-linked characteristic is expressed depends upon which X chromosome (and, consequently, which allele for a particular trait) was inactivated in the embryonic cell from which the population of cells descended. A striking example is provided by color blindness, a human sex-linked characteristic that we will consider in more detail in Chapter 19. Women who are heterozygous for color blindness are sometimes color blind in one eye but not the other.

In cats, the alleles for black or yellow coat color are carried on the X chromosome. Male cats, having only a single X chromosome with one or the other of these alleles, are either black or yellow. Tortoiseshell cats have coats with patches of both black and yellow. As you would expect, they are almost always female—neatly fitting the predictions of the Lyon hypothesis.

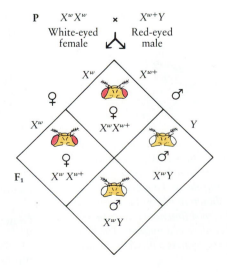

13–6 *Offspring of a cross between a white-eyed female fruit fly and a red-eyed male fruit fly, illustrating what happens when a recessive allele is carried on an X chromosome. The F₁ females, with one X chromosome from the mother and one from the father, are all heterozygous (X^wX^{w+}) and so will be red-eyed. But the F₁ males, with their single X chromosome received from the mother carrying the recessive (w) allele, will all be white-eyed because the Y chromosome carries no gene for eye color. Thus the recessive allele on the X chromosome inherited from the mother will be expressed in the male offspring.*

13–7 *Many human characteristics are governed by simple Mendelian inheritance of dominant alleles. Among these are dimples and cleft chin, as illustrated by the actor Kirk Douglas and his sons.*

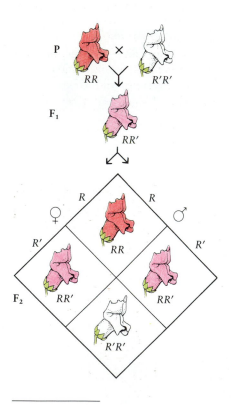

13–8 *A cross between a red (RR) snapdragon and a white (R'R') snapdragon. This looks very much like the cross between a purple- and a white-flowering pea plant shown in Figure 11–8, but there is a significant difference because in this case neither allele is dominant. The flower of the heterozygote is a blend of the two colors.*

BROADENING THE CONCEPT OF THE GENE

At the same time that the early work on *Drosophila* was proceeding in Morgan's laboratory, experiments and observations in other laboratories were also confirming and extending Mendel's principles. Many additional dominant and recessive alleles were demonstrated in a variety of other organisms, including humans (Figure 13–7). Alleles, even sex-linked ones, segregate and assort independently according to the principles of Mendel. However, as the golden age of genetics progressed, new studies, while confirming Mendel's work in principle, showed that the patterns of inheritance are not always as simple and direct as Mendel's reported results had indicated. (This fact is not surprising when you recall that Mendel had carefully selected certain traits for study, using only those that showed clear-cut differences.) The phenotypic effects of a particular gene, it was found, are influenced not only by the alleles of that gene present in the organism, but also by other genes and by the environment. And many, indeed most traits are influenced by more than one gene, just as most genes can influence more than a single trait. Some examples follow.

Allele Interactions

Incomplete Dominance and Codominance

Dominant and recessive characteristics are not always as clear-cut as in the seven traits studied by Mendel in the pea plant. Some characteristics appear to blend. For instance, as Bateson and Punnett showed in 1906, a cross between a homozygous red-flowering snapdragon *(RR)* and a homozygous white-flowering snapdragon *(R'R')* produces heterozygotes that are pink, a phenotype intermediate between those of the homozygotes (Figure 13–8). This phenomenon is known as **incomplete dominance.** As we shall see in Chapter 15, it is a result of the combined effects of gene products. When the heterozygous pink snapdragons are allowed to self-pollinate, red and white characteristics sort themselves out once again, showing that the alleles themselves, as Mendel had asserted, remain discrete and unaltered.

In other cases, alleles may act in a **codominant** manner, with heterozygotes expressing not an intermediate phenotype but rather both homozygous phenotypes simultaneously. A familiar example is found in the human blood type AB, in which the distinctive characteristics of red blood cells of both type A and type B are expressed in the phenotype.

Multiple Alleles

Although any individual diploid organism can have only two alleles of any given gene, it is possible that more than two forms of a gene—**multiple alleles**—may be present in a population of organisms. Multiple alleles result from different mutations of a single gene.

The members of the "set" of alleles may have different dominance relationships with one another. For instance, coat color in rabbits is determined by a series of four alleles: C (wild type, or agouti), c^{ch} (light gray, or chinchilla), c^h (albino with black extremities, or Himalayan), and c (albino). Different combinations of any two of these four possible alleles produce different coat colors (Figure 13–9).

In humans, the three alleles—A, B, and O—that determine the principal blood groups (to be discussed in more detail in Chapter 39) are probably the best known example of multiple alleles.

Gene Interactions

In addition to the interactions that occur between alleles of the same gene, interactions also occur among the alleles of different genes. Indeed, most of the

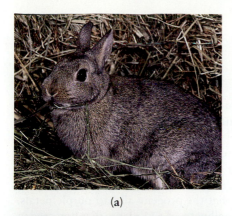

(a)

(b)

(c)

(d)

13–9 *Coat color in rabbits is determined principally by a single gene, of which four different alleles are known. Any individual rabbit, of course, carries only two alleles for this gene in its body cells. Different combinations of alleles produce (a) the wild-type, or agouti, rabbit (CC, Cc^{ch}, Cc^h, or Cc), (b) the chinchilla rabbit (c^{ch}c^{ch}, c^{ch}c^h, or c^{ch}c), (c) the Himalayan rabbit (c^{h}c^h or c^{h}c), and (d) the albino rabbit (cc).*

characteristics (both structural and chemical) that constitute the phenotype of an organism are the result of the interaction of many distinct genes.

Appearance of Novel Phenotypes

Sometimes, when one trait is affected by two (or more) different genes, a completely novel phenotype may appear. For instance, as demonstrated by Bateson and Punnett, comb shape in chickens is determined by two different genes, rose and pea, each with two alleles (*R, r* and *P, p*). *RR* or *Rr* results in rose comb, whereas *rr* produces single comb. *PP* or *Pp* produces pea comb, and *pp* produces single comb. However, when *R* and *P* occur together in the same individual, a novel phenotype, walnut comb, results. Thus, four different types of combs are possible depending on the interaction of the alleles of these two genes (Figure 13–10).

(a)

(b)

(c)

(d)

13–10 *The four types of combs observed in chickens are (a) single comb, (b) rose comb, (c) pea comb, and (d) walnut comb. These comb types are determined by two different genes, of which the alleles are R, r and P, p. Single comb occurs in chickens homozygous recessive for both genes (rrpp). Rose comb results from the presence of at least one dominant R allele coupled with two recessive p alleles (Rrpp or RRpp). Pea comb, by contrast, is produced by at least one dominant P allele coupled with two recessive r alleles (rrPp or rrPP). When at least one dominant allele of each gene is present in the same individual, a novel phenotype, walnut, is produced. Genotypically, chickens with walnut comb may be RRPP, RRPp, RrPP, or RrPp.*

269

Epistasis

In other cases, a novel phenotype does not appear when genes interact. Instead, different genes interact so that one gene interferes with or modifies the effect of the other. This type of interaction is called **epistasis** ("standing upon"). If gene *A* masks the effects of gene *B*, then *A* is said to be epistatic to *B*.

A classic example of epistatic gene interaction was reported by Bateson. Scientists in his laboratory crossed two pure-breeding white-flowered varieties of sweet pea *(Lathyrus odoratus)* and found that the progeny all had purple petals. When these F_1 plants were allowed to self-pollinate, of 651 plants that flowered in the F_2 generation, 382 had purple petals and 269 were white. At first, these figures may seem meaningless, but if you examine them closely, you will see that they fit a 9:7 ratio. In a cross demonstrating independent assortment of two nonallelic genes in a ratio of 9:3:3:1, 9/16, you will remember, is the proportion of offspring that show the effects of the two dominant alleles. So we can conclude that only a plant that has received at least one dominant allele from each gene (that is, allele *P* and allele *C)* can make the purple pigment (Figure 13–11).

In this case, either gene in the recessive homozygous condition is epistatic to, or hides, the effects of the other gene. When gene *C* is homozygous recessive *(cc)*, flowers will be white even if a dominant *P* is present (as in the phenotypes of *ccPp* and *ccPP).* Similarly, when gene *P* is homozygous recessive *(pp)*, flowers are also white (as in the phenotypes of *Ccpp* and *CCpp).*

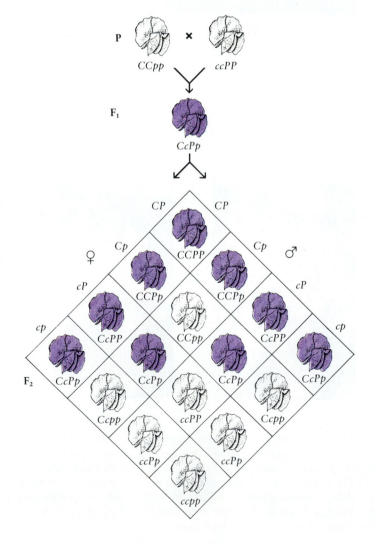

13–11 *Epistasis in sweet peas. When two different varieties of white-flowered sweet pea plants are crossed, all of the* F_1 *plants have purple flowers. In the* F_2 *generation, the ratio of purple- to white-flowered plants is 9:7. Purple color is due to the presence of both dominant alleles* P *and* C; *the homozygous recessive of either gene masks, or is epistatic to, the effects of the other gene.*

The phenomenon of epistasis indicates once again that what we see at the phenotypic level is actually the product of a very complicated series of events that influence one another during the development and lifetime of the organism. No gene works alone. The expression of all genes is in some way influenced by many other genes.

Genes and the Environment

The expression of a gene is always the result of its interaction with the environment. To take a common example, a seedling may have the genetic capacity to be green, to flower, and to fruit, but it will never turn green if it is kept in the dark, and it may not flower and fruit unless certain precise environmental requirements are met.

The water buttercup, *Ranunculus peltatus,* is a more striking example. It grows with half the plant body submerged in water. Although the leaves are genetically identical, the broad, floating leaves differ markedly in both form and physiology from the finely divided leaves that develop under water (Figure 13–12).

Temperature often affects gene expression. Primrose plants that are red-flowered at room temperature are white-flowered when raised at temperatures above 30°C (86°F). Similarly, as we noted in Chapter 8, Himalayan rabbits are white at high temperatures and black at low temperatures. In addition, Siamese cats raised at room temperature are black in their cooler peripheral areas, such as ears, nose, and tail tip.

These are extreme examples of a universal verity: The phenotype of any organism is the result of interaction between genes and environment.

Expressivity and Penetrance

When the expression of a gene is altered by environmental factors or other genes, two outcomes are possible. First, the degree to which a particular genotype is expressed in the phenotype of an individual may vary. This variable **expressivity** is seen for polydactyly, the presence of extra fingers and toes, which is caused by a dominant allele. Often, there is great variability in expressivity among members of a family, with the result that some individuals have extra digits on both hands and feet, while others may have only a portion of an extra toe on one foot.

Second, the proportion of individuals that show the phenotype ascribed to a particular genotype may be less than expected; the genotype shows incomplete **penetrance.** For example, individuals known to carry the allele for polydactyly may have absolutely normal hands and feet.

Examples of variable expressivity and incomplete penetrance abound among human genetic characteristics, often making it difficult to analyze the patterns of inheritance for certain genetic diseases or abnormalities.

Polygenic Inheritance

Some traits, such as size or height, shape, weight, color, metabolic rate, and behavior, are not the result of interactions between one, two, or even several genes; instead, they are the cumulative result of the combined effects of many genes. This phenomenon is known as **polygenic inheritance.**

A trait affected by a number of genes, or polygenes, does not show a clear difference between groups of individuals—such as the differences tabulated by Mendel. Instead, it shows a gradation of small differences, which is known as **continuous variation.** If you make a chart of differences among individuals for any trait affected by a number of genes, you get a curve such as that shown in Figure 13–13.

13–12 *The water buttercup. Leaves growing above water are broad, flat, and lobed. The genetically identical underwater leaves are thin and finely divided.*

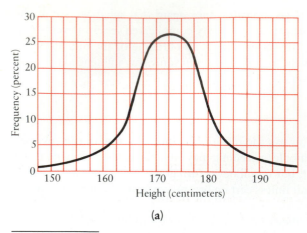

(a)

1	0	0	1	5	7	7	22	25	26	27	17	11	17	4	4	1
4:10	4:11	5.0	5:1	5:2	5:3	5:4	5:5	5:6	5:7	5:8	5:9	5:10	5:11	6:0	6:1	6:2

(b)

13–13 (a) *Height distribution of males in the United States. Height is an example of polygenic inheritance; that is, it is affected by a number of genes. Such genetic traits are characterized by small gradations of difference. A graph of the distribution of such traits takes the form of a bell-shaped curve, as shown, with the mean, or average, usually falling in the center of the curve. The larger the number of genes involved, the smoother the curve.* (b) *A company of student recruits at the Connecticut Agricultural College about 80 years ago. The number of men in each group and their height (in feet and inches) are shown below the photograph.*

Males in the United States are taller, on the average, than they were fifty years ago, due to better nutrition and other environmental factors. However, the shape of the curve is the same; in other words, the great majority fall within the middle range, and the extremes in height are represented by only a few individuals. Some of these height variations are produced by environmental factors, such as diet, but even if all the men in a population were maintained from birth on the same type of diet, there would still be a continuous variation in height in the population. This is due to genetically determined differences in hormone production, bone formation, and numerous other factors.

Table 13–1 illustrates a simple example of polygenic inheritance, color in wheat kernels, which is controlled by two genes, the four alleles of which exhibit cumulative quantitative effects. Human skin color is believed to be under a similar kind of genetic control (although involving more than two genes), as are many other traits. In fact, most normal human traits are believed to be polygenic.

TABLE 13-1 The Genetic Control of Color in Wheat Kernels (Polygenic inheritance*)

Parents:	$R_1R_1R_2R_2$ (Dark red)	$\times$	$r_1r_1r_2r_2$ (White)		
F_1:	$R_1r_1R_2r_2$ (Medium red)				
F_2:	**Genotype**			**Phenotype**	
1	$R_1R_1R_2R_2$			Dark red	
2 } 4	$R_1R_1R_2r_2$			Medium-dark red	
2	$R_1r_1R_2R_2$			Medium-dark red	
4	$R_1r_1R_2r_2$			Medium red	15 red
1 } 6	$R_1R_1r_2r_2$			Medium red	to
1	$r_1r_1R_2R_2$			Medium red	1 white
2 } 4	$R_1r_1r_2r_2$			Light red	
2	$r_1r_1R_2r_2$			Light red	
1	$r_1r_1r_2r_2$			White	

* Two genes are involved, each with two alleles: R_1 and r_1 for gene 1, and R_2 and r_2 for gene 2.

Pleiotropy

It is also possible for a single gene to affect more than one characteristic, a phenomenon known as **pleiotropy.** The frizzle trait in fowl (Figure 13–14) is an example of pleiotropy. In these animals, a change in feather formation due to the abnormal action of a single gene leads to drastic changes in many other aspects of their physiology. Similarly, the gene for white coat color in cats has a pleiotropic effect on the eye and ear. Cats that are all white and have blue eyes are often deaf. Some white cats have one blue eye and one yellow-orange eye. These too are deaf, but only on the side where the blue eye is present.

In rats, a single mutation affecting a gene that produces a protein involved in the formation of cartilage causes a whole complex of congenital deformities. These include thickened ribs, a narrowing of the tracheal passage (through which air moves to and from the lungs), blocked nostrils, a blunt snout, a loss of elasticity in the lungs, thickening of the heart muscle, and, needless to say, a greatly increased mortality. Since cartilage is one of the most common structural substances of the body, the widespread consequences of such a mutation are not difficult to understand. In fact, it is very likely that Mendel's allele for wrinkled peas, for example, also affected other structural characteristics of the pea plant. Similarly, the higher mortality rate of white-eyed fruit flies was a result of pleiotropic effects.

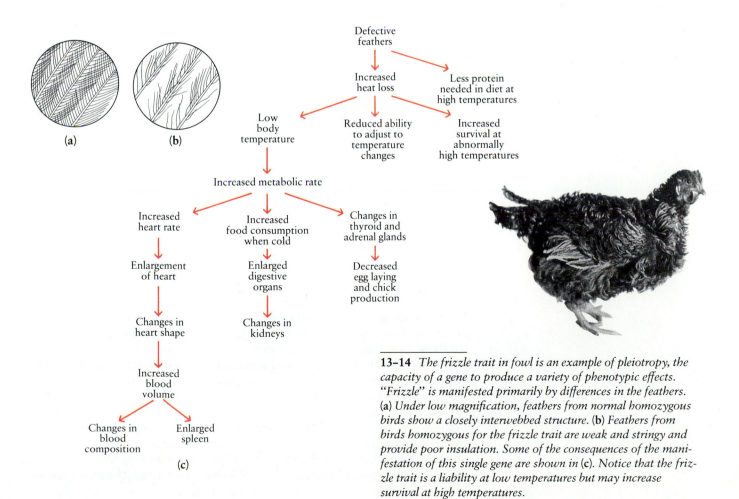

13–14 *The frizzle trait in fowl is an example of pleiotropy, the capacity of a gene to produce a variety of phenotypic effects. "Frizzle" is manifested primarily by differences in the feathers.* **(a)** *Under low magnification, feathers from normal homozygous birds show a closely interwebbed structure.* **(b)** *Feathers from birds homozygous for the frizzle trait are weak and stringy and provide poor insulation. Some of the consequences of the manifestation of this single gene are shown in* **(c).** *Notice that the frizzle trait is a liability at low temperatures but may increase survival at high temperatures.*

GENES AND CHROMOSOMES

Linkage

Mendel showed that certain pairs of alleles, such as those for round and wrinkled peas, assort independently of other pairs, such as those for yellow and green peas. However, as we noted previously, the alleles of two different genes will always assort independently if the genes are on different pairs of homologous chromosomes. If the alleles of the two genes are on the same pair of homologous chromosomes, then segregation of the alleles of one gene will not be independent of the segregation of the alleles of the other gene. In other words, if the alleles of two different genes are on the same chromosome, they should both be transmitted to the same gamete at meiosis. Genes that tend to stay together because they are on the same pair of homologous chromosomes are said to be **linked,** or in the same **linkage group.**

In 1927, an important research tool became available when H. J. Muller, one of Morgan's collaborators, found that exposure to x-rays greatly increases the rate at which mutations occur in *Drosophila*. Other forms of radiation, such as ultraviolet light, and certain chemicals were also shown to act as **mutagens,** or agents that produce mutations. As increasing numbers of mutants were found in Columbia University's *Drosophila* collection, as a result of the application of Muller's discovery, the mutations began to fall into four linkage groups, in accord with the four pairs of chromosomes visible in the cells. Indeed, in all organisms that have been studied in sufficient genetic detail, the number of linkage groups and the number of pairs of chromosomes have been the same, providing further support for Sutton's hypothesis that genes are on the chromosomes.

Recombination

Large-scale studies of linkage groups soon revealed some unexpected difficulties. For instance, most fruit flies have light tan bodies and long wings, both of which are dominant characteristics. When individuals homozygous for these characteristics were bred with mutant fruit flies having black bodies and short wings (both recessive characteristics), all the F_1 offspring had light tan bodies and long wings, as would be expected. Then the F_1 generation was inbred. Two outcomes seemed possible:

1. The genes for body color and wing length would be assorted independently, giving rise to Mendel's 9:3:3:1 ratio in the phenotypes and indicating that the genes for these two traits were on different chromosomes.

2. The genes for the two traits would be linked. In this case, 75 percent of the flies would be tan with long wings and 25 percent, homozygous for both recessives, would be black with short wings.

In the case of these particular traits, the results closely resembled the second possibility, but they did not conform exactly. In a few of the offspring, the genes for these traits seemed to assort independently; that is, some few flies appeared that were tan with short wings, and some that were black with long wings. How could this be? Somehow genes that were presumed to be on the same chromosome had become separated.

To find out what was happening, Morgan tried a testcross, breeding a member of the F_1 generation with a homozygous recessive. If black and tan, long and short, assorted independently—that is, if they were on different chromosomes—25 percent of the offspring of this cross should be black with long wings, 25 percent tan with long wings, 25 percent black with short wings, and 25 percent tan with

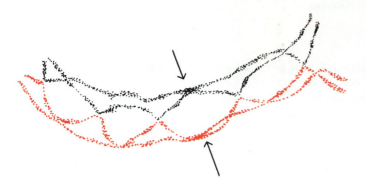

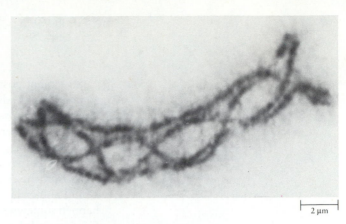

2 μm

13-15 *Homologous chromosomes of a grasshopper, as seen in prophase I. All four chromatids are visible. Crossing over—that is, the exchange of genetic material—has probably occurred at* *the points at which these chromatids intersect, the chiasmata. The arrows indicate the position of the centromeres.*

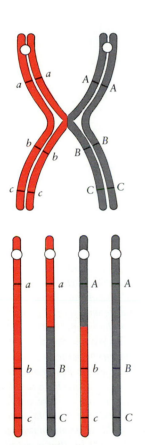

13-16 *Crossing over takes place when breaks occur in the chromatids of homologous chromosomes during prophase I, when the chromosomes are paired. The broken end of each chromatid joins with the chromatid of a homologous chromosome. In this way, alleles are exchanged between chromosomes. The white circles symbolize centromeres.*

short wings. On the other hand, if the genes for color and wing size were on the same chromosome and so moved together, half of the testcross offspring should be tan with long wings and half should be black with short wings. But actually, as it turned out, over and over, in counts of hundreds of fruit flies resulting from such crosses, 41.5 percent were tan with long wings, 41.5 percent were black with short wings, 8.5 percent were tan with short wings, and another 8.5 percent were black with long wings.

Morgan was convinced by this time that genes are located on chromosomes. It now seemed clear that the genes for the two traits, body color and wing length, were located on a single pair of homologous chromosomes, since the characteristics did not show up in the 1:1:1:1 ratio of independently assorted alleles. The only way in which the observed figures could be explained, Morgan reasoned, was if alleles could sometimes be exchanged between homologous chromosomes, that is, be recombined.

As we noted in Chapter 12, it now has been established that exchange of portions of homologous chromosomes—crossing over—takes place at the beginning of meiosis (Figure 13–15). If crossing over takes place between the positions at which two different genes are located on the same pair of homologues, then the alleles of the two genes can become separated as chromatids of the two homologues break and rejoin with each other (Figure 13–16).

Mapping the Chromosome

With the discovery of crossing over, accumulating evidence clearly supported not only the premise that genes are carried on chromosomes, but also that they must be positioned at particular spots, or **loci** (singular, locus), on the chromosomes. It followed that the alleles of any given gene must occupy corresponding loci on homologous chromosomes. Otherwise, the exchange of sections of chromosomes would result in genetic chaos rather than in an exact exchange of alleles.

As other traits were studied, it became clear that the percentage of recombinations between any two genes, such as those for body color and wing length, was different from the percentage of recombinations between two other genes, such as those for body color and leg length. In addition, as Morgan's experiments had shown, these percentages were fixed and predictable. It occurred to A. H. Sturtevant, who was an undergraduate working in Morgan's laboratory at this time, that the percentage of recombinations probably had something to do with the physical distances between the gene loci, or, in other words, with their spacing along the chromosome. This concept opened the way to the "mapping" of chromosomes.

Sturtevant postulated (1) that genes are arranged in a linear order on chromosomes, like beads on a string; (2) that genes that are close together will be separated by crossing over less frequently than genes that are farther apart; and (3) that it should therefore be possible, by determining the frequencies of recombinations, to plot the sequence of the genes along the chromosome and the relative distances between them. In Figure 13-16, for example, you can see that in a crossover, the chance that a strand would break and rejoin with its homologous strand somewhere between B and C should be less likely than if this were to happen somewhere between A and B, simply because the distance between B and C is less and there is less room (fewer chances) for crossovers to occur. Similarly, the chance of a crossover between A and B is less than the chance of a crossover between A and C.

In 1913, Sturtevant began constructing chromosome maps using data from crossover studies in fruit flies. As a standard unit of measure, he arbitrarily defined one map unit as equal to the distance that would give (on the average) one recombinant organism per 100 fertilized eggs (1 percent recombination). Genes with 10 percent recombination would be 10 map units apart; those with 8 percent recombination would be 8 map units apart (Figure 13-17). The fewer map units between genes, the less likely they were to be separated, while genes more than 50 map units apart on the same chromosome assorted independently.

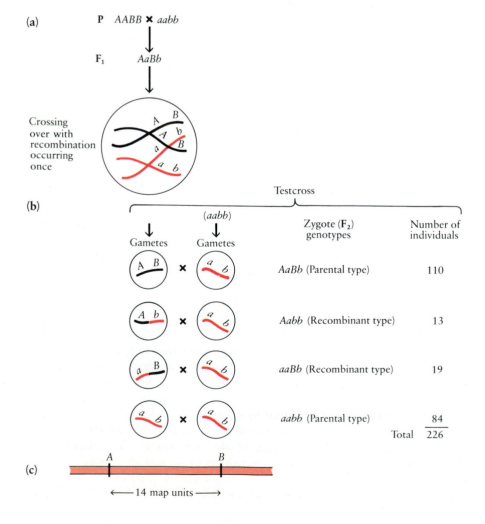

(a) **P** *AABB* ✕ *aabb*

F₁ *AaBb*

Crossing over with recombination occurring once

Testcross

(b)

(*aabb*)

Gametes   Gametes

Zygote (**F₂**) genotypes	Number of individuals
AaBb (Parental type)	110
Aabb (Recombinant type)	13
aaBb (Recombinant type)	19
aabb (Parental type)	84
Total	226

(c)   *A*   *B*

← 14 map units →

13-17 *Determining the map distance between two genes on the same chromosome.* (**a**) *When an individual homozygous dominant* (AABB) *for two genes located on the same pair of homologous chromosomes is crossed with one that is homozygous recessive* (aabb), *the F₁ offspring will all be heterozygous for both genes* (AaBb). *If crossing over occurs during meiosis in the heterozygote, alleles on the chromatids of the two homologues can be exchanged, and four different types of gametes can result from the recombination: parental-type gametes AB and ab, along with recombinant-type gametes Ab and aB.* (**b**) *The heterozygote is then mated to a homozygous recessive individual (a test-cross).* (**c**) *The number of recombinant offspring divided by the total number of offspring yields a recombination percentage (32/226 = 0.14) that is defined as the map distance between the genes. Genes A and B are 14 map units apart.*

13-18 *Mapping a chromosome. Alleles A and B recombine with a and b in 4 percent of the offspring, and alleles A and C with a and c in 9 percent of the offspring. Use as the map unit the distance that will give (on the average) one recombinant per 100 fertilized eggs. (a) Start with the highest recombination percentage and establish the relative positions of A and C on the chromosome. (b) A and B, as you know from the data, are 4 units apart. Thus B could theoretically be either to the left of A or to the right. (c) However, if it were to the left, B and C would be 13 units apart, a distance that does not conform to the data. B must therefore be between A and C.*

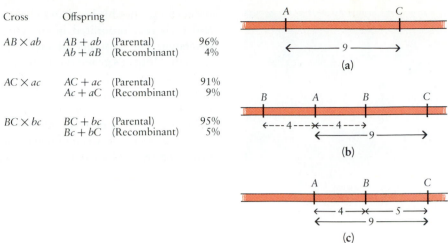

Cross	Offspring		
$AB \times ab$	$AB + ab$	(Parental)	96%
	$Ab + aB$	(Recombinant)	4%
$AC \times ac$	$AC + ac$	(Parental)	91%
	$Ac + aC$	(Recombinant)	9%
$BC \times bc$	$BC + bc$	(Parental)	95%
	$Bc + bC$	(Recombinant)	5%

By correlating recombination frequencies (that is, crossover percentages) with the relative distance between genes, Sturtevant and other geneticists located a variety of genes on the chromosome maps of *Drosophila* (Figure 13–19). These map distances, however, do not necessarily accurately reflect physical distances because breaking and rejoining of chromosome segments are more likely to occur at some sites on the chromosomes than at others.

These important studies confirmed that not only are genes located on chromosomes, as Sutton had hypothesized, but they also have fixed positions in a linear sequence.

ABNORMALITIES IN CHROMOSOME STRUCTURE

Recombination does not affect the order of genes on a chromosome. However, it is possible, under certain circumstances, for chromosomes to break apart and rejoin in a different orientation on the same chromosome or even join with another chromosome. The outcome for both possibilities is a change in the sequence of genes on the affected chromosomes.

Cytological evidence for such changes in chromosome structure, giving rise to changes in gene order and altered hereditary patterns, came from study of the giant chromosomes discovered in 1933 in the salivary glands of *Drosophila* larvae. In *Drosophila*, as in many other insects, certain cells do not divide during the larval stages of the insect. In such cells, however, the chromosomes continue to replicate, over and over again, but since daughter chromosomes do not separate from one another after replication, they simply become larger and larger until they are composed of thousands of copies. Also, salivary gland homologous chromosomes pair tightly along their entire lengths, adding to their size; for example, a salivary gland cell of *Drosophila*, in which $2n = 8$, appears to have only four giant chromosomes. As you can see in Figure 13–20, when stained, these giant chromosomes are characterized by very distinctive dark and light bands.

These banding patterns became another useful tool for geneticists, enabling them to detect structural changes in the chromosomes themselves. In the giant chromosomes, geneticists can actually locate the positions where changes in chromosome structure have occurred by observing changes in the banding pattern. Sometimes a whole segment is lost; usually the effect of such a loss—known as **deletion**—is lethal. In some cases, the "lost" segment becomes incorporated into its homologue, in which the segment then appears twice; this phenomenon is known as **duplication**. Sometimes a portion is transferred from one chromosome to another, nonhomologous chromosome, a process known as **translocation**. In

Normal   **Mutant**

Long aristae — 0 — Short aristae

Long wings — 13.0 — Dumpy wings

— 31.0 —

Long legs   Short legs

Red eyes — 54.5 — Purple eyes

— 67.0 —

Long wings   Vestigial wings

Red eyes —104.0— Brown eyes

13-19 *A portion of one chromosome map of* Drosophila melanogaster, *showing relative positions of some of the genes on chromosome 2, as calculated by the frequency of recombinations. As you can see, more than one gene may affect a single trait, such as eye color.*

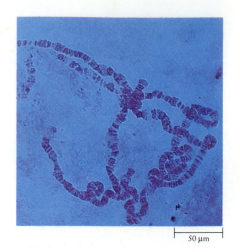

50 μm

13-20 *Chromosomes from the salivary gland of a* Drosophila *larva. These chromosomes are 100 times larger than the chromosomes in ordinary body cells, and their details are therefore much easier to see. (This micrograph was taken with a light microscope, not an electron microscope.) Because of the distinctive banding patterns, it is possible in some cases to associate genes with specific regions in particular chromosomes. Each visible band represents dozens of genes.*

other cases, a double break in a chromosome occurs and a segment is turned 180° and then reincorporated in the chromosome; this phenomenon is known as **inversion.** The existence of such chromosomal aberrations had been hypothesized by Morgan's group on the basis of mapping studies. The studies of the giant chromosomes of *Drosophila* confirmed the existence of these chromosomal aberrations (Figure 13–21), and it became possible to assign genes to physical locations on the *Drosophila* chromosome.

Perhaps more important, almost a quarter of a century after the *Drosophila* work had begun, it was no longer just supposition that little, seemingly homogeneous, particles of chromatin could serve as the repositories of the mysteries of heredity.

SUMMARY

The rediscovery of Mendel's work in 1900 was the catalyst for many new discoveries in genetics, leading to the identification of chromosomes as the carriers of heredity and to the modification and extension of some of Mendel's conclusions.

Strong support for the hypothesis that genes are on the chromosomes came from studies by Morgan and his group on the fruit fly *Drosophila melanogaster*. Because it is easy to breed and maintain, *Drosophila* has been used in a wide variety of genetic studies. It has four pairs of chromosomes; three pairs (the autosomes) are structurally the same in both sexes, but the fourth pair, the sex chromosomes, is different. In fruit flies, as in many other species (including humans), the two sex chromosomes are *XX* in females and *XY* in males.

At the time of meiosis, the sex chromosomes, like the autosomes, segregate. Each egg cell receives an *X* chromosome, but half the sperm cells receive an *X* chromosome and half receive a *Y* chromosome. Thus, in fruit flies, humans, and many (but not all) other organisms, it is the paternal gamete that determines the sex of the offspring.

In the early 1900s, breeding experiments with *Drosophila* showed that certain traits are sex-linked, that is, their genes are carried on the sex chromosomes.

13-21 *A chromosomal inversion occurs when a segment of a chromosome breaks off, is turned 180°, and rejoins by the "wrong" ends.* **(a)** *An inversion results in a reversal of the sequence of genes, as shown in the chromosome on the right.* **(b)** *When one member of a homologous pair contains a large inverted segment, that chromosome must loop inside the other for close pairing to occur.* **(c)** *In giant chromosomes, such loops are greatly magnified, making it possible to readily identify regions with inverted segments.*

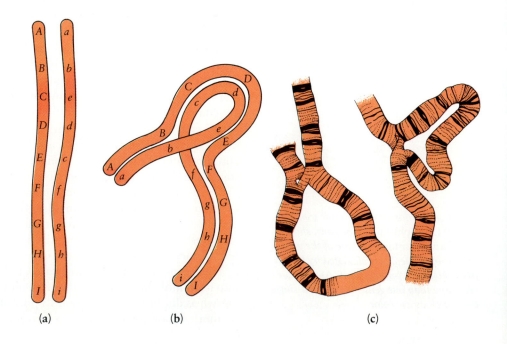

(a)   (b)   (c)

Because the X chromosome carries genes that are not present on the Y chromosome, a single recessive allele in the male, if carried on the X chromosome, will result in a recessive phenotype since no other allele is present. By contrast, a female heterozygous for a sex-linked trait will show the dominant characteristic.

Many traits are inherited according to the patterns revealed by Mendel. However, in others—perhaps the majority—the patterns are more complex. These complexities are caused by interactions among alleles, interactions among genes, and interactions with the environment.

Although many alleles interact in a dominant-recessive manner, some show varying degrees of incomplete dominance and codominance. In a population of organisms, a single gene may occur as multiple alleles, as the result of a series of different mutations of that gene. However, only two alleles can be present in any particular diploid individual.

Different genes can also interact with one another. Novel phenotypes may result from these interactions, or genes may affect one another in an epistatic manner, such that one hides the effect of the other. As a result, the expected phenotypic ratio is altered. Moreover, variable expressivity or reduced penetrance often results from the effects of other genetic influences, environmental conditions, or both.

The phenotypic expression of many traits is influenced by a number of genes. This phenomenon is known as polygenic inheritance. Such traits typically show continuous variation, as represented by a bell-shaped curve. Conversely, a single gene can affect two or more superficially unrelated traits; this property of a gene is known as pleiotropy.

Some genes assort independently in breeding experiments, and others tend to remain together. Genes that do not assort independently (because they are on the same chromosome) are said to be linked; a linkage group consists of a pair of homologous chromosomes.

Alleles are sometimes exchanged between homologous chromosomes as a result of crossing over in meiosis. Such recombinations can take place because: (1) the genes are arranged in a fixed linear array along the length of the chromosomes, and (2) the alleles of a given gene are at the corresponding sites (loci) on homologous chromosomes. Chromosome maps, showing the relative positions of gene loci along the chromosomes, have been developed from recombination data provided by breeding experiments.

Genetic studies have shown that chromosome breaks other than those resulting in crossovers may sometimes occur. A portion of a chromosome may be lost, or deleted, it may be duplicated, it may be translocated to a nonhomologous chromosome, or it may be inverted. Studies of the giant chromosomes of *Drosophila* larvae provided visual confirmation of these changes, as well as the final, conclusive evidence that the chromosomes are the carriers of the genetic information.

QUESTIONS

1. Distinguish among the following: sex chromosome/autosome; heterogametic/homogametic; incomplete dominance/codominance/epistasis; polygenic inheritance/pleiotropy; variable expressivity/incomplete penetrance; inversion/deletion/duplication/translocation.

2. Draw a diagram similar to Figure 13–2 indicating sex determination in a robin. Note that in birds the sex chromosomes are labeled Z and W; the genotype for the homogametic sex is ZZ and for the heterogametic sex is ZW.

3. With regard to sex-linked traits, the heterogametic sex, usually the male, is said to be hemizygous. What does this term mean? What two types of individuals can be used as testcross organisms for sex-linked traits?

4. The genes for coat color in cats are carried on the X chromosome. Black (b) is the recessive and yellow (B) is the dominant. What coat colors would you expect in the offspring of a cross between a black female and a yellow male? What coat colors would you expect in the sons of a tortoiseshell female?

5. Judging from the colors of the tortoiseshell cat (page 267), at what stage in development does the X chromosome become inactivated?

6. The so-called "blue" (really gray) Andalusian variety of chicken is produced by a cross between the black and white varieties. Only a single pair of alleles is involved. What color chickens (and in what proportions) would you expect if you crossed two blues? If you crossed a blue and a black? Explain.

7. In snapdragons, the allele that produces tall stems is completely dominant to the allele for dwarf stems, while the allele that produces red flowers is only partially dominant to that for white flowers. Describe the phenotype (height and flower color) of the F_1 plants resulting from a cross between a homozygous tall, red-flowered plant and a homozygous dwarf, white-flowered plant. If one of these F_1 plants self-pollinates, what will be the appearance and proportions of phenotypes in the resulting F_2 generation? Which two of these phenotypes will breed true?

8. In chickens, as we have seen, two pairs of alleles determine comb shape. RR or Rr results in rose comb, whereas rr produces single comb. PP or Pp produces pea comb, and pp produces single comb. When R and P occur together, they produce a new type of comb: walnut. What would be the genotype of the F_1 generation resulting from $RRpp \times rrPP$? The phenotype? If F_1 hybrids were crossbred, what would be the probable distribution of genotypes? Of phenotypes? (Illustrate this cross with a Punnett square.)

9. In Duroc-Jersey pigs, coat color is determined by two genes, R and S. The homozygous recessive condition, $rrss$, produces a white coat. The presence of at least one copy each of R and S produces red. The presence of one or the other allele (either R or S) produces a new phenotype, sandy. Give the phenotypes of the following genotypes:

RRSS	rrss
RrSs	rrSs
RRSs	rrSS
RrSS	RRss

10. Mating a red Duroc-Jersey boar to sow A (white) gave pigs in the ratio of 1 red: 2 sandy: 1 white. Mating this same boar to sow B (sandy) gave 3 red: 4 sandy: 1 white. When this boar was mated to sow C (sandy), the litter had equal numbers of red and sandy piglets. Using the information presented in Question 9, give the possible genotypes of the boar and the three sows.

11. You and a geneticist are looking at a mahogany-colored Ayrshire cow with a newly born red calf. You wonder if it is male or female, and the geneticist says it is obvious from the color which sex the calf is. She explains that in Ayrshires the genotype AA is mahogany and aa is red, but the genotype Aa is mahogany in males and red in females. What is she trying to tell you—that is, what sex is the calf? What are the possible phenotypes of the calf's father?

12. In one strain of mice, skin color is determined by five different pairs of alleles. The colors range from almost white to dark brown. Would it be possible for some pairs of mice to produce offspring darker or lighter than either parent? Explain.

13. The size of an egg laid by one variety of hens is determined by three pairs of alleles; hens with the genotype $AABBCC$ lay eggs weighing 90 grams, and hens with the genotype $aabbcc$ lay eggs weighing 30 grams. Each of the alleles A, B, or C adds 10 grams to the weight of the egg. When a hen from the 90-gram strain is mated with a rooster from the 30-gram strain, the hens of the F_1 generation lay eggs weighing 60 grams. If a hen and rooster from this F_1 generation are mated, what will be the weight of the eggs laid by hens of the F_2?

14. Height and weight in animals follow a distribution similar to that shown in Figure 13–13. By inbreeding large animals, breeders are usually able to produce some increase in size among their stock. But after a few generations, increase in size characteristically stops. Why?

15. In Jimson weed, the allele that produces violet petals is dominant over that for white petals, and the allele that produces prickly capsules is dominant over that for smooth capsules. A plant with white petals and prickly capsules was crossed with one that had violet petals and smooth capsules. The F_1 generation was composed of 47 plants with white petals and prickly capsules, 45 plants with white petals and smooth capsules, 50 plants with violet petals and prickly capsules, and 46 plants with violet petals and smooth capsules. What were the genotypes of the parents?

16. A diploid organism has 42 chromosomes per cell. How many linkage groups does it have?

17. Segregation of alleles can occur at either of two stages of meiosis. Name the two stages and explain what happens in each of them.

18. Does crossing over necessarily result in a recombination of alleles? Explain.

19. In a series of breeding experiments, a linkage group composed of genes A, B, C, D, and E was found to show approximately the recombination frequencies in the chart below. Using Sturtevant's standard unit of measure, "map" the chromosome.

		Gene				
		A	B	C	D	E
	A	—	8	12	4	1
	B	8	—	4	12	9
Gene	C	12	4	—	16	13
	D	4	12	16	—	3
	E	1	9	13	3	—

Recombinations per 100 fertilized eggs

C H A P T E R **14**

The Chemical Basis of Heredity: The Double Helix

By the early 1940s, the existence of genes and the fact that they were in chromosomes were no longer in doubt. But what were the genes? What did they really do? A turning point in genetics came when scientists focused on the question of how it was possible for these little lumps of matter—the chromosomes—to be the bearers of what they had come to realize must be an enormous amount of extremely complex information.

THE CHEMISTRY OF HEREDITY

The chromosomes, like all the other parts of a living cell, are composed of atoms arranged into molecules. As we noted previously, some scientists, a number of them eminent in the field of genetics, thought it would be impossible to understand the complexities of heredity in terms of the structure of "lifeless" chemicals. Others thought that if the chemical structure of the chromosomes was understood, we could then come to understand how chromosomes function as the bearers of the genetic information. This thinking marked the beginning of the vast range of investigations that we know as "molecular genetics."

The Language of Life

Early chemical analyses of the hereditary material revealed that the eukaryotic chromosome consists of both deoxyribonucleic acid (DNA) and protein, in about equal amounts. Thus, both were candidates for the role of the genetic material. Proteins seemed the more likely choice because of their greater chemical complexity. (As you will recall from Chapter 3, proteins are polymers of amino acids, of which there are 20 different types found in living cells; DNA, by contrast, is a polymer formed from only four different types of nucleotides.) Speculative thinkers in the field of biology were quick to point out that the amino acids, the number of which is so provocatively close to the number of letters in our own alphabet, could be arranged in a variety of different ways. The amino acids were seen as making up a sort of language—"the language of life"—that spelled out the directions for all the many activities of the cell. Many prominent investigators, particularly those who had been studying proteins, believed that the genes themselves were proteins. They thought that the chromosomes contained master models of all the proteins that would be required by the cell and that enzymes and other proteins active in cellular life were copied from these master models. This was a logical hypothesis, but, as it turned out, it was wrong.

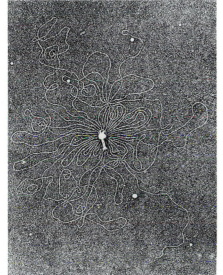

0.5 µm

14–1 *Molecular genetics is concerned with the chemical basis of heredity. The genetic information has been shown to be contained in a large, complex molecule known as deoxyribonucleic acid (DNA). This electron micrograph shows a bacteriophage, a virus that attacks bacterial cells. It is surrounded by its genetic material, a long, continuous molecule of DNA. (The molecule shown here, however, has been broken apart at one point—note the free ends at the top and bottom.) In the center of the micrograph is the outer coat of the virus, made of protein, from which the DNA has been released.*

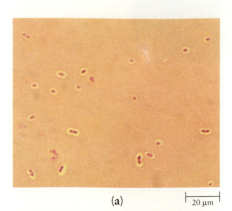

(a)   20 µm

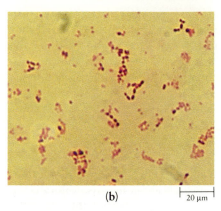

(b)   20 µm

14-2 (a) *Encapsulated and* (b) *nonencapsulated forms of pneumococci. The capsule is made up of polysaccharides deposited outside the cell wall. The encapsulated form, which is resistant to phagocytosis by white blood cells, produces pneumonia; the mutant, nonencapsulated form is harmless.*

THE DNA TRAIL

Sugar-Coated Microbes and the Transforming Factor

To trace the beginning of the other hypothesis—the one that ultimately proved correct—it is necessary to go back to 1928 and pick up an important thread in modern biological history. In that year, an experiment was performed that seemed at the time to have little relevance to the field of genetics. Frederick Griffith, a public health bacteriologist in England, was studying the possibility of developing vaccines against *Streptococcus pneumoniae,* a type of bacterium that causes one form of pneumonia. In those days, before the development of antibiotics, bacterial pneumonia was a serious disease, the grim "captain of the men of death."

As Griffith knew, these bacteria, commonly called pneumococci, come in either virulent (disease-causing) forms with polysaccharide capsules or nonvirulent (harmless) forms without capsules (Figure 14-2). The production of the capsule and its composition are both genetically determined—that is, they are inherited properties of the bacteria. (It is now known that the nonencapsulated pneumococcus is a mutant form; in Griffith's time, however, the term mutant was not applied to bacteria.) Griffith was interested in finding out whether injections of heat-killed virulent pneumococci, which do not cause disease, could be used to immunize against pneumonia. In the course of various experiments, he performed one that gave him very puzzling results. He injected mice simultaneously with heat-killed virulent bacteria and with living nonvirulent bacteria, each of which was harmless—but all the mice died. When Griffith performed autopsies on them, he found their bodies filled with living encapsulated (and therefore virulent) bacteria (Figure 14-3). Had the dead virulent cells come back to life, or had something been passed from them to the living, nonvirulent cells that, in turn, endowed the living cells with the capacity to make capsules and therefore to be virulent?

Within the next few years, it was shown that the same phenomenon could be reproduced in the test tube and these questions could be answered. It was found that extracts from the killed encapsulated bacteria, when added to the living harmless bacteria, could convert them to the virulent type with the capacity to make capsules. Furthermore, once converted, they could transmit this characteristic to their progeny. This phenomenon was known as **transformation,** and the "something" in the extract that caused the conversion was called the **transforming factor.**

In 1943, after almost a decade of patient chemical isolation and analysis, O. T. Avery and his coworkers at Rockefeller University demonstrated that the transforming factor was DNA. Subsequent experiments showed that a variety of genetic factors could be passed from bacterial cells of one strain to bacterial cells of another, similar strain by means of isolated DNA.

The Nature of DNA

DNA had first been isolated by a German physician named Friedrich Miescher in 1869—in the same remarkable decade in which Darwin published *The Origin of Species* and Mendel presented his results to the Brünn Natural History Society. The substance Miescher isolated was white, sugary, slightly acidic, and contained phosphorus. Since he found it only in the nuclei of cells, he called it "nuclein." This name was later amended to nucleic acid, and then, still later, to deoxyribonucleic acid, to distinguish it from a similar chemical also found in cells, ribonucleic acid (RNA).

Almost 50 years later, in 1914, another German, Robert Feulgen, discovered that DNA had an unusually strong attraction for a red dye called fuchsin.

14–3 *Discovery of the transforming factor, a substance that can transmit genetic characteristics from one cell to another, resulted from studies of pneumococci, which are pneumonia-causing bacteria. One strain of these bacteria has polysaccharide capsules (protective outer layers); another does not. The capacity to make capsules and cause disease is an inherited characteristic, passed from one bacterial generation to another as the cells divide. (a) Injection into mice of encapsulated pneumococci killed the mice. (b) The nonencapsulated strain produced no infection. (c) If the encapsulated strain was heat-killed before injection, it too produced no infection. (d) If, however, heat-killed encapsulated bacteria were mixed with live nonencapsulated bacteria and the mixture was injected into mice, the mice died. (e) Blood samples from the dead mice revealed live encapsulated pneumococci. Something had been transferred from the dead bacteria to the live ones that endowed them with the capacity to make polysaccharide capsules and cause pneumonia. This "something" was later isolated and found to be DNA.*

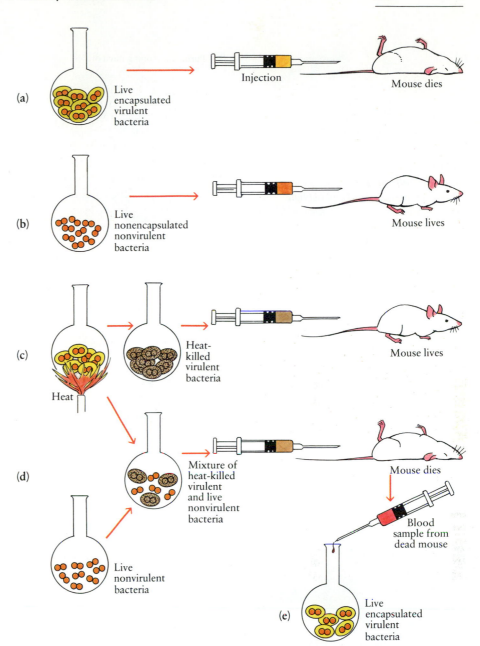

However, he considered this finding so unimportant that he did not trouble to report it for a decade. Feulgen staining, as it was called when it finally made its way into use, revealed that DNA was present in all cells and was characteristically located in the chromosomes.

For the next few decades, however, there was no particular interest in DNA since no role had been postulated for it in cellular metabolism. During the 1920s, most of the work on its chemistry was carried out in a single laboratory by the eminent biochemist P. A. Levene. He showed that DNA could be broken down into a five-carbon sugar, a phosphate group, and four nitrogenous bases—adenine and guanine (the purines) and thymine and cytosine (the pyrimidines). From the proportions of these components he made two deductions, one correct and one incorrect:

1. Each nitrogenous base is attached to a molecule of sugar, which is, in turn, attached to a phosphate group to form a single molecule, a nucleotide (Figure 14–4). This deduction was correct.

(a)

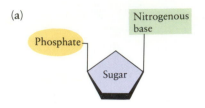

14–4 (a) *A nucleotide is made up of three different components: a nitrogenous base, a five-carbon sugar, and a phosphate group.* (b) *The four types of nucleotides found in DNA. Each nucleotide consists of one of four possible nitrogenous bases, a deoxyribose sugar, and a phosphate group.*

(b) Purine-containing nucleotides   Pyrimidine-containing nucleotides

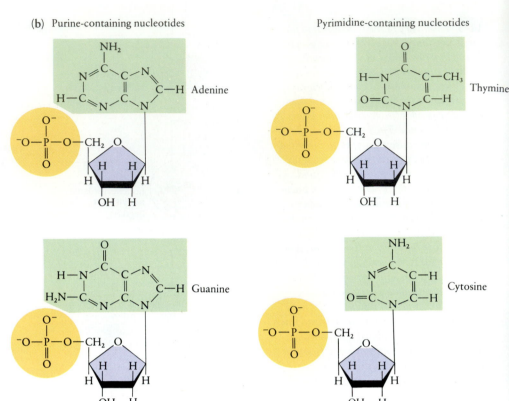

14–5 *Max Delbrück and Salvador Luria at the Cold Spring Harbor Laboratory of Quantitative Biology in 1953. They shared the Nobel Prize with A. D. Hershey in 1969 for "their discoveries concerning the replication mechanism and the genetic structure of viruses."*

2. Since, in all the samples he measured, the proportions of the nitrogenous bases were approximately equal, Levene concluded that all four nitrogenous bases must be present in nucleic acid in equal quantity. Furthermore, he hypothesized that these molecules must be grouped in clusters of four—a tetranucleotide, he called it—that repeated over and over again along the length of the molecule. Although this deduction was incorrect, it dominated scientific thinking about the nature of DNA for more than a decade.

Because Levene's "tetranucleotide theory" was given great weight by his renown as a biochemist, biologists were generally slow to recognize the importance of Avery's demonstration that the transforming factor in bacteria is DNA. This was partly because bacteria, which are, of course, prokaryotes, were considered "lower" and "different" and partly because the DNA molecule—made up of only four components—seemed too simple for the enormously complex task of carrying the hereditary information. Avery, like Mendel before him, was a traveler bearing an odd tale that did not fit.

The Bacteriophage Experiments

In 1940, Max Delbrück and Salvador Luria, both of whom had left Europe in the mass intellectual exodus of the 1930s, initiated a series of studies with another "fit material," destined to become as important to genetic research as the garden pea and the fruit fly. The fit material was a group of viruses that attack bacterial cells and are therefore known as **bacteriophages** ("bacteria eaters"), or phage for short. Every known type of bacterial cell is preyed upon by its own type of bacterial virus, and many bacteria are host to many different kinds of viruses. Delbrück, Luria, and the group that joined them in these studies agreed to concentrate on a series of seven related viruses that attack *Escherichia coli*, the familiar bacterium

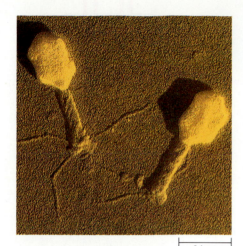

├─ 0.1 μm ─┤

14-6 *T4 bacteriophages, as revealed by a transmission electron micrograph of a replica prepared by "shadowing" (page 98). Notice their highly distinctive "tadpole" shape. Each bacteriophage consists of a head, which appears hexagonal in electron micrographs, and a complex tail assembly. These bacteriophages attach to* E. coli *cells by means of the thin fibers extending from the tail assembly.*

that inhabits the healthy human intestine. These viruses were numbered T1 through T7, with the T standing simply for "type." As it turned out, most of the early work was done on T2 and T4, which became known as the T-even bacteriophages.

These viruses were inexpensive to work with, easy to maintain in the laboratory, and demanded little space or equipment. Furthermore, they were phenomenal at reproducing themselves. Twenty-five minutes from the time a single virus infected a bacterial cell, that cell would burst open, releasing a hundred or more new viruses, all exact copies of the original virus. Another advantage (which was not discovered until after the research was begun) was that this group of bacteriophages has a highly distinctive shape (Figure 14–6), and so can be readily identified with the electron microscope.

According to electron-microscope studies of infected *E. coli* cells (broken open at regular intervals after infection), the bacteriophages do not multiply like bacteria. Except for a few fragments, they disappear the moment after infection, and for the first 10 to 11 minutes of the infection cycle, not a single virus can be seen within the bacterial cell. Then, depending on when the cell is opened during the course of the infection, increasing numbers of completed bacteriophages can be seen and, mixed with them, odds and ends that appear to be bits of incomplete bacteriophages.

Chemical analysis of the bacteriophages revealed that they consist quite simply of DNA and of protein, the two leading contenders in the 1940s for the role of the genetic material. The chemical simplicity of the bacteriophage offered geneticists a remarkable opportunity. The viral genes—the hereditary material that directed the synthesis of new viruses within the bacterial cell—had to be carried either on the protein or on the DNA. If it could be determined which of the two it was, then the chemical identity of the gene would be known.

In 1952, a set of simple but ingenious experiments were carried out by Alfred D. Hershey and his colleague, Martha Chase.* They prepared two separate samples of viruses, one in which the DNA was labeled with a radioactive isotope of phosphorus, ^{32}P, and the other in which the protein was labeled with a radioactive isotope of sulfur, ^{35}S. Each type of virus was produced by growing the *E. coli* host on a medium that contained the appropriate radioactive isotope. After a cycle of multiplication, the newly formed viruses all contained some of the radioactive isotope in place of the common nonradioactive isotope. If you recall the chemical structure of nucleic acids and proteins, you will note that DNA contains phosphorus but no sulfur, while the amino acid components of proteins contain no phosphorus, although two amino acids (methionine and cysteine) contain sulfur. Thus ^{32}P and ^{35}S can serve as specific radioactive labels that distinguish DNA from protein.

One culture of bacteria was infected with ^{32}P-labeled phage and another with ^{35}S-labeled phage (Figure 14–7). After the infection cycle had begun, the cells were agitated in a blender and then spun down in a centrifuge to separate them from any viral material remaining outside the cells. The two samples—one containing extracellular material and the other intracellular material—were then tested for radioactivity. Hershey and Chase found that the ^{35}S had remained outside the bacterial cells with the empty viral coats and the ^{32}P had entered the cells, infected them, and caused the production of new virus progeny. It was therefore concluded that the genetic material of the virus is DNA rather than protein.

---

* We are including the names of the scientists involved in these experiments, not only to give credit where it is due, but also because the names have become synonymous with the work. What we are describing now are the Hershey-Chase experiments.

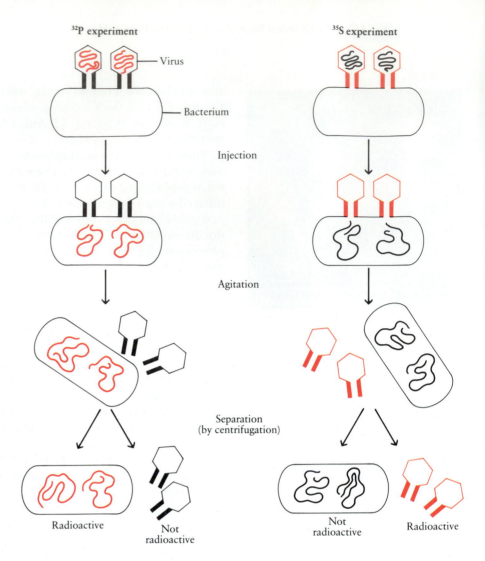

Virus

Bacterium

Injection

Agitation

Separation
(by centrifugation)

Radioactive

Not
radioactive

Not
radioactive

Radioactive

14–7 *A summary of the Hershey-Chase experiments demonstrating that DNA is the hereditary material of a virus. Radioactively labeled molecules are shown in color.*

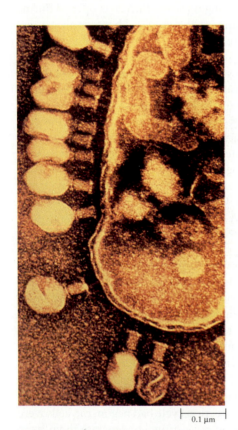

0.1 μm

14–8 *Electron micrograph of T4 bacteriophages attacking a cell of* E. coli. *The viruses attach to the bacterial cell by their tail fibers; viral DNA, contained within the head of the virus, is injected through the tail and into the cell. As you can see, the heads of some of the viruses are empty, indicating that the injection process has already occurred. A complete cycle of virus infection takes only about 25 minutes. At the end of that period about a hundred new virus particles are released from the cell.*

Electron micrographs have now confirmed that the T4 bacteriophage attaches to the bacterial cell wall by its tail fibers. They also indicate that the phage injects its DNA into the cell, leaving the empty protein coat on the outside (Figure 14–8). In short, the protein is just a container for the bacteriophage DNA. It is the DNA of the bacteriophage that enters the cell and carries the complete hereditary message of the virus particle, directing the formation of new viral DNA and new viral protein.

Further Evidence for DNA

The role of DNA in transformation and in viral replication formed very convincing evidence that DNA is the genetic material. Two other lines of experimental work also helped to lend weight to the argument. First, Alfred Mirsky, in a long series of careful studies conducted at Rockefeller University, showed that, in general, the somatic cells of any given species contain equal amounts of DNA and the gametes contain just half as much DNA as is found in the somatic cells. This is consistent with the observed result of meiosis, in which the diploid chromosome number is reduced to the haploid number.

Chargaff's Results

A second important series of contributions was made by Erwin Chargaff of Columbia University's College of Physicians and Surgeons. Chargaff analyzed the purine and pyrimidine content of the DNA of many different kinds of living things and found that, in contradiction to Levene's conclusions (page 283), the nitrogenous bases do *not* always occur in equal proportions. The proportions of the four nitrogenous bases are the same in all cells of all individuals of a given species, but they vary from one species to another. Therefore variations in base composition could very well provide a "language" in which the instructions controlling cell growth could be written. Some of Chargaff's results are reproduced in Table 14–1. Can you, by examining these figures, notice anything interesting about the proportions of purines and pyrimidines?

TABLE 14–1 Percentage Composition of DNA in Several Species

	PURINES		PYRIMIDINES	
SOURCE	ADENINE	GUANINE	CYTOSINE	THYMINE
Human being	30.4%	19.6%	19.9%	30.1%
Ox	29.0	21.2	21.2	28.7
Salmon sperm	29.7	20.8	20.4	29.1
Wheat germ	28.1	21.8	22.7	27.4
E. coli	24.7	26.0	25.7	23.6
Sea urchin	32.8	17.7	17.3	32.1

The Hypothesis Is Confirmed

Taken together, all of the studies we have traced thus far provided convincing evidence that DNA is the genetic material. Nonetheless, a critical question remained unanswered: *How* is the genetic information contained in the DNA? The answer to this question was to be found in the structure of the DNA molecule itself.

To fulfill its biological role, the genetic material has to meet at least four requirements:

1. It must carry the genetic information from parent cell to daughter cell and from generation to generation. Further, it must carry a great deal of information. Consider how many instructions must be contained in the set of genes that directs, for example, the development of an elephant, or a tree, or even a *Paramecium.*

2. It must contain information for producing a copy of itself, for it is copied with every cell division and with great precision.

3. It must be chemically stable; otherwise, it would not carry identical information from generation to generation and offspring would not resemble their parents.

4. On the other hand, it must be capable of mutation. When a gene changes, that is, when a "mistake" is made, the "mistake" must be copied as faithfully as was the original. This is a most important property, for without the capacity to replicate "errors," there would be no genetic variation. As a consequence, there would also be no evolution by natural selection.

It was when the DNA molecule was found to have the size, the configuration, and the complexity necessary to meet these requirements that DNA became universally accepted as the genetic material. The scientists primarily responsible for working out the structure of DNA were James Watson and Francis Crick, and their feat is one of the milestones in the history of science.

THE WATSON-CRICK MODEL

In the early 1950s, a young American scientist, James Watson, went to Cambridge, England, on a research fellowship to study problems of molecular structure. There, at the Cavendish Laboratory, he met physicist Francis Crick. Both were interested in DNA, and they soon began to work together to solve the problem of its molecular structure. They did not do experiments in the usual sense but rather undertook to examine all the data about DNA and to unify them into a meaningful whole.

The Known Data

By the time Watson and Crick began their studies, quite a lot of information on the subject had already accumulated:

1. The DNA molecule was known to be very large, and also very long and thin, and to be composed of nucleotides containing the nitrogenous bases adenine, guanine, thymine, and cytosine.

2. According to Levene's interpretation of his data, these nucleotides were assembled in repeating units of four.

3. Linus Pauling, in 1950, had shown that a protein's component chains of amino acids are often arranged in the shape of a helix and are held in that form by hydrogen bonds between successive turns of the helix (see page 75). Pauling had suggested that the structure of DNA might be similar.

4. X-ray diffraction photographs of DNA (Figure 14–9) from the laboratories of Maurice Wilkins and Rosalind Franklin at King's College, London, showed patterns that almost certainly reflected the turns of a giant helix.

5. Also crucial were the data of Chargaff indicating, as you perhaps noticed in Table 14–1, that (within experimental error) the amount of adenine is the same as the amount of thymine, and the amount of guanine is the same as the amount of cytosine: A = T and G = C.

Building the Model

From these data, some of them contradictory, Watson and Crick attempted to construct a model of DNA that would fit the known facts and explain the biological role of DNA. In order to carry the vast amount of genetic information, the molecules should be heterogeneous and varied. Also, there must be some way for them to replicate readily and with great precision so that faithful copies could be passed from cell to cell and from parent to offspring, generation after generation.

On the other hand, Watson and Crick could not be sure that the chemical structure of DNA would actually reveal its biological function. After all, this idea had never really been tested rigorously. "In pessimistic moods," Watson has recalled, "we often worried that the correct structure might be dull—that is, that it would suggest absolutely nothing."

It turned out, in fact, to be unbelievably "interesting." By piecing together the various data, they were able to deduce that DNA is an exceedingly long, entwined double helix.

If you were to take a ladder and twist it into a helix, keeping the rungs perpendicular, you would have a crude model of the DNA molecule (Figure 14–10). The two rails, or sides, of the ladder are made up of alternating sugar and phosphate molecules. The perpendicular rungs of the ladder are formed by the nitrogenous bases—adenine (A), thymine (T), guanine (G), and cytosine (C). Two bases form each rung, with each base covalently bonded to a sugar-phosphate

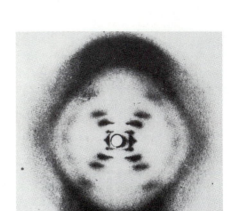

(a)

(b)

14–9 (a) *The critical x-ray diffraction photograph of DNA, taken by Rosalind Franklin. The reflections crossing in the middle indicate that the molecule is a helix. The heavy dark regions at the top and bottom are due to the closely stacked bases perpendicular to the axis of the helix.* (b) *Rosalind Franklin, photographed while vacationing in France in 1950 or 1951. She died of cancer in 1958, at the age of 37.*

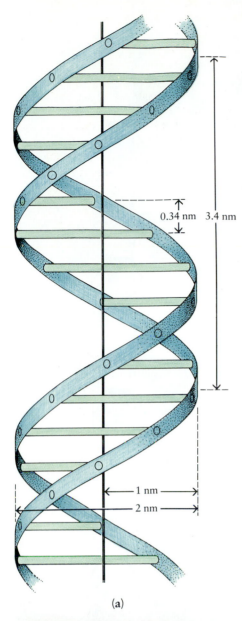

(a)

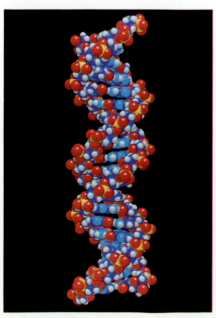

(b)

14–10 (a) *The double-stranded helical structure of DNA, as first presented in 1953 by Watson and Crick. The framework of the helix is composed of the sugar-phosphate units of the nucleotides. The rungs are formed by the four nitrogenous bases adenine and guanine (the purines) and thymine and cytosine (the pyrimidines). Each rung consists of two bases. Knowledge of the distances between the atoms was crucial in establishing the structure of the DNA molecule. The distances were determined from x-ray diffraction photographs of DNA taken by Rosalind Franklin and Maurice Wilkins. (b) A computer-generated space-filling model of DNA. The paired bases form the rungs connecting the sugar-phosphate side rails. In this model, the atoms are color-coded as follows: white = hydrogen (H), red = oxygen (O), yellow = phosphorus (P), dark blue = carbon (C), and turquoise = nitrogen (N).*

unit. The paired bases meet across the helix and are joined together by hydrogen bonds, the relatively weak bonds that Pauling had demonstrated in his studies of protein structure.

The distance between the two sides, or railings, according to x-ray measurements, is 2 nanometers. Two purines in combination would take up more than 2 nanometers, and two pyrimidines would not reach all the way across. But if a purine paired in each case with a pyrimidine, there would be a perfect fit and the molecule would be the same width along its entire length. The paired bases—the "rungs" of the ladder—would therefore always be purine-pyrimidine combinations.

As Watson and Crick analyzed the data, they assembled actual tin-and-wire models of the molecules (see essay, page 291), testing where each piece would fit into the three-dimensional puzzle. As they worked with the models, they realized that the nucleotides along any one strand of the double helix could be assembled in any order: for example, TTCAGTACATTGCCA, and so on (Figure 14–11a). Since a DNA molecule may be thousands of nucleotides long, there is a possibility for great variety in the sequence of bases, and variety is one of the primary requirements for the genetic material. Note also that the strand has direction: each phosphate group is attached to one sugar at the 5′ position (the fifth carbon in the sugar ring) and to the other sugar at the 3′ position (the third carbon in the sugar ring). Thus, the strand has a 5′ end and a 3′ end.

The most exciting discovery came, however, when Watson and Crick set out to construct the matching strand. They encountered another interesting and important restriction. Not only could purines not pair with purines and pyrimidines not pair with pyrimidines, but because of the structures of the bases, adenine could pair only with thymine, forming two hydrogen bonds, (A═T), and guanine only with cytosine, forming three hydrogen bonds (G≡C). The paired bases were **complementary.** Look at Table 14–1 again and see how well these chemical requirements explain Chargaff's data.

The double-stranded structure of a DNA molecule is shown in Figure 14–11b. As you can see, the two strands run in opposite directions; that is, the direction from the 5′ end to the 3′ end of each strand is opposite. The strands are said to be **antiparallel.** Although the nucleotides along one chain of the double helix can occur in any order, their sequence then determines the order of nucleotides in the other chain. This is necessarily the case, since the bases are complementary (G with C and A with T). Thus, for example, the complementary strand of (5′)-TTCAGTACATTGCCA-(3′) must have the nucleotide sequence (3′)-AAGTCATGTAACGGT-(5′).

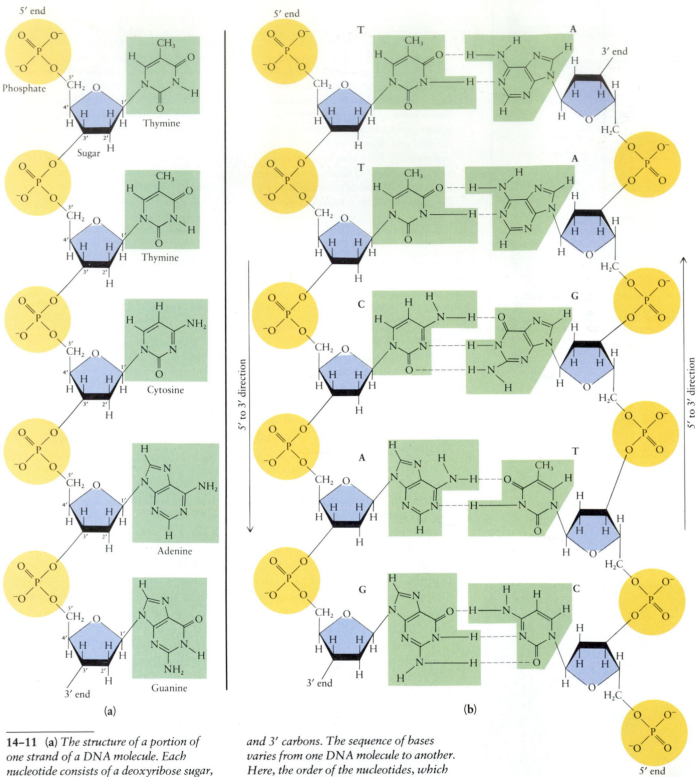

14–11 **(a)** *The structure of a portion of one strand of a DNA molecule. Each nucleotide consists of a deoxyribose sugar, a phosphate group, and a purine or pyrimidine base. Note the repetitive sugar-phosphate-sugar-phosphate sequence that forms the backbone of the molecule. Each phosphate group is attached to the 5' carbon of one sugar subunit and to the 3' carbon of the sugar subunit in the adjacent nucleotide. The DNA strand thus has a 5' end and a 3' end, determined by these 5'* *and 3' carbons. The sequence of bases varies from one DNA molecule to another. Here, the order of the nucleotides, which is customarily written in the 5' to 3' direction, is TTCAG. **(b)** The double-stranded structure of a portion of a DNA molecule. The strands are held together by hydrogen bonds (represented here by dashes) between the bases. Notice that adenine and thymine can form two hydrogen bonds, whereas guanine and cytosine can form three. Because of these bonding require-* *ments, adenine can pair only with thymine and guanine only with cytosine. Thus the order of bases along one strand determines the order of bases along the other strand. The strands are antiparallel; that is, the direction from the 5' to the 3' end of one strand is opposite to that of the other.*

Who Might Have Discovered It?

Then there is the question, what would have happened if Watson and I had not put forward the DNA structure? This is "iffy" history which I am told is not in good repute with historians, though if a historian cannot give plausible answers to such questions I do not see what historical analysis is about. If Watson had been killed by a tennis ball I am reasonably sure I would not have solved the structure alone, but who would? Olby has recently addressed himself to this question. Watson and I always thought that Linus Pauling would be bound to have another shot at the structure once he had seen the King's College x-ray data, but he has recently stated that even though he immediately liked our structure it took him a little time to decide finally that his own was wrong. Without our model he might never have done so. Rosalind Franklin was only two steps away from the solution.

Watson (left) and Crick in 1953 with one of their models of DNA. "DNA, you know, is Midas' gold," said Maurice Wilkins, with whom they shared the Nobel Prize. "Everyone who touches it goes mad."

She needed to realise that the two chains must run in opposite directions and that the bases, in their correct tautomeric forms, were paired together. She was, however, on the point of leaving King's College and DNA, to work instead on TMV [tobacco mosaic virus] with Bernal. Maurice Wilkins had announced to us, just before he knew of our structure, that he was going to work full time on the problem. Our persistent propaganda for model building had also had its effect (we had previously lent them our jigs to build models but they had not used them) and he proposed to give it a try. I doubt myself whether the discovery of the structure could have been delayed for more than two or three years.

There is a more general argument, however, recently proposed by Gunther Stent and supported by such a sophisticated thinker as Medawar. This is that if Watson and I had not discovered the structure, instead of being revealed with a flourish it would have trickled out and that its impact would have been far less. For this sort of reason Stent had argued that a scientific discovery is more akin to a work of art than is generally admitted. Style, he argues, is as important as content.

I am not completely convinced by this argument, at least in this case. Rather than believe that Watson and Crick made the DNA structure, I would rather stress that the structure made Watson and Crick. After all, I was almost totally unknown at the time and Watson was regarded, in most circles, as too bright to be really sound. But what I think is overlooked in such arguments is the intrinsic beauty of the DNA double helix. It is the molecule which has style, quite as much as scientists. The genetic code was not revealed all in one go but it did not lack for impact once it had been pieced together. I doubt if it made all that difference that it was Columbus who discovered America. What mattered much more was that people and money were available to exploit the discovery when it was made. It is this aspect of the history of the DNA structure which I think demands attention, rather than the personal elements in the act of discovery, however interesting they may be as an object lesson (good or bad) to other workers.

FRANCIS CRICK: "The Double Helix: A Personal View," *Nature,* vol. 248, pages 766–769, 1974.

DNA REPLICATION

An essential property of the genetic material is the ability to provide for exact copies of itself. Does the Watson-Crick model satisfy this requirement? In their published account, Watson and Crick wrote, "It has not escaped our notice that the specific pairing we have postulated immediately suggests a possible copying mechanism for the genetic material." Implicit in the double and complementary structure of the DNA helix is a method by which it can reproduce itself. At the time of chromosome replication, the molecule "unzips" down the middle, the paired bases separating at the hydrogen bonds. As the two strands separate, they act as **templates,** or guides; each directs the synthesis of a new complementary strand along its length, using the raw materials in the cell (Figure 14–12). This mechanism of DNA replication, hypothesized by Watson and Crick on the basis of their model of DNA structure, is called **semiconservative replication,** since half the molecule is conserved. Each old strand forms a template for the production of a new one. If a T is present on the old strand, only an A can fit into place in the new strand; a G will pair only with a C, and so on. In this way, each strand forms a copy of the original partner strand, and two exact replicas of the molecule are produced. The age-old question of how hereditary information is duplicated and passed on, generation after generation, had apparently been answered.

A Confirmation of Semiconservative Replication

The Watson-Crick model of DNA replication, however, was not the only possible mechanism. Matthew Meselson and Franklin W. Stahl, working at the California Institute of Technology, devised an elegant experiment to choose among three possible models (Figure 14–13).

In designing their experiment, they took advantage of the availability of a heavy isotope of nitrogen (^{15}N) and an extremely sensitive method of separating macromolecules on the basis of density. The method, which had been devised by Meselson while he was a graduate student, involves placing a solution of cesium chloride (CsCl) in a tube and spinning it in an ultracentrifuge. The small, dense CsCl molecules form a continuous density gradient, less concentrated at the top of the tube and more concentrated at the bottom. When DNA molecules are centrifuged in this solution, they will form a band at the point in the gradient at which the DNA and the CsCl solution have equal densities (Figure 14–14a). CsCl was selected for this procedure because the range of densities in the gradient it forms includes that of DNA.

Meselson and Stahl grew *E. coli* for several generations in a medium in which the sole nitrogen source contained ^{15}N, the heavy isotope of nitrogen. At the end of this period, the DNA of the bacterial cells contained a large proportion of heavy nitrogen. Although the density of this DNA was only about 1 percent greater than that of normal DNA, it formed a separate and distinct band in the cesium chloride gradient (Figure 14–14b and c).

They then placed a sample of cells containing heavy nitrogen in a medium containing ^{14}N; the cells were left in this medium only long enough for the DNA to replicate once (as determined by a doubling of the number of cells). A sample of DNA from these cells was spun in the ultracentrifuge (Figure 14–14d). Then, a second sample of cells was grown in the ^{14}N medium for two generations. Their DNA was also ultracentrifuged (Figure 14–14e).

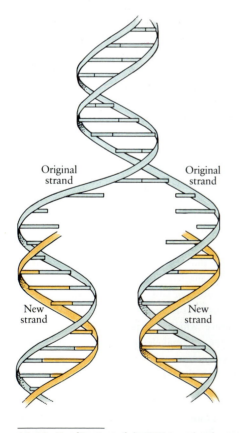

Original strand

Original strand

New strand

New strand

14–12 *Replication of the DNA molecule, as predicted by the Watson-Crick model. The strands separate down the middle as the paired bases separate at the hydrogen bonds. Each of the original strands then serves as a template along which a new,* complementary strand forms from nucleotides available in the cell. Subsequent research has resulted in modification of some of the details of this process, as we shall see shortly, but the underlying principle is unchanged.

(a)

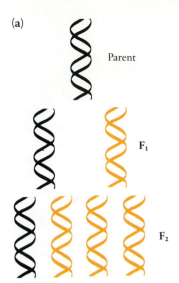

Parent

F₁

F₂

(b)

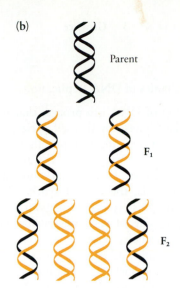

Parent

F₁

F₂

(c)

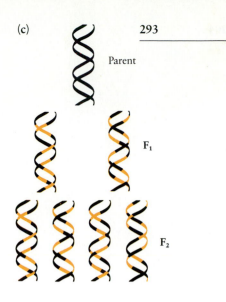

Parent

F₁

F₂

14-13 *Three possible mechanisms of DNA replication. In this diagram the original strands are shown in black and the newly replicated strands are shown in color. (a) Conservative replication. Each of the two strands of parent DNA is replicated, without strand separation. In the first generation, one daughter is all old DNA and one daughter is all new. The second generation contains one helix com-*

posed of two old strands and three made up entirely of new strands. (b) Semiconservative replication. The two parental strands separate, and each forms a template for a new strand. In the first generation, each daughter is half old and half new. The second generation comprises two hybrid DNAs (half old, half new) and two DNAs made up entirely of new strands.

(c) Dispersive replication. During replication, parent chains break at intervals, and replicated segments are combined into strands with segments from parent chains. All daughter helixes are part old, part new. The Meselson-Stahl experiment (Figure 14–14) was undertaken to determine which of these three possibilities was correct. Watson and Crick had predicted (b).

Each sample of DNA contained more light DNA, as would be expected, because newly formed DNA had to incorporate the available ^{14}N. Moreover—and this was of crucial importance—the density of the first generation DNA was exactly halfway between that of heavy parent DNA and that of ordinary light DNA, as it should be if each molecule contained one old (heavy) strand and one new (light) strand. The second generation contained one-half half-heavy DNA and one-half light DNA, which again, exactly and ingeniously, confirmed the Watson-Crick hypothesis of semiconservative replication (Figure 14–13b).

14-14 *The Meselson-Stahl experiment. (a) Normal light DNA forms a precisely located band when ultracentrifuged in a density gradient of cesium chloride. (b) E. coli cells cultured in a medium containing heavy nitrogen (^{15}N) accumulate a heavy DNA, which forms a separate and distinct band. (c) When a mixture of heavy and light DNA is centrifuged in a cesium chloride density gradient, the two types of DNA separate into two distinct bands. (d) When cells grown in a medium containing heavy nitrogen (^{15}N) are permitted to multiply for one generation in a medium containing ordinary light nitrogen (^{14}N), their DNA forms a band in the cesium chloride density gradient that is located midway between the bands of the heavy and light DNAs. (e) When cells containing heavy DNA are grown for two generations in the ^{14}N medium, their DNA forms two bands in the density gradient—one band of light DNA and one band of half-heavy DNA. The column on the right shows the investigators' interpretations. As you can see, this experiment confirmed the Watson-Crick hypothesis of semiconservative replication.*

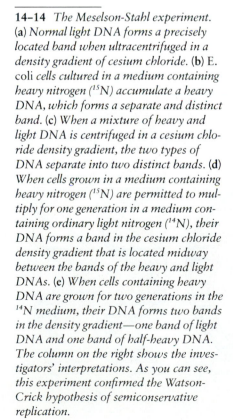

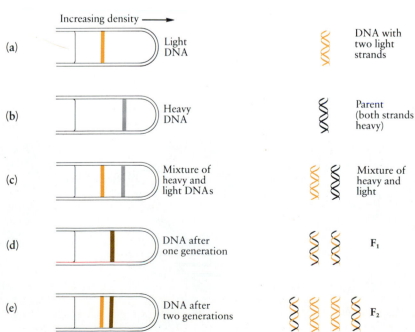

Increasing density →

(a) Light DNA — DNA with two light strands

(b) Heavy DNA — Parent (both strands heavy)

(c) Mixture of heavy and light DNAs — Mixture of heavy and light

(d) DNA after one generation — F₁

(e) DNA after two generations — F₂

The Mechanics of DNA Replication

Replication of DNA is a process that occurs only once in each cell generation, during the S phase of the cell cycle (page 146); it is the essential event in the replication of the chromosomes. In most eukaryotic cells, DNA replication leads ultimately to mitosis, but in the primary spermatocytes and oocytes it leads instead to meiosis. It is a remarkably rapid process; for example, in humans and other mammals, the rate of synthesis is about 50 nucleotides per second. In prokaryotes, it is even faster—about 500 nucleotides per second.

The principle of semiconservative replication, in which each strand of the DNA double helix serves as a template for the formation of a new strand, is relatively simple and easy to understand. However, the actual process by which the cell accomplishes replication is considerably more complex. Like other biochemical reactions of the cell, DNA replication requires a number of different enzymes, each catalyzing a particular step of the process. The identification of the principal enzymes, their precise functions, and the sequence of events in replication has required a number of years and the efforts of many scientists working in different laboratories. Although our understanding is still incomplete, the general outlines of the process are now clear.

Initiation of DNA replication always begins at a specific nucleotide sequence known as the **origin of replication.** It requires special initiator proteins and, in addition, enzymes known as **helicases.** These enzymes break the hydrogen bonds linking the complementary bases at the replication origin, opening up the helix so replication can occur. However, as the strands of the helix separate, the adjacent portions of the double helix are in danger of becoming more and more tightly coiled, or supercoiled. Other enzymes, the **topoisomerases,** break and reconnect one or both strands of the helix, allowing swiveling to occur and thus relieving strain on adjacent portions of the molecule.

Once the two strands of the DNA double helix are separated, additional proteins, known as single-strand binding proteins, attach to the individual strands, holding them apart and preventing kinking. This makes possible the next stage, the actual synthesis of the new strands, catalyzed by a group of enzymes known as **DNA polymerases.**

14–15 *An overview of DNA replication. The two strands of the DNA molecule* (a) *separate at the origin of replication as a result of the action of initiator proteins and enzymes known as helicases.* (b, c) *The two replication forks move away from the origin of replication in opposite directions, forming a replication bubble that expands bidirectionally.* (d) *When synthesis of the new DNA strands is complete, the two double-stranded chains separate into two new double helixes. The DNA has been replicated semiconservatively; that is, each new double helix consists of one old strand and one new strand.*

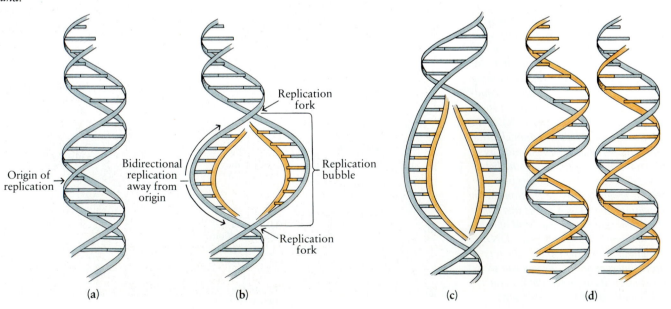

(a)   (b)   (c)   (d)

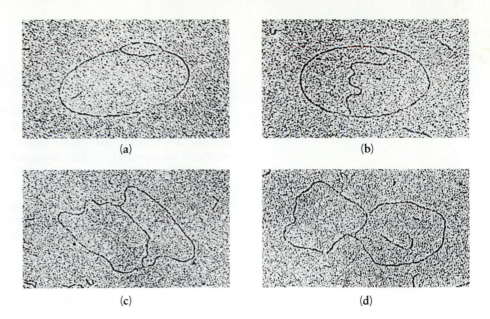

(a)   (b)

(c)   (d)

14–16 *Replication of the circular* E. coli *chromosome, as revealed by the electron microscope. The process begins in a specific sequence of nucleotides, the origin of replication, and proceeds in opposite directions from that point. (a) In the early stages of bidirectional replication, the enlarging replication bubble resembles an eye. (b) Subsequently, the replicating chromosome takes on the shape of the Greek letter theta. (c) Bidirectional replication continues until the entire chromosome has been replicated (d). The two newly formed chromosomes then separate from each other (see Figure 7–2, page 145).*

If replicating DNA is viewed under the electron microscope, the region of synthesis appears as an "eye," or replication bubble. At either end of the bubble, where the old strands are being separated by helicase and the complementary strands are being synthesized, the molecule appears to form a Y-shaped structure. This is known as a **replication fork**. Replication is bidirectional, with two replication forks moving in opposite directions away from the origin (Figure 14–15).

In prokaryotes, there is a single replication origin, located within a specific nucleotide sequence about 300 base pairs long. As the replicated strands begin to separate, a structure forms that resembles the Greek letter theta (θ), eventually giving rise to two circular DNAs (Figure 14–16). In eukaryotes, by contrast, there are many replication origins; replication proceeds along the linear chromosomes as each bubble expands bidirectionally until it meets an adjacent bubble (Figure 14–17). The entire length of the chromosome is replicated as these bubbles merge.

14–17 *Replication of eukaryotic chromosomes is initiated at multiple origins. The individual replication bubbles spread until ultimately they meet and join. In this electron micrograph of an embryonic cell of* Drosophila melanogaster, *the centers of the replication bubbles are indicated by the arrows.*

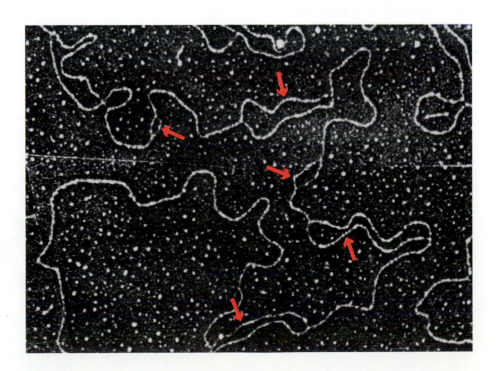

RNA Primers and the Direction of Synthesis

In order for synthesis of a new, complementary strand of DNA to occur, it is not enough that the old strand—serving as the template for the new strand—be present. There must also be the beginning of the new strand. This beginning is provided by a **primer,** formed of nucleotides of ribonucleic acid (RNA). As you will recall from Chapter 3 (page 80), RNA is a closely related nucleic acid in which the five-carbon sugar is ribose; in addition, it contains the nitrogenous base uracil instead of thymine. RNA nucleotides can form hydrogen bonds with the nucleotides of a DNA strand, following a similar principle of complementarity. Guanine pairs with cytosine, the adenine of RNA pairs with the thymine of DNA, and the uracil of RNA pairs with the adenine of DNA.

In eukaryotes, RNA primers are typically about 10 nucleotides long. Their synthesis on the exposed single strand of DNA is catalyzed by an enzyme known as **RNA primase.** With the RNA primers in place, DNA polymerases begin to synthesize new complementary DNA strands along the template strands, adding nucleotides one by one to the growing strands.

The first DNA polymerases discovered were found to synthesize new DNA strands only in the 5′ to 3′ direction; that is, incoming nucleotides were added only at the 3′ end of the chain. This posed a major problem. Because of the antiparallel structure of the DNA double helix, replication of the two new DNA strands on the two arms of the Y-shaped replication fork seemed to require not only synthesis in the 5′ to 3′ direction but also in the 3′ to 5′ direction. For a number of years, researchers sought unsuccessfully to identify another DNA polymerase that would function in the 3′ to 5′ direction. The cell's solution to this problem was eventually revealed by the Japanese scientist Reiji Okazaki, who found that, although the 5′ to 3′ strand is synthesized continuously as a single unit, the 3′ to 5′ strand is synthesized discontinuously, as a series of fragments, each synthesized in the 5′ to 3′ direction (Figure 14–18). The strand that is synthesized continuously is known as the **leading strand,** and the strand synthesized as a series of fragments is known as the **lagging strand.** The fragments that

14–18 *A close-up of the leading and lagging strands near the replication fork. The addition of nucleotides to both of these strands is catalyzed by the enzyme DNA polymerase, which works only in the 5′ to 3′ direction. In order for DNA polymerase to begin adding nucleotides, an RNA primer must be present, hydrogen bonded to the template strand; synthesis of the RNA primer is catalyzed by RNA primase. (The RNA primer for the leading strand is located at the origin of replication, not visible in this diagram.) The leading strand is synthesized continuously in the 5′ to 3′ direction. The lagging strand, by contrast, is synthesized discontinuously, in the form of Okazaki fragments. These fragments are synthesized in the 5′ to 3′ direction, which is, however, opposite to the overall direction of replication of this strand. When an Okazaki fragment has become long enough to encounter the RNA primer ahead of it, other enzymes replace the RNA nucleotides of the primer with DNA nucleotides. DNA ligase then connects the fragment with the adjacent newly synthesized fragment of the strand.*

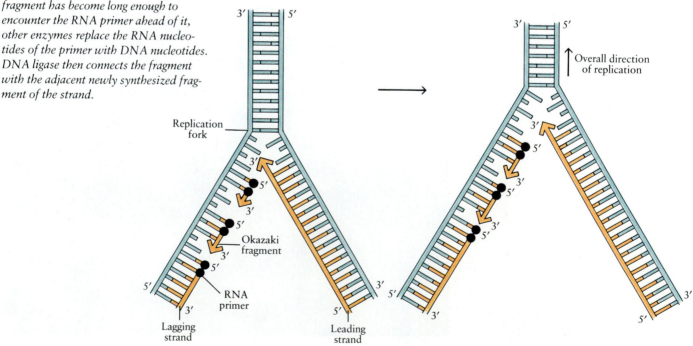

form the lagging strand, the **Okazaki fragments,** are typically 1,000 to 2,000 nucleotides long in prokaryotes and 100 to 200 nucleotides long in eukaryotes.

The role of the RNA primers, on both the leading and lagging strands, is to provide correctly paired strands of nucleotides with exposed 3′ OH groups to which the DNA polymerase can begin attaching DNA nucleotides in sequence. DNA polymerase, beginning at an RNA primer, moves along a DNA strand, adding nucleotides to the 3′ end of the growing strand, until it encounters the 5′ end of another strand of RNA primer. On the lagging strand, in which synthesis moves away from the replication fork, this primer represents the beginning of another Okazaki fragment. On the leading strand, in which synthesis moves in the same direction as the replication fork, this primer represents the beginning of an Okazaki fragment of the next replication bubble. When DNA polymerase makes contact with the 5′ end of an RNA primer, other enzymes are activated; these enzymes remove the RNA nucleotides and replace them with DNA nucleotides. Another enzyme, **DNA ligase,** then connects the newly synthesized DNA segment to the growing DNA strand by catalyzing the condensation reaction that bonds adjacent phosphate and sugar groups.

This complex process, which occurs prior to every cell division, is summarized in Figure 14–19.

Proofreading

One of the essential features of DNA replication is that DNA polymerase can add nucleotides to the 3′ end of a strand only if the nucleotides previously added to that strand are correctly paired with their complementary nucleotides on the template strand. As we have seen, the function of the RNA primer is to provide a correctly paired sequence of nucleotides at which DNA synthesis can begin. In the

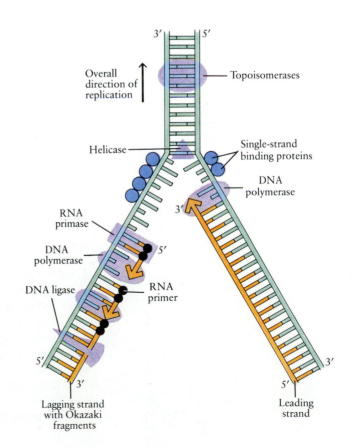

14–19 *A summary of DNA replication. The two strands of the DNA double helix separate, and new complementary strands are synthesized, using the original strands as templates. Separation of the two strands of the original double helix is due to the action of helicases, enzymes that operate at the replication forks. Supercoiling is prevented by the topoisomerases, which catalyze the formation and resealing of nicks in one or both strands ahead of the replication forks. The separated DNA strands are stabilized by single-strand binding proteins. Synthesis of RNA primers is catalyzed by the enzyme RNA primase. Building onto the RNA primers, the DNA polymerases add nucleotides in a 5′ to 3′ direction. Synthesis of the leading strand is continuous (as in Figure 14–18, its RNA primer is not visible). Along the lagging strand, 5′ to 3′ synthesis is discontinuous. Okazaki fragments are synthesized in the 5′ to 3′ direction, and, following replacement of the RNA primer of the previous Okazaki fragment by DNA nucleotides, the fragment is joined to the growing strand by the enzyme DNA ligase.*

course of synthesis, however, errors are sometimes made, and the wrong nucleotide is added to the newly forming strand—that is, the nucleotide added to the strand is not complementary to the nucleotide on the template strand. When this occurs, DNA polymerase backtracks, removing nucleotides until it encounters a correctly paired nucleotide. At that point, the enzyme stops its reverse movement and resumes moving in the 5′ to 3′ direction, adding nucleotides to the growing strand as it goes. This ability to remove incorrectly paired nucleotides provides an important proofreading that ensures the accuracy of DNA replication.

In addition to the proofreading that occurs during DNA replication, other enzymes constantly monitor the DNA double helixes of the cell. Whenever an incorrectly paired nucleotide is encountered, DNA repair enzymes move in to snip it out and replace it with the correct nucleotide. This process is essential in maintaining the integrity of the genetic machinery.

The Energetics of DNA Replication

The nucleotides required for DNA synthesis are assembled by the same biosynthetic pathways as the nucleotides required for other functions of the cell. They are assembled not in the form of the monophosphates shown in Figure 14–4 but rather as triphosphates; that is, adenine is provided as deoxyadenosine triphosphate (dATP), guanine as the analogous deoxyguanosine triphosphate (dGTP), and so forth. The $P \sim P$ groups of the triphosphates provide the energy to power the reactions catalyzed by DNA polymerase. As each nucleotide is attached to the growing DNA strand, the two "extra" phosphates are removed and released. Almost immediately, another enzyme breaks the bond between the two phosphates, releasing them as inorganic phosphates.

Why does the cell do it this way? The release of the energy-yielding $P \sim P$ group and the subsequent breaking of the bond between the two phosphates seem like a waste of carefully stored chemical energy. Is the cell really so profligate?

Measurement of the energy changes involved reveals an interesting point. The reaction in which the activated nucleotide is attached to the growing DNA strand is only slightly exergonic. Therefore, it could go in either direction. Under certain equilibrium conditions, the DNA strand could come apart about as fast as it was synthesized, with the enzyme working both ways. However, with removal of the $P \sim P$ fragment and its immediate degradation, the reaction becomes highly exergonic. Thus, the reverse reaction—which would necessitate reforging the $P \sim P$ group—becomes highly endergonic and so, for all practical purposes, does not occur. This is yet another example of the way in which the living cell exercises tight control over its biochemical activities.

DNA AS A CARRIER OF INFORMATION

You will recall that a necessary property of the genetic material is the capacity to carry information. The Watson-Crick model showed that the DNA molecule is able to do this. The information is carried in the sequence of the bases, and *any* sequence of bases is possible. Since the number of paired bases ranges from about 5,000 for the simplest known virus up to an estimated 5 billion in the 46 human chromosomes, the number of possible variations is astronomical. The DNA from a single human cell—which if extended in a single thread would be almost 2 meters long—can contain information equivalent to some 600,000 printed pages of 500 words each, or a library of about a thousand books. Obviously, the DNA structure can well account for the endless diversity among living things.

SUMMARY

Classical genetics had been concerned with the mechanics of inheritance—how the units of heredity are passed from one generation to the next and how changes in the hereditary material are expressed in individual organisms. In the 1930s, new questions arose and geneticists began to explore the nature of the gene—its structure, composition, and properties.

During the 1940s, many investigators believed that genes were proteins, but others were convinced that the hereditary material was deoxyribonucleic acid (DNA). Important, although not widely accepted, evidence for the genetic role of DNA was presented by Avery in his experiments to identify the transforming factor of pneumococci. Confirmation of Avery's hypothesis came from studies with bacteriophages (bacterial viruses) showing that DNA and not protein is the genetic material of the virus.

Further support for the genetic role of DNA came from two more sets of data: (1) Almost all somatic cells of any given species contain equal amounts of DNA, and (2) the proportions of nitrogenous bases are the same in the DNA of all cells of a given species, but they vary in different species.

In 1953, Watson and Crick proposed a structure for DNA. The DNA molecule, according to their model, is a double-stranded helix, shaped like a twisted ladder. The two sides of the ladder are composed of repeating subunits consisting of a phosphate group and the five-carbon sugar deoxyribose. The "rungs" are made up of paired nitrogenous bases, one purine base pairing with one pyrimidine base. There are four bases in DNA—adenine and guanine, which are purines, and thymine and cytosine, which are pyrimidines. Adenine (A) can pair only with thymine (T), and guanine (G) only with cytosine (C). The four bases are the four "letters" used to spell out the genetic message. The paired bases are joined by hydrogen bonds.

When the DNA molecule replicates, the two strands come apart, separating at the hydrogen bonds. Each strand acts as a template for the formation of a new complementary strand from nucleotides available in the cell. The semiconservative (one strand conserved) nature of this process was confirmed by studies using heavy isotopes.

DNA replication begins at a particular nucleotide sequence on the chromosome, the origin of replication. It proceeds bidirectionally, by way of two replication forks that move in opposite directions. Helicases unwind the double helix at each replication fork, and single-strand binding proteins stabilize the separated strands. Topoisomerases relieve supercoiling of the helix ahead of the replication forks by nicking, which allows the chains to swivel freely, followed by resealing. A strand of RNA primer, correctly base-paired to the template strand, is required before replication can begin; attachment of the nucleotides that form the primer is catalyzed by the enzyme RNA primase. Addition of DNA nucleotides to the strand is catalyzed by DNA polymerases. These enzymes synthesize new strands in the 5' to 3' direction only, adding nucleotides, one by one, to the 3' end of the growing strand. Replication of the leading strand is continuous, but replication of the lagging strand is discontinuous. On the lagging strand, segments known as Okazaki fragments are synthesized in the 5' to 3' direction. DNA ligase catalyzes the condensation reaction that links adjacent Okazaki fragments together. In the course of DNA synthesis, DNA polymerase proofreads, backtracking when necessary to remove nucleotides that are not correctly paired with the template strand.

The nucleotides incorporated into the growing DNA strands are supplied in the form of triphosphates. The energy required to power replication is supplied by removal of the two "extra" phosphates and degradation of the $P \sim P$ bond.

With the elucidation of the structure of the DNA double helix by Watson and Crick, the role of DNA as the carrier and transmitter of the genetic information was universally accepted. With the discovery of the complex—and extremely precise—mechanism by which the living cell replicates its DNA, the question of how the hereditary information is faithfully transmitted from parent cell to daughter cell, generation after generation, was answered.

QUESTIONS

1. Distinguish among the following: purine/pyrimidine; origin of replication/replication bubble/replication fork; helicases/topoisomerases; RNA primase/DNA polymerases; leading strand/lagging strand; RNA primers/Okazaki fragments/DNA ligase.

2. What are the steps by which Griffith demonstrated the existence of the transforming factor? Can you think of any implications of Griffith's discovery for modern medicine?

3. What characteristics of bacteriophages make them a useful experimental tool?

4. One of the chief arguments for the erroneous hypothesis that proteins constitute the genetic material was that proteins are heterogeneous. Explain why the genetic material must have this property. What feature of the Watson-Crick model of DNA structure is important in this respect?

5. When the structure of DNA was being worked out, it became apparent that one purine base must be paired with a pyrimidine base, and that the other purine base must be paired with the other pyrimidine base. The evidence for this requirement came from two types of data. What were the data, and how did they indicate this structural requirement?

6. Further consideration of the structures of the four nitrogenous bases indicated that adenine could pair only with thymine, and cytosine only with guanine. What feature of the structures of the bases imposed this requirement on the structure of the DNA molecule?

7. Suppose you are talking to someone who has never heard of DNA. How would you support an argument that DNA is the genetic material? List at least five of the strong points in such an argument.

8. Suppose the Meselson-Stahl experiment were extended to the third generation. What would be the proportion of light to half-heavy DNA?

9. Eukaryotic cells are grown for a number of generations in a medium containing thymine labeled with ^3H. They are then removed from the radioactive medium, placed in an ordinary medium, and allowed to divide. Studies of the distribution of the radioactive isotope are made after each generation to determine the presence or absence of radioactive material in the chromatids. Before the cells are placed in the nonradioactive medium, all the chromatids contain ^3H. After one generation in the nonradioactive medium, all of the chromatids still contain radioactive ^3H. Assuming each chromatid contains a single DNA molecule, explain the results. Is this consistent with the Watson-Crick hypothesis? What would be the distribution of the ^3H after two generations in the nonradioactive medium? Why?

10. When 5-bromodeoxyuridine (BrdU) is added to the medium in which cells are cultured, it is used in place of thymidine (thymine plus deoxyribose). All DNA synthesized after the cells have been placed in the medium will contain BrdU. After the cells have divided once in the BrdU-containing medium, the chromatids all look alike, as you can see in micrograph (a) below. Both sister chromatids of each chromosome contain parent strands of original DNA combined with new strands of BrdU-substituted DNA. The original DNA stains darkly, so both sides are dark. Micrograph (b) shows the appearance of the chromosomes after the second cell division in the BrdU-containing medium. As you will notice, one sister chromatid is dark, and the other is light. Does this confirm or refute the results of the Meselson-Stahl experiment? Explain your answer.

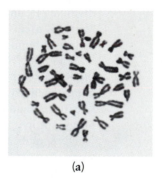 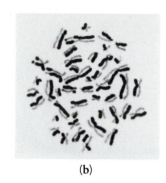

(a)   (b)

11. In the DNA replication bubble diagrammed below, label all 5′ and 3′ ends on each nucleotide strand and identify the leading and lagging strands. Describe what will happen next.

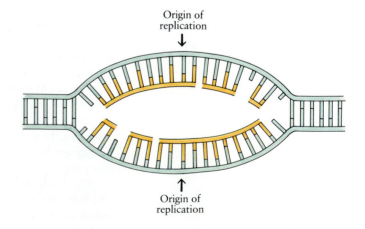

Origin of replication
↓

↑
Origin of replication

The Genetic Code and Its Translation

The Watson-Crick model, which both established the structure of DNA and indicated the mechanism for its replication, led to the virtually universal acceptance of DNA as the repository of the hereditary information. But how does a DNA molecule—"a lump of inert matter"—embody the instructions that specify a bacterium or a fruit fly or a student of biology? In the same year (1953) that their model was first published, Watson and Crick speculated—quite correctly as it turned out—that "it . . . seems likely that the precise sequence of the bases is a code which carries the genetical information." And so began an avalanche of studies to decipher the code and, even more interesting, to discover how the instructions it contains are decoded, transmitted, and carried out. These studies had their beginnings in a concept first proposed early in this century, shortly after Mendel's work was rediscovered.

GENES AND PROTEINS

Inborn Errors of Metabolism

In 1908, an English physician, Sir Archibald Garrod, presented a series of lectures in which he set forth a new concept of human diseases, which he called "inborn errors of metabolism." Garrod postulated that certain diseases that are caused by the body's inability to perform particular chemical processes are hereditary in nature.

One such disease was described in 1649:

> The patient was a boy who passed black urine and who, at the age of fourteen years, was submitted to a drastic course of treatment which had for its aim the subduing of the fiery heat of his viscera, which was supposed to bring about the condition in question by charring and blackening his bile. Among the measures prescribed were bleedings, purgation, baths, a cold and watery diet, and drugs galore. None of these had any obvious effect, and eventually the patient, who tired of the futile and superfluous therapy, resolved to let things take their natural course. None of the predicted evils ensued, he married, begat a large family, and lived a long and healthy life, always passing urine black as ink.

Sir Archibald hypothesized that this condition, alkaptonuria, is the result of an enzyme deficiency and is hereditary in nature. Implicit in his hypothesis was the idea that genes act by influencing the production of enzymes.

One Gene–One Enzyme

By the 1940s, biologists had come to realize that all of the biochemical activities of the living cell, including the multitude of synthetic reactions that produce all of its constituent molecules—carbohydrates, lipids, and proteins—depend upon different specific enzymes. Even the synthesis of enzymes depends on enzymes.

$\vdash$ 5 mm $\dashv$

15-1 Neurospora, *the red bread mold, growing on a corn tortilla. Colonies of unidentified black and gray molds are also present.* Neurospora *is another "fit material" that has played a key role in the history of genetics. Studies of* Neurospora *mutants by George Beadle and Edward Tatum in the 1940s provided the first definitive demonstration that genes contain the information specifying particular protein molecules.*

Further, it was becoming clear that the specificity of the different enzymes is a result of their primary structure, the linear sequence of amino acids in the molecule.

Meanwhile, George Beadle, a geneticist, was working with the eye-color mutants of *Drosophila* that had been discovered in Morgan's laboratory (page 265). As a result of his studies, Beadle formulated the hypothesis that each of the various eye colors observed in the mutants is the result of a change in a single enzyme in a biosynthetic pathway. To test this notion—that genes control enzymes—on a broader scale, he teamed up in 1941 with Edward L. Tatum, a biochemist. Instead of picking a genetic characteristic and working out its chemistry, they decided to begin with step-by-step chemical reactions—controlled by enzymes—to see if mutations affected these reactions.

The organism they chose for their studies was the red bread mold *Neurospora crassa* (Figure 15–1), which has since become almost as famous a research tool in genetics as the fruit fly. This fungus has several obvious advantages for genetic research:

1. Its life cycle is brief.

2. It can be grown in vast quantities in the laboratory.

3. Throughout most of its life cycle, it is haploid. As a consequence, when a mutation occurs, its effects are detectable immediately. The lack of a homologous chromosome rules out the masking of any mutation by a dominant allele.

4. Meiosis takes place in saclike reproductive structures known as asci (singular, ascus), in which the products of meiosis are not only packaged but also are replicated by mitosis and lined up neatly for inspection (Figure 15–2).

5. Many chromosome mapping studies had already been done with *Neurospora*, which facilitated further genetic analysis.

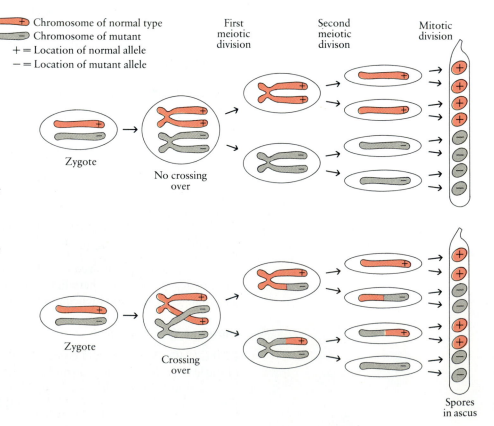

15–2 *Meiosis in* Neurospora crassa, *showing only one of its seven chromosomes. The zygotes shown here were produced by crossing a mutant strain with a normal strain (lacking that particular mutation). As a result of meiosis and the single mitotic division that follows it, eight spores are produced, lined up in a single, narrow spore case, the ascus. Four spores will be normal, and four will be mutants. The order in which the spores are lined up in the ascus depends on whether crossing over has or has not occurred.*

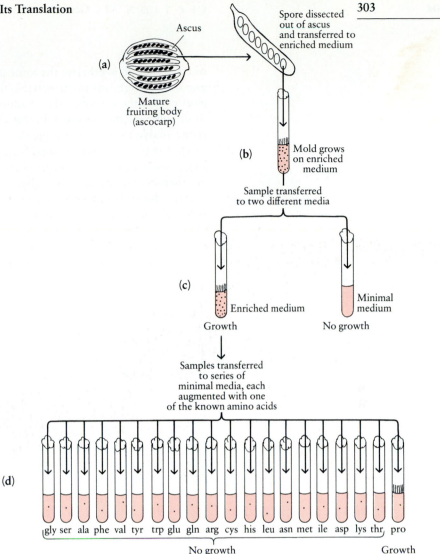

15-3 *How Beadle and Tatum tested the mutants of* Neurospora. *In these experiments, they were able to show that a change in a single gene results in a change in a single enzyme.* (a) *Asci are removed from the fruiting bodies (ascocarps) of* Neurospora, *and the spores are dissected out.* (b) *Each spore is transferred to an enriched medium, containing all that* Neurospora *normally needs for growth plus supplementary amino acids.* (c) *A fragment of the mold is tested for growth on the minimal medium. If no growth is observed on the minimal medium, it may mean that a mutation has occurred that renders this mutant incapable of making a particular amino acid.* (d) *Subcultures of molds that grow on the enriched medium but not on the minimal medium are tested for their ability to grow on minimal media supplemented with only one of the amino acids. In the example shown here, a mold that has lost its capacity to synthesize the amino acid proline is unable to survive in a medium that lacks that amino acid. Further tests are then made to discover which enzymatic step in the synthesis of proline has been impaired.*

In the figure labels: Ascus — Spore dissected out of ascus and transferred to enriched medium; Mature fruiting body (ascocarp); (b) Mold grows on enriched medium; Sample transferred to two different media; (c) Enriched medium — Growth; Minimal medium — No growth; Samples transferred to series of minimal media, each augmented with one of the known amino acids; (d) gly ser ala phe val tyr trp glu gln arg cys his leu asn met ile asp lys thr pro — No growth — Growth.

Another important feature of *Neurospora*, from the point of view of the investigators, was the fact that it can be grown on a very simple medium—a minimal medium—containing any one of several sugars as a carbon and energy source, one vitamin (biotin), and a few minerals. This undemanding fungus is able to make for itself all the amino acids, other vitamins, polysaccharides, and other substances essential for its growth and reproduction.

The synthesis of an amino acid or a vitamin requires a series of chemical reactions, each of which is catalyzed by a particular enzyme. If, as a result of a mutation, *Neurospora* were to lose any one of the enzymes involved, for example, in making the amino acid arginine, it could no longer grow on the minimal medium. The mutant could, however, grow on a medium supplemented with arginine.

Beadle and Tatum x-rayed *Neurospora* spores to increase the mutation rate, allowed them to germinate and grow on a medium enriched with all of the amino acids, and then crossed these irradiated strains with normal strains. The spores produced as a result of these crosses were then used in the experimental tests, ensuring that any changes observed were indeed genetic ones. On the basis of their experiments, summarized in Figure 15–3, Beadle and Tatum were able to demonstrate that different mutations resulted in loss of the ability to synthesize different amino acids. Moreover, as a result of the genetic analyses they performed, they showed that, in a number of cases, several different mutations could result in loss of the ability to synthesize the same amino acid. This indicated that the different mutations were affecting enzymes catalyzing different steps of the reaction pathway for the synthesis of that amino acid. For example, as we saw on page 173, three different mutations, affecting three different enzymes, can render *Neurospora* unable to synthesize arginine.

On the basis of their studies, Beadle and Tatum set forth the then-daring (and later, Nobel Prize–winning) proposal that a single gene specifies a single enzyme; in other words, one gene–one enzyme.

This formulation turned out to be an oversimplification, however, for although enzymes are indeed proteins, not all proteins are enzymes. Some proteins, for instance, are hormones, like insulin, and others are structural proteins, like collagen. These proteins, too, are specified by genes. This led to an expansion of the original concept, but did not modify it in principle. "One gene–one enzyme," as it was first formulated, was simply amended to "one gene–one protein." Subsequently, with the realization that many proteins consist of more than one polypeptide chain, it was modified once more, to the less memorable but more precise "one gene–one polypeptide chain." (As we shall see, this, too, has now been further amended.)

The Structure of Hemoglobin

What causes the change or loss of function in an enzyme or protein specified by a gene that has undergone a mutation? Linus Pauling was one of the first to see some of the implications of the work of Beadle and Tatum with regard to this question. Perhaps, Pauling reasoned, human diseases involving hemoglobin, such as sickle cell anemia, could be traced to a variation from the normal protein structure of the hemoglobin molecule. To test this hypothesis—which, at this stage, was pure speculation—he took samples of hemoglobin from people with sickle cell anemia (a homozygous recessive condition), from others heterozygous for the allele, and from still others homozygous for the normal allele. To try to detect differences in these proteins, Pauling used a technique known as electrophoresis, in which organic molecules dissolved in a solution are separated in a weak electric field.

As we noted in Chapter 3, individual amino acids may have a positive charge, a negative charge, or no electric charge at all. Therefore, a substitution of one amino acid for another may change the total charge of the protein molecule, and the normal and the variant protein molecules will, as a result, move at different rates in an electric field.

Figure 15–4 shows the results of Pauling's experiment. A person who has sickle cell anemia makes a different sort of hemoglobin than a person who does not have the disease. A person who is heterozygous (carrying one copy of the allele for sickling and one copy of the allele for normal hemoglobin) makes both kinds of hemoglobin molecules; however, enough normal molecules are produced to prevent anemia. (Notice that the terms "dominant" and "recessive" are beginning to have a less definite meaning.)

A few years later, Vernon Ingram, of Cambridge University in England, was able to show that the actual difference between the normal and the sickle cell hemoglobin molecules is a change in a single amino acid out of 300, as we noted previously (page 80).

The Virus Coat

Additional evidence that DNA specifies the structure of proteins came from the studies of bacteriophages described in Chapter 14. You will recall that the introduction of viral DNA into a bacterial cell results in the production not only of more viral DNA, but also of the proteins of the virus coat. Clearly, the viral DNA carries the information for the synthesis of the coat proteins.

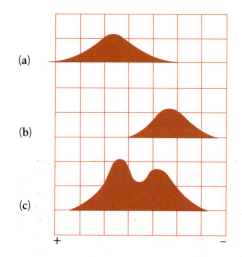

(a)

(b)

(c)

+   −

15–4 *Results from electrophoresis of (a) normal hemoglobin, (b) the hemoglobin of a person with sickle cell anemia, and (c) the hemoglobin of a person who is heterozygous for the sickle cell allele. The height of the curves indicates the amount of hemoglobin at each point. Because of slight differences in electric charge, normal and sickle cell hemoglobins move at different rates in an electric field. The normal hemoglobin is more negatively charged; hence it has moved closer to the positive pole than has the sickle cell hemoglobin. The hemoglobin from the heterozygote separates into the two different positions, indicating that it consists of both normal and sickle cell hemoglobins.*

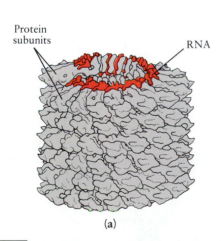

Deoxyribose   Ribose

(a)

Thymine   Uracil

(b)

15–5 *Chemically, RNA is very similar to DNA, but there are two differences in its nucleotides. (a) One difference is in the sugar component; instead of deoxyribose, RNA contains ribose, in which a hydroxyl group replaces a hydrogen on the 2' carbon. (b) The other difference is that instead of thymine, RNA contains the closely related pyrimidine uracil (U). Uracil, like thymine, pairs only with adenine.*

A third, and very important, difference between the two nucleic acids is that most RNA is single-stranded and does not form a regular helical structure.

FROM DNA TO PROTEIN: THE ROLE OF RNA

As a result of all of these studies, there was general agreement that the DNA molecule is a code that contains instructions for biological structure and function. Moreover, these instructions are carried out by proteins, which also contain a highly specific biological "language." As we saw in Chapter 3, the linear sequence of amino acids in a polypeptide chain determines the three-dimensional structure of the completed protein molecule, and it is the three-dimensional structure that determines function. The question thus became one of translation. How did the order of bases in DNA specify the sequence of amino acids in a protein molecule?

The search for the answer to this question led to ribonucleic acid (RNA), the close chemical relative of DNA that we have encountered previously. (The chemical differences between the two molecules are reviewed in Figure 15–5.) There were several clues that RNA might play a role in the translation of genetic information from DNA into a sequence of amino acids. First, there was some circumstantial evidence provided by eukaryotic cells. Unlike DNA, which is found primarily in the nucleus, RNA is found mostly in the cytoplasm, and it is there that most protein synthesis takes place. Embryologists noted that the cells of developing embryos of many different kinds contain high levels of RNA. Second, both prokaryotic and eukaryotic cells making large amounts of protein have numerous ribosomes. A rapidly growing *Escherichia coli* cell, for instance, contains about 15,000 ribosomes, constituting about one-half of the total mass of the cell. And ribosomes are two-thirds RNA and one-third protein.

Additional evidence came from experiments with viruses. When a bacterial cell is infected by a DNA-containing bacteriophage, RNA is synthesized from the viral DNA before viral protein synthesis begins. Also, some viruses contain no DNA—only RNA and protein; tobacco mosaic virus (TMV) is an example (Figure 15–6). When a tobacco leaf is infected with RNA purified from TMV, new viruses are produced, protein coats and all. In other words, RNA as well as DNA seemed to contain information about proteins.

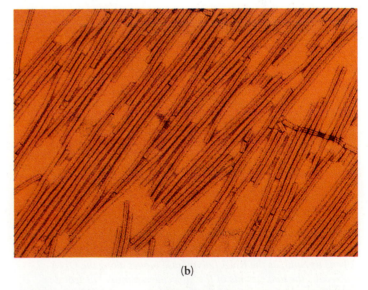

Protein subunits   RNA

(a)   (b)

15–6 *(a) Diagram of a very small portion of tobacco mosaic virus (TMV). This virus has a central core not of DNA but rather of RNA (ribonucleic acid). Its outer coat is composed of 2,150 identical pro-* *tein molecules, each consisting of 158 amino acids. If the RNA is separated from its protein coat and rubbed into scratches on a tobacco leaf, new TMV particles are formed, complete with new protein coats.* *Studies with TMV were among the first to suggest that RNA also could direct the assembly of proteins. (b) When viewed under the electron microscope, TMV particles appear as rod-shaped structures.*

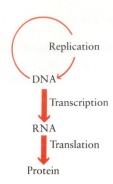

15-7 *The central dogma of molecular genetics: information flows from DNA to RNA to protein. Replication of the DNA occurs only once in each cell cycle, during the S phase prior to mitosis or meiosis. Transcription and translation, however, occur repeatedly throughout the interphase portions of the cell cycle.*

The Central Dogma

Based on early evidence, Crick proclaimed what he called the "central dogma," illustrated in Figure 15–7. DNA specifies RNA, which, in turn, specifies proteins. Note that the arrows go in only one direction. The genotype (DNA) determines the phenotype by dictating the composition of proteins. However, proteins do not alter the genotype—that is, proteins do not send instructions back to the DNA.

Crick called his proposal "dogma" because, at the time, there was little supporting evidence for it. A diversity of experiments have since, however, showed it to be true, almost without exception.*

The direction of information flow from DNA to RNA to proteins provided an important confirmation of Darwin's theory of evolution, in which natural selection acts on inherited variations—and a refutation of Lamarck's contention (page 4) that acquired characteristics could be inherited.

RNA as Messenger

As it turned out, not one but three kinds of RNA play roles as intermediaries in the steps that lead from DNA to protein. At this point in our story, we shall describe just one of them: **messenger RNA** (mRNA).

Messenger RNA molecules are copies (transcripts) of DNA sequences. Unlike DNA molecules, however, RNA molecules are usually single-stranded. Each new mRNA molecule is copied, or transcribed, from one of the two strands of DNA (the template strand) by the same base-pairing principle that governs DNA replication (Figure 15–8). Like a strand of DNA, each RNA molecule has a 5' end and a 3' end. As in the synthesis of DNA, the ribonucleotides, which are present in

---

* As we shall see in Chapter 17, the principal exception to the central dogma is a process known as reverse transcription, in which the information encoded by some RNA viruses is transcribed into DNA.

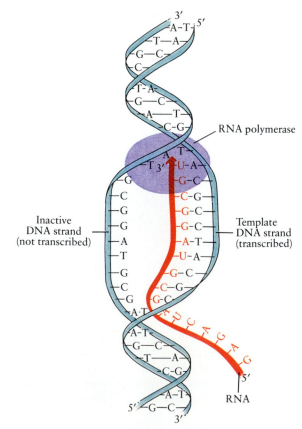

15-8 *A schematic representation of RNA transcription. At the point of attachment of the enzyme RNA polymerase, the DNA opens up, and, as the RNA polymerase moves along the DNA molecule, the two strands of the molecule separate. Nucleotide building blocks are assembled into RNA in a 5' to 3' direction as the enzyme reads the template DNA strand in a 3' to 5' direction. Note that the RNA strand is complementary—not identical—to the template strand from which it is transcribed; its sequence is, however, identical to that of the inactive (untranscribed) DNA strand, except for the replacement of thymine (T) by uracil (U).*

the cell as triphosphates, are added, one at a time, to the 3′ end of the growing RNA chain. The process, known as **transcription,** is catalyzed by the enzyme RNA polymerase. This enzyme operates in the same manner as DNA polymerase, moving in a 3′ to 5′ direction along the template DNA strand, synthesizing a new complementary strand of nucleotides—in this case, ribonucleotides—in a 5′ to 3′ direction. Thus, the mRNA strand is antiparallel to the DNA template strand from which it is transcribed.

Messenger RNA is the working copy of the genetic information. Incorporating the instructions encoded in DNA, mRNA dictates the sequence of amino acids in proteins.

THE GENETIC CODE

The identification of mRNA as the working copy of the genetic instructions still left the big question unresolved. Scientists from many disciplines were intrigued by the puzzle of the genetic code. One of these was George Gamow, the astronomer who was the father of the "big bang" theory of cosmology (page 23).

Proteins contain 20 different amino acids (Figure 3–18, page 73), but DNA and RNA each contain only four different nucleotides. As Gamow pointed out, if a single nucleotide "coded" for one amino acid, only four amino acids could be specified by the four bases. If two nucleotides specified one amino acid, there could be a maximum number, using all possible arrangements, of 4 × 4, or 16—still not quite enough to code for all 20 amino acids. Therefore, following the code analogy, at least three nucleotides in sequence must specify each amino acid. This would provide 4 × 4 × 4, or 64, possible combinations, or **codons**—clearly, more than enough.

The three-nucleotide, or triplet code, was widely adopted as a working hypothesis. Its existence, however, was not actually demonstrated until the code was finally broken, a decade after Watson and Crick first presented their DNA structure. The scientists who performed the initial, crucial experiments toward breaking the code were Marshall Nirenberg and his colleague Heinrich Matthaei, both of the National Institutes of Health.

Breaking the Code

Messenger RNA, then newly discovered, gave Nirenberg the tool he needed. He broke apart *E. coli* cells, extracted their contents, and added to them radioactively labeled amino acids and crude samples of RNA from a variety of cell sources. All of the RNA samples stimulated protein synthesis; the amounts of protein produced were small but measurable. In other words, the material extracted from the *E. coli* cells would start producing protein molecules even when the RNA "orders" it received were from a "complete stranger." Even the RNA from tobacco mosaic virus, which naturally multiplies only in cells of the leaves of tobacco plants, could be read as an mRNA by the machinery of the bacterial cell.

Nirenberg and Matthaei then tried an artificial RNA. Perhaps if the cell-free extracts could read a foreign message and translate it into protein, they could read a totally synthetic message, one dictated by the scientists themselves. Severo Ochoa of New York University had developed an enzymatic process for linking ribonucleotides into a long strand of RNA. With this process, carried out in a test tube, he had produced an RNA molecule that contained only one nitrogenous base, uracil, repeated over and over again. It was called "poly-U."

Nirenberg and Matthaei prepared 20 different test tubes, each of which contained cell-free extracts of *E. coli* that included ribosomes, ATP, the necessary enzymes, and all of the amino acids. In each test tube, one of the amino acids, and only one, carried a radioactive label. Synthetic poly-U was added to each test tube.

The Elusive Messenger

The cytoplasm of cells that are synthesizing proteins is full of RNA. This observation was the major clue that RNA plays a role in directing the assembly of proteins. Although the existence of RNA molecules that carry the genetic information from DNA to protein was hypothesized, confirmation of the hypothesis required the detection and isolation of the messenger molecules. The problem was complicated not by a shortage of RNA but by the fact that most of any cell's RNA is bound into ribosomes, and, as we shall see shortly, ribosomal RNA is not heterogeneous. It was therefore an unlikely candidate for a carrier of genetic information. This paradox puzzled molecular biologists for almost a decade.

Escherichia coli and its phages once again provided the tools of discovery. As you know, the genetic material of the bacteriophages is DNA, and their coats are made of protein. When these viruses infect a bacterial cell, new coat proteins are synthesized. If the messenger hypothesis were true, new RNA should be formed between the time of cell infection and the making of new virus particles. To test the hypothesis, *E. coli* cells were infected with phage and then exposed briefly to uracil labeled with radioactive carbon. Short-lived RNA molecules with radioactive labels were detected in the cells; they were associated with ribosomes but they were not a part of the ribosomes. Were they the long-sought messengers?

To answer this second question it was necessary to show that the radioactive RNA was complementary to the phage DNA. The method used was simple but ingenious. If dissolved DNA molecules are heated gently, the hydrogen bonds break apart and the two strands of the double helix separate. Subsequently, when the solution is slowly cooled, complementary strands pair up again and the hydrogen bonds reform. If the newly formed, radioactive RNA were complementary to the phage DNA, the investigators reasoned, it too should form a duplex with that DNA. As a control, the radioactive RNA molecules were first mixed in a beaker with a solution of *E. coli* DNA and heated; when this mixture was cooled, no duplexes containing radioactivity could be detected. No hybrid RNA-DNA molecules had formed. Then the radioactive RNA molecules were mixed with a solution of phage DNA; when this mixture was heated and then cooled, the results were clear-cut. Duplexes containing radioactivity had formed, indicating that the RNA had bound to its complementary strand of viral DNA. The messenger had been disclosed.

Hybridization of DNA with DNA and of RNA with DNA has since become an enormously powerful tool in both molecular genetics and evolutionary taxonomy. It is used in a great variety of studies, ranging from the detection of specific genes responsible for particular human diseases (page 397) to the resolution of evolutionary enigmas (page 422).

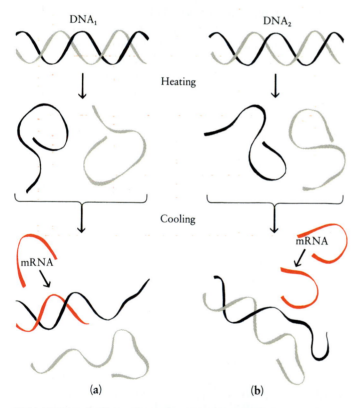

RNA-DNA hybrids can be used to show the complementarity in nucleotide sequence between an RNA molecule and the DNA molecule from which it was transcribed. (a) The formation of a hybrid molecule between a radioactive RNA molecule (color) and its template DNA strand (black). (b) If the same RNA molecule is mixed with unrelated DNA, no hybrid molecules containing radioactivity are formed.

In 19 of the test tubes, no radioactive polypeptides were produced, but in the twentieth tube, to which radioactive phenylalanine had been added, the investigators were able to detect newly formed, radioactive polypeptide chains. When the polypeptides were analyzed, they were found to consist only of phenylalanines, one after another. Nirenberg and Matthaei had dictated the message "uracil . . . uracil . . . uracil . . . uracil . . . uracil . . . uracil . . . ," and a clear answer had come back, "phenylalanine . . . phenylalanine. . . ." The experiment not only defined the first code word (UUU = phe) but also made available a method for defining the others.

As a result of further experiments (see essay on page 310), the mRNA codons for all of the amino acids were worked out (Figure 15–9). Of the 64 possible triplet combinations, 61 specify particular amino acids and three are termination codons. With 61 combinations coding for 20 amino acids, you can see that there must be more than one codon for many of the amino acids; thus the genetic code is said to be **degenerate**.*

15–9 *The genetic code, consisting of 64 triplet combinations (codons) and their corresponding amino acids. The codons shown here are the ones that would appear in the mRNA molecule. Of the 64 codons, 61 specify particular amino acids. The other three codons are stop signals, which cause the chain to terminate.*

Since 61 triplets code for 20 amino acids, there are "synonyms," as many as six different codons for leucine, for example. Most of the synonyms, as you can see, differ only in the third nucleotide. Each codon, however, specifies only one amino acid.

Second letter

First letter (5' end)

	U	C	A	G	
U	UUU UUC } phe UUA UUG } leu	UCU UCC UCA UCG } ser	UAU UAC } tyr UAA stop UAG stop	UGU UGC } cys UGA stop UGG trp	U C A G
C	CUU CUC CUA CUG } leu	CCU CCC CCA CCG } pro	CAU CAC } his CAA CAG } gln	CGU CGC CGA CGG } arg	U C A G
A	AUU AUC } ile AUA AUG met	ACU ACC ACA ACG } thr	AAU AAC } asn AAA AAG } lys	AGU AGC } ser AGA AGG } arg	U C A G
G	GUU GUC GUA GUG } val	GCU GCC GCA GCG } ala	GAU GAC } asp GAA GAG } glu	GGU GGC GGA GGG } gly	U C A G

Third letter (3' end)

PROTEIN SYNTHESIS

With a knowledge of the genetic code, we can turn our attention to the question of how the information encoded in the DNA and transcribed into mRNA is subsequently translated into a specific sequence of amino acids in a polypeptide chain. The answer to this question is now understood in great detail. The basic principles of protein synthesis are the same in both prokaryotic and eukaryotic cells, but there are some differences in detail, which will be described in subsequent chapters. Here we shall focus on the process as it takes place in prokaryotes, particularly *E. coli.*

Instructions for protein synthesis are encoded in sequences of nucleotides in the DNA molecule. Semiconservative replication of DNA transmits these instructions from parent cell to daughter cell and from generation to generation. Thus, each new cell and each new organism inherits the necessary information for synthesizing the specific proteins that determine its particular structure and functions.

---

* Degeneracy, in this context, does not imply a moral judgment. It is a term used by physicists to describe multiple states that amount to the same thing. The word persists in biology as a testimony to the roles of physicists Delbrück, Wilkins, Crick, Gamow, and others in the research that ultimately led to the breaking of the genetic code.

AGA-GAG-AGA

Among the many experiments that contributed to breaking the genetic code were those of H. G. Khorana at the University of Wisconsin. Khorana synthesized an artificial messenger in which two nucleotides were repeated over and over again in a known sequence: AGAGAGAGAG, UCUCUCUCUC, ACACACACAC, and UGUGUGUGUG. Each of these RNA chains, when used as a messenger in the cell-free system, produced polypeptide chains of alternating amino acids. Poly-AG produced arginine and glutamic acid over and over again; poly-UC, serine and leucine; poly-AC, threonine and histidine; and poly-UG, cysteine and valine. This is, of course, what you would expect from a triplet code. A poly-AG message would be read AGA. . .GAG. . .AGA. . ., for instance.

Khorana also synthesized artificial messengers in which three nucleotides were repeated over and over again. These messengers could produce three different polypeptides, each consisting of only one amino acid, repeated over and over. Which polypeptide was produced depended on where the reading process began.

These studies provided the first clear demonstration (1) that mRNA is read sequentially (that is, one

In Khorana's experiments, an artificial mRNA in which two nucleotides alternated over and over produced a polypeptide chain of alternating amino acids. An artificial mRNA with three different nucleotides produced three different polypeptides, each consisting of only one type of amino acid.

codon after another); (2) that how it is read depends on the reading frame—that is, the nucleotide at which translation starts; and (3) that the codon consists of an uneven number of nucleotides, lending support to the triplet hypothesis.

As we have seen, these instructions are transcribed—that is, copied—into an mRNA molecule following the same base-pairing rules that govern DNA replication; the only difference is that in mRNA uracil substitutes for thymine. Specific nucleotide sequences of the DNA, called **promoters,** are the start signals for RNA synthesis, and others, called **terminators,** are the stop signals for RNA synthesis. The RNA is transcribed in only one direction—5' to 3'—and along only one strand of the DNA duplex. The mRNA molecules are long—500 to 10,000 nucleotides—and single-stranded. These molecules, as we have noted previously, are the working copies used in protein synthesis to determine the amino acid sequences.

The synthesis of proteins requires, in addition to mRNA molecules, two other types of RNA: **ribosomal RNA** (rRNA) and **transfer RNA** (tRNA). These molecules differ both structurally and functionally from mRNA. In most cells, ribosomal RNA is by far the most abundant type, a fact that hampered the search for mRNA, which typically has only a very transitory existence in an *E. coli* cell. Ribosomes consist of two subunits (Figure 15–10) and are, by weight, about two-thirds RNA and one-third protein. In the ribosomes of *E. coli*, the smaller (30S) subunit has one type of rRNA, 1,542 nucleotides in length (commonly referred to as 16S rRNA), and a single molecule each of 21 different proteins. The larger (50S) subunit has two types of rRNA, one consisting of 120 nucleotides (5S rRNA) and the other of 2,904 nucleotides (23S rRNA), and 34 different proteins.

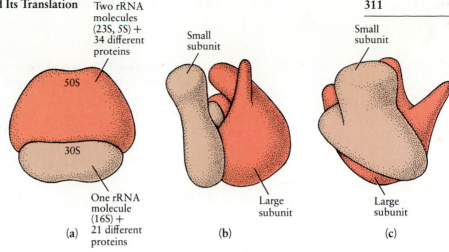

15–10 *Ribosomes, in both prokaryotes and eukaryotes, consist of two subunits, one large and one small. Each subunit is composed of specific rRNA and protein molecules. Studies of rates of sedimentation in the ultracentrifuge have revealed that the size and density of both the subunits and the whole ribosome differ in prokaryotes and eukaryotes. (a) The subunits of the E. coli ribosome have sedimentation values of 50S and 30S; these values, which are a function of both molecular weight and the shape of the molecule, are not additive. The prokaryotic 30S and 50S ribosomal subunits combine to form a ribosome that has a sedimentation value of only 70S.*

(b, c) Two views of the three-dimensional structure of the E. coli ribosome, as revealed by electron micrographs.

The smaller subunit has a binding site for the mRNA molecule; in *E. coli* and other prokaryotes, the leading (5′) end of the mRNA molecule attaches to this binding site even as the rest of the molecule is still being transcribed. The larger subunit has two binding sites for the third type of RNA, transfer RNA.

Transfer RNA molecules are, in effect, the dictionary by which the language of nucleic acids is translated into the language of proteins. These molecules are comparatively small, ranging from 75 to 85 nucleotides in length. There are more than 20 different kinds in every cell, at least one for each of the kinds of amino acids found in proteins. Each tRNA molecule has two important attachment sites. One such site, known as the **anticodon,** binds to the codon on the mRNA molecule. The other, on the 3′ end of the tRNA molecule, attaches to a particular amino acid. Thus, tRNA molecules provide the crucial link between nucleic acids and proteins, the two languages of the living cell.

All tRNA molecules have approximately the same cloverleaf shape shown in Figure 15–11. The 3′ end of the molecule—the one that attaches to the amino acid—always terminates in a (5′)-CCA-(3′) sequence. The sequence of the other nucleotides, however, varies according to the particular type of tRNA. Attachment of tRNA molecules to their amino acids is brought about by a group of enzymes known as aminoacyl-tRNA synthetases. There are at least 20 different aminoacyl-tRNA synthetases, one or more for each amino acid. Each of these enzymes has a binding site for a particular amino acid and for its matching tRNA molecule.

15–11 *(a) The structure of a tRNA molecule. Such molecules consist of about 80 nucleotides linked together in a single chain. The chain always terminates in a (5′)-CCA-(3′) sequence. An amino acid can link to its specific tRNA at this end. Some nucleotides are the same in all tRNAs; these are shown in gray. The other nucleotides vary according to the particular tRNA. The symbols D, γ, ψ, and T represent unusual modified nucleotides characteristic of tRNA molecules.*

Some of the nucleotides are hydrogen-bonded to one another, as indicated by the dashed lines. In some regions, the unpaired nucleotides form loops. The loop on the right in this diagram, known as the TψC loop, is thought to play a role in binding the tRNA molecule to the surface of the ribosome. Three of the unpaired nucleotides in the loop at the bottom of the diagram (indicated in color) form the anticodon. They serve to "plug in" the tRNA molecule to an mRNA codon.

(b) The molecule folds over on itself, producing this three-dimensional structure. This is a photograph of a model based on x-ray analysis.

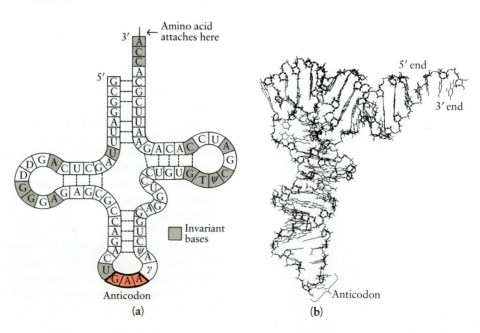

The enzymatic reaction linking an amino acid to its tRNA molecule takes place in two steps. In the first, which supplies the energy required for the reaction, an ATP molecule is cleaved, two phosphates are released, and a complex is formed that consists of an amino acid, a molecule of AMP (adenosine monophosphate), and the enzyme. This amino acid–AMP–enzyme complex remains intact until it encounters the appropriate tRNA molecule (out of the 20 or more different types in the cell). Work is presently underway to identify the portion of the tRNA molecule that is recognized by the enzyme and to determine the means by which recognition occurs.

The second step of the reaction takes place when the amino acid–AMP–enzyme complex meets its tRNA molecule. The AMP molecule is released from the enzyme, a bond is formed between the amino acid and the 3′ end of the tRNA molecule, and then the amino acid–tRNA complex is also released. Subsequently, when the tRNA molecule is hydrogen-bonded to the mRNA molecule, anticodon to codon, it thus brings the specified amino acid into place. Only then is the bond broken between the tRNA and the amino acid, as a new bond is formed—a peptide bond that links the newly arrived amino acid to the growing polypeptide chain. Simultaneously, the tRNA molecule is released, once again free to pick up another molecule of its amino acid and to repeat the cycle.

Translation

The synthesis of proteins is known as **translation,** since it is the transfer of information from one language (nucleotides) to another (amino acids). It takes place in three stages: initiation, elongation, and termination (Figure 15–12).

The first stage, **initiation,** begins when the smaller ribosomal subunit attaches to a strand of mRNA near its 5′ end, exposing its first, or initiator, codon. (As we noted earlier, in *E. coli* the 3′ end of the mRNA is still bound to the DNA helix, with transcription continuing even as translation begins at the 5′ end.) Next, the first tRNA comes into place to pair with the initiator codon of mRNA. This initiator codon, which is usually (5′)-AUG-(3′), pairs in an antiparallel fashion with the tRNA anticodon (3′)-UAC-(5′). The incoming initiator tRNA, which binds to the AUG codon, carries a modified form of the amino acid methionine, *N*-formylmethionine, or fMet, as its amino acid (Figure 15–13). This fMet will be the first amino acid in the newly synthesized polypeptide chain, but it may later be removed. The combination of the small ribosomal subunit, mRNA, and the initiator tRNA is known as the **initiation complex.** The larger ribosomal subunit then attaches to the smaller subunit, and the initiator tRNA becomes locked into the P (peptide) site of the larger subunit—one of two sites for binding tRNA molecules. The energy for this step is provided by the hydrolysis of guanosine triphosphate (GTP).

At the beginning of the **elongation** stage, the second codon of the mRNA is positioned opposite the A (aminoacyl) site of the large subunit. A tRNA with an anticodon complementary to the second mRNA codon plugs into the mRNA molecule and, with its amino acid, occupies the A site of the ribosome. When both the P and A sites are occupied, an enzyme, peptidyl transferase, which is part of the larger subunit of the ribosome, forges a peptide bond between the two amino acids, attaching the first amino acid (fMet) to the second. The first tRNA is released. The ribosome moves one codon down the mRNA chain; consequently, the second tRNA, to which is now attached fMet and the second amino acid, is transferred from the A to the P position. A third tRNA–amino acid moves into the A position opposite the third codon on the mRNA, and the step is repeated. The P position accepts the tRNA bearing the growing polypeptide chain; the A position accepts the tRNA bearing the new amino acid that will be added to the chain. As the ribosome moves along the mRNA chain, the initiator portion of the

15–12 *Three stages in protein synthesis.* (a) *Initiation. The smaller ribosomal subunit attaches to the 5' end of the mRNA molecule. The first tRNA molecule, bearing the modified amino acid fMet, plugs into the AUG initiator codon on the mRNA molecule. The larger ribosomal subunit locks into place, with the tRNA occupying the P (peptide) site. The A (aminoacyl) site is vacant. The initiation complex is now complete.*

(b) *Elongation. A second tRNA with its attached amino acid moves into the A site, and its anticodon plugs into the mRNA. A peptide bond is formed between the two amino acids brought together at the ribosome. At the same time, the bond between the first amino acid and its tRNA is broken. The ribosome moves along the mRNA chain in a 5' to 3' direction, and the second tRNA, with the dipeptide attached, is moved to the P site from the A site as the first tRNA is released from the ribosome. A third tRNA moves into the A site, and another peptide bond is formed. The growing peptide chain is always attached to the tRNA that is moving from the A site to the P site, and the incoming tRNA bearing the next amino acid always occupies the A site. This step is repeated over and over until the polypeptide is complete.*

(c) *Termination. When the ribosome reaches a termination codon (in this example, UGA), the polypeptide is cleaved from the last tRNA and the tRNA is released from the P site. The A site is occupied by a release factor that triggers the dissociation of the two subunits of the ribosome.*

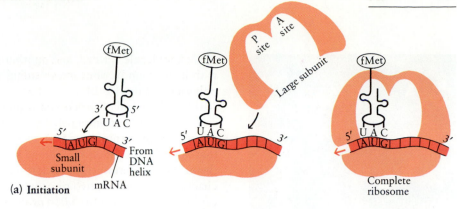

(a) **Initiation**

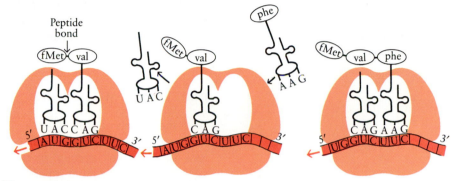

(b) **Elongation**

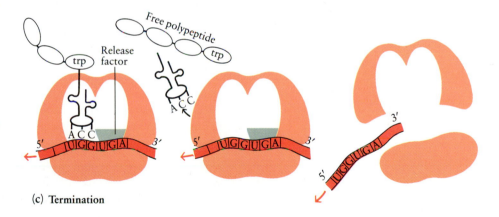

(c) **Termination**

15–13 *The structures of methionine and N-formylmethionine (fMet), the modified amino acid carried by the initiator tRNA in prokaryotes. Note that the attachment of the formyl group (shown in color) prevents a peptide bond from forming on the "wrong" (amino) end of the amino acid and ensures that its peptide bond with the second amino acid in the sequence will occur at the carboxyl end. The mRNA codon for both methionine and fMet is (5')-AUG-(3'). It is not known how AUG can code for fMet in the initiator position and methionine elsewhere in the polypeptide chain. Two different tRNAs are involved, but the anticodons are the same.*

0.1 µm

15–14 *Clusters of ribosomes reading the same mRNA strand. Such groups are called polyribosomes, or polysomes.*

mRNA molecule is freed, and another ribosome can form an initiation complex with it. A group of ribosomes reading the same mRNA molecule is known as a **polysome** (Figure 15–14).

Toward the end of the coding sequence of the mRNA molecule is a codon that serves as a **termination** signal. Three termination codons are known (UAG, UAA, and UGA), and often more than one is present. No tRNAs exist with anticodons that "match" these codons, and so no tRNAs will enter the A site in response to them. When a termination codon is reached, translation stops, the polypeptide chain is freed, and the two ribosomal subunits separate. It is estimated that *E. coli* can synthesize as many as 3,000 proteins, each different and each assembled in this same way.

The elucidation of the details of this precise and elegant process of translation was an awe-inspiring achievement. Even more awe-inspiring is the knowledge that at this very moment a similar process is taking place in virtually every cell of our own bodies.

REDEFINING MUTATIONS

With the process of protein synthesis in mind, let us consider some of the broader implications of the genetic code and its translation. For example, take another look at sickle cell anemia, in the light of Figure 15–9. Normal hemoglobin contains glutamic acid at a particular position; sickle cell hemoglobin contains valine in the same position. The difference between the mRNA codons for glutamic acid and valine is a single nucleotide. In mRNA, GAA or GAG specifies glutamic acid (glu), and GUU, GUC, GUA, or GUG specifies valine (val). So the difference between the two is the replacement of one adenine by one uracil. In the template strand of DNA from which the mRNA was transcribed, the difference is the replacement of one thymine by one adenine in a sequence of nucleotides that, since it dictates a polypeptide that contains more than 150 amino acids, must contain more than 450 nucleotides. In other words, the tremendous functional difference—literally a matter of life and death—can be traced to a single "misprint" in over 450 nucleotides.

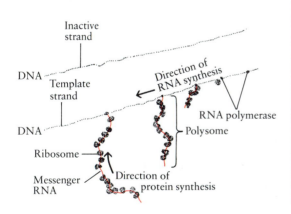

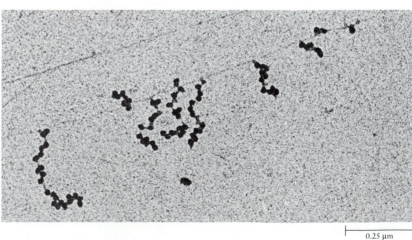

0.25 µm

15–15 *A bacterial gene in action. In the micrograph, you can see several different mRNA strands (shown in color in the diagram) being transcribed simultaneously from the same DNA template. The longest one, at the left, was the first one synthe-sized. As each mRNA strand peels off the DNA molecule, ribosomes attach to the mRNA, translating its encoded information into protein. You can also see molecules of RNA polymerase, the enzyme that catalyzes the transcription of RNA from DNA. The RNA polymerase molecule at the far right is approximately at the point where transcription begins. The protein molecules are not visible in this micrograph.*

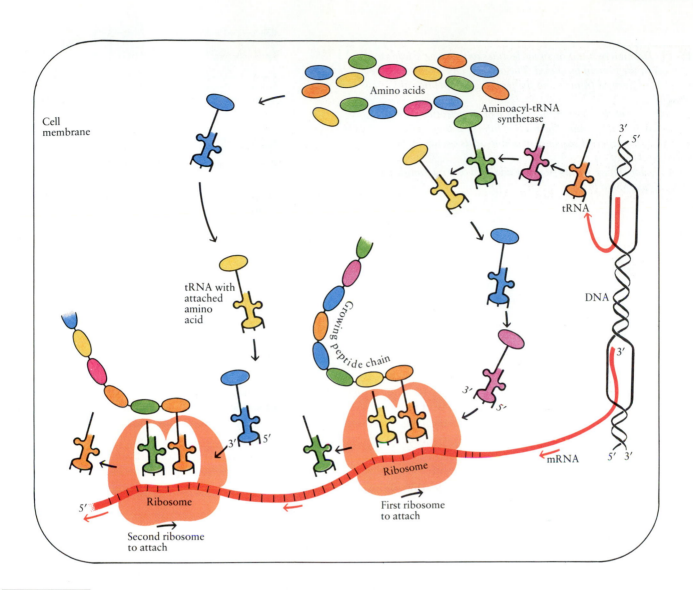

15-16 *A summary of protein synthesis in a bacterial cell. At least 32 different kinds of tRNA molecules are transcribed from the DNA of the bacterial cell. These molecules are so structured that each can be attached at one end (by an aminoacyl-tRNA synthetase) to a specific amino acid. Each contains an anticodon that is complementary to an mRNA codon for that particular amino acid.*

The process begins when an mRNA strand is transcribed from a DNA template. At the point of attachment to the ribosome, the matching tRNA molecule, with its amino acid, plugs in momentarily to the codon in the mRNA. As the ribosome moves along the mRNA strand, a tRNA linked to its particular amino acid fits into place and the first tRNA molecule

is released, leaving behind its amino acid, now enzymatically linked to the second amino acid by a peptide bond. As the process continues, the amino acids are brought into line one by one, following the exact order originally dictated by the DNA from which the mRNA was transcribed.

De Vries, almost 90 years ago, defined mutation in terms of characteristics appearing in the phenotype. In the light of current knowledge, the definition is somewhat different: A mutation is a change in the sequence or number of nucleotides in the nucleic acid of a cell. Mutations that occur in gametes or the cells that give rise to gametes are transmitted to future generations. Mutations that occur in somatic cells are transmitted to the daughter cells produced by mitosis and cytokinesis.

15–17 *Deletion or addition of nucleotides within a gene leads to changes in the protein produced. The original DNA molecule, the mRNA transcribed from it, and the resulting polypeptide are shown in (a).*

In (b) we see the effect of the deletion of a nucleotide pair (T-A), as indicated by the arrow. The reading frame for the gene is altered, and a different sequence of amino acids occurs in the polypeptide.

A similar change results from the addition of a nucleotide pair (brown), as shown in (c).

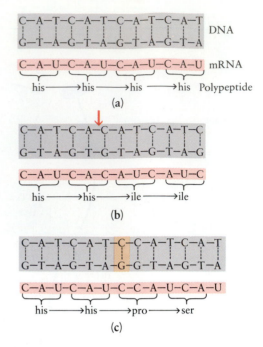

Most mutations involve only a single nucleotide substitution and are called **point mutations.** As in the case of sickle cell anemia, such a substitution can lead to changes in the protein produced by a gene. Other changes in the amino acid sequence of a protein can result from deletion or addition of nucleotides within a gene (Figure 15–17). When this occurs, the reading frame of the gene may shift—that is, the way in which the nucleotides are grouped into triplets changes resulting in the production of an entirely new protein. These **frame-shifts,** as they are known, almost invariably lead to "bad" proteins.

UNIVERSALITY OF THE GENETIC CODE

In the decades since the genetic code was broken, the DNA and proteins of many more organisms have been examined. The evidence is now overwhelming: for virtually all organisms, from *E. coli* to *Homo sapiens,* the genetic code is universal.* It evolved early, remained constant, and determines the underlying unity of all living things.

SUMMARY

Genetic information is coded in the sequence of nucleotides in molecules of DNA, and these, in turn, determine the sequence of amino acids in molecules of protein. That genes exert their effects by influencing the production of specific protein molecules was demonstrated by the experiments of Beadle and Tatum with *Neurospora* mutants. As a result of their experiments, they formulated the principle of "one gene–one enzyme," subsequently amended to "one gene–one polypeptide." Studies of normal and sickle cell hemoglobin molecules demonstrated that a change in a single amino acid of a polypeptide chain can cause a dramatic change in the function of the resulting protein.

---

* The only apparent exceptions to this statement, as we shall see in Chapter 18, involve mitochondria. These organelles contain their own DNA, transcribe their own mRNA, rRNA, and tRNA molecules, and carry out some protein synthesis. In several instances, the mitochondrial code differs from that carried in the chromosomes of both prokaryotes and eukaryotes.

The way in which DNA is translated into protein has been worked out in considerable detail. The information is transcribed from one strand of the DNA (the template strand) into a long, single strand of RNA (ribonucleic acid). This type of RNA molecule is known as messenger RNA, or mRNA. An enzyme, RNA polymerase, catalyzes the transcription process. The mRNA is synthesized in the 5′ to 3′ direction, following the principles of base pairing first suggested by Watson and Crick. It is therefore complementary to the template DNA strand. Each group of three nucleotides in the mRNA molecule is the codon for a particular amino acid.

The genetic code has been deciphered, that is, it is now known which amino acid is called for by a given mRNA codon. Of the 64 possible triplet combinations of the four-letter nucleotide code, 61 combinations have been identified with one of the 20 amino acids that make up protein molecules. The other three triplets serve as "stop signals," terminating protein synthesis. The code is universal.

Protein synthesis—translation—takes place at the ribosomes. A ribosome is formed from two subunits, one large and one small, each consisting of characteristic ribosomal RNAs (rRNAs) complexed with specific proteins. Also required for protein synthesis are another group of RNA molecules, known as transfer RNA (tRNA), each of which is folded into a cloverleaf configuration. These small molecules can carry an amino acid on one end and have a triplet of bases, the anticodon, on a central loop at the opposite end of the molecule. The tRNA molecule is the adapter that pairs the correct amino acid with each mRNA codon during protein synthesis. There is at least one kind of tRNA molecule for each kind of amino acid found in cells. Enzymes known as aminoacyl-tRNA synthetases catalyze the binding of each amino acid to its specific tRNA molecule.

In *E. coli* and other prokaryotes, even as the 3′ end of an mRNA strand is being transcribed, ribosomes are attaching near its 5′ end. At the point where the strand of mRNA is in contact with a ribosome, tRNAs are bound temporarily to the mRNA strand. This bonding takes place by complementary base pairing between the mRNA codon and the tRNA anticodon. Each tRNA molecule carries the specific amino acid called for by the mRNA codon to which the tRNA attaches. Thus, following the sequence originally dictated by the DNA, the amino acid units are brought into line one by one and, as peptide bonds form between them, are linked into a polypeptide chain.

Mutations are now defined as changes in the sequence or number of nucleotides in the nucleic acid of a cell or organism. Point mutations may take the form of substitutions of one nucleotide for another, or deletions or additions of nucleotides.

QUESTIONS

1. Distinguish among the following: mRNA/tRNA/rRNA; code/codon/anticodon; transcription/translation; P site/A site; initiation/elongation/termination.

2. A person heterozygous for the allele for sickle cell anemia makes the variant hemoglobin molecules but does not suffer from anemia. In what respect is this situation different from the definitions of "dominant" and "recessive" given in Chapter 11?

3. Define the "central dogma" and describe its importance in evolutionary theory.

4. Most of the bacterial DNA codes for mRNA, and most of the RNA produced by the cell is mRNA. Yet analysis of the RNA content of a cell reveals that, typically, rRNA is about 80 percent of the cellular RNA and tRNA makes up most of the rest. Only about 2 percent is normally mRNA. How do you explain these findings? What do you think the functional explanation might be?

5. Even before the genetic code was deciphered, it was believed to be degenerate. Explain why.

6. Transcription and translation in prokaryotes are "linked" pro-

cesses but, as we shall see in Chapter 18, this is not the case in eukaryotes. On the basis of your knowledge of cell structure, propose a likely explanation.

7. Given the details of protein synthesis, what further amendment would you make to the principle "one gene–one polypeptide"?

8. In a hypothetical segment of one strand of a DNA molecule, the sequence of bases is (3′)-AAGTTTGGTTACTTG-(5′). What would be the sequence of bases in an mRNA strand transcribed from this DNA segment? What would be the sequence of amino acids coded by the mRNA? Does it matter at what point on the template strand the transcription from DNA to mRNA begins? Explain your answer.

9. Suppose you have the peptide arg-lys-pro-met, and you know that the tRNA molecules used in its synthesis had the following anticodons:

    (3′)-GGU-(5′)
    (3′)-GCU-(5′)
    (3′)-UUU-(5′)
    (3′)-UAC-(5′)

Determine the DNA nucleotide sequence for the template strand of the gene that codes for this peptide.

10. Deletion or addition of nucleotides within a gene leads to

changes in the protein produced. The original DNA molecule, the mRNA transcribed from it and the resulting polypeptide are:

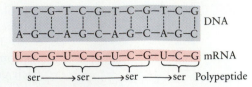

Deleting the second T-A pair yields the following DNA molecule:

How is the resulting amino acid sequence altered?
How does the addition of a C-G pair to the original molecule,

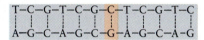

affect the amino acid sequence?

CHAPTER 16

The Molecular Genetics of Prokaryotes and Viruses

As we have seen in the two previous chapters, many of the important early advances in molecular genetics resulted from experiments with prokaryotes and viruses. Investigations with pneumococci, *Escherichia coli*, bacteriophages, and TMV contributed to the identification of DNA as the genetic material, to the breaking of the genetic code, and to the elucidation of both the principles and the details of transcription and translation. This work, however, is not just of historical interest. Studies with these and other viruses and bacteria have laid the foundation for much of the current work in genetics.

Biologists are now able to manipulate genes in ways never before imagined—modifying and recombining portions of DNA molecules from different sources and inserting these altered molecules into other cells where they are expressed. This new technology, known as **recombinant DNA,** has generated an avalanche of studies and information, some of which will be discussed in the following chapters. In order to understand this rapidly advancing frontier of genetics, however, we must look more closely both at the ways in which bacteria and viruses modify, recombine, and exchange genetic material entirely on their own, without any human intervention whatsoever, and at the ways in which they regulate the expression of their genes.

THE *E. COLI* CHROMOSOME

The chromosome of *E. coli* is a single, continuous (circular) thread of double-stranded DNA, approximately 1 millimeter long when fully extended but only 2 nanometers in diameter (Figure 16–1). It contains some 4.7 million base pairs. The

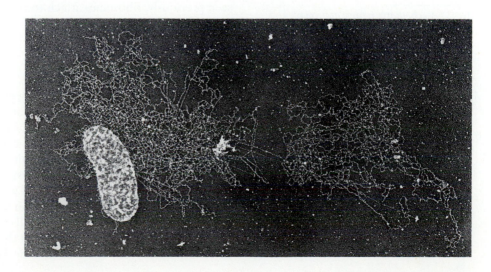

16–1 *An* E. coli *cell that has been lysed (broken apart), releasing its circular chromosome. Even when freed from the cell, the chromosome appears as a highly folded structure, composed of many loops. In the intact cell, the chromosome, which is some 500 times longer than the cell itself, is tightly packed into the nucleoid region.*

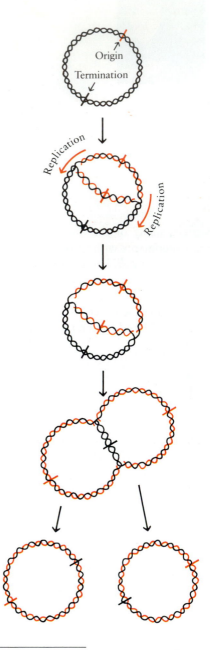

16-2 *A schematic representation of bidirectional (theta) replication of the DNA of the prokaryotic chromosome. The points of origin and termination of replication are indicated in red and black, respectively, and the newly synthesized DNA strand is shown in red. The replication forks are moving away from the origin in opposite directions.*

bacterial cell itself is less than 2 micrometers long, about 1/500th the extended length of its chromosome. Within the cell, the chromosome is compacted into an irregularly shaped body known as the nucleoid (see Figure 4–8, page 91).

As we saw in Chapter 14, bidirectional replication of the chromosome in *E. coli* and other bacteria begins at one specific nucleotide sequence known as the origin of replication. As the two replication forks move away from the origin in opposite directions, DNA polymerase adds nucleotides, one by one, to the 3′ ends of both the leading strands and the Okazaki fragments of the lagging strands. When the circular bacterial chromosome is replicating, it forms a structure resembling the Greek letter θ (theta); hence, its replication is known as **theta replication** (Figure 16–2).

TRANSCRIPTION AND ITS REGULATION

Transcription in *E. coli* and other prokaryotes takes place, as we saw in the last chapter, by the synthesis of a molecule of mRNA along a template strand of DNA. The process begins when the enzyme RNA polymerase attaches to the DNA at a specific site known as the promoter. The RNA polymerase molecule binds tightly to the promoter and causes the DNA double helix to open, initiating transcription. The growing RNA strand remains hydrogen-bonded to the DNA template briefly—only about 10 or 12 ribonucleotides are bonded to the DNA at any one time—and then it peels off in a single strand.

A segment of DNA that codes for a polypeptide (a protein) is known as a **structural gene.** Often structural genes coding for polypeptides with related functions occur together in sequence on the bacterial chromosome. Such functional groups might include, for instance, two polypeptide chains that together constitute a particular enzyme or three enzymes that work in a single enzymatic pathway. Groups of genes coding for such molecules are typically transcribed into a single mRNA strand. Thus, a group of polypeptides that are needed by the cell at the same time can be synthesized simultaneously, a simple and efficient inventory-control system.

The newly synthesized mRNA molecule (Figure 16–3) has a short "leader" sequence at its 5′ end, part of which may assist in binding mRNA to the ribosome. The coding region of the molecule is a linear sequence of nucleotides that precisely dictates the linear sequence of amino acids in particular polypeptide chains. There may be several stop and start codons within the mRNA molecule, marking the end of one structural gene and the beginning of the next. An additional nucleotide sequence at the 3′ end is known as a "trailer." As we have seen previously, ribosomes attach to the mRNA molecule even before transcription is complete, initiating the sequence of events shown in Figure 15–12 (page 313).

The Need for Regulation

In the course of their long evolutionary history, *E. coli* and other prokaryotes have evolved ways to maximize their utilization of nutrients for cellular growth. If a bacterial cell could be said to have a purpose or function, it would be to grow and multiply as rapidly as possible. And, bacteria are excellent at achieving this; a culture of *E. coli* cells, for example, can double in number every 20 minutes.

One reason for *E. coli*'s effectiveness in using nutrients is its versatility; it can make at least 1,700 enzymes and other proteins, enabling it to utilize a wide range of potential nutrients. A second reason is that the cell is highly efficient in its synthetic activities. It does not make all of its possible proteins all of the time, but only when they are needed and only in the amounts needed. For example, cells of *E. coli* supplied with the disaccharide lactose as a carbon and energy source

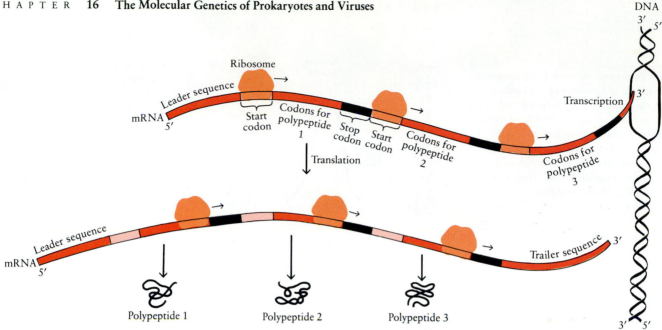

16-3 *In prokaryotes, transcription often results in an mRNA molecule that contains coding sequences for several different polypeptide chains, with the sequences separated by stop and start codons. In this* *diagram, the stop and start codons are adjacent, but sometimes they are separated by as many as 100 to 200 nucleotides. The 5' end of the mRNA molecule has a short leader sequence, and the 3' end* *has a trailer sequence; neither of these sequences codes for protein. Translation generally begins at the leading end of the mRNA while the rest of the molecule is still being transcribed.*

require the enzyme beta-galactosidase to split the disaccharide (Figure 16–4). Cells growing on lactose have approximately 3,000 molecules of beta-galactosidase per cell. In the absence of lactose, there is an average of one molecule of the enzyme per cell. In short, the presence of lactose induces the production of the enzyme molecules needed to break it down (Figure 16–5a). Such enzymes are said to be **inducible**.

Conversely, the presence of a particular nutrient may inhibit the transcription of a group of structural genes. *E. coli*, like other bacteria, can synthesize each of its amino acids from ammonia and a carbon source. The structural genes for the enzymes needed for the biosynthesis of the amino acid tryptophan, for instance,

16-4 *Lactose (milk sugar) is an important energy source for* E. coli. *The splitting of lactose into galactose and glucose requires the enzyme beta-galactosidase. Normal* E. coli *cells synthesize beta-galactosidase only when lactose is present in the medium in which they are growing.*

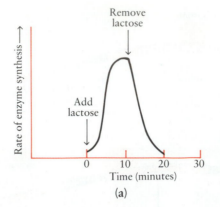
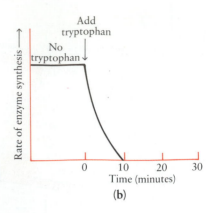

(a)   (b)

16–5 *Inducible and repressible enzymes.* (**a**) *The rate of synthesis of beta-galactosidase, an inducible enzyme produced by E. coli, increases dramatically when lactose is added to the surrounding growth medium. As long as lactose is abundant in the medium, enzyme production continues at its maximum rate. However, when lactose is removed from the medium, the rate of synthesis of beta-galactosidase immediately plummets.* (**b**) *In the absence of an essential substance, such as the amino acid tryptophan, the enzymes required for its production are synthesized at a maximum rate. If, however, tryptophan is added to the medium, synthesis of these enzymes is rapidly repressed.*

are grouped together and are transcribed into a single mRNA molecule. This mRNA is produced continuously in growing cells—unless tryptophan is present. In the presence of tryptophan, production of the enzymes ceases (Figure 16–5b). Such enzymes, the synthesis of which is reduced by the presence of the products of the reactions they catalyze, are said to be **repressible**.

Mutants of *E. coli* sometimes occur that are unable to regulate enzyme production. These cells produce beta-galactosidase even in the absence of lactose, for example, or the enzymes that synthesize tryptophan even when tryptophan is present. These and similar mutants are generally at a disadvantage because they are squandering their energies and resources. Normal *E. coli* cells rapidly outmultiply them.

Although regulation of protein synthesis could theoretically take place at many points in the biosynthetic process, in prokaryotes it occurs mostly at the level of transcription. Regulation involves interactions between the chemical environment of the cell and special regulatory proteins, coded by regulatory genes. These proteins can work either as negative controls, repressing mRNA transcription, or as positive controls, enhancing transcription. The fact that mRNA is translated into protein so immediately (before transcription is even completed) and broken down so rapidly further increases the efficiency of this strategy of regulation.

The Operon

It was the detection of the mutants described above that led to our current understanding of the regulation of transcription in prokaryotes. This understand-

16–6 *François Jacob, André Lwoff, and Jacques Monod in Paris in 1966, shortly after they shared the Nobel Prize for their discoveries concerning the genetics of prokaryotes. Jacob and Monod were honored for their work on the operon model of genetic regulation, and Lwoff for his work on a phenomenon known as lysogeny, which we shall consider later in this chapter. Most of the research of these three scientists was carried out at the Pasteur Institute, beginning at the time of World War II. During that war Jacob served in the French army, while Lwoff and, most notably, Monod were active in the French Resistance.*

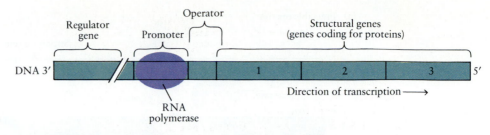

16–7 *A schematic representation of an operon. An operon consists of a promoter, an operator, and structural genes (that is, genes that code for proteins, often enzymes that work sequentially in a particular reaction pathway). The promoter, which precedes the operator, is the binding site for RNA polymerase. The operator is the site at which a repressor protein can bind; it may overlap the promoter, the first structural gene (as shown here), or both. Another gene involved in operon function is the regulator, which codes for the repressor. Although the regulator may be adjacent to the operon, in most cases it is located elsewhere on the bacterial chromosome.*

ing rests upon a model, known as the **operon** model, proposed some years ago by the French scientists François Jacob and Jacques Monod, who shared the Nobel Prize in 1965 with their colleague André Lwoff. According to the model formulated by Jacob and Monod, groups of genes coding for proteins with related functions are arranged in units known as operons. An operon (Figure 16–7) comprises the promoter, the structural genes, and another DNA sequence known as the **operator.** The operator is a sequence of nucleotides located between the promoter and the structural gene or genes; the operator may overlap the promoter, the adjacent structural gene, or both.

Transcription of the structural genes often depends on the activity of still another gene, the **regulator,** which may be located anywhere on the bacterial chromosome. This gene codes for a protein called the **repressor,** which binds to the operator. When a repressor is bound to the operator, it obstructs the promoter. As a consequence, RNA polymerase either cannot bind to the DNA molecule or, if bound, cannot begin its movement along the molecule. The result in either case is the same: no mRNA transcription occurs. However, when the repressor is removed, transcription may begin. Evidence for the existence of the regulator gene was derived from studies of *E. coli* cells that could not stop making beta-galactosidase. In these cells, a mutation in the regulator gene for the lactose *(lac)* operon provided the essential clue that such a gene existed in normal cells.

The capacity of the repressor to bind to the operator and thus to block protein synthesis depends, in turn, on another molecule that functions as an effector. Depending on the operon, an effector can either activate or inactivate the repressor for that particular operon (Figure 16–8). For example, when lactose is present in the growth medium, the first step in its metabolism produces a closely related sugar, allolactose, that binds to and inactivates the repressor, removing it

16–8 *In an operon system, the synthesis of proteins is regulated by interactions involving either a repressor and an inducer or a repressor and a corepressor. (a) In inducible systems, such as the lac operon, the repressor molecule is active until it combines with the inducer (in this case, allolactose). (b) In repressible systems, such as the trp operon, the repressor is not active until it combines with the corepressor.*

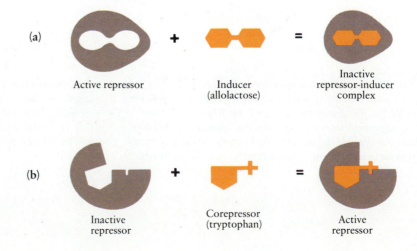

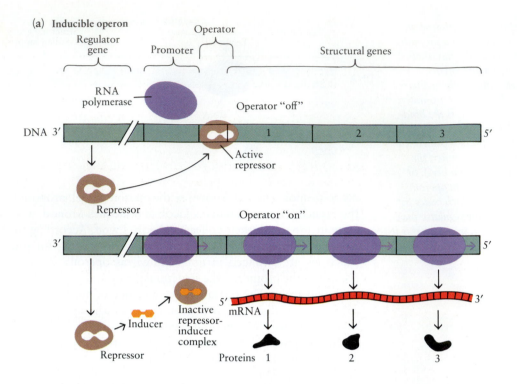

(a) **Inducible operon**

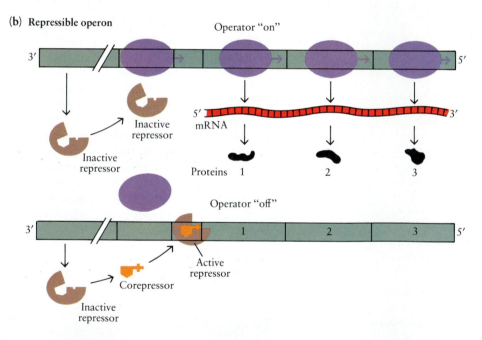

(b) **Repressible operon**

16–9 *Inducible and repressible operons are both turned off by repressor proteins that are coded by regulator genes. The repressor binds to DNA at the operator and so prevents RNA polymerase from initiating transcription. (a) In inducible operons, the inducer counteracts the effect of the repressor by binding with it and maintaining it in an inactive form. Thus, when the inducer is present,* the repressor can no longer attach to the operator, permitting transcription and translation to proceed.

(**b**) *In repressible operons, the repressor can bind to the operator only when it is combined with a corepressor. Thus transcription and translation proceed until a corepressor is produced.*

from the operator of the *lac* operon. As a consequence, RNA polymerase can begin its movement along the DNA molecule, transcribing the structural genes of the operon into mRNA (Figure 16–9a). In the case of the tryptophan *(trp)* operon, the presence of the amino acid activates the repressor, which then binds to the operator and blocks the synthesis of the unneeded enzymes (Figure 16–9b). Both allolactose and tryptophan—as well as the molecules that interact with the repressors of other operons—are allosteric effectors (page 176), exerting their effects by causing a change in the configuration of the repressor molecule.

Some 75 different operons have now been identified in *E. coli*, comprising 260 structural genes. Some are, like the *lac* operon, inducible, while others are, like the *trp* operon, repressible. Note, however, that both inducible and repressible systems are examples of negative control, since both involve repressors that turn off transcription.

The CAP–Cyclic AMP System

Catabolite activator protein, or CAP, is a regulatory protein that exerts positive control on the operon. Like the operon itself, the CAP system was initially investigated in relation to lactose metabolism and is now known to be of much wider significance. CAP combines with a molecule known as cyclic AMP (cAMP), and this combination binds to the promoter region of the operon. Only then, when the CAP-cAMP complex is bound to the promoter, does maximum transcription take place (Figure 16–10). As we have seen, the operon is under the negative control of the repressor: no transcription takes place unless the repressor is removed. It is also under the positive control of the CAP-cAMP complex, which enhances transcription when it is bound to the operon.

16–10 *Negative and positive regulation of the* lac *operon.* (a) *In the* lac *operon (and other operons regulated by the CAP-cAMP system), the promoter includes two distinct regions: a binding site for the CAP-cAMP complex and an entry site for RNA polymerase molecules. In order for RNA polymerase to bind efficiently to the promoter, the CAP-cAMP complex must be in place on its binding site.* (b) *In the absence of the inducer (allolactose), the repressor binds to the operator, which, in the* lac *operon, overlaps the first structural gene. Although RNA polymerase can bind to the promoter, it cannot move past the repressor to begin transcription.* (c) *In the presence of the inducer, the repressor is inactivated and can no longer attach to the operator. If, under these circumstances, the CAP-cAMP complex is in place at its binding site, RNA polymerase molecules immediately begin transcription of mRNA molecules that direct the synthesis of three proteins: the enzyme beta-galactosidase, a transport protein that brings lactose from the external medium into the cell, and the enzyme transacetylase, which transfers an acetyl group from acetyl CoA (page 194) to galactose.*

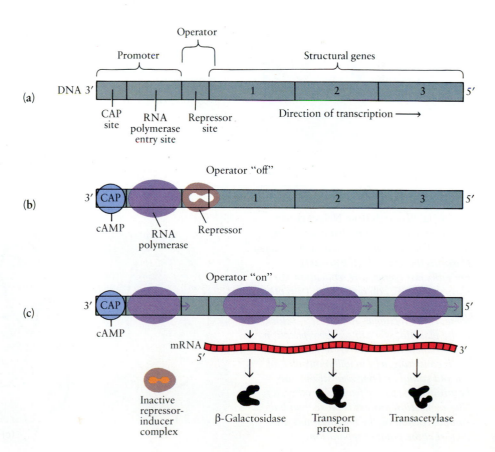

Discovery of this control system came about because of the observation that *E. coli* will not use lactose as an energy source if glucose is present. In other words, in the presence of glucose, the *lac* operon remains repressed, even though lactose is present in the cell. The intermediary in this regulatory process is cAMP. When the supply of glucose in the cell decreases, the level of cAMP increases, more CAP-cAMP complexes form and become available to bind to the *lac* operon, more of the proteins coded by the *lac* operon are produced, and more lactose is broken down. The process by which a decrease in the concentration of glucose leads to an increase in the concentration of cAMP remains a mystery.

These mechanisms are, in themselves, further examples of the precision with which the living cell regulates its biochemical activities. Their manipulation is, as we shall see in the next chapter, an essential component in the scientific trick of inducing bacterial cells to synthesize mammalian proteins of medical importance, such as human insulin.

PLASMIDS AND CONJUGATION

Although the bacterial chromosome contains all of the genes necessary for growth and reproduction of the cell, virtually all types of bacteria have been found to carry additional DNA molecules known as **plasmids.** Plasmids, which are much smaller than the bacterial chromosome, may carry as few as two genes or as many as thirty. Certain plasmids can move into and out of the bacterial chromosome; a plasmid that is incorporated into the chromosome is known as an **episome.**

Like the bacterial chromosome, plasmids are circular and self-replicating (Figure 16–11). Some plasmids replicate in synchrony with the chromosome, and each daughter cell has only one copy of the plasmid. Other plasmids replicate asynchronously, with the result that the cell may contain multiple copies. In the case of some small plasmids, as many as 50 copies have been detected in a single cell. Alternatively, if the plasmid replicates less frequently than the chromosome, some daughter cells may not receive any copies of the plasmid. The DNA of an episome, as you would expect, replicates when the chromosome itself replicates.

About a dozen different kinds of plasmids have been described in *E. coli* alone. Two of the most important types are sex factor, or F, plasmids, and drug resistance, or R, plasmids.

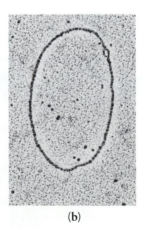

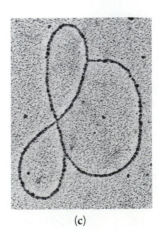

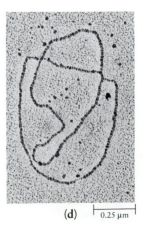

(a) |⎯⎯⎯⎯| 0.5 μm

(b)   **(c)**   **(d)** |⎯⎯| 0.25 μm

16–11 (a) *Plasmids from* Neisseria gonorrhoeae, *the bacterium that causes gonorrhea. The two pairs of connected plasmids (indicated by the arrows) are probably just completing replication.* (b) *A plasmid from* E. coli *replicating. Its replication, like that of the bacterial chromosome, is bidirectional. At almost 2 o'clock in the plasmid, you can see the replication "eye," where the two DNA double helixes (each consisting, as you will remember, of one old strand and one new strand) are beginning to separate from each other.* (c) *The replication process is more than half-completed, and* (d) *the two plasmids are almost at the point of separation.*

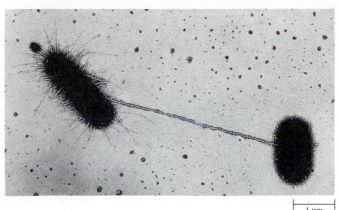

16–12 *Electron micrograph of conjugating* E. coli *cells. The elongated F⁺ (male) cell at the top of the micrograph is connected to the more rotund F⁻ (female) cell by a long pilus. Genes on the F plasmid are responsible for the production of these specialized pili, which are necessary for conjugation. Numerous shorter pili are visible on the F⁺ cell.*

⊢ 1 μm

The F Plasmid

The first plasmid to be recognized as such was the F (for fertility) factor of *E. coli*. This F factor, or F plasmid as it is sometimes called, contains some 25 genes, many of which control the production of F pili. F pili are long, rod-shaped protein structures that extend from the surface of cells containing the F plasmid, which are known as male (donor), or F⁺, cells. Cells that lack the F plasmid are known as female (recipient), or F⁻, cells. The F⁺ cells can attach themselves to F⁻ cells by the pili (Figure 16–12) and transfer the F plasmid to them through cytoplasmic bridges. Transfer of the F factor gives the recipient cells the capacity to produce F pili and to transfer the F plasmid (that is, the recipient cells become F⁺ cells). In a mixed bacterial culture, all F⁻ cells quickly become F⁺ cells. In bacteria, maleness is thus a highly contagious condition, a phenomenon fortunately without parallel in humans. Transfer of DNA from one cell to another by cell-to-cell contact is known as **conjugation.**

The transfer of an F plasmid is diagrammed in Figure 16–13. Note that it involves a mode of replication, known as **rolling-circle replication,** that differs significantly from the bidirectional (theta) replication of the bacterial chromosome (see Figure 16–2).

The F factor, like many other plasmids, can become integrated into the bacterial chromosome. A bacterial cell that contains the F factor as part of its chromosome—that is, as an episome—is known as an Hfr (high frequency of recombination) cell. An Hfr cell has an astonishing property: when it attaches to an F⁻ cell, the replicating bacterial chromosome itself (or a portion thereof) can be transferred from the Hfr cell to the F⁻ cell. In other words, the genes in the bacterial chromosome can be passed from one cell to another, resulting in a new gene combination in the recipient cell. Conjugation is thus, in effect, a form of sexual recombination.

In the Hfr cell at conjugation, a break occurs in the F-factor sequence in the chromosome, and rolling-circle replication begins (Figure 16–14). Leading by its 5′ end, a single strand of DNA passes from the Hfr cell to the F⁻ cell. Recombination may then occur between the chromosome of the recipient cell and portions

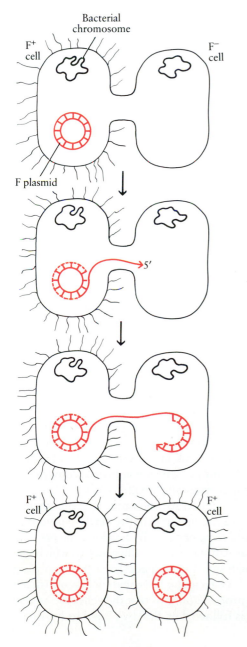

Bacterial chromosome

F⁺ cell   F⁻ cell

F plasmid

5′

F⁺ cell   F⁺ cell

16–13 *Transfer of an F plasmid from an F⁺ cell to an F⁻ cell via rolling-circle replication. In these diagrams, the plasmid is shown greatly enlarged; in reality, it is much smaller than the bacterial chromosome and contains far fewer nucleotide pairs.*

A single strand of DNA moves into the recipient cell, where its complementary strand is synthesized (dashed lines). As the DNA strand is transferred, the donor

strand "rolls" counterclockwise, exposing the unpaired nucleotides. These serve as a template for the synthesis of a complementary DNA strand (dashed lines). As a result, the plasmid in the donor cell continues to be a circle of double-stranded DNA. The transferred plasmid converts the recipient cell to an F⁺ cell. Transfer of DNA by cell-to-cell contact is known as conjugation.

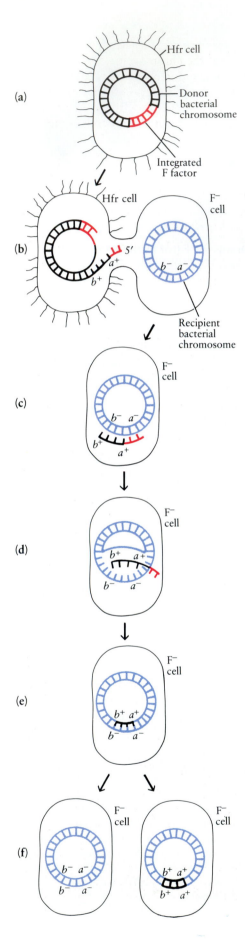

(a) Hfr cell — Donor bacterial chromosome — Integrated F factor

(b) Hfr cell — F⁻ cell — Recipient bacterial chromosome

(c) F⁻ cell

(d) F⁻ cell

(e) F⁻ cell

(f) F⁻ cell — F⁻ cell

16–14 *Transfer of a portion of the bacterial chromosome during conjugation.* **(a)** *An F⁺ cell is converted into an Hfr cell when an F plasmid becomes inserted into its chromosome.* **(b)** *A break occurs in the F-factor sequence in the chromosome, and rolling-circle replication begins. Leading by its 5′ end, a single strand of the DNA, containing a portion of the F-factor sequence followed by genes a⁺ and b⁺, moves into the recipient F⁻ cell. In this example, only a portion of the chromosome is transferred before the cells separate from each other.* **(c)** *The single strand of transferred DNA is complementary to the part of the recipient chromosome that carries the same genes. The "match" is not exact, however, since genes a⁻ and b⁻ are alternate forms of genes a⁺ and b⁺. They differ in nucleotide sequence as a result of mutations that have caused, in this example, a⁻ and b⁻ to be nonfunctional—that is, they do not result in the synthesis of products a and b.* **(d)** *Recombination occurs between the donor DNA and the recipient chromosome. The displaced recipient DNA strand and the noncomplementary portion of the donor strand (that is, the portion containing part of the F-factor sequence) are degraded by enzymes. The structure shown here exists only briefly.* **(e)** *The chromosome containing the integrated recipient DNA immediately replicates. This allows each of the "almost-matched" strands to act as a template for the formation of a new complementary strand in which all bases can be paired exactly.* **(f)** *The daughter cell containing genes a⁻ and b⁻ will not be able to synthesize products a and b, but the daughter cell that contains transferred genes a⁺ and b⁺ will be able to synthesize these products. This offers a means by which conjugation can be demonstrated. Note that the donor cell remains Hfr, and the recipient cell is still F⁻, as are its daughter cells.*

of the chromosome of the donor cell, with new material replacing old in the recipient cell's chromosome. Such recombinations can be detected by working with suitable auxotrophic strains of F⁻ bacteria—that is, mutant strains unable to synthesize particular molecules, which must be supplied in the growth medium for the bacteria to survive. Gaining the ability to synthesize those particular molecules demonstrates that the deficient F⁻ cells have received the necessary genes from the normal donor cells.

Chromosome Mapping

Studies of conjugation revealed that the genes of a bacterium are arranged in a regular linear order around the circular chromosome. The essential laboratory tools in these conjugation studies were a kitchen blender and a timer. Stop here a moment and see if you can figure out the role of these two comparatively humble instruments in mapping the bacterial chromosome. Here are two clues: (1) movement of a strand of DNA from the donor cell into the recipient cell during conjugation proceeds at a constant rate, and (2) transmission of a copy of the entire *E. coli* chromosome requires about an hour and a half at 37°C.

During the approximately 90 minutes required for the transfer of a complete chromosome copy from an Hfr cell to an F⁻ recipient cell, the newly synthesized strand of DNA is making its way into the recipient cell. By breaking the physical contact between the conjugating cells at various times in the process (which can be accomplished by whirling them at high speed in a blender) and then analyzing which genes have been transferred to the F⁻ recipient, it is possible to construct a map of the chromosome. The mapping process confirms that the genes are carried in a linear array on the chromosome: *A* is followed by *B*, *B* by *C*, and so on, down to *XYZ*.

As different Hfr strains were studied, it was found that insertion sites for the F plasmid differed from strain to strain and that the chromosome could be transferred in either direction, depending on the orientation of the inserted F plasmid. It was these studies that also first gave a clue that the *E. coli* chromosome is circular; that is, *Y* followed *X*, and *Z* followed *Y*, but next would come *A* and then *B*. The circularity of the chromosome—the fact that it has no ends—has now been confirmed by electron microscopy.

R Plasmids

In 1959, a group of Japanese scientists discovered that resistance to certain antibiotics and other antibacterial drugs can be readily transferred from one bacterial cell to another. Under experimental conditions, 100 percent of a population of drug-sensitive cells can become resistant within an hour after being mixed with suitable drug-resistant bacteria. It was subsequently found that the genes conveying drug resistance are often carried on plasmids, which have come to be known as R plasmids.

Resistance genes can also be transferred from one R plasmid to another. A single plasmid may collect as many as 10 resistance genes, making the cell it inhabits (and any cell to which it is transferred) resistant to as many as 10 different antibiotics (Figure 16–15). Resistance genes can also be transferred from plasmids to the bacterial chromosome, to viruses, and, most disturbing of all, to bacteria of other species. Thus, the usually innocuous *E. coli* can pick up R plasmids by conjugation and transfer them to *Shigella*, a bacterium capable of causing a sometimes fatal form of dysentery. Infectious drug resistance has now been found among an increasing number of types of pathogens, including those responsible for typhoid, gastroenteritis, plague, undulant fever, meningitis, and gonorrhea.

Typically, only a few copies of these large plasmids exist in a single cell. They are passed from mother to daughter cells at cell division, are transferred by conjugation, or, in another example of bacterial transformation (page 282), they may be simply passed from cell to cell through the cell membranes.

Drug resistance in bacterial cells is often the result of the synthesis of enzymes that break down the drug or that set up a new enzymatic pathway, circumventing the effects of the drug. Thus, resistance may depend on the synthesis of specific enzymes in high concentrations. The fact that resistance genes are on plasmids may allow for many copies of those genes to be produced very rapidly within a single cell.

16–15 *This plasmid, known as R6, carries genes that confer resistance to six different drugs, including the antibiotics tetracycline, neomycin, and streptomycin.*

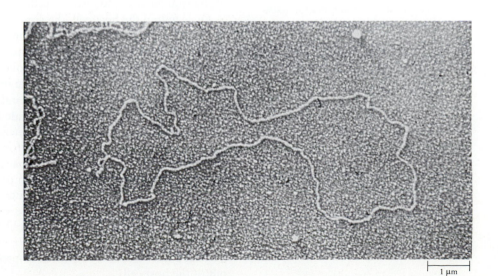

1 μm

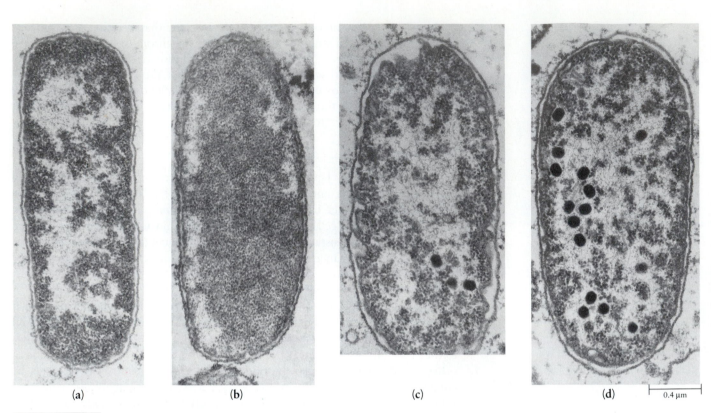

(a)   (b)   (c)   (d)   0.4 µm

16–16 *Stages in the replication of T4 bacteriophage in cells of* E. coli. *(a) As the viral infection begins, the DNA of the bacterial cell is visible in the electron-transparent nucleoid region. (b) After 5 minutes, the bacterial DNA has changed in appearance and has moved toward the cell membrane. (c) After 15 minutes, the bacterial DNA has disappeared, replaced by vacuoles containing threads of replicating phage DNA; transcription from phage DNA is occurring simultaneously. You can also see proteins assembling themselves into the distinctive outer coat of the bacteriophage. (d) After 30 minutes, many phages, both complete and incomplete, are present.*

VIRUSES

Viruses consist essentially of a molecule of nucleic acid enclosed in a protein coat, or **capsid.** They contain no cytoplasm or ribosomes or other cellular machinery. However, as we saw in Chapter 14, they can move from cell to cell and, within a host cell, utilize its enzyme systems and organelles to replicate their nucleic acid and synthesize new coat proteins. The coat may consist of one protein molecule repeated over and over, as in tobacco mosaic virus (page 305), or a number of different kinds of proteins, as in the T-even bacteriophages with their complex tail assemblies (page 285). The composition of the protein coat determines the attachment of the virus to the membrane of the host cell and the subsequent entry of the viral nucleic acid into the cell.

Once within the host cell, the viral nucleic acid directs the production of new viruses (Figure 16–16). This is accomplished using the raw materials of the cell—such as nucleotides, amino acids, and the cell's ATP and other energy sources—and also the cell's metabolic machinery. Thus viruses are obligate parasites; they cannot multiply outside the host cell.

The nucleic acid of a virus—the viral chromosome—may be either DNA or RNA, single-stranded or double-stranded, circular or linear. Viral chromosomes vary greatly in size from some 5,400 nucleotides for a small, single-stranded DNA bacteriophage, φX174, to 180,000 for the T-even bacteriophages. The viral chromosome always codes for the coat protein or proteins, and also for one or more enzymes involved in replication of the viral chromosome. These enzymes ensure the rapid replication of viral nucleic acid in preference to the nucleic acid of the host cell. The viral chromosome also codes for an enzyme or enzymes that, once the new virus particles are assembled, enable them to lyse the host cell and escape. The infection cycle is complete when the viral nucleic acid molecules are packaged into the newly synthesized protein coats and the virus particles break out of the host cell.

16–17 *When certain types of viruses—known as temperate bacteriophages—infect bacterial cells, one of two events may occur.* **(a)** *The viral DNA may enter the cell and set up an infection, or lytic, cycle, or* **(b)** *the viral DNA may become part of the bacterial chromosome, replicating with it and being passed on to daughter cells. Bacteria harboring such viruses are known as lysogenic, because from time to time, such a virus, called a prophage, becomes activated and sets up a new lytic cycle.*

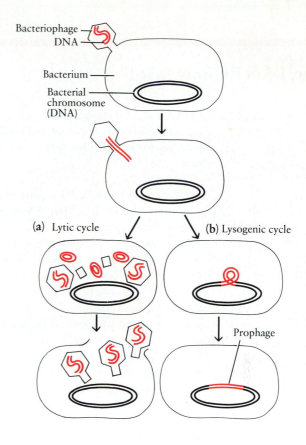

Viruses as Vectors

The genetic makeup of the DNA of bacterial cells can be altered, as we have seen, by the introduction of DNA from other bacterial cells. Both transformation (page 282) and conjugation can result in recombination between donor and host DNA. Similarly, the transfer of plasmids and their insertion into the chromosome of a bacterial cell can change the DNA composition of the recipient cell. Viruses can also play a role as vectors (carriers) that move pieces of DNA from one bacterium to another.

Temperance and Lysogeny

Early in the study of bacteriophages, it was noted that a virus infection could suddenly erupt in a colony of apparently uninfected bacterial cells. Such cells were termed **lysogenic,** because of their capacity to generate a cycle of cell lysis that spread through neighboring cells. The cause of lysogeny, it was discovered, was the capacity of certain viruses to set up a long-term relationship with their host cell, remaining latent for many cellular generations before initiating an infection cycle. Such viruses became known as **temperate bacteriophages.** The DNA of temperate phages, like that of the F plasmid, may become integrated at specific sites in the host chromosome, replicating along with the chromosome. Such integrated bacteriophages are known as **prophages.** Prophages break loose from the host chromosome spontaneously about once in every 10,000 cell divisions, triggering a lytic cycle (Figure 16–17). In the laboratory, lysis may be triggered by ultraviolet light (see essay), x-rays, or other agents that damage nucleic acids.

Temperate phages resemble plasmids in that (1) they are autonomously replicating molecules of DNA, and (2) they may become integrated into the bacterial cell chromosome. They differ from plasmids both in their capacity to manufacture a protein coat and thus to exist (though not to replicate) outside the cell and in their capacity to lyse the host cell.

"Sir, I Am Entirely Lysed"

André Lwoff, who appears in the photograph on page 322, wrote the following account of his discovery of a technique for inducing lysogenic bacteria to undergo lysis:

Our aim was to persuade the totality of the bacterial population to produce bacteriophage. All our attempts—a large number of attempts it was—were without result.... Yet I had decided that extrinsic factors must induce the formation of bacteriophage. Moreover, the hypothesis had been published already [1949], and when one publishes an hypothesis, one is sentenced to hard labor....

Our experiments consisted in inoculating exponentially growing bacteria into a given medium and following bacterial growth by measuring optical density [that is, the turbidity of the culture, which provided an indirect measure of the number of intact bacterial cells in the culture]. Samples were taken every fifteen minutes, and the technicians reported the results. They (the technicians, that is) were so involved that they had identified themselves with the bacteria, or with the growth curves, and they used to say, for example: "I am exponential," or "I am slightly flattened...."

So negative experiments piled up, until after months and months of despair, it was decided to irradiate the bacteria with ultraviolet light. This was not rational at all, for ultraviolet radiations kill bacteria and bacteriophages, and on a strictly logical basis the idea still looks illogical in retrospect. Anyhow, a suspension of lysogenic bacilli was put under the UV lamp for a few seconds.

The Service de Physiologie Microbienne is located in an attic, just under the roof of the Pasteur Institute, with no proper insulation. The thermometer sometimes rises in a manner that leaves no conclusion other than that the temperature is high. It was a very hot summer day and the thermometer was unusually high. After irradiation, I collapsed in an armchair, in sweat, despair, and hope. Fifteen minutes later, Evelyne Ritz, my technician, entered the room and said: "Sir, I am growing normally." After another quarter of an hour, she came again and reported simply that she was normal. After fifteen more minutes, she was still growing. It was very hot and more desperate than ever. Now sixty minutes had elapsed since irradiation; Evelyne entered the room again and said very quietly, in her soft voice: "Sir, I am entirely lysed." So she was: the bacteria had disappeared! As far as I can remember, this was the greatest thrill—molecular thrill—of my scientific career.

From André Lwoff, "The Prophage and I," in *Phage and the Origins of Molecular Biology,* a collection of essays compiled in 1966 by the Cold Spring Harbor Laboratory and dedicated to Max Delbrück on his sixtieth birthday.

Transduction

The process known as **transduction** is the transfer of cellular DNA from one host cell to another by means of viruses. In the course of a lytic cycle, as we have noted, viruses exploit the resources of the host cell. During the lytic cycle of many viruses, the host DNA becomes fragmented; when these viruses leave the cell, some of them may contain DNA fragments from the host chromosome. Since the amount of DNA that can be packaged within the protein coat is limited, such viruses lack some or all of their own necessary genetic information. Although they may be able to infect a new host cell, they are not be able to complete a lytic cycle. However, the genes they carry from their previous host may become incorporated into the chromosome of the new host. Depending on the genes that are transferred, these recombinations may be detected. This process is called general transduction (Figure 16–18a) because virtually any gene can be transferred by this mechanism.

16–18 *The two types of viral transduction. (a) General transduction occurs when a nontemperate bacteriophage infects a bacterial cell. The viral DNA enters the bacterial cell and undergoes a lytic cycle. In the course of this cycle, the DNA of the host cell is broken apart, and some of the fragments are accidentally incorporated into newly formed virus particles. When released, a virus particle containing bacterial DNA may infect another bacterial cell. Although such a virus is defective and unable to set up a lytic cycle, the bacterial DNA it has introduced may recombine with the DNA of the new host cell.*

(b) Specialized, or restricted, transduction occurs when a temperate bacteriophage infects a bacterial cell and enters a lysogenic cycle. The viral DNA is incorporated into the host chromosome, where it may remain as a prophage for many generations. When the prophage leaves the bacterial chromosome, it often takes with it a piece of the bacterial DNA. In this case, only DNA adjacent to the insertion site of the prophage is picked up with the viral DNA. The linked bacterial and viral DNA is replicated and incorporated into new virus particles that are released from the cell when it is lysed. These particles infect other bacterial cells, and the genes of the first host cell may recombine with those of the new host cell. Viral DNA may also become integrated into the DNA of the new host cell.

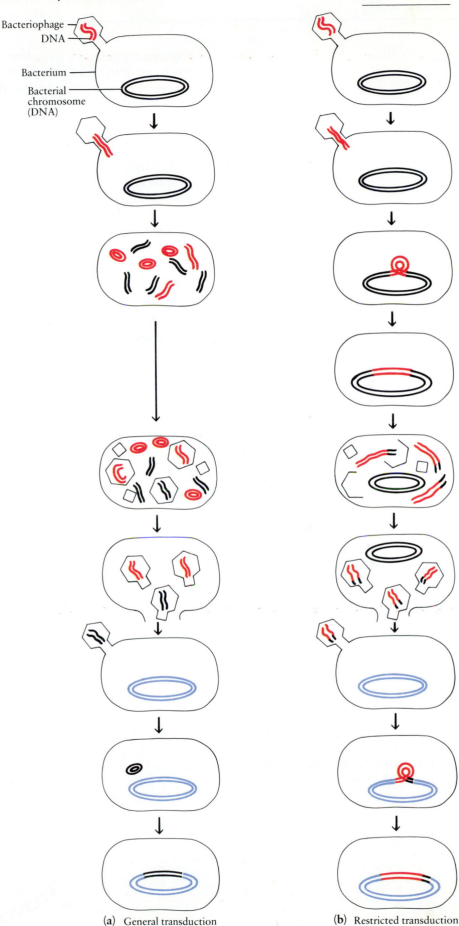

(a) General transduction

(b) Restricted transduction

When prophages break loose from a host chromosome to initiate a lytic cycle, they may, similarly, take a fragment of the host chromosome with them. The chromosome of each newly formed phage then consists of both host DNA and viral DNA. In this situation, the host DNA is not picked up at random, as is the case with nontemperate phages, but is, quite specifically, restricted to the portions of the host chromosome adjoining the insertion site of the prophage. Hence, this process is known as specialized, or restricted, transduction (Figure 16–18b).

Transduction resembles conjugation in that it involves the transfer of bacterial genes from one bacterial cell to another. It differs from conjugation in that in transduction the genes are carried by viruses.

Introducing Lambda

Lambda, the best studied of the temperate bacteriophages, has several interesting and instructive features. When the double-stranded DNA of the virus is packaged in its protein coat, it is linear—that is, it has two free ends. Once the viral chromosome is released into the host cell, however, it forms a circle. This closing of the circle takes place because a single strand of 12 nucleotides protrudes on the 5′ end of each strand of the DNA molecule (Figure 16–19). These strands are exactly complementary to one another and are said to be "sticky." Hydrogen bonds form between the complementary base pairs, joining the ends of the molecule together. Such a cohesive nucleotide sequence, which has since been found in other DNAs, is known in shorthand as a COS region.

When the DNA of lambda's circular chromosome is replicated, it is rolled out in numerous copies, joined in a single long molecule. A special enzyme then cleaves the molecule over and over at the COS sites and packages the individual lambda chromosomes into the waiting protein coats.

Lambda's integration into the *E. coli* chromosome takes place because the *E. coli* chromosome contains a short nucleotide sequence identical to a sequence on the lambda chromosome. Attachment of the bacteriophage chromosome to the

16–19 (a) *A portion of the lambda chromosome, showing the nucleotide sequences of the single-stranded "sticky" ends.* (b) *In the host cell, the DNA molecule forms a circle when these single strands come together and their complementary bases are paired. The enzyme DNA ligase catalyzes the condensation reaction that links the phosphate subunit at each 5′ end to the deoxyribose subunit at each 3′ end, closing the "breaks."*

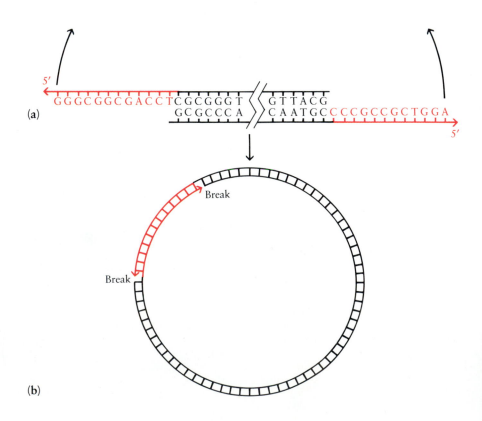

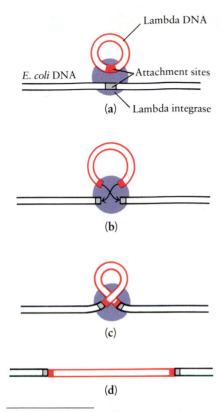

16–20 *Integration of lambda DNA into the chromosome of E. coli. (a) Lambda integrase recognizes the attachment sites on the DNA molecules of lambda and E. coli and brings them into close proximity. (b) This enzyme catalyzes the breakage of both circular DNA helixes and (c) their joining together, resulting in the incorporation of the lambda prophage into the E. coli chromosome (d).*

bacterial chromosome is brought about by a special enzyme that recognizes both sequences, brings the two circular DNA helixes together, and initiates the cutting and sealing reactions (Figure 16–20). This enzyme, lambda integrase, is coded by the lambda chromosome. When lambda leaves the bacterial chromosome to begin a lytic cycle, another enzyme frees it from the bacterial chromosome by making a staggered cut that leaves "sticky" ends protruding. These ends rapidly rejoin one another, forming a circular chromosome once more.

As we shall see in the next chapter, specific recognition sites, staggered cuts, and "sticky" ends, the everyday equipment of lambda, were to prove equally essential to molecular geneticists.

TRANSPOSONS

More recently, another type of movable genetic element has been found; it is known as a **transposon.** Like episomes and prophages, transposons are segments of DNA that are integrated into the chromosomal DNA. However, they differ from episomes and prophages in that they contain a gene that codes an enzyme, transposase, that catalyzes their insertion into a new site. Also, at each end they have a repeated nucleotide sequence. This sequence may consist of direct repeats —such as ATTCAG and ATTCAG—or of inverted repeats—such as ATTCAG and GACTTA. The repeated sequences are typically 20 to 40 nucleotides in length. At the time of insertion, the target site on the host chromosome—the site at which the transposon becomes inserted—is duplicated. The target sequence, which is 5 to 10 base pairs in length, then flanks the transposon (Figure 16–21). In some cases, the transposons do not actually move—that is, they do not disappear from their initial site when they appear at a new location. Instead the original, parental transposon gives rise to a new copy that becomes inserted elsewhere.

Two kinds of transposons are known, simple and complex. Simple transposons, also called **insertion sequences,** are only about 600 to 1,500 base pairs in length and do not carry any genes beyond those essential for the process of transposition. At least six different simple transposons have been found in *E. coli.* They are detectable because they cause mutations. If one of these transposons becomes inserted into a gene, it inactivates it. Simple transposons also contain promoter sequences, which may lead to the inappropriate initiation of transcription of previously inactive genes of the host chromosome. Simple transposons appear to have no function but to duplicate themselves; thus they are among that group of molecules known as "selfish DNA."

16–21 *Insertion of a transposon into recipient DNA. (a) The nucleotide sequence at which insertion occurs is known as the target site. (b) Staggered cuts are made in the target site, and (c) the transposon is attached to the protruding ends of the cuts. (d) As the gaps are filled by complementary nucleotides, identical repeats are formed on the two sides of the inserted transposon. These are often used as "landmarks" in the identification of DNA sequences that have been transposed.*

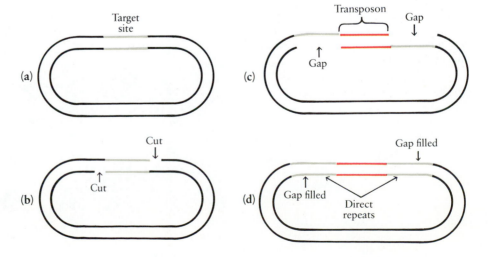

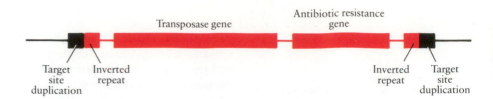

Transposase gene   Antibiotic resistance gene

Target site duplication   Inverted repeat   Inverted repeat   Target site duplication

16–22 *The structure of a complex transposon. The transposable element shown here consists of the gene that codes for transposase, a gene that codes for an enzyme that confers resistance to an antibiotic, and, at each end, inverted repeats. The duplicated target site sequence is part of the host chromosome.*

Complex transposons are much larger and carry genes that code for additional proteins (Figure 16–22). As is the case with simple transposons, complex transposons may cause mutations, but they are also detectable because of their gene products. Genes that are part of a complex transposon can move from place to place on a chromosome or from chromosome to chromosome, and are therefore known as "jumping genes." Drug-resistance genes are often part of transposons, and so can be transferred readily from plasmid to plasmid and from plasmid to bacterial chromosome to plasmid again. Complex transposons are often found to have simple transposons flanking them—one at each end—which suggests that complex transposons might have come about by two simple transposons jumping at the same time, taking with them everything in between.

RECOMBINATION STRATEGIES

We have now described four different ways that new, information-carrying DNA can be introduced into a bacterial cell: transformation (page 282), which is the uptake of fragments of DNA; conjugation, which is the direct transfer of DNA from one cell to another; viral infection, with the injection of viral nucleic acid; and transduction, the transfer by viruses of nonviral genetic material from one cell to another.

We have also, as you may have noticed, described two different ways in which genetic recombination can take place. One involves exchange between homologous segments of DNA. When two such segments of double-stranded DNA are aligned with one another, exchanges occur between the molecules in such a way that genes may be transferred from one molecule to the other. This phenomenon occurs in eukaryotic cells during meiosis as a result of crossing over; it also takes place during conjugation, transformation, and transduction in bacterial cells.

16–23 *Transposons can carry genes conferring antibiotic resistance from R plasmid to R plasmid and also into and out of the bacterial chromosome. This micrograph shows three transposons that carry a gene conferring resistance to the antibiotic ampicillin; they are from a gonorrhea-causing bacterium.*

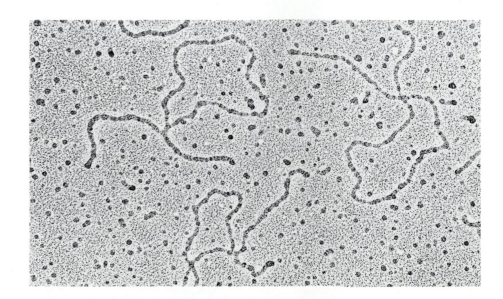

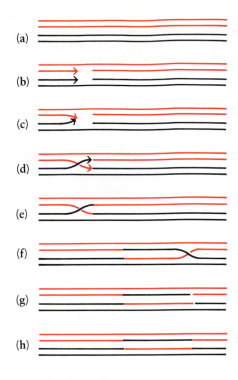

16–24 *The "single-strand switch" model of genetic recombination between two homologous strands of DNA.* (a) *The homologous parental DNAs are indicated in black and color.* (b) *One strand of each DNA molecule is broken,* (c, d) *switched over to the other molecule, and* (e) *joined to the opposite switched strand.* (f) *The exchange of strands between DNAs proceeds along the chromosome, and* (g) *at a specific point the switched strands are again broken and* (h) *resealed, completing the exchange and recombination of the genes.*

Several models of how homologous recombination takes place have been proposed; one of them, the "single-strand switch," is shown in Figure 16–24.

A second kind of recombination involves the insertion of movable (and removable) genetic elements. These elements can enter or leave the DNA of a chromosome without the occurrence of homologous recombination. The F plasmid, lambda phage, and transposons are all examples of this phenomenon.

Genetic recombination clearly was not invented by late twentieth-century molecular biologists; it has undoubtedly been taking place for billions of years, long before *Homo sapiens* was even a twinkle in the eye of evolution.

SUMMARY

The essential genetic information of prokaryotes, of which *E. coli* is the best-studied example, is coded in a circular double-stranded molecule of DNA. It is tightly packed in the bacterial cell in the nucleoid region. Replication begins at a particular site on the chromosome and proceeds bidirectionally (theta replication).

A principal means of genetic regulation in bacteria is the operon system. An operon is a linear sequence of genes coding for a group of functionally related proteins plus the promoter and operator. The structural genes of the operon are transcribed as a single mRNA molecule. Transcription from the operon is controlled by the promoter and operator sequences, which are adjacent to the structural genes and bind specific proteins. The promoter contains the binding site for RNA polymerase and may contain a binding site for the CAP-cyclic AMP complex. The operator is the binding site for a repressor, a protein coded by another gene, the regulator, which may be located some distance away on the bacterial chromosome. The operator overlaps the promoter, the first structural gene, or both; when the repressor is attached to the DNA molecule at the operator, RNA polymerase cannot initiate transcription of mRNA. When the repressor is not present, RNA polymerase can attach to the DNA and begin its movement along the chromosome, permitting transcription and protein synthesis to take place.

The *lac* operon is an example of an inducible operon. It is turned from "off" to "on" when an inducer binds to and inactivates the repressor. Other operons, such as the *trp* operon, are repressible. These are turned from "on" to "off" by the action of a corepressor that binds to an inactive repressor. This activates the repressor and it binds to the operator. Both induction and repression are forms of negative regulation.

Positive regulation of some operons is provided by the binding of the CAP-cAMP complex. For example, when glucose is present in the cell, cyclic AMP levels are low, and the CAP-cAMP complex does not form. As glucose is depleted, cAMP levels rise, and CAP-cAMP complexes form and then bind to the promoter. With lactose present (and the repressor thus inactivated) and the CAP-cAMP complex in place, RNA polymerase also binds to the promoter, and transcription from the operon takes place.

In addition to the genes carried on the bacterial chromosome, the bacterial cell contains other genes carried in plasmids, which are much smaller, also circular, double-stranded DNA molecules. Most plasmids can be transferred from cell to cell. Such transfer of DNA by cell-to-cell contact is known as conjugation. Some plasmids can become reversibly integrated into the bacterial chromosome, in which case they are known as episomes. Plasmids often carry genes for drug resistance; as many as 10 such genes have been located on a single plasmid.

The F (fertility) factor of *E. coli* is a plasmid present in donor F$^+$ (male) cells that can be transferred to recipient F$^-$ (female) cells; such cells then become F$^+$ and can transfer the F factor. When the F factor becomes integrated into the chromosome of an *E. coli* cell (making it an Hfr cell), part or (rarely) all of the chromosome can be transferred to another *E. coli* cell. At the time of transfer, the chromosome replicates by the rolling-circle mechanism, and a single-stranded copy of the DNA enters the recipient cell linearly so that the bacterial genes enter the cell one after another in a fixed sequence. The complementary strand is then synthesized from nucleotides available in the recipient cell. Because the rate at which the bacterial genes enter the recipient cell is constant at a given temperature, separation of conjugating cells at regular intervals provides a means for mapping the bacterial chromosome.

Viruses consist of either DNA or RNA wrapped in a protein coat. Within a host cell, the viral nucleic acid can make use of the cell's metabolic resources to synthesize more viral nucleic acid molecules and more viral proteins. Packaged in their protein coats, the virus particles can then break out of the cell to start a new infection cycle.

The DNA of some viruses, known as temperate viruses, can become integrated into the host chromosome where, like an episome, it replicates along with the chromosome. When integrated into a host chromosome, the DNA of a bacterial virus is known as a prophage. From time to time, prophages break loose from the chromosome and set up a new infection (lytic) cycle.

Viruses can serve as vectors, transporting genes from cell to cell, a process known as transduction. General transduction occurs when host DNA, fragmented in the course of the viral infection, is incorporated into new virus particles that carry these fragments to a new host cell. Specialized transduction occurs when a prophage, on breaking away from the host chromosome, carries with it—as part of the viral chromosome—host genes, which are then transported to a new host cell.

Lambda is a temperate bacteriophage of *E. coli*. The lambda chromosome is linear when it is in the viral protein coat, but when it is released into the cytoplasm of *E. coli*, it forms a circle. The closing occurs because of the existence of "sticky" ends—single-stranded complementary DNA sequences on either end of the molecule. Lambda becomes integrated into the bacterial chromosome at a specific attachment site bearing a nucleotide sequence identical to a sequence in lambda itself.

Transposons are movable genetic elements that differ from plasmids and viruses in several respects: (1) they carry a gene for the enzyme transposase, which catalyzes their integration into the host chromosome; (2) a repeated sequence, either direct or inverted, is present at each end of the transposon; (3) the target sequence on the host chromosome is duplicated when the transposon is inserted, with the result that a transposon is flanked on each end by the target sequence. Transposons may cause mutations by interfering with the normal expression of host-cell genes. Simple transposons contain only genes involved in their transposition; complex transposons carry additional structural genes.

Genetic recombinations can take place either because of exchanges between homologous sequences of DNA—substitution of one sequence for another—or by insertion of additional, new DNA into a recipient chromosome. Exchanges of host-cell genes in transformation, conjugation, and transduction all take place by the former mechanism (substitution of one sequence of DNA for another similar sequence). The second type of genetic recombination (the insertion of additional, new DNA) is characteristic of transposons, of the viral DNA of prophages, and of plasmids, such as the F factor, that can be added to or subtracted from the bacterial chromosome.

QUESTIONS

1. Distinguish among the following: inducible enzyme/repressible enzyme; operon/promoter/operator; inducer/repressor/co-repressor; transformation/conjugation/transduction; plasmid/virus/transposon; F plasmid/R plasmid; F$^+$ cell/F$^-$ cell/Hfr cell; theta replication/rolling-circle replication; prophage/lysogenic bacterium; general transduction/specialized transduction.

2. Compounds can be used by cells in two different ways. One type is broken down (usually as an energy source). Another is used as a building block for a larger molecule. Which type of compound would you expect to function as a corepressor? As an inducer? Do the examples given in the text conform to your expectations?

3. A culture of bacterial cells is grown in a medium in which both glucose and lactose are present in fixed amounts as the sole carbon sources. Describe the series of events that take place in the operon as the sugars are metabolized.

4. Gene expression theoretically can be regulated at the level of transcription, translation, or activation of the protein. In the latter case, the polypeptide produced by protein synthesis is in an inactive form. It undergoes some structural modification (catalyzed by enzymes) before it can perform its function in the cell. What would be the advantages of each type of regulation, in terms of the cell? Under what circumstances might one type be more useful than another? Which is the more economical?

5. How did the mapping experiments with *E. coli* demonstrate that the bacterial chromosome is both linear and circular?

6. You are trying to map the DNA of a strange new bacterium. It is a circular duplex similar to that of *E. coli*. It contains an F plasmid and can transfer DNA during conjugation. After allowing different strains of the bacterium to conjugate, two by two, for different lengths of time, the process is interrupted. In each experiment, you test whether a certain gene (for example, met$^-$) that was previously nonfunctional in the recipient strain is now functional in that strain—that is, you want to determine if transfer of the gene from the donor chromosome, followed by recombination with the recipient chromosome, has occurred. To do this, you plate the recipient bacteria onto a minimal medium and see if they grow. Only cells containing genetic recombinants will grow. From the results given below, determine the order of genes on the chromosome.

			GROWTH ON MINIMAL MEDIUM			
DONOR		RECIPIENT	5 MIN	10 MIN	15 MIN	20 MIN
phe$^+$met$^-$	$\times$	phe$^-$met$^+$	no	no	yes	yes
met$^+$leu$^-$	$\times$	met$^-$leu$^+$	yes	yes	yes	yes
leu$^+$phe$^-$	$\times$	leu$^-$phe$^+$	no	yes	yes	yes
leu$^-$phe$^+$	$\times$	leu$^+$phe$^-$	no	no	yes	yes
met$^-$pro$^+$	$\times$	met$^+$pro$^-$	no	no	no	yes

7. As bacterial conjugation indicates, it is possible to separate the production of new genetic combinations from reproduction. Why do you think these two processes are combined in eukaryotic cells?

8. Describe the two possible outcomes of infection of a bacterial cell by a temperate virus.

Recombinant DNA: The Tools of the Trade

Larger

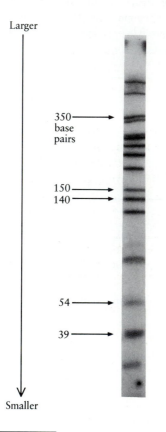

350 base pairs →

150 →
140 →

54 →

39 →

Smaller

17-1 *Electrophoresis, the technique used by Pauling to separate normal and sickle cell hemoglobins (page 304) can also be used to separate fragments of DNA. In electrophoresis, the electrical field can separate molecules not only on the basis of their charge (the feature utilized by Pauling) but also on the basis of their size. Smaller molecules move faster than larger ones. As shown in this photograph of an actual electrophoretic gel, DNA fragments containing different numbers of base pairs can be cleanly separated from one another. The separated, purified fragments can subsequently be washed out of the gel unharmed. This separation procedure is important in many aspects of recombinant DNA work.*

The revelation of the many ways in which cells process, add, delete, and transfer genetic information opened the way for molecular biologists to carry out their own genetic manipulations. This field of human technological activity is often referred to simply as recombinant DNA—a term that overlooks the fact that the recombining of DNA was, by all evidence, going on long before the first amoeba appeared, much less a curious primate.

Recombinant DNA technology has made possible further investigations of the structure and function of genes, especially the otherwise inaccessible genes of eukaryotes. The results of some of these studies will be described in the next chapter. It has also suddenly and dramatically opened the way to new understandings of human genetics (the subject of the final chapter of this section), leading to the accurate diagnosis of many human genetic diseases, and, very likely, someday, the successful treatment of such diseases.

In this chapter we shall describe some of the basic techniques of recombinant DNA—the tools of the trade. The most important ones are (1) methods for obtaining specific, uniform DNA sequences—that is, segments of DNA molecules of a size suitable for analysis and manipulation; (2) DNA cloning, which makes it possible to produce such segments of DNA in large quantities; (3) nucleic acid hybridization, a method for identifying specific segments of DNA and RNA and for estimating similarities between nucleic acids from different sources; and (4) DNA sequencing, the determination of the exact order of nucleotides in a DNA segment—in effect, allowing a direct "reading" of the encoded genetic information.

ISOLATION OF SPECIFIC DNA SEGMENTS

When researchers first confronted the size and complexity of the DNA of even the simplest virus, the possibility of ever deciphering the encoded genetic information seemed virtually hopeless. Chemical analysis depends upon being able to obtain uniform samples of a manageable size and, for a while, that seemed impossible. However, as has occurred repeatedly in recombinant DNA research, the tools were provided by the organisms themselves. In this case, the tools were **restriction enzymes,** which are synthesized by certain bacterial cells, and another enzyme, known as **reverse transcriptase,** which is encoded by the nucleic acid of certain RNA viruses.

Restriction Enzymes: gDNA

Restriction enzymes were discovered in the early 1970s in a number of bacterial species, including *Escherichia coli.* The function of these enzymes is to cleave foreign DNA—that is, DNA from other bacterial strains or from viruses. For example, if DNA from *E. coli* strain B is introduced into cells of *E. coli* strain C, it is broken into fragments by restriction enzymes of the strain C cells.

The essential feature of restriction enzymes, which has made them indispensable for recombinant DNA technology, is that they cleave the DNA not at random but only at very specific nucleotide sequences, some four to eight base pairs in length. These sequences are referred to as **recognition sequences,** since they are "recognized" by specific restriction enzymes. The bacteria—for example, *E. coli* cells of strain C—protect their own DNA from their restriction enzymes by adding a methyl group ($-CH_3$) to one or more of the nucleotides in the recognition sequences of their DNA. This methylation, which occurs during DNA replication, is accomplished by special enzymes that are commonly found with the restriction enzymes. For example, one restriction enzyme of *E. coli*, called *Eco*RI, cleaves DNA only at the sequence GAATTC. Cells that produce *Eco*RI also produce a specific methylating enzyme that adds a methyl group to one of the adenines in the GAATTC sequence, thus protecting their own DNA from recognition and cleavage.

A second important feature of restriction enzymes, as shown in Figure 17–2, is that not all of them make straight cuts through both strands of the DNA molecule; some, including *Eco*RI, cut through the strands a few nucleotides apart, leaving "sticky" ends, as in lambda (page 334). As we have seen, such "sticky" ends can rejoin when hydrogen bonds form spontaneously between complementary bases and the enzyme DNA ligase forges a sugar-phosphate bond linking the ends of each strand together. Also—and this is most important—these "sticky" ends can join with any other segment of DNA that has been cleaved by the same restriction enzyme and, as a result, has complementary "sticky" ends. This discovery marked the beginning of recombinant DNA technology.

More than 200 different restriction enzymes have now been isolated, making it possible to cleave a DNA molecule at any one of more than 90 recognition sequences, thus producing multiple uniform fragments of DNA molecules. The DNA fragments produced by restriction enzymes acting on the DNA of a cell or an organism are known as **genomic DNA,** or gDNA. Because of the number of different restriction enzymes and recognition sequences, it is now possible to splice together segments of gDNA from a limitless variety of sources.

It is interesting to note that certain viruses have enzymes analogous to the restriction enzymes and methylating enzymes of bacteria. For instance, the DNA of the T-even bacteriophages codes for enzymes that break the DNA of *E. coli* cells into fragments; the fragments are then recycled by the bacteriophages to make new bacteriophage DNA. These viral enzymes, like restriction enzymes, cleave DNA only at specific nucleotide sequences. The bacteriophages protect their own DNA from cleavage by methylation of the cytosines that occur within the sequences recognized by the enzymes.

17–2 *The DNA nucleotide sequences recognized by three widely used restriction enzymes:* (a) *Hpa*I, (b) *Eco*RI, *and* (c) *Hind*III. *Recognition sequences are frequently, as for these enzymes, six base pairs long, and, when read in the 5' to 3' direction, the two strands of the sequence are identical. Eco*RI *and* Hind*III* cleave *the DNA so that "sticky" ends result. Restriction enzymes are generally obtained from bacteria: Hpa*I *is from* Hemophilus parainfluenzae, Eco*RI *is from* E. coli, *and* Hind*III is from* Hemophilus influenzae.

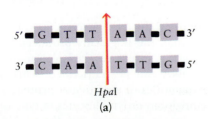

*Hpa*I
(a)

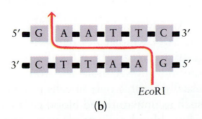

*Eco*RI
(b)

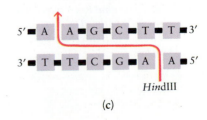

*Hind*III
(c)

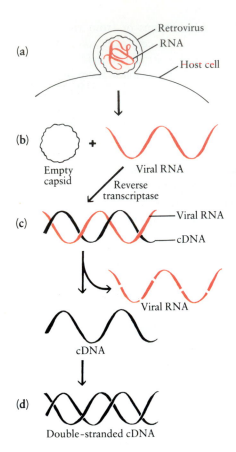

(a)

Retrovirus

RNA

Host cell

(b)

Empty
capsid

+

Viral RNA

Reverse
transcriptase

(c)

Viral RNA

cDNA

Viral RNA

cDNA

(d)

Double-stranded cDNA

17–3 *Infection of an animal cell by a retrovirus. (**a**) The capsid of a retrovirus is typically surrounded by an outer lipoprotein envelope formed from elements of the cell membrane of its previous host. This envelope can fuse with the cell membrane of a new host, allowing the virus to enter the cell. (**b**) Once the retrovirus has gained entry to the cell, the viral RNA is released from the capsid and (**c**) transcribed into a single strand of complementary DNA (cDNA). (**d**) Synthesis of the matching DNA strand follows immediately, producing a double-stranded molecule of cDNA. These reactions, as well as the degradation of the original viral RNA molecule, are all catalyzed by reverse transcriptase.*

As we shall see in the next chapter, the double-stranded cDNA can become integrated into the host-cell chromosome. Subsequently, new viral RNA molecules are transcribed from the cDNA, as are mRNA molecules that direct the synthesis of viral proteins.

Reverse Transcriptase: cDNA

A second method of obtaining specific DNA segments for cloning and manipulation became available with the unexpected discovery of a type of animal virus known as a **retrovirus**. Eukaryotic cells, including the cells of plants and animals, are host to both DNA and RNA viruses. In the case of DNA viruses, the viral DNA is both replicated, forming more viral DNA, and transcribed into messenger RNA, directing the synthesis of viral proteins. With most RNA viruses, the RNA is similarly replicated, forming new viral RNA, and also serves as messenger RNA. However, some RNA viruses have a different method of replication: in these viruses, the RNA serves first as a template for synthesizing DNA, using a viral enzyme, reverse transcriptase (Figure 17–3). Both new viral RNA and mRNA for the synthesis of viral proteins are subsequently transcribed from this DNA.

The discovery of retroviruses, as you might expect, has led to a revision in the "central dogma" of molecular genetics (Figure 17–4). These viruses are also of great interest to medical scientists; a number of retroviruses have been shown to cause cancer in animals, and the virus responsible for the devastating disease AIDS (to be discussed in Chapter 39) is a retrovirus. In addition, retroviral reverse transcriptase has proved to be a valuable tool in recombinant DNA studies.

DNA molecules synthesized by reverse transcriptase from an RNA template are known as **complementary DNA,** or cDNA. cDNA molecules can be spliced into other DNA molecules by means of artificial "sticky" ends. These are constructed by the addition of a single strand of a single nucleotide—TTTTTT, for instance—to the end of the cDNA sequence. The cDNA molecule can then combine with any other DNA molecule to which a complementary single strand has been added—AAAAAA, for instance.

17–4 *The revision of the "central dogma" of molecular genetics required by the discovery of retroviruses. Note that although information can flow from RNA to DNA, as well as from DNA to RNA, the essential feature of the "central dogma" remains intact: information does not flow from the phenotype (protein) to the genotype (nucleic acid).*

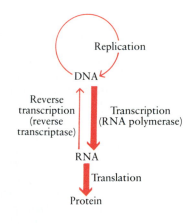

Replication

DNA

Reverse
transcription
(reverse
transcriptase)

Transcription
(RNA polymerase)

RNA

Translation

Protein

An advantage of using reverse transcriptase to produce DNA segments is that the cDNA molecules represent genes, rather than the fragments produced by restriction enzymes. A disadvantage, however, is that it requires the isolation of a specific mRNA molecule from which the cDNA can be transcribed. This is usually feasible only in cells producing large quantities of a particular protein—such as immature red blood cells making hemoglobin or lymphocytes (a type of white blood cell) producing antibodies—a serious limitation on the cDNA molecules that can be prepared.

17-5 *Preparation of a synthetic oligonucleotide. (a) Organic molecules that function as blocking groups are added chemically to either the 5′ or 3′ ends of mononucleotides. Some blocking groups can be removed by treatment with an acid, and others by treatment with a base. Typically, one type of blocking group is used on the 3′ end and another type on the 5′ end. (b) The blocking groups prevent a condensation reaction at the "wrong" ends of the nucleotides, but allow the desired reaction to occur. In this example, treatment with acid then removes the blocking group from the 5′ end of the newly formed dinucleotide, (c) allowing it to undergo a condensation reaction with another nucleotide that is blocked at the 5′ end but free to react at the 3′ end. The result is a trinucleotide (d). These steps are repeated over and over until the desired oligonucleotide is synthesized.*

Synthetic Oligonucleotides

Scientists have also developed methods for synthesizing short sequences of DNA and RNA in the laboratory. These molecules, called **synthetic oligonucleotides** (from *oligo*, meaning "few"), are a third source of uniform DNA or RNA segments.

As shown in Figure 17–5, the preparation of synthetic oligonucleotides depends on the capacity to selectively block either the 3′ end or the 5′ end of nucleotides. This ensures that the condensation reactions forming the sugar-phosphate bonds occur in the proper sequence to link the nucleotides in the desired order. When this procedure was first developed, synthesis of a segment containing from 12 to 20 nucleotides required several days. It has now become possible to attach one end of the chain to a resin bead in a column through which nucleotides and the appropriate chemical reagents can be passed sequentially, leading to automation of the procedure. As a result, longer oligonucleotides can be synthesized, in a much shorter period of time.

CLONES AND VECTORS

The next requirement for the detailed study of DNA was a methodology for obtaining gDNA, cDNA, or synthetic oligonucleotide molecules in large quantities. The machinery for duplicating DNA was already present in *E. coli* and other bacterial cells. What were needed were vectors that could carry the DNA molecules of interest into these cells and initiate replication. Again, prokaryotes and viruses provided the solution.

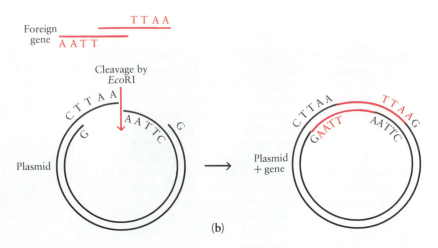

(b)

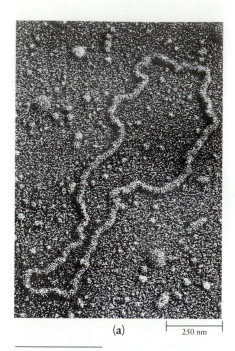

(a)

250 nm

17–6 (a) *The plasmid pSC101.* (b) *The restriction enzyme EcoRI cleaves this plasmid at the sequence GAATTC, leaving "sticky" ends exposed. These ends, consisting of TTAA and AATT sequences, can join with any other segment of DNA that has been cleaved by the same enzyme. Thus it is possible to splice a foreign gene into the plasmid. (In this diagram, the length of the GAATTC sequences is exaggerated and the lengths of the other portions of both the foreign gene and the plasmid are compressed.)*

When such plasmids, incorporating a foreign gene, are released into a medium in which bacteria are growing, they are taken up by some of the bacterial cells. As these cells multiply, the plasmids also replicate; the result is an increasing number of cells, all making copies of the same plasmid. The plasmids can then be separated from the other contents of the cells and treated with EcoRI to release multiple copies of the cloned gene.

Plasmids as Vectors

Not long after the discovery of the restriction enzyme *Eco*RI, Stanley Cohen and his coworkers from the Stanford University Medical School isolated a small plasmid of *E. coli*, designated pSC101 (note the initials of its discoverer), that makes the bacteria resistant to the antibiotic tetracycline (Figure 17–6). pSC101 has only one GAATTC sequence in its entire molecule, and, as a consequence, it is cleaved at only one site by *Eco*RI. The insertion of a small segment of foreign DNA into the plasmid does not affect either the uptake of the plasmid by *E. coli*, its capacity to make the recipient cells tetracycline-resistant, or its ability to replicate. Typically, the plasmid replicates several times in its host cell, producing about 10 new plasmids per cell. Because the pSC101 plasmids containing the foreign DNA segment are larger than those that do not contain it, after replication they can be readily isolated and collected. Treatment of the isolated plasmids by *Eco*RI releases the foreign DNA, which can then be separated from the pSC101 DNA by electrophoresis (see Figure 17–1).

The discovery that pSC101 could be used as a vector for foreign DNA opened the way to the production of uniform, identical segments of DNA in large enough amounts to be analyzed by biochemical means. Such multiple copies are known as **clones,** a term also applied to genetically identical bacteria and other organisms produced asexually from a single parent cell or organism.

Lambda and Cosmids

Plasmids are useful as vectors because they multiply rapidly and are easily taken up by bacteria through the cell membrane. Their chief shortcoming is that the life of a plasmid is very competitive, and those that multiply the most rapidly have an evolutionary advantage. And, as you might expect, the larger the plasmid, the more time it requires in order to replicate. Thus, although short sequences of foreign DNA are tolerated, longer segments tend to be eliminated over the generations. Plasmids are reliable vectors only for segments of up to about 4,000 base pairs in length.

Specially modified strains of bacteriophage lambda can be used to clone larger DNAs, segments of up to 20,000 base pairs in length. Such preparations are made by removing the central section of the phage DNA, using strains of lambda that have appropriately located recognition sites for restriction enzymes. This central section contains genes that are involved solely with the integration of lambda into the bacterial host chromosome (see Figure 16–20, page 335); they are not required for infection of the cell or multiplication within it. This large section of the lambda genome is replaced with foreign DNA. If the foreign DNA is about the same length as the deleted DNA—that is, some 20,000 base pairs—it can be introduced into the *E. coli* cell and will multiply as the virus undergoes its usual cycles of lytic infection.

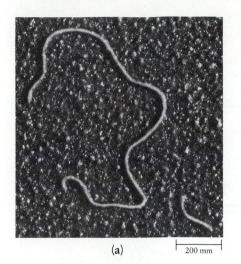

(a) 200 mm

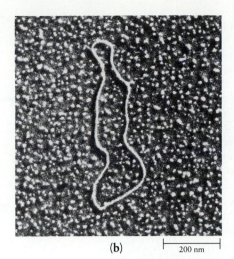

(b) 200 nm

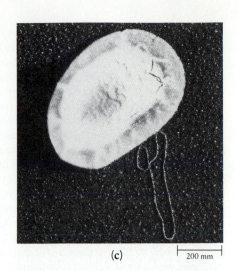

(c) 200 mm

17–7 *Insertion of a foreign gene into a bacterial cell.* (a) *A plasmid has been cut open with a restriction enzyme, leaving two "sticky" ends. A small segment of foreign DNA (lower right) also has "sticky" ends that can join with the ends of the plasmid by base pairing.* (b) *With the aid of DNA ligases, the foreign DNA has been spliced into the plasmid.* (c) *The plasmid, now containing the foreign DNA, is about to enter a bacterial cell. (Huntington Potter and David Dressler,* LIFE *Magazine, 1980, Time, Inc.)*

In order to clone still larger DNA segments, biologists have taken further advantage of the unusual and marvelous features of lambda, in this case, its cohesive ends. As we noted in the last chapter, the COS regions of lambda are the recognition sites for the enzyme that cuts the newly synthesized DNA into individual viral chromosomes and packs them into their protein coats. As it turns out, all that is required for the packaging to take place are two COS regions 35,000 to 40,000 base pairs apart. Using restriction enzymes, biologists are now able to construct DNA segments flanked by appropriately placed COS regions; these segments are then conveniently packaged into lambda protein coats by the viral enzyme. The protein coat gains them entry into the bacterial cell; once inside, the introduced DNA segments, like the normal lambda chromosome, assume a circular form (see Figure 16–19, page 334) and begin to multiply like plasmids. These vectors are appropriately known as **cosmids**.

NUCLEIC ACID HYBRIDIZATION

One of the earliest—and still one of the most useful—methods for studying DNA and RNA molecules is nucleic acid hybridization. This technique takes advantage of the base-pairing properties of nucleic acids. If DNA is heated, the hydrogen bonds holding the two strands together are broken and the strands separate; the three-dimensional structure of the molecule is lost, and the molecule is said to be denatured. When the solution is cooled, the hydrogen bonds re-form, reconstituting the double helix.

When denatured DNAs from different sources are mixed together, they undergo random collisions. If two strands with nearly complementary sequences (the matching need not be exact) find each other, they will form a hybrid double helix. The extent to which the segments from two samples reassociate and the speed with which they do so provide an estimate of the similarity between their nucleotide sequences. As we shall see in Chapter 20, this method is being used with increasing success to determine the evolutionary relationships among organisms.

When denatured DNA is mixed with single-stranded RNA, DNA-RNA hybrids can be formed. As we saw in Chapter 15 (page 308), this technique made possible the first identification and isolation of mRNA. Now, mRNA molecules are routinely used to identify and isolate corresponding DNA segments, and vice versa.

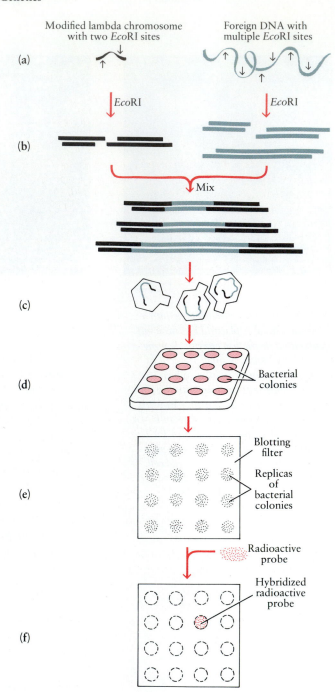

17–8 *The use of a radioactive probe to locate a DNA segment of interest.* (**a**) *Multiple copies of the DNA of a vector, such as a modified lambda chromosome, are cut with a suitable restriction enzyme, such as EcoRI. Foreign DNA containing the segment of interest is cut with the same restriction enzyme.* (**b**) *The restriction fragments of the two DNAs are mixed under conditions that allow the foreign DNA to become incorporated into the lambda chromosomes.* (**c**) *The resulting DNA molecules are enclosed in protein capsids, in a reaction catalyzed by an enzyme obtained from lambda. This produces phages capable of infecting bacterial cells.* (**d**) *Colonies, each consisting of a few identical bacterial cells, are established in small wells on a culture plate and infected with the phages. On the average, only one phage is added to each colony. After the cells have multiplied a number of times, which allows the vectors to multiply also, a replica of the bacterial colonies is obtained by blotting them with a specially prepared filter.* (**e**) *Chemical treatments release, denature, and attach the DNA to the filter. The filter is then incubated with a single-stranded radioactive probe, complementary to the DNA segment of interest.* (**f**) *After incubation, all single-stranded nucleic acid is washed from the filter, and the colony replica containing radioactive hybrid molecules is identified. The corresponding original colony is then treated to release the vectors, which can be purified and cloned in other bacteria to produce multiple copies of the DNA of interest.*

Radioactive Probes

Before a particular DNA or mRNA segment of interest can be cleaved, cloned, sequenced, or otherwise manipulated, it must first be located and isolated. One of the most important tools in accomplishing such location and isolation is an application of nucleic acid hybridization, using radioactive probes. These probes are short segments of single-stranded DNA or RNA labeled with a radioactive isotope. They may be mRNA or synthetic oligonucleotides or cloned fragments of gDNA or cDNA. Such probes may be used in a variety of ways.

For example, suppose that you are interested in studying a particular gene. One procedure that you could follow to locate and isolate DNA containing the gene is illustrated in Figure 17–8. First, a restriction enzyme is used to fragment DNA molecules known to contain the gene, and the fragments are incorporated into

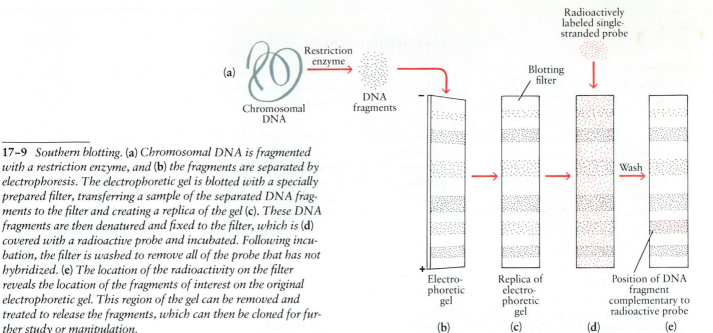

17-9 *Southern blotting.* (a) *Chromosomal DNA is fragmented with a restriction enzyme, and* (b) *the fragments are separated by electrophoresis. The electrophoretic gel is blotted with a specially prepared filter, transferring a sample of the separated DNA fragments to the filter and creating a replica of the gel* (c). *These DNA fragments are then denatured and fixed to the filter, which is* (d) *covered with a radioactive probe and incubated. Following incubation, the filter is washed to remove all of the probe that has not hybridized.* (e) *The location of the radioactivity on the filter reveals the location of the fragments of interest on the original electrophoretic gel. This region of the gel can be removed and treated to release the fragments, which can then be cloned for further study or manipulation.*

appropriate vectors. Some of the vectors carry fragments containing the gene of interest (or, more likely, portions of that gene), while others carry other fragments, which are, for your present purposes, not of particular interest. The vectors are then introduced into a series of identical bacterial colonies, each containing a very small number of cells. Approximately one vector is added to each colony, with the result that—if the vector is taken up by a cell—as each colony multiplies, an increasing number of copies of that one vector will be present. Using a special type of blotting filter, a replica of the colonies is made by transferring a portion of each colony to the filter, which can be treated in such a way that the DNA is released, denatured, and chemically attached to the filter. Next, the replica is exposed to a radioactive probe, previously prepared from, for example, a segment of corresponding mRNA. The probe is allowed to hybridize with any complementary DNA it encounters, and then all of the single-stranded nucleic acid is washed out of the filter; any double-stranded molecules that have formed remain in place on the filter. The colony replicas that are radioactive after this treatment identify the original colonies containing the vectors with the DNA segment you wish to study. Once the colony or colonies are identified, the vectors bearing the desired fragment can be collected, introduced into other bacterial cells, and cloned to produce large quantities of that DNA segment.

An analogous technique, illustrated in Figure 17–9, is known as Southern blotting (named after molecular biologist E. M. Southern, who developed it). DNA is first digested with one or more restriction enzymes. The resulting restriction fragments are separated by size on an electrophoretic gel and then blotted onto a filter. The filter thus contains a replica of the fragments on the gel, just as the filter in the previously described method contained a replica showing the positions of the bacterial colonies. The DNA fragments on the filter are then denatured, attached, and exposed to a radioactive probe, which, following the treatment outlined above, reveals the location of fragments containing a sequence complementary to the probe. Once the positions of the fragments of interest are identified, those fragments can be purified from the electrophoretic gel and used for subsequent study or manipulation. Southern blotting provides a very sensitive method for detecting specific nucleotide sequences of interest.

DNA SEQUENCING

The development of techniques for cleaving DNA molecules into smaller pieces and cloning them into multiple copies now makes it possible, in principle, to determine the nucleotide sequence of any isolated DNA molecule. One of the most important features of restriction enzymes is that different enzymes cleave DNA molecules at different sites (Figure 17–10). Cleavage of a DNA molecule with one restriction enzyme produces one particular set of short DNA fragments; cleavage of an identical DNA molecule with a different restriction enzyme produces a different set of short DNA fragments. The fragments of each set can be separated from each other by electrophoresis on the basis of their lengths and then cloned into multiple copies for use in sequencing studies.

One of the earliest methods for determining the nucleotide sequence of short DNA segments—and also one of the easiest to understand—was worked out by Allan Maxam and Walter Gilbert. It uses radioactive labeling techniques and electrophoresis. Multiple copies of single-stranded fragments are labeled on one end by removing a phosphate and substituting a radioactive phosphate in its place. Then, the mixture of identically labeled DNA molecules is separated into four portions. Each portion is treated chemically in such a way that one base—C, for example—and no other, is damaged and removed from the molecule, breaking the strand at that site. Crucial to the method is the fact that the chemical treatment is regulated so that not all the C's are damaged and those that are damaged are hit at random. Take, for example, a hypothetical 10-nucleotide sequence in which the base at the 5′ end bears a radioactive label, as indicated by the color:

$$5'\ 3'$$
$$\text{G}-\text{A}-\text{T}-\text{C}-\text{A}-\text{G}-\text{C}-\text{T}-\text{A}-\text{G}$$

Random excision of the C's produces the following mixture:

$$\text{G}-\text{A}-\text{T}-\text{C}-\text{A}-\text{G}$$
$$\text{T}-\text{A}-\text{G}$$
$$\text{G}-\text{A}-\text{T}$$
$$\text{A}-\text{G}-\text{C}-\text{T}-\text{A}-\text{G}$$
$$\text{A}-\text{G}$$

Of these fragments, two are radioactively labeled at the 5′ end.

17–10 (a) *Simian virus 40, or SV40, has an icosahedral (20-sided) protein coat enclosing its chromosome, a circular molecule of double-stranded DNA. SV40, originally isolated from monkeys, is of particular interest because it can produce cancers in baby hamsters and other laboratory animals.* (b) *Because the SV40 chromosome has only one GAATTC sequence, EcoRI cleaves the SV40 DNA at only one point, which becomes the reference point for other analyses and manipulations.* (c) *The restriction enzyme HindIII cleaves the SV40 DNA at six points, yielding six fragments.* (d) *Electrophoresis separates the fragments according to size. Lane 1 shows uncut DNA, and lane 2 shows SV40 DNA cleaved by HindIII.*

The use of different restriction enzymes to cleave DNA molecules, such as the SV40 chromosome, into different sets of specific fragments is an essential component in determining the complete nucleotide sequence of the molecules.

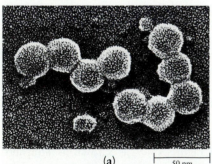

(a)   50 nm

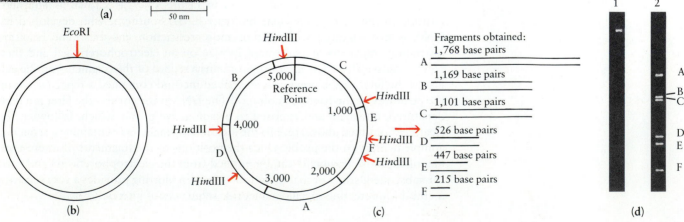

(b)

Fragments obtained:

A 1,768 base pairs

B 1,169 base pairs

C 1,101 base pairs

D 526 base pairs

E 447 base pairs

F 215 base pairs

(c)

Lane 1   Lane 2

(d)

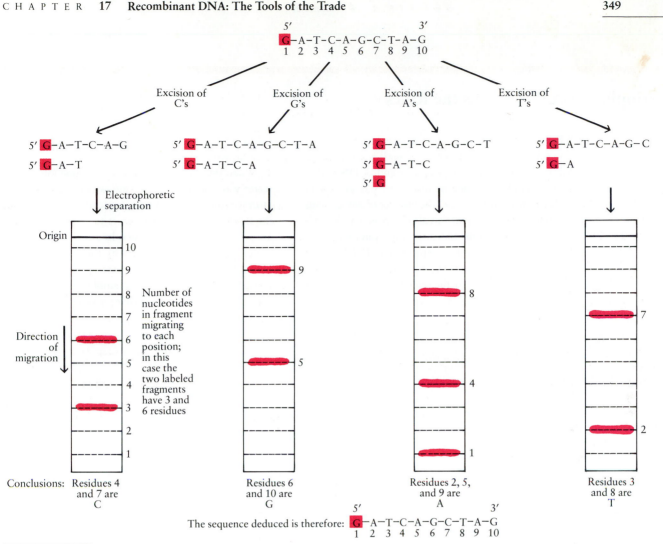

17–11 *The sequencing of a segment of a DNA molecule by the method of Maxam and Gilbert. The single-stranded segment (present in multiple copies) is radioactively labeled at the 5' end. The solution containing the labeled DNA is divided into four portions, each of which is subjected to a different chemical treatment to break the molecule at only one of the four bases. The resulting fragments are then separated by electrophoresis on lanes that are calibrated to indicate the positions occupied by fragments of different lengths. Thus the location of the radioactivity reveals the number of nucleotides contained in the labeled fragments. By combining the information gained from each procedure, the sequence of the complete segment can be read directly off the lanes.*

The mixture of nucleotide fragments is then placed on a gel and subjected to electrophoresis, which separates them on the basis of their lengths. The shorter fragments move farther through the gel than do the longer fragments. The positions to which the two radioactive fragments move, which is revealed by their radioactive label, can be used to determine the number of nucleotides the labeled fragments contain. This same procedure, repeated for all four bases, results in a sequence ladder from which the order of nucleotides can be read off directly (Figure 17–11).

Because the sets of fragments produced by different restriction enzymes overlap, the information obtained from sequencing the fragments can be pieced together like a puzzle to reveal the entire sequence of the DNA molecule or isolated gene.

Another, widely used method for determining the nucleotide sequences of short DNA segments, using enzymatic rather than chemical procedures, was devised by Frederick Sanger. Using this method, Sanger provided the first complete sequence of a genome, that of the bacteriophage φX174 (see essay on the next page). Gilbert and Sanger were honored with the Nobel Prize in 1980 for their accomplishments in nucleic acid sequencing. It was Sanger's second; his first, received 22 years earlier, was for the first sequencing of a protein, insulin.

Today, the sequencing of genes from sources that range from the smallest viruses to human cells is going forward in hundreds of laboratories around the world.

Bacteriophage φX174 Breaks the Rules

The first complete nucleotide sequence to be worked out was that for the DNA of a small bacteriophage known as φX174. The single-stranded DNA that forms the chromosome of this virus was known to code for nine proteins, and the number of amino acids in each of these proteins was also known. This knowledge raised a serious and interesting question, even before the analysis was complete. The DNA (which contains 5,375 nucleotides) was insufficient to code for the nine proteins by the triplet code hypothesis. It simply was not long enough. Were the concepts established by a quarter-century of intensive effort to be overthrown by a submicroscopic particle?

When the nucleotide sequence became known, its secret was revealed. The investigators had originally assumed that each gene was physically separate along the DNA molecule. However, it turns out that there are pairs of overlapping genes. In other words, different genes in φX174 are coded by the same regions of DNA but using different reading frames. More recently, several additional examples of overlapping genes have been found in other viruses.

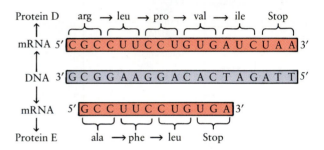

A segment of the DNA of φX174, showing a portion of the genes for proteins D and E. Both are coded by the same stretch of DNA; their reading frames, however, are different. The gene for protein E is completely contained within the gene for D, an entirely unrelated protein.

BIOTECHNOLOGY

Early in the course of recombinant DNA research, biologists realized that if the segments of DNA coding for certain proteins (particularly those of medical or agricultural importance) could be transferred into bacteria and expressed, the bacteria could function as "factories," providing a virtually limitless source of the proteins.

The first synthesis of a mammalian protein in a bacterial cell was reported by Keiichi Itakura and his associates at the City of Hope Medical Center. They selected the gene for the hormone somatostatin because it was a small protein (only 14 amino acids) and it could be detected in very small amounts.

The amino acid sequence of somatostatin is alanine-glycine-cysteine-lysine-asparagine-phenylalanine-phenylalanine-tryptophan-lysine-threonine-phenylalanine-threonine-serine-cysteine. Knowing this sequence, the investigators determined what one of the possible sequences of nucleotides in the DNA could be (you can do this, too) and then prepared a synthetic gene, including an initiation codon, bonding together the nucleotides one by one (see page 343).

Then, using a restriction enzyme, the synthetic gene was spliced into plasmids carrying genes for drug resistance, and the plasmids were supplied to *E. coli* cells. Some cells took up the plasmids (as evidenced by the fact that they were now drug-resistant), but there was no evidence of somatostatin synthesis. Itakura's

17–12 *The somatostatin project. A gene for somatostatin, synthesized artificially, was fused to the beta-galactosidase gene in a bacterial plasmid. Introduced into* E. coli, *this plasmid directs the synthesis of a hybrid protein that begins as beta-galactosidase but ends as somatostatin. Cyanogen bromide cleaves the protein at methionine, thus releasing the hormone intact (together with many fragments of beta-galactosidase, from which it can be separated).*

Reading in a clockwise direction, the nucleotide sequence shown in this diagram is the 5′ to 3′ sequence of the inactive strand of the DNA double helix. It is complementary to the template strand from which the mRNA is transcribed, and, with the exception of the substitution of thymine for uracil, it is identical to the 5′ to 3′ sequence of the transcribed mRNA molecule.

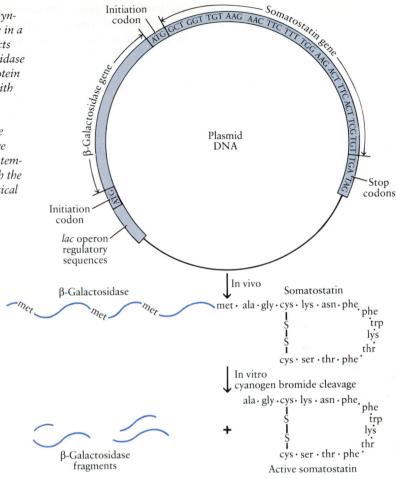

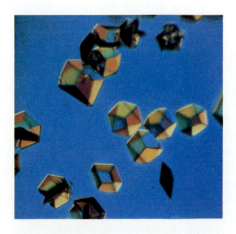

17–13 *Crystals of human insulin, known as "Humulin," produced by bacteria that have been modified by genetic engineering.*

group needed some way to turn the gene on. They inserted the regulatory sequences for the *lac* operon into the plasmid upstream from the somatostatin gene. When they turned the operon on, they were able to detect somatostatin, but only in very small amounts. Cells degrade foreign proteins, and the hormone was being destroyed almost as fast as it was being produced. Finally, to protect the newly synthesized somatostatin from the bacterial enzymes, the scientists spliced the somatostatin gene onto the beta-galactosidase gene, the first structural gene of the *lac* operon. At the point of the splice, for reasons that we shall see shortly, they retained the initiation codon of the synthetic gene; as you will recall, this is also the codon for methionine.

These plasmids were reintroduced into host cells, and, at last, clones of bacteria were obtained that were able to synthesize the hybrid beta-galactosidase-somatostatin protein. The protein was then isolated and treated with the chemical cyanogen bromide, which cleaved it exactly at the methionine insert, releasing the somatostatin (Figure 17–12). The somatostatin, tested in laboratory animals, was found to have the biological activity of the natural hormone. In other words, it worked!

More recently, genes for other medically useful proteins have been successfully introduced into bacterial cells and have functioned in protein synthesis. One example is the gene for human insulin (Figure 17–13). Another is the gene for somatotropin, or growth hormone, which is used to treat some forms of dwarfism in children. Somatotropin had previously been available only in very small amounts, being extracted from human pituitary glands. Several cases of contamination of these pituitary extracts by viruses causing human neurological diseases made the bacterial synthesis of this hormone of crucial medical importance.

On the economic front, the bacterial synthesis of other proteins is of increasing importance. For example, the enzyme rennin, extracted from calves' stomachs and used by the dairy industry for cheesemaking, has been produced by recombinant DNA technology. More recently, scientists at Cornell University have succeeded in inducing bacteria to synthesize the enzyme cellulase, which is produced in nature by certain fungi. This enzyme converts cellulose, which is indigestible by most organisms, into glucose, a food molecule of major importance.

Vaccines against viral disease are another important product of the new biotechnology. All viruses, as you know, consist of nucleic acid wrapped in a protein coat. It is the exterior proteins of the capsid that determine whether or not the virus can attach itself to and enter a target cell. In the animal bloodstream, these proteins of the virus, recognized by cells of the immune system as foreign, evoke the formation of antibodies, molecules that play a crucial role in future immunity against the virus. Most vaccines are made using killed or altered forms of the virus particles. Vaccines produced from synthetic protein coats alone are safer, since without the viral nucleic acid, no contamination of the vaccine by infectious particles can occur.

GENE TRANSFER: THE CASE OF THE GLOWING TOBACCO PLANT

Agrobacterium tumefaciens is a common soil bacterium that infects plants, producing a lump, or tumor, of tissue known as a crown gall (Figure 17–14). Even if the bacteria are destroyed with antibiotics, the gall, once started, continues to grow. Investigations have revealed that the cause of the gall is not *Agrobacterium tumefaciens* itself but rather a large (200,000 base pairs) plasmid of the bacterium. A portion of this plasmid, which is known as Ti (for tumor-inducing), becomes integrated into the DNA of the host plant cell. Crown gall disease is the only naturally occurring genetic recombination yet recorded between prokaryotic and eukaryotic cells.

To try to understand how Ti exerts its effects, recombinant DNA technology has been used to examine the genes of the Ti plasmid. Three of its genes, it has been found, direct the synthesis of plant hormones that act directly on the gall cells to promote their growth. One or more additional genes subvert the cell's machinery to produce unusual amino acids, called opines, which can be used by the gall cells but not by normal cells. Moreover, opines act as molecular aphrodisiacs, increasing bacterial conjugation and thus promoting the spread of the Ti plasmid to uninfected bacterial cells. In effect, Ti takes over and directs the activities of both of its hosts—the bacterial cells and the plant cells—to promote its own multiplication.

The Ti plasmid has attracted considerable attention not only because of its remarkable powers but also because of its potential as a vector to ferry useful genes into crop plants. Among the genes that are candidates for such transfer are those conferring resistance to major plant diseases and those required for C_4 photosynthesis (page 223) and for nitrogen fixation (page 664).

In a notable recent experiment, investigators from the University of California at San Diego isolated the gene for the enzyme luciferase from fireflies. The substrate of luciferase is a protein called luciferin; in the presence of oxygen, luciferin plus luciferase plus ATP produces bioluminescence, as seen in the flash of the firefly. The luciferase gene was cloned in *E. coli* and then spliced into the chromosome of a plant virus, which provided a regulatory sequence for the gene.

17–14 *Crown galls growing on a tobacco stem.*

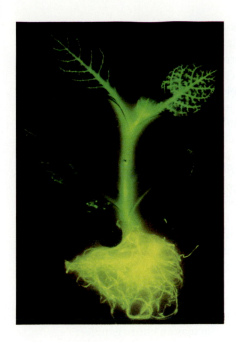

17–15 *Genes for the production of the enzyme luciferase are inserted into cells isolated from the normal tobacco plant* (Nicotiana tabacum), *using* Agrobacterium *as a vector. After the undifferentiated callus cells develop into a mature plant, the cells that have incorporated the luciferase gene into their DNA are luminescent in the presence of luciferin, ATP, and oxygen.*

The altered viral chromosome was then inserted into Ti plasmids, the plasmids were transferred to the bacteria, and the bacteria were incubated with tobacco leaf cells. The cells formed a mass of tissue, known as a callus, from which, in a suitable growth medium, new plants were produced. The new plants were watered with a solution containing luciferin. You can see the results in Figure 17–15: the plants shone!

One of the technical problems encountered in attempts at gene transfer is knowing whether a particular gene has actually been introduced into a new host cell and, if transferred, whether it is directing the synthesis of protein. Luciferase clearly provides an extraordinary signal that gene transfer has taken place, and experiments are now underway to combine it with other genes that are candidates for transfer. In the meantime, these experiments are a glowing example of the ingenuity of both molecular geneticists and the Ti plasmid.

SUMMARY

Recombinant DNA technology includes four basic techniques: (1) methods for obtaining specific, uniform DNA segments of a size suitable for analysis and manipulation; (2) DNA cloning, which makes it possible to produce identical DNA segments in large quantities; (3) nucleic acid hybridization, a method for identifying specific segments of DNA and RNA and for estimating similarities between nucleic acids from different sources; and (4) DNA sequencing, the determination of the exact order of nucleotides in a DNA segment. Another important tool used in recombinant DNA work is electrophoresis, which provides a means of separating DNA segments from one another on the basis of their size.

Specific, uniform DNA segments may be produced by cleaving DNA molecules with restriction enzymes, by transcribing mRNA into DNA with the enzyme reverse transcriptase, or through the laboratory synthesis of oligonucleotides. Restriction enzymes are found in nature in bacterial cells. They cleave DNA molecules at specific recognition sequences that are typically four to eight nucleotides in length. Their function in the bacterial cells in which they occur is the degradation of foreign DNA molecules. The DNA of the bacterial cell is protected from its own restriction enzymes by the methylation of nucleotides at the recognition sequences. Some restriction enzymes produce straight cuts through the DNA molecule. Others cut unevenly, leaving "sticky" ends that can then join by complementary base pairing with other fragments produced by the same enzyme. This makes it possible to combine DNA segments from different sources. The DNA segments produced by cleavage with restriction enzymes are known as genomic DNA, or gDNA.

Reverse transcriptase is an enzyme produced by certain RNA viruses, known as retroviruses. When these viruses infect a host cell, reverse transcriptase catalyzes the synthesis of DNA from the viral RNA template; viral mRNA, coding for viral proteins, is transcribed from this DNA, as is the viral RNA to be packaged into new virus particles. In the laboratory, reverse transcriptase can be used to synthesize DNA from an RNA template, such as an mRNA molecule. DNA segments produced in this manner are known as complementary DNA, or cDNA.

Clones, in molecular genetics, are multiple copies of the same DNA sequence. The sequences to be cloned are introduced into bacterial cells by means of vectors. Plasmids and bacteriophages, particularly bacteriophage lambda, are used as vectors; cosmids are synthetic vectors that combine the cohesive ends (COS regions) of lambda with the DNA segment to be cloned. Once in the bacterial cell, the vector and the foreign DNA it carries are replicated, and the multiple copies can be harvested from the cells.

Nucleic acid hybridization techniques depend upon the capacity of a single strand of RNA or DNA (which can be released from the double helix by heating) to combine, or hybridize, with another strand with a complementary nucleotide sequence. The greater the similarity between the nucleotide sequences of the two strands, the more rapid and more complete the hybridization. This technique makes possible a range of procedures, including estimating the evolutionary affinities of different organisms, determining relationships between DNAs and transcribed RNAs, and the use of radioactive probes to identify specific nucleotide sequences of interest. Radioactive probes can be used, for example, to identify bacterial colonies in which a particular DNA sequence, introduced by a vector, is being cloned, or to identify segments of interest on an electrophoretic gel.

DNA sequencing is the determination, nucleotide by nucleotide, of the sequence of bases in a molecule of DNA. Two principal techniques of sequencing are in current use, one involving enzymatic methods and the other chemical methods. Sequencing depends on the availability of multiple copies of uniform DNA segments, cloned from the DNA fragments produced by restriction enzymes. By combining sequencing information for sets of short segments produced by different restriction enzymes, molecular biologists can determine the complete sequence of a long DNA segment (such as an entire gene).

Techniques have now been developed for incorporating specific genes into suitable vectors, introducing them into bacterial cells, and inducing the bacterial cells to synthesize the proteins coded by the genes. Human insulin and human growth hormone are two medically important proteins now produced in this way.

The Ti plasmid of *Agrobacterium tumefaciens,* the cause of crown gall disease in plants, is being used as a vector for introducing genes into plant cells. Incorporation of the gene for the enzyme luciferase into the Ti plasmid makes it possible to determine visually if the genes carried by the plasmid have been successfully transferred to the plant cells and are being expressed.

QUESTIONS

1. Distinguish between the following: gDNA/cDNA; restriction enzyme/methylating enzyme; retrovirus/reverse transcriptase.

2. Identify the four techniques that form the basis of recombinant DNA technology and give a brief description of each.

3. What is electrophoresis? Why is it of such great value in recombinant DNA studies?

4. How is the DNA of a bacterial cell protected from the action of its own restriction enzymes? Why does this protection not extend to foreign DNA introduced into the cell?

5. Describe the role of "sticky" ends in recombinant DNA technology. What enzyme is required to complete the recombination?

6. Compare and contrast the features of plasmids, bacteriophage lambda, and cosmids as vectors. What are the advantages of the Ti plasmid as a vector for introducing genes into plant cells?

7. Suppose you treated a DNA molecule with a particular restriction enzyme and obtained five fragments, which you separated and cloned into multiple copies. Using the multiple copies, you then sequenced the five fragments. What would you do next to establish the order of the five fragments in the original molecule?

8. Suppose you wish to locate on the chromosome the gene coding for a small protein molecule. You know the amino acid sequence of the protein, and you have the technical skill to synthesize an artificial mRNA molecule with any nucleotide sequence you choose. How would you go about locating the gene? Suppose you wish to separate the gene from the rest of the chromosome. How would you proceed?

9. Why, in the somatostatin project, did the investigators link the synthetic somatostatin gene to regulatory elements of the *lac* operon? Why did they also include the first structural gene of the *lac* operon (the gene coding for beta-galactosidase) in the plasmid that was introduced into the *E. coli* cells?

10. *E. coli* cells used in recombinant DNA studies are "disabled"; that is, they lack the capacity to synthesize key components of their cell walls, for example, or to make a nitrogenous base, such as thymine. Consequently, they are able to survive only in an enriched laboratory medium. Why is such a precaution taken by molecular biologists?

CHAPTER 18

The Molecular Genetics of Eukaryotes

The discovery, early in the history of molecular genetics, that the genetic code is apparently universal—the same in *Escherichia coli, Homo sapiens,* and all other organisms—is awesome evidence that all living things are descended from a common ancestor. It originally tempted molecular biologists to think that the eukaryotic chromosome would turn out to be simply a large-scale version of the *E. coli* chromosome. This has not proved to be the case. As studies of the molecular genetics of eukaryotes have progressed, increasingly aided by the tools of recombinant DNA technology, the gulf between eukaryotes and prokaryotes has widened rather than narrowed. It is now clear that, at the molecular level, there are many important differences in the genetics of eukaryotes and prokaryotes—some expected and some very surprising. As we shall see, these differences include (1) a far greater quantity of DNA in the eukaryotic cell; (2) a great deal of repetition in this DNA, with much of it lacking any apparent function; (3) a close association of the DNA with proteins that play a major role in chromosome structure; and (4) considerably more complexity in the organization of the protein-coding sequences of the DNA and the regulation of their expression.

THE EUKARYOTIC CHROMOSOME

DNA is an "exquisitely thin filament," in the words of E. J. DuPraw, who calculated that a length sufficient to reach from the earth to the sun would weigh

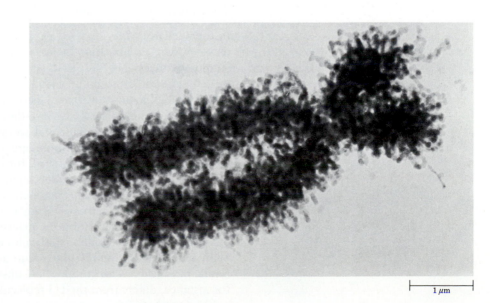

18–1 *In eukaryotic cells, the DNA of the genome is always found in association with protein. Chromosomes are made up of a protein-DNA combination called chromatin. At mitosis and meiosis, the chromatin condenses and becomes visible under the microscope. Each chromatid, according to present evidence, contains a single molecule of double-stranded DNA and is, by weight, about 60 percent protein. Shown here is human chromosome 12 at metaphase of mitosis.*

1 µm

18–2 *Computer-generated diagrams of B-DNA and Z-DNA. B-DNA is the form described by Watson and Crick. It consists of two strands of nucleotides twisted around each other, with the backbones forming a smooth, right-handed helix. In Z-DNA, the backbones of the nucleotide strands form a zig-zag, left-handed helix. The two forms are reversible, suggesting that the changes in structure might reflect some regulatory role.*

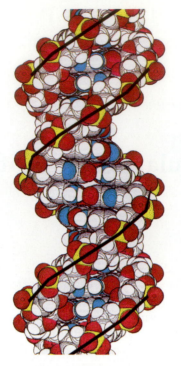

B-DNA   Z-DNA

only half a gram. The DNA of each eukaryotic chromosome is believed to be in the form of a single molecule. In a human chromosome, each of these molecules is believed to be from 3 to 4 centimeters long. Each diploid cell, with its 46 chromosomes, thus contains about 2 meters of DNA, and the entire human body contains some 25 billion kilometers of DNA double helix.

Double-stranded DNA is always a helix, and the helix is usually the tightly coiled right-handed helix, known as the B form, first described by Watson and Crick. However, x-ray diffraction studies have shown that DNA can assume other helical conformations: A-DNA, which is also a right-handed helix, less tightly coiled than the B form; and Z-DNA, which is a left-handed helix (Figure 18–2). The left-handed coiling results in zigs and zags of the sugar-phosphate backbone, hence the name Z-DNA. It is hypothesized that changes in the configuration of the DNA molecule may affect the binding of proteins to the molecule, and that this, in turn, affects gene expression.

Structure of the Chromosome

In the nucleus of the eukaryotic cell, the DNA is always found combined with proteins. This combination, as we noted in Chapter 5, is known as chromatin ("colored threads") because of its staining properties. Chromatin is more than half protein, and the most abundant proteins, by weight, belong to a class of small polypeptides known as histones. Histones are positively charged (basic) and so are attracted to—and, in turn, attract—the negatively charged (acidic) DNA. They are always present in chromatin and are synthesized in large amounts during the S phase of the cell cycle. The histones are primarily responsible for the folding and packaging of DNA. In a human cell, for instance, the approximately 2 meters of DNA are packed into 46 cylinders that, when condensed at metaphase, have a combined length of only 200 nanometers.

There are five distinct types of histones, known as H1, H2A, H2B, H3, and H4. They are present in enormous quantities—about 30 million molecules of H1 per cell, and about 60 million molecules of each of the other four types per cell. With the exception of H1, the amino acid sequences of the histones are very similar in widely diverse groups of organisms. The H3 molecule of the garden pea, for instance, differs from the H3 molecule of the cow by only four amino acids out of a total of 135.

18-3 (a) *Chromatin that has been decondensed to reveal the beadlike nucleosomes. The distance between nucleosomes is about 10 to 11 nanometers, and the diameter of each bead is about 7 nanometers. The core of each nucleosome is composed of about 140 base pairs of DNA and an assembly of eight histone molecules. The strand of linker DNA between the nucleosome cores contains another 30 to 60 base pairs.*

(b) *The structure of a nucleosome. The negatively charged DNA coils twice around the protein core, composed of eight positively charged histone molecules. An H1 histone molecule (also positively charged) binds to the outer surface of the nucleosome.*

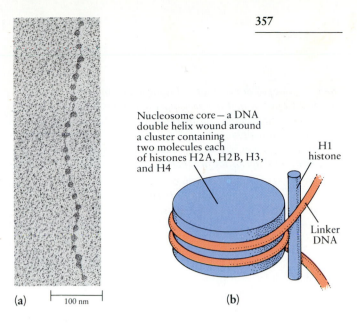

Nucleosome core — a DNA double helix wound around a cluster containing two molecules each of histones H2A, H2B, H3, and H4

H1 histone

Linker DNA

(a)   100 nm

(b)

The fundamental packing unit of chromatin is the **nucleosome** (Figure 18–3), which is composed of a core of two molecules each of histones H2A, H2B, H3, and H4—eight molecules in all—around which the DNA filament is wrapped twice, like thread around a spool. Each nucleosome contains, in addition to the eight histone molecules, about 140 pairs of nucleotides, and the strand of DNA between the nucleosomes contains another 30 to 60 nucleotide pairs. The fifth type of histone, H1, lies on this strand, outside the nucleosome core. When a fragment of DNA is tied up in a nucleosome, it is about one-sixth the length it would be if fully extended.

In electron micrographs, such as Figure 18–3a, the nucleosomes and the in-between, linking strands of DNA resemble beads on a string. In fact, electron micrographs provided the first clue to this remarkable structure, which has since been analyzed in detail by biochemical techniques.

Chromatin isolated at the next level of condensation is shown in Figure 18–4a. The details of this structure (which is 30 nanometers in diameter and is thus known as the 30-nanometer fiber) are not known. Two models that have been proposed are shown in Figure 18–4b and c.

18-4 (a) *Electron micrograph of a chromatid strand that is more tightly condensed than the beads-on-a-string form shown in Figure 18–3a. Because of its diameter, this form of chromatin is known as a 30-nanometer fiber.* (b, c) *Two proposed models for the way in which the beads-on-a-string form is packed into the 30-nanometer fiber.*

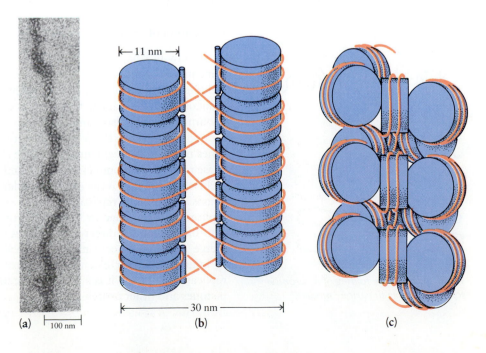

← 11 nm →

(a)   100 nm

← 30 nm →

(b)

(c)

18–5 *The looped domains of a chromosome can be revealed by chemical treatment that removes most of the histone proteins. In this electron micrograph of a single chromatid of an insect chromosome at mitotic anaphase, the DNA can be seen looping out from a nonhistone protein scaffolding. This scaffolding is essentially the same size and shape as the original anaphase chromosome.*

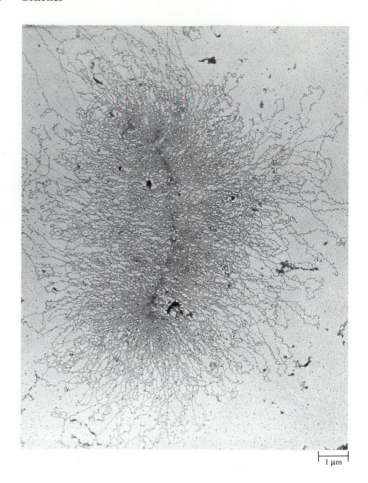

1 μm

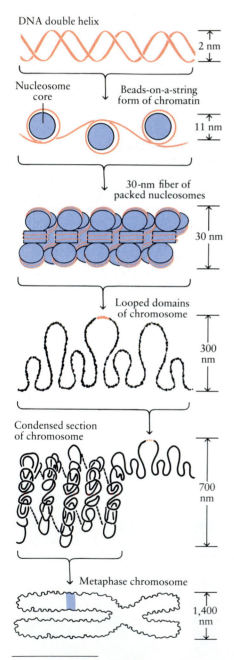

DNA double helix

2 nm

Nucleosome core

Beads-on-a-string form of chromatin

11 nm

30-nm fiber of packed nucleosomes

30 nm

Looped domains of chromosome

300 nm

Condensed section of chromosome

700 nm

Metaphase chromosome

1,400 nm

18–6 *Stages in the folding of a chromosome, according to various models.*

Further condensation occurs as the 30-nanometer fiber forms a series of loops, known as **looped domains** (Figure 18–5). Each looped domain coils until, ultimately, clusters of neighboring looped domains condense into the compact chromosomes that become visible during mitosis and meiosis (Figure 18–6).

Other proteins associated with the chromosome are the enzymes concerned with DNA and RNA synthesis, along with regulatory proteins, plus a large number and variety of unidentified molecules. Unlike the histones, these proteins vary from one cell type to another, both in their abundance and in their identity.

Replication of the Chromosome

As we saw in Chapter 14, replication of the DNA of eukaryotes is the same, in principle, as replication of the DNA of prokaryotes. Nucleotides in the energy-yielding form of triphosphates are assembled along a template DNA strand in the semiconservative manner first proposed by Watson and Crick and subsequently confirmed by Meselson and Stahl (page 293). As in prokaryotes, DNA polymerases operate only in the 5' to 3' direction; the 3' to 5' strand is synthesized as a series of Okazaki fragments, which are then joined together by the enzyme DNA ligase to form the complementary strand (see Figure 14–19, page 297).

In the comparatively small, circular prokaryotic chromosome, replication begins at a single replication origin and proceeds bidirectionally along two replication forks (page 295). In eukaryotic chromosomes, there are many replication origins, and bidirectional synthesis takes place until the replication forks merge (page 295). Replication is much slower in eukaryotes than in prokaryotes; in human cells, for example, the rate is about 50 base pairs per second per replication fork. As it is synthesized, eukaryotic DNA becomes complexed with histones and other proteins.

REGULATION OF GENE EXPRESSION IN EUKARYOTES

As we saw in Chapter 16, regulation of gene expression in prokaryotes largely involves the fine tuning of the metabolic machinery of the cell in response to changes in available nutrients in the environment. In eukaryotes, especially multicellular eukaryotes, the problems of regulation are very different. A multicellular organism usually starts life as a fertilized egg, the zygote. The zygote divides repeatedly by mitosis and cytokinesis, producing many cells. At some stage these cells begin to differentiate, becoming, for example, muscle cells, nerve cells, blood cells, intestinal cells, and so forth. Each cell type, as it differentiates, begins to produce characteristically different proteins that distinguish it from other types of cells. This is nicely illustrated by mammalian red blood cells. In the early stages of fetal life, developing red blood cells synthesize one type of fetal hemoglobin; red blood cells produced at later stages contain a second type of fetal hemoglobin; then, sometime after the birth of the organism, the developing red blood cells begin to produce the alpha and beta chains characteristic of adult hemoglobin. Thus the genes are expressed in a carefully controlled sequence, one after the other. The DNA segments that code for these hemoglobin molecules are expressed only in developing red blood cells.

There is evidence, however, that all of the genetic information originally present in the zygote is also present in every diploid cell of the organism (Figure 18-7). In other words, the DNA segments that code for hemoglobin (both the fetal types and the adult type) are present in skin cells and heart cells and liver cells and nerve cells and, indeed, in every one of the nearly 200 different types of cells in the body. Similarly, the DNA sequence that codes for the hormone insulin is present not only in the specialized cells of the pancreas that manufacture insulin but also in all the other cells. Since each type of cell produces only its characteristic proteins—and not the proteins characteristic of other cell types—it becomes apparent that differentiation of the cells of a multicellular organism depends on the inactivation of certain groups of genes and the activation of others.

Condensation of the Chromosome and Gene Expression

Many lines of evidence indicate that the degree of condensation of the DNA of the chromosome, as shown by chromatin staining, plays a major role in the regulation of gene expression in eukaryotic cells. Staining reveals two types of chromatin: **euchromatin,** the more open chromatin, which stains weakly, and **heterochromatin,** the more condensed chromatin, which stains strongly (Figure 18-8). During interphase, heterochromatin remains condensed, but euchromatin becomes dispersed. Transcription of DNA to RNA takes place only during interphase, when the euchromatin is dispersed.

Some regions of heterochromatin are constant from cell to cell and are never expressed. An example is the highly condensed chromatin located in the centromere region of the chromosome. This region, which does not code for protein, is believed to play a structural role in the movement of the chromosomes during mitosis and meiosis. Similarly, little or no transcription takes place from Barr bodies (page 267), which are X chromosomes that are tightly condensed and irreversibly inactivated.

Other regions of condensed chromatin, by contrast, vary from one type of cell to another within the same organism, reflecting, it is believed, the biosynthesis of different proteins by different types of cells. Also, as a cell differentiates during embryonic development, the proportion of heterochromatin to euchromatin increases as the cell becomes more specialized.

Additional evidence linking the degree of chromosome condensation to gene expression comes from studies of the giant chromosomes of insects (see page 277).

18-7 *A number of experiments have shown that early development does not result in permanent inactivation of genes or the loss of functional DNA. In experiments by J. B. Gurdon, illustrated here, nuclei were removed from intestinal cells of a tadpole and implanted into egg cells in which the nucleus had been destroyed. In some cases, the egg developed normally, indicating that the tadpole intestinal cell nucleus contained all the information required for all the cells of the organism. In other experiments, F. C. Steward demonstrated that under certain conditions, a single differentiated cell of a carrot can be persuaded to reconstitute an entire carrot plant.*

Labels in figure: Cell from tadpole intestine; Micropipette containing nucleus from tadpole cell; Egg from another frog—nucleus has been destroyed by ultraviolet radiation; Nucleus from tadpole cell is implanted in enucleated egg; Egg develops into frog with the same characteristics as original, nucleus-donating tadpole

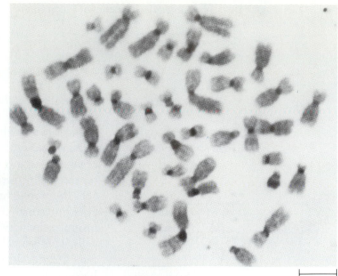

18-8 *Human chromosomes at metaphase, stained to distinguish the more tightly condensed heterochromatin from the less tightly condensed euchromatin. Notice the deeply stained heterochromatin in the region of the centromere.*

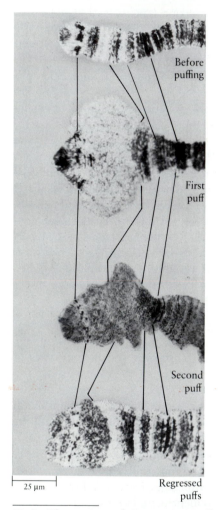

Before puffing

First puff

Second puff

25 μm

Regressed puffs

18-9 *Observations of chromosome puffs support the concept that the DNA is somehow unwound to make it available for RNA transcription. These puffs were observed in chromosomes of the Brazilian gnat, which, like the fruit fly, has giant chromosomes in some of its cells. Puffs occur normally but can also be induced experimentally. These puffs occurred in response to a hormone that causes molting. As the micrographs indicate, the puffs occurred sequentially along one chromosome.*

At various stages of larval growth in insects, it is possible to observe diffuse thickenings, or "puffs," in various regions of these chromosomes (Figure 18–9). The puffs are open loops of DNA, and studies with radioactive isotopes indicate that these loops are sites of rapid RNA synthesis. When ecdysone, a hormone that produces molting in insects, is injected, the puffs occur in a definite sequence that can be related to the developmental stage of the animal. For example, in one species of *Drosophila*, ecdysone initiates three new puffs and causes increases in 18 other puffs within 20 minutes after it is injected; during this same period, 12 other puffs decrease in size. After 4 to 6 hours, five additional puffs can be seen. The looping out of the DNA occurs before RNA synthesis is initiated. The mechanism of this spinning out of the chromosomal DNA is still poorly understood.

Methylation and Gene Expression

Once the DNA helix is formed, specific enzymes add methyl groups to nucleotides of cytosine. Methylation in eukaryotes is hypothesized to inhibit gene expression and so to provide a form of gene regulation. Methylcytosine is found almost exclusively at the complementary sequences (5')-C-G-(3') and (3')-G-C-(5'). In birds and mammals, 50 to 70 percent of cytosines in such sequences are methylated. Z-DNA (page 356) contains a large proportion of methylated cytosines, and Z-DNA is believed to be inactive. On the other hand, some eukaryotes —insects, for example—have no methylated bases in their DNA and yet they regulate the expression of their genes with no apparent loss of efficiency.

Regulation by Specific Binding Proteins

In eukaryotes, as in prokaryotes, transcription is also regulated by proteins that bind to specific sites on the DNA molecule. To date, only a few of these proteins and their binding sites have been identified. However, it is increasingly clear that this level of transcriptional control is far more complex in eukaryotes, particularly multicellular eukaryotes, than in prokaryotes. A gene in a multicellular organism appears to respond to the sum of many different regulatory proteins, some tending to turn the gene on and others to turn it off. The sites at which these regulatory proteins bind may be hundreds or even thousands of base pairs away from the promoter sequence at which RNA polymerase binds and transcription begins. This, as you might expect, adds to the difficulty of identifying the regulatory molecules and also of understanding exactly how they exert their effects. Recent research suggests that changes in the activities of some of these regulators are linked with the development of cancers, a matter we shall explore later in this chapter.

The DNA of the Energy Organelles

Two important generalizations have emerged from our study of genetics in the preceding chapters. One is that the genome of the offspring of sexually reproducing organisms is a new combination of maternal and paternal genes. The second, emphasized in the last three chapters, is that the genetic code is universal. As it turns out, neither of these generalizations is quite true.

The exceptions are found in the eukaryotic cell's energy organelles, the mitochondria and the chloroplasts. These organelles contain their own DNA, which, like that of prokaryotes, is not associated with histones. This DNA is replicated within the organelle, and new mitochondria and chloroplasts are formed by simple division, much as in *E. coli*. Because they are self-replicating and because their DNA resembles that of prokaryotes, mitochondria and chloroplasts are hypothesized to have evolved from prokaryotes that became parasites in primitive eukaryotic cells early in evolutionary history, a topic that we shall explore further in Chapter 22.

The DNA of mitochondria and chloroplasts is transcribed and translated, although most of the proteins of these organelles are coded by the nuclear DNA and are synthesized in and imported from the cytoplasm. The relatively small DNA molecule (16,569 base pairs) of human mitochondria has now been sequenced, making possible a variety of studies. Comparison of this DNA sequence with the RNAs and the few proteins made by the mitochondria has

revealed that the genetic code is not, strictly speaking, universal. For instance, in human mitochondria, UGA codes for tryptophan and not termination, and AUA codes for methionine and not isoleucine (see page 309). Also, mitochondria have many fewer tRNAs than either *E. coli* or eukaryotic cells, not enough to translate all of the possible codons by conventional base pairing. The origin of these discrepancies remains unknown and puzzling, for despite the differences in the mitochondrial code, the code employed in the translation of the information encoded in the DNA of the nucleus is the same in all eukaryotic organisms in which it has been checked and is the same as the code employed in prokaryotic cells.

In about two-thirds of all plant species, the male gamete contributes neither chloroplasts nor mitochondria to the zygote. Such non-Mendelian inheritance was first noted some 80 years ago in plants in which deficient chloroplasts produced mottled leaves —but only if the defect was present in the plant in which the egg cell developed. Similarly, in humans and other animal species, the sperm contributes almost no cytoplasm to the fertilized egg and, consequently, all of the mitochondria are maternal in origin. As we shall see in Chapter 50, the maternal transmission of the mitochondrion—coupled with the capacity to rapidly sequence the mitochondrial DNA of different individuals—is shedding new light on the evolution of *Homo sapiens*.

THE EUKARYOTIC GENOME

Examination of the DNA of eukaryotic cells revealed four major surprises. First, with some few exceptions, the amount of DNA per cell is the same for every diploid cell of any given species (which is not surprising), but the variations among different species are enormous. *Drosophila* has about 1.4×10^8 base pairs per haploid genome, only about 70 times more than *E. coli*. Humans (with approximately 3.5×10^9 base pairs) have 25 times as much as *Drosophila*, somewhat more than a mouse, but about the same amount as a toad (3.32×10^9 base pairs). The largest amount of DNA found so far has been located in a salamander with 8×10^{10} base pairs per haploid genome.

Second, in every eukaryotic cell, there is what appears to be a great excess of DNA, or at least of DNA whose functions are unknown. It is estimated that in eukaryotic cells less than 10 percent of all the DNA codes for proteins; in humans, it may even be as little as 1 percent. By contrast, prokaryotes, as we have seen, use their DNA very thriftily; viruses even more so. Except for regulatory or signal sequences, virtually all of their DNA is expressed.

Third, almost half of the DNA of the eukaryotic cell consists of nucleotide sequences that are repeated hundreds, even millions, of times. This was a particularly startling discovery. In *E. coli,* long the model for molecular geneticists, each chromosomal DNA molecule typically contains only one copy of any given gene. (The principal exceptions are the genes coding for the ribosomal RNAs.) Moreover, according to Mendelian genetics, a gene should be present only twice per diploid eukaryotic cell, not in a multitude of copies.

Introns

The fourth surprise—and perhaps the most unexpected of all—was that the protein-coding sequences of eukaryotic genes are usually not continuous but are instead interrupted by noncoding sequences. These noncoding interruptions within the gene are known as intervening sequences, or **introns,** and the coding sequences, the sequences that are expressed, are called **exons.**

Introns were discovered in the course of hybridization experiments, when investigators found that there was not a perfect match between eukaryotic messenger RNA molecules and the genes from which they were transcribed. The nucleotide sequences of the genes were much longer than their complementary mRNA molecules that were found in the cytoplasm. Subsequently, introns and exons were actually visualized in electron micrographs (Figure 18–10).

18–10 *This electron micrograph reveals the results of an experiment in which a single strand of DNA containing the gene coding for ovalbumin was hybridized with the messenger RNA for ovalbumin. The complementary sequences of the DNA and mRNA are held together by hydrogen bonds; there are eight such sequences, the exons labeled L and 1 through 7 in the accompanying diagram. Some segments of the DNA do not have corresponding mRNA segments and so loop out from the hybrid; these are the seven introns, labeled A through G. Only the exons are translated into protein.*

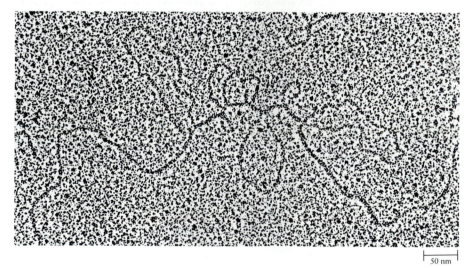

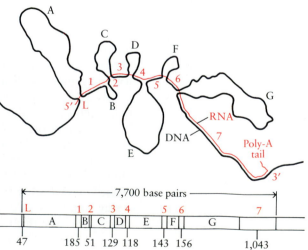

It is now known that most, but not all, structural genes of multicellular eukaryotes contain introns. The introns are transcribed onto RNA molecules and excised before translation. The number of introns per gene varies widely. For example, the gene for ovalbumin, a protein found in large quantities in vertebrate egg cells, has seven introns. By contrast, the mammalian gene for beta globin, one of the polypeptides of the hemoglobin molecule, has only two introns, one large and one small. In chickens, the gene for collagen, a very common protein (Figure 3-23, page 77), has 50 introns. Introns have also been found in genes coding for transfer RNAs and ribosomal RNAs, and even in some viruses.

In general, the more complex an organism and the more recently it has evolved, the larger and more abundant are its introns. It is not known which came first—continuous genes lacking introns or interrupted genes containing introns. It has been suggested that perhaps the latter came first, but that in bacteria and other unicellular organisms that are highly selected for rapid growth, any unneeded DNA has been eliminated in the course of their evolution.

The Function of Introns

Are introns accidents, or do they have a function? One suggestion for their continued existence in multicellular eukaryotes is that they promote recombination; crossing over during meiosis is more likely in genes containing introns than in genes lacking introns, just because of the distances involved. There are also indications that, in some cases, different exons code for different structural and functional segments, or domains, of the finished protein (Figure 18-11). For example, the central exon of the beta-globin gene codes for the domain of the polypeptide that holds the heme group, and the other two exons code for domains of the molecule that fold around this central portion (see Figure 3-27, page 79). It is hypothesized that new combinations of such domains, brought about by the reshuffling of exons, might foster the rapid evolution of new proteins.

18-11 *An attractive hypothesis: Exons code for discrete functional regions in the protein for which the entire gene codes.*

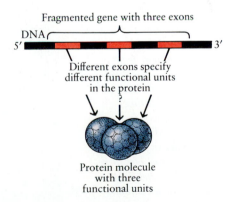

Fragmented gene with three exons

DNA

Different exons specify different functional units in the protein

Protein molecule with three functional units

Classes of DNA: Repeats and Nonrepeats

The repetitious nature of much of the DNA in the eukaryotic cell was first revealed in hybridization studies carried out before the discovery of restriction enzymes. In these studies, DNA was broken by chemical means into fragments about 1,000 base pairs in length, denatured, and allowed to reassociate. When *E. coli* DNA is treated in this way, hybridization takes place at a uniform rate; every strand present has an equal chance of finding a partner strand because each is present in equal numbers. However, if eukaryotic DNA is treated in this way, up to 30 percent of the DNA, depending on the species, reassociates very rapidly, indicating that multiple copies of the same sequence are present in each genome. This class of DNA has come to be called simple-sequence DNA. Another fraction—known as intermediate-repeat DNA—reassociates more slowly. A third fraction forms hybrids at a still slower rate, indicating that only one or a few copies of each sequence are present. This latter fraction contains most of the protein-coding genes.

Simple-Sequence DNA

Simple-sequence DNA proved easy to analyze because it consists of short sequences, as its name implies, arranged in tandem, head to tail. These sequences are typically 5 to 10 base pairs in length, although a few are as long as 200 to 300 base pairs.

Simple-sequence DNA is present in enormous quantities. Half of the DNA in one species of crab, for instance, is made up of ATATATAT and so on; *Drosophila virilis* has ACAAACT repeated 12 million times. About 10 percent of the DNA of the mouse, and about 20 to 30 percent of human DNA, is made up of short, highly repetitive sequences. It is interesting to note that repetitive sequences, especially those in which purines (A or G) and pyrimidines (T or C) alternate, are especially likely to form Z-DNA.

Simple-sequence DNA is thought by many investigators to be vital to chromosome structure. Long blocks of short repetitive sequences have been found around the centromere (Figure 18–12) and, indeed, may *be* the centromere. More recently, the tips of all human chromosomes have been found to consist of some 1,500 to 6,000 nucleotides in which a simple sequence—TTAGGG—is repeated over and over. This same simple, repeated sequence has also been found at the chromosome tips in a wide range of other mammals, birds, reptiles, and even protists. The "caps" formed by this repeated sequence are hypothesized to play a role in chromosome integrity and stability.

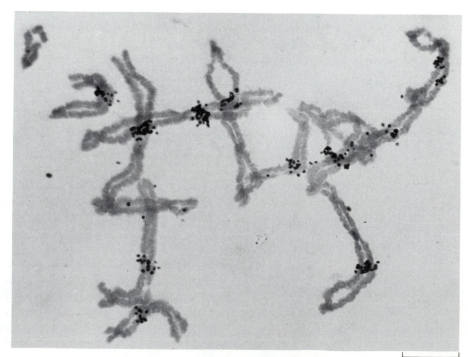

18–12 *Salamander chromosomes in which the concentration of simple-sequence DNA in the centromere regions was revealed by the use of a radioactive probe. The probe—RNA labeled with tritium—was prepared in a test tube, using as the template simple-sequence DNA that had been separated from other chromosomal DNA in a density gradient. When the salamander chromosomes were exposed to the radioactively labeled RNA, the regions of the chromosomes complementary to the RNA hybridized with it. After the RNA that had not hybridized was washed from the preparation, a photographic emulsion was placed over it. The dark spots indicate the location of the radioactivity. These chromosomes are in meiotic metaphase I.*

10 μm

Intermediate-Repeat DNA

In the hybridization experiments described earlier, intermediate-repeat sequences reassociated more slowly than simple sequences and more rapidly than single-copy DNA. About 20 to 40 percent of the DNA of multicellular organisms consists of this class. Intermediate-repeat DNA differs in several features from simple-sequence DNA. First, the sequences are longer, generally about 150 to 300 nucleotides. Second, they are similar but not identical to one another (hence they are sometimes referred to as "families"). Third, with the exception of the rRNA and histone genes, they are scattered throughout the genome. Fourth, some of the intermediate-repeat sequences, though only a small proportion, have known functions.

18–13 *A single gene codes for three types of rRNA molecules (18S, 5.8S, and 28S) found in the eukaryotic ribosome. This gene occurs in multiple copies, repeated in tandem (head to tail). (a) Electron micrograph of five copies of this gene separated by nontranscribed spacer sequences. (b) An enlargement of one of the genes and its transcribed RNA precursor molecules (the fine fibrils perpendicular to the DNA molecule). Enzymatic cleavage of the RNA precursor molecules and removal of the transcribed spacer sequences yields individual 18S, 5.8S, and 28S rRNA molecules. (c) A map of the three-rRNA gene. The portions of the DNA coding for the three rRNA molecules are shown in color.*

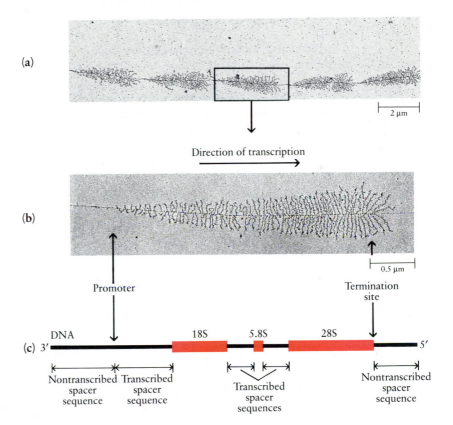

Among the most thoroughly studied intermediate-repeat sequences are the genes coding for histones and for the ribosomal RNAs (Figure 18–13). The histone genes are present in multiple copies (from 50 to 500) in the cells of all multicellular eukaryotes. Cells of multicellular eukaryotes, which may contain some 10 million ribosomes per cell, also have from 50 to 5,000 copies of the rRNA genes. The rRNA genes occur in tandem, head to tail; the chromosome regions in which they are located form the structure we recognize as the nucleolus (see essay).

Most intermediate-repeat sequences are more mysterious in nature. One of the most common families of intermediate repeats, for example, is the *Alu* sequence. Although all members of this family are not identical, they typically contain a recognition sequence for a restriction enzyme known as *Alu*I. About 5 to 10 percent of the entire human genome is made up of *Alu* sequences, often located within introns.

The Nucleolus

As you know, the most prominent feature in the nucleus of a cell in interphase is the nucleolus. During mitosis and meiosis, however, the nucleolus disappears, only to reappear following telophase. Because of its prominence in the cell and its puzzling behavior, the nucleolus has long been an object of scrutiny by cytologists. (A review of observations on the nucleolus published in 1898 contained some 700 references.) Although some details are still missing, its structure and function are now known in broad outline.

Structurally, the nucleolus is not actually a distinct entity, but rather a cluster of loops of chromatin, often from different chromosomes. For example, 10 of the 46 human chromosomes contribute chromatin loops to the nucleolus. The loops that form the basic structure of the nucleolus are the DNA segments that contain copies of the gene coding for three of the four types of rRNA molecules found in eukaryotic ribosomes. During the condensation of the chromosomes at the beginning of meiosis or mitosis, these loops are reeled back into their respective chromosomes, and the nucleolus disappears.

Functionally, the nucleolus is a ribosome factory in which rRNA molecules are transcribed from the chromatin loops and ribosomal subunits are assembled. Ribosomal proteins, themselves synthesized on ribosomes in the cytoplasm of the cell, are transported into the eukaryotic nucleus and assembled into subunits. The nearly completed ribosomal subunits, containing both rRNAs and protein, are then shipped back into the cytoplasm where, after a few finishing touches, they begin to perform their essential functions in the assembly of amino acids into proteins.

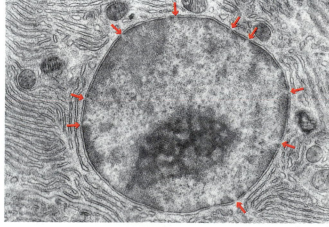

Electron micrograph of the nucleus of a type of pancreatic cell that produces and exports many of the enzymes used in digestion. The dark body in the center is the nucleolus, where the RNAs of the ribosomes are synthesized and the ribosomal subunits are assembled. You can see partially formed ribosomal subunits around its periphery. Notice also the nuclear envelope, with its many nuclear pores (arrows). Surrounding the nucleus are membranes of the endoplasmic reticulum, as well as a few mitochondria.

1 μm

Single-Copy DNA

The rest of the genome (anywhere from 50 to 70 percent, depending on the species) is made up of sequences that are not repeated or are only repeated a few times. With the exception of the histone genes, all of the known protein-coding genes belong to this fraction of the DNA. However, only a small proportion of the single-copy DNA—perhaps as little as 1 percent of the total—appears to be translated into protein. Transcription units, which consist of exons plus introns, are separated by great distances of nontranscribed spacer DNA. Moreover, introns are often longer than exons; for example, in vertebrates, introns may make up more than 80 percent of the DNA within transcription units.

"Protein-coding genes," in the words of molecular biologist James Darnell, "seem to be islands floating in a sea of meaningless DNA."

18–14 *One important group of globin genes is the beta globin family, which is located in a long nucleotide sequence in human chromosome 11. The epsilon (ε) gene is expressed early in embryonic life, followed by the two gamma (γ) genes (which code for the gamma chains that combine with alpha chains to produce fetal hemoglobin). The delta (δ) gene, found only in primates, and the beta (β) gene are expressed in adult life. This gene family also includes two pseudogenes— DNA sequences similar to the structural genes, but which are not expressed.*

Gene Families

As we have seen, some genes occur in multiple identical copies. These include the genes coding for the ribosomal RNAs and also the genes coding for histones, the only multiple repeats that are known to code for proteins. Both of these families of genes code for molecules needed in large quantities.

Other protein-coding genes are found in gene families made up of similar but not identical genes. The best studied of these is the globin family. Adult hemoglobin, as we saw in Chapter 3, is a complex of four polypeptide chains—two alpha (α) globin chains and two beta (β) globin chains—each carrying an identical heme group. In humans, the beta branch of the globin gene family is clustered together on one chromosome (chromosome 11). The cluster contains five different protein-coding genes, spaced out along the chromosome (Figure 18–14). They differ slightly in their nucleotide sequences, and all consist of three exons and two introns, one large and one small, in the same position in each gene.

As we noted previously, these genes are expressed one after the other in the course of embryonic development, producing polypeptide molecules that differ very slightly. Combined with alpha chains, they form hemoglobins with higher affinities for oxygen than the alpha-beta hemoglobin of the mother; hence the developing fetus can compete successfully with the mother for oxygen, wresting O_2 molecules from the heme in her bloodstream.

Figure 18–15 summarizes the evolutionary steps that are believed to have led to the present-day human globin gene family. The ancestral molecule is believed to have resembled myoglobin, which is a relatively small protein (153 amino acids) found in muscle cells; myoglobin consists of a single polypeptide chain and holds a single oxygen-binding heme group. It is believed that the ancestral gene—coding for the ancestral protein—was accidentally duplicated several times in the course of evolutionary history and that these duplicates were preserved by unequal cross-over events. Once the genes were duplicated, mutations led to their divergence, eventually giving rise to the present family of genes. A corresponding divergence of function occurred in the proteins for which the genes coded, leading to modern myoglobin and to the forerunners of the alpha and beta chains of hemoglobin.

18–15 *Proposed evolutionary relationships among some of the globin genes. The solid dots represent duplications of genes. The divergence of the genes following duplication is the result of mutations.*

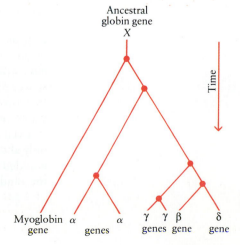

In addition to the protein-coding genes shown in Figure 18–14, there are two other nucleotide sequences that closely resemble members of the beta group but are not expressed. These sequences, known as beta pseudogenes, are believed to be copies that have been disabled by their accumulated mutations. It remains to be seen—if there are human descendants and molecular biologists in future millennia—whether these will be discarded in the course of subsequent evolution or if, as the result of further mutations, they will eventually become active family members.

Several other gene families are known that are also made up of closely related genes coding for proteins with slightly different properties. The actin family, for example, codes for the various forms of the contractile protein actin (page 121), a component of the cytoskeleton (page 110) and also one of the principal components of animal muscle fibers. Slightly different forms of actin are present in the mammalian fetus and in the adult, as is the case with beta globin; also, slightly different forms of actin are present in different types of muscle, for example, skeletal muscle and cardiac muscle.

TRANSCRIPTION AND PROCESSING OF mRNA IN EUKARYOTES

Transcription in eukaryotes is the same, in principle, as in prokaryotes. It begins with the attachment of a special enzyme, an RNA polymerase, to a particular nucleotide sequence, the promoter, on one strand of the DNA double helix. This strand then functions as a template for the assembly of ribonucleotides, as shown in Figure 15–8 (page 306). The transcribed RNA molecules (rRNAs, tRNAs, and mRNA) then play their various roles in the translation of the encoded genetic information into protein.

Despite this basic similarity, there are some significant differences between transcription in prokaryotes and eukaryotes. One difference is that eukaryotic genes are not grouped in operons in which two or more structural genes are transcribed onto a single RNA molecule, as they often are in prokaryotes. In eukaryotes, each structural gene is transcribed separately, and its transcription is under separate controls.

There are also differences in the enzymes involved in transcription. Most notably, in prokaryotes, a single RNA polymerase catalyzes the biosynthesis of the three types of RNA—messenger, transfer, and ribosomal. In eukaryotes, there are three different RNA polymerases: one transcribes the genes that will be translated into proteins, a second transcribes the genes for the large ribosomal RNAs, and the third transcribes a variety of small RNAs, including the tRNAs and the small RNAs of the ribosome.

mRNA Modification and Editing

In prokaryotes, as you will recall, ribosomes attach to an mRNA molecule and begin its translation into protein even before transcription is completed. In eukaryotes, however, transcription and translation are separated in both time and space. After transcription is completed in the nucleus, the mRNA transcripts are extensively modified before they are transported to the cytoplasm—the site of translation.

Even before transcription is completed, while the newly forming RNA strand is only about 20 base pairs long, a "cap" of an unusual nucleotide, 7-methylguanine, is added to the 5′ end of the messenger. This cap, it is now known, is necessary for the binding of the mRNA to the ribosome. After transcription has been com-

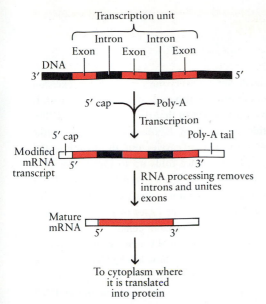

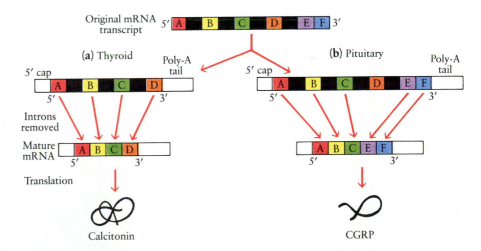

18-16 *A summary of the stages in the processing of mRNA transcribed from the structural genes of eukaryotes. The genetic information encoded in the DNA is transcribed into an RNA copy. This copy is then edited, with the addition of the 5' cap and a poly-A tail, the excision of the introns, and the splicing together of the exons. The mature mRNA then goes to the cytoplasm, where it is translated into protein.*

pleted and the molecule released from the DNA template, special enzymes add a string of adenine nucleotides to the 3' end of the molecule. This added segment, known as the poly-A tail, may contain as many as 200 nucleotides. Although its function in the cell has not been determined, its utility to molecular biologists is clear: by synthesizing a poly-T strand and anchoring it to a suitable support, they can catch mRNA molecules by their poly-A tails.

Before the modified mRNA molecules leave the nucleus, the introns are excised, and the exons spliced together to form a single, continuous molecule (Figure 18–16). The exact way that excision and splicing take place is not known, but it must be very exact, since the slightest error would cause a frame-shift in the transcribed message (see page 316).

Several instances have now been found in which identical mRNA transcripts are processed in more than one way. Such alternative splicing can result in the formation of more than one functional polypeptide from RNA molecules that were originally identical (Figure 18–17). In such cases, an intron may become an exon, or vice versa. Thus, as you can see, the more that is learned about eukaryotic DNA and its expression, the more difficult it becomes to define "gene" or "intron" or "exon."

The mRNAs that are transported to the cytoplasm are associated with proteins in ribonucleoprotein particles (mRNPs). The associated proteins may aid in transporting the mRNA molecules through pores in the nuclear envelope and may also help to bind the mRNAs to ribosomes.

18-17 *Alternative splicing of identical mRNA transcripts results in the synthesis of different polypeptides from the information encoded by a single gene. For example, when the mRNA transcript shown here is processed in (a) the thyroid gland, the transcript is cut and the poly-A tail is added to the 3' end of exon D. The introns are then removed, and the mature mRNA molecule is translated into the peptide hormone calcitonin. (b) In the pituitary gland, however, the poly-A tail is added to the 3' end of exon F. Five introns are removed from this transcript, including segment D, which in the thyroid was retained as an exon. The mature mRNA, composed of five exons, is translated into a different hormone, known as calcitonin-gene-related protein (CGRP).*

RNA and the Origin of Life

The discovery of the role of DNA in heredity launched a debate among biologists as to whether life had its molecular beginnings in protein, the original candidate, or in DNA. DNA was a likely choice because it is the repository of the genetic information and provides the template for its own precise replication; proteins have neither of these qualifications. Proponents of proteins, however, have noted that virtually all of the chemical reactions of the cell depend on the catalytic activities of proteins. Even the so-called self-replicating properties of DNA are actually protein-dependent.

An important clue toward the resolution of this "chicken-and-egg" dilemma has come from an unexpected source. T. C. Cech and his coworkers at the University of Colorado were studying the excision of introns and the splicing together of exons. This process must be carried out with exquisite precision; a mistake of one nucleotide could render the entire molecule nonfunctional. By a happy coincidence, the biological system the investigators were using was the unicellular protist *Tetrahymena*. In order to isolate the catalysts required for the reaction, Cech and his coworkers set up two cell-free systems. One contained not only an RNA molecule from which an intron was to be excised but also proteins that were potential catalysts; the other system, the control, was protein-free. The intron was neatly excised in the first system,

as expected, but, to everyone's surprise, the excision and splicing process also took place in the control. It was subsequently shown that the intron itself—a 400-nucleotide sequence of RNA—has an enzyme-like catalytic activity that carries out the excision and splicing. This sequence folds up to form a complex surface that functions like an enzyme. Although RNA catalysts are not common, other examples have now been found both in other types of reactions and in exon-splicing in other types of cells.

The discovery that RNA can act as a catalyst makes it easier to imagine how life had its beginnings. According to Bruce M. Alberts, "One suspects that a crucial early event was the evolution of an RNA molecule that could catalyze its own replication." These molecules then diversified into a collection of catalysts that could, for example, assemble ribonucleotides in RNA synthesis or accumulate lipid-like molecules to form the first primitive cell membranes. Gradually, other RNAs evolved and assembled the first proteins, which, because they were better catalysts, gradually took over the enzymatic functions. In the third step, DNA appeared on the scene, and its more stable double-stranded structure became the ultimate repository of the genetic information. Thus, researchers speculate, the catalytic intron can be regarded as a living fossil, a provocative clue to the events of almost 4 billion years ago.

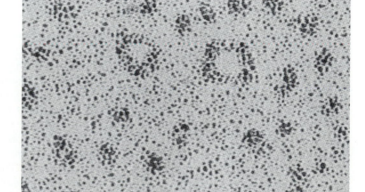

The small circles visible in this electron micrograph are introns removed from a transcribed RNA molecule of the protist Tetrahymena. *These introns have the capacity to catalyze their own excision and to splice together the exons.*

GENES ON THE MOVE

As we saw in Chapter 16, studies of the prokaryotic chromosome revealed, unexpectedly, the existence of a number of genetic elements—transforming factors, plasmids, bacteriophages, and transposons—that move into and out of the bacterial genome, affecting both structure and function. Analysis of the eukaryotic chromosome has revealed that it too is subject to rearrangements, deletions, and additions. We shall discuss a few examples, notably antibody-coding genes, viruses, and transposons.

Antibody-Coding Genes

Antibodies are complex globular proteins produced in large quantities by specialized white blood cells (lymphocytes) in response to the presence of foreign molecules. A substance that evokes the production of antibodies is known as an **antigen;** virtually all foreign proteins and most foreign polysaccharides can act as antigens. An antibody recognizes and combines with its particular antigen in much the same way, and as specifically, as an enzyme combines with its substrate. Antibodies immobilize or destroy foreign proteins, virus particles, bacterial cells, and other invaders. The problem with antibodies, from the geneticist's perspective, is that a single organism—a mouse, for instance—is capable of making at least 10 million different kinds of antibodies, and there are not enough protein-coding genes in the entire mammalian genome to account for this many different proteins.

Analyses of the amino acid sequences of antibody molecules have shown that each is made up of two heavy (long) polypeptide chains and two light (short) chains (Figure 18–18). Each type of chain has a constant region that is characteristic of the species of organism and the type of antibody, and each has variable regions. The variable regions are responsible for the highly specific reaction between antigen and antibody. These variable regions consist of only about 100 amino acids.

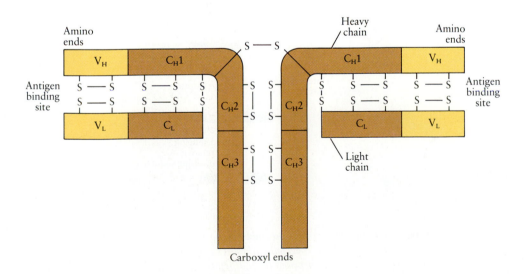

18–18 *Structure of an antibody molecule. It has two light (L) and two heavy (H) polypeptide chains, each of which has a variable (V) region (yellow) and a constant* (C) *region (brown). The polypeptide chains are connected to each other by disulfide bridges. The variable regions of the light and heavy chains form a complex three-dimensional structure that is the portion of the antibody molecule that recognizes and binds to foreign antigens.*

More than 20 years ago it was hypothesized that the constant and variable regions of antibody molecules might be encoded by separate genes. Variability could thus be generated by combining a single constant sequence with different variable sequences. With the discovery of restriction enzymes and the refinement of hybridization techniques, it became possible to test this hypothesis. Studies conducted by Susumu Tonegawa, now of the Massachusetts Institute of Technology, demonstrated conclusively that separate DNA segments code for the variable portions of the antibody molecules. Detailed analysis has shown that the DNA for the variable regions of the heavy chain consists of at least 400 different variable (V) sequences, about 12 diversity (D) sequences, and four joining (J) sequences, which can be assembled in many millions of different ways.

Moreover, by comparing the nucleotide sequences of mature mouse lymphocytes with those of embryonic mouse cells, Tonegawa was able to show that segments that code for the variable regions of an antibody molecule are actually moved into a new place on the chromosome during the differentiation of a lymphocyte (Figure 18–19). This phenomenon—the rearrangement of gene fragments in somatic cells to produce functional genes—is known to occur only in the immune system. Tonegawa's work, for which he received the 1987 Nobel Prize, laid the foundation for a virtual avalanche of new discoveries concerning the cells of the immune system, about which we shall have a great deal more to say in Chapter 39.

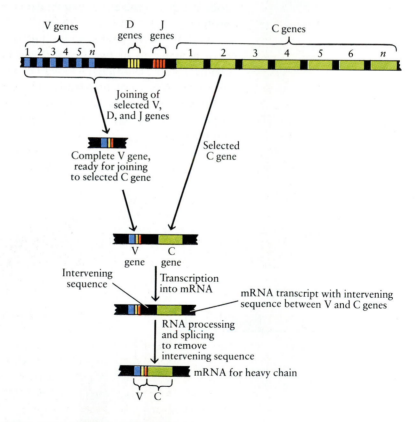

18–19 *A schematic representation of the assembly of the genes coding for an antibody heavy chain. Selected variable (V), diversity (D), and joining (J) genes are transposed from different regions of the chromosome to form a complete variable gene that then joins to one of the constant (C) genes. The intervening sequence between the variable and constant sequences is removed from the RNA transcript to yield the finished mRNA molecule.*

18–20 *Susumu Tonegawa, the first Japanese scientist to receive the Nobel Prize, was inundated with congratulatory telephone calls at his Massachusetts home when the award was announced in October of 1987. For his nine-month-old son, Hidde, however, the focus of attention was the photographer and his equipment.*

Viruses

The viruses of eukaryotes, like those of prokaryotes, consist essentially of nucleic acid enclosed in a protein capsid. Like the bacteriophages described in Chapter 16, eukaryotic viruses may be either DNA or RNA. Moreover, like lambda and other temperate bacteriophages, certain eukaryotic viruses can also become integrated into the chromosomal DNA of the host cell. When integrated, these viruses are known as **proviruses.** They are, in effect, mobile genetic elements. In eukaryotes, such viruses are of two general types: DNA viruses (analogous to the temperate bacteriophages) and RNA retroviruses.

A number of DNA viruses are known that can, depending on the type of cell they infect, either initiate an infection cycle, damaging the cell, or insert themselves into the chromosomal DNA of the host cell. An example is simian virus 40, or SV40 (Figure 17–10, page 348). SV40 is a virus of monkeys that was first discovered in cells, growing in tissue culture, that were being used for the development of polio vaccines. SV40 was subsequently found to cause cancers in newborn hamsters, though not in the monkeys that are its normal hosts. The cancers are caused by specific growth-promoting proteins produced in the cells by the viral genes. In short, SV40 can introduce new, functional genes into the DNA of the host cell, as can a number of other DNA viruses.

The second group of eukaryotic viruses that can become integrated into host cell chromosomes are the RNA retroviruses. The integration of an RNA virus into a DNA chromosome poses some special problems that are, as we saw in the previous chapter, solved by the enzyme reverse transcriptase. Molecules of reverse transcriptase are carried within the capsid of an RNA retrovirus, along with its RNA. Once within the host cell, the viral RNA is copied by reverse transcriptase to produce, after a complex series of events, a double-stranded DNA molecule. In the course of these events, reverse transcriptase also directs the duplication of sequences at the ends of the virus, producing repeated sequences known as long terminal repeats (LTRs). LTRs are distinctive features of retroviruses.

Once integrated into a host chromosome, the DNA derived from the viral RNA utilizes the RNA polymerases and other resources of the host cell to produce new viral RNA and protein molecules, which are packaged into new virus particles. Depending on the site of insertion into the host chromosome, DNA derived from a retrovirus may cause mutations by interfering with the expression of host cell genes, either inhibiting them or releasing them from repression. Characteristically, however, most retroviral insertions do not damage or destroy their host cells but become permanent additions to the host cell genome. If germ cells (the cells destined to become eggs and sperm) are infected with such a retrovirus, its genetic information will be transmitted to the next generation.

In mice, it is estimated that from 0.5 to 1.0 percent of the total DNA is of retroviral origin. Moreover, retroviruses are so efficient at promoting their own transcription that as much as 10 percent of the total mRNA in a cell may be of retroviral origin.

Eukaryotic Transposons

Transposons in prokaryotes, as you will recall, are genetic elements—nucleotide sequences—that can move, either directly or in the form of replicas, from one place to another in the bacterial genome. Analogous transposable elements have also been identified in eukaryotic cells. In fact, they were first reported in plants some 40 years ago (see essay). More recently, transposons have been identified in yeast and in *Drosophila*, and there is evidence that many of the repeated DNA sequences in these and other organisms originated as transposons.

Eukaryotic transposons resemble their bacterial counterparts in structure (Figure 18–21), and, like bacterial transposons, they can cause mutations when they become inserted into structural genes or promoter regions. They differ from bacterial transposons in one significant and interesting feature: in eukaryotes, many transposons are first copied into RNA—and then back into DNA—before their insertion into a new location in the chromosomal DNA. This discovery was surprising, for it was previously thought that reverse transcription was unique to retroviruses. Moreover, DNA sequencing has uncovered pseudogenes (nonfunctional genes) that lack introns and, in some cases, even have poly-A tails—clear evidence that they were originally copied from messenger RNA rather than DNA. If this were a mystery story, this would be the devastating clue incriminating the perpetrator—the enzyme reverse transcriptase.

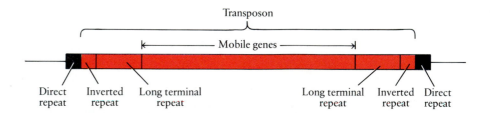

18–21 *Elements typically found in eukaryotic transposons. The central region of the transposon is a DNA sequence, often several thousand base pairs in length, that contains the gene coding for transposase (the enzyme that inserts the transposon into the host chromosome by cutting and splicing the host DNA). This segment may also carry other protein-coding genes. The insertion sequence, which also contains a promoter sequence, consists of a long terminal repeat, usually several hundred base pairs in length and the same at both ends of the transposon. The inverted repeats are also identical at each end, except that they are mirror images; if one, for example, is AATG, the other will be GTAA. Inverted repeats are the recognition sequences for transposases. The direct repeats, outside the transposon itself, are copies of a short segment (5 to 10 base pairs) of host DNA.*

"It Was Fun . . ."

Forty years ago, Barbara McClintock, working in the Cold Spring Harbor Laboratories on Long Island, New York, was studying the genetics of corn *(Zea mays)*. Working with corn kernels (each of which is an embryonic corn plant), she was performing genetic analyses of color differences and other variations, similar to those carried out in Morgan's laboratory with *Drosophila.* In the course of these studies, she encountered instances of unexplained sudden gene inactivation. On the basis of mapping and cytological studies, she was able to deduce that these changes in gene function were not due to mutations but rather came about as a consequence of the movement of genetic elements—"controlling elements," she called them—from place to place on the chromosome. These elements, she reported, actually "jumped" from one site to another on a chromosome and even from one chromosome to another.

Her findings, first published in 1951, were largely ignored, another odd traveler's tale that did not fit into the scheme of things as then understood. "Fiercely independent, beholden to no one," in the words of James Watson, she stubbornly pursued her research, sometimes working without pay. "It was fun," she is reported to have said, "I could hardly wait to get up in the morning."

In the last decade, with the discovery of a host of movable genetic elements, McClintock has received a

Barbara McClintock, photographed in 1983 with one of her colleagues, Stephen Dellaporta, at the Cold Spring Harbor Laboratories.

barrage of accolades. In 1981, at the age of 79, she was given eight separate awards, including a lifetime grant, and was hailed as a scientific prophet. In 1983, she became a Nobel laureate. Her life hasn't changed much, however; she is still at work in her laboratories in Cold Spring Harbor.

Although nucleotide sequences that resemble defective genes for reverse transcriptase have been identified in the genome of some eukaryotes, there is, thus far, no evidence of reverse transcriptase activity in cells uninfected by retroviruses. It is, however, widely accepted that the action of reverse transcriptase in the course of retroviral infections has played an important role in the evolution of the eukaryotic genome and, in particular, in the evolution of transposons.

GENES, VIRUSES, AND CANCER

Cancer is a disease in which cells escape the factors, still largely unknown, that regulate normal cell growth. As a consequence, the cells multiply out of control, crowding out, invading, and destroying other tissues. Cancer is often considered a group of diseases rather than a single disease because, with few exceptions, any one of the 200 or more cell types in the human body can become malignant. The behavior of the cells and the prognosis of the illness depend on the type of cells that have become malignant.

Three lines of evidence have long linked the development of cancer with changes in the genetic material. First, once a cell has become cancerous, all of its daughter cells are cancerous; in other words, cancer is an inherited property of cells. Second, gross chromosomal abnormalities, such as deletions and translocations, are often visible in cancer cells. Third, most carcinogens—agents known to cause cancer, such as x-rays, ultraviolet radiation, tobacco smoke, and a variety of chemicals—are also mutagens.

As long ago as 1911, a cancer-causing virus, the Rous sarcoma virus, was isolated from chicken tumors. (In those days, a virus was defined simply as "a cell-free extract that produces a disease when it is injected into a suitable host.") Even though other cancer-causing viruses, particularly viruses affecting laboratory mice, were gradually discovered, a viral theory of cancer was slow to emerge. For one thing, viruses could not be shown to be important as causes of human cancer. (Even today, after years of searching, only a few rare human cancers have been linked to viruses.) For another, the "viral theory" of cancer seemed to be at odds with the "mutation theory." In addition, the fact that most of the known cancer-causing viruses, including the Rous sarcoma virus, are RNA viruses rather than DNA viruses also seemed to make these two hypotheses incompatible.

Eventually, however, evidence emerged that viruses, like mutagens, can bring about changes in the cell's genetic makeup and that, furthermore, all known cancer-causing viruses are viruses that introduce information into host cell chromosomes. These include both DNA viruses, like SV40, and RNA retroviruses. The discovery of the role of reverse transcriptase forged the crucial link between retroviruses and the chromosomes of eukaryotic cells. The Rous sarcoma virus has been shown to be a retrovirus, and DNA segments produced by reverse transcriptase from its RNA have been located in host cell chromosomes.

Recombinant DNA techniques have enabled molecular biologists to study some of the changes in eukaryotic chromosomes that lead to cancer. These studies are usually carried out in cells growing in tissue culture. When such cells are exposed to a cancer-causing agent, such as a virus, they may undergo characteristic changes in their growth patterns and in their shape. Such cells are said to be **transformed** (Figure 18–22). Transformed cells can produce cancers when they are transplanted into laboratory animals. (Note that transformation has two meanings in biology: one is the introduction of new characteristics into a cell by means of DNA from another cell, as in the pneumococcus experiments of some 60 years ago; the other is the induction of cancer.)

Studies of transformed cells have uncovered a group of genes known as **oncogenes** (from the Greek word *onkos*, meaning "tumor"). Oncogenes closely resemble normal genes of the eukaryotic cells in which they are found. According to the oncogene hypothesis, cancer is caused when something goes wrong in the expression of these normal cellular genes, as a result of mutations in the genes themselves, changes in gene regulation, or both. Thus, viruses can cause cancer in three different ways. First, simply by their presence in the chromosome, viruses may disrupt the function of normal genes. Second, viruses may encode proteins needed for viral replication that also affect the regulation of cellular genes. Third, and most interesting of all, viruses may serve as vectors of oncogenes. In fact, oncogenes were first discovered when genetic analyses of cancer-causing retroviruses revealed the presence of genes that were not required by the viruses for their own multiplication. It was subsequently found that the nucleotide sequences of these genes not only closely resemble those of normal genes of the host cell but also cause malignant transformation of the cells.

With these discoveries, the "viral theory" and the "mutation theory" of cancer are no longer regarded as incompatible but rather as mutually supportive. About

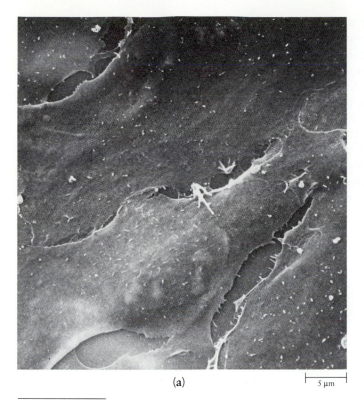

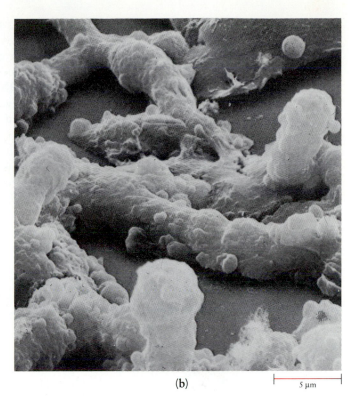

(a)   ├─ 5 μm ─┤   (b)   ├─ 5 μm ─┤

18–22 *Scanning electron micrographs of (a) normal cells growing in tissue culture and (b) the same type of cells after transformation with a cancer-causing virus. Observe that the cancer cells not only show striking surface changes but also have piled up on top of one another. The normal cells are inhibited by cell-to-cell contact and stop multiplying (see page 147), whereas the cancer cells do not.*

50 oncogenes have been discovered so far. Their gene products that have been identified all seem to be regulatory proteins of some sort, involved with the control of either cell growth or cell division. Thus, this work not only is bringing us closer to the control of one of our oldest and ugliest enemies but also is yielding new information on the fundamental question of the regulation of cell growth.

TRANSFERS OF GENES BETWEEN EUKARYOTIC CELLS

The most extravagant hope with regard to the applications of recombinant DNA technology is that at some future time it may be possible to correct genetic defects by substituting "good" genes for "bad" ones. This is an enormously complex undertaking. It requires, first, the preparation of a gene that will be taken up by a eukaryotic cell, become incorporated into a chromosome, and be expressed there—but this is only the beginning. The new gene must be established in a large number of cells of the appropriate type (blood cells should not produce somatostatin, for example) and be subject to the complicated—and as yet largely unknown—regulatory controls of the normal gene.

To Cells in Test Tubes

The first stage of the project has proved easier than expected; foreign genes will undergo recombination in eukaryotic cells growing in test tubes. In the first such experiment, SV40 virus was used as a vector to insert a rabbit gene for the beta-globin polypeptide into monkey cells. The recipient cells produced rabbit beta globin. The advantage of using a virus as a vector is that not only can viruses gain access to target cells but also they characteristically possess strong promoters, with the result that the gene is efficiently expressed.

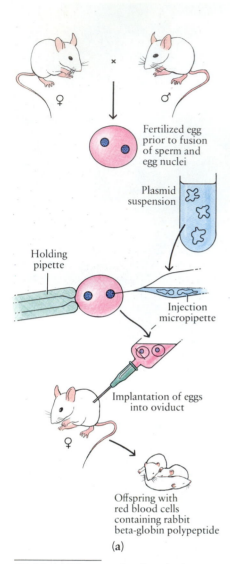

18–23 (a) *The procedure by which Gordon and Ruddle inserted the gene for rabbit beta globin into mice. The gene was spliced into plasmids, which were then injected into fertilized eggs before the egg and sperm nuclei fused. After injection, the fertilized eggs were implanted in a female mouse, who gave birth. The rabbit beta-globin polypeptide was present in the red blood cells of the offspring, and hybridization techniques revealed that the gene had been incorporated into their DNA.*

(b) Injection of the plasmid suspension into a fertilized egg. The diameter of the micropipette tip is only about 0.5 micrometer, but Gordon says that the injection "is equivalent to your being speared by a telephone pole." Nevertheless, a significant number of the eggs survive.

Subsequently, it was found that exposing cells in tissue culture to purified DNA precipitated with calcium ion (Ca^{2+}) stimulates the uptake of DNA; apparently some of the cells—about one in a million—actually phagocytize the precipitated granules and incorporate the DNA. To identify those cells that have taken up the foreign DNA, a marker is needed. To solve this problem, a line of mouse cells has been developed that lacks the enzyme thymidine kinase (TK). These TK⁻ cells are exposed to DNA molecules containing both the gene for thymidine kinase and other genes. Cells that have taken up the DNA molecules can be identified by their ability to grow in a medium in which the TK⁻ cells cannot.

Studies such as these demonstrate that eukaryotic cells have, in principle, the same capacities for incorporating foreign DNA as do prokaryotic cells. Moreover, they make possible further analysis of the regulation of expression of eukaryotic genes.

To Fertilized Mouse Eggs

Foreign genes have also been introduced into fertilized eggs and expressed in the organisms that developed from the eggs. Jon W. Gordon and Frank H. Ruddle of Yale University were the first to successfully insert a DNA sequence into the fertilized eggs of mice. When the eggs were injected with a rabbit beta-globin gene, the mice derived from the eggs were found to contain the rabbit beta globin in their red blood cells (Figure 18–23). The fact that the gene was expressed only in the red blood cells—and not in other tissues of the mouse—indicated that it had been incorporated in the "right" place and so had come under cellular control mechanisms. The gene was also passed, in a Mendelian distribution, to subsequent generations.

Using the same technique, Ralph Brinster of the University of Pennsylvania and Richard Palmiter of the University of Washington combined the human gene for the growth hormone somatotropin with the regulatory portion of a mouse gene and injected it into fertilized mouse eggs. The resultant "transgenic" mice—mice receiving the transferred human gene—grew to twice the normal size (Figure 18–24) indicating that the human gene was incorporated into the mouse genome and was producing growth hormone. Because the DNA was injected into the egg cells, it was found in all cells, including the germ cells, and thus could be passed on to the next generation.

This type of procedure has now been carried out with a number of cloned genes. From 10 to 30 percent of the eggs survive the manipulation and, of these, the foreign gene functions in up to 40 percent.

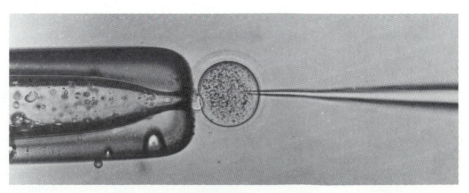

(b)

18–24 *The two female mice shown here are litter mates, approximately 24 weeks old. The fertilized egg from which the mouse on the left developed was injected with a gene consisting of the promoter and regulator sequences of a mouse gene combined with the structural gene for human growth hormone. Following integration of the new gene into the genome of the female mouse, it is passed on to her offspring. On the average, mice that express the new gene grow two to three times as fast as mice lacking the gene and, as adults, they are twice the normal size. (Those unable to see the utility of a giant mouse may be interested to learn that a similar procedure has now been successfully performed in fish.)*

To *Drosophila* Embryos

More recently, in a most elegant experiment, Allan C. Spradling and Gerald M. Rubin at the Carnegie Institution of Washington used naturally occurring transposons of *Drosophila* to ferry genes into embryonic fruit flies. They first demonstrated that if the transposons were injected into early embryos, they could become incorporated into the cells destined to become gametes. They then spliced a gene for red eyes into the transposons, cloned the transposons bearing the inserted gene, and injected them into early embryos of mutant brown-eyed flies. About 8 percent of the injected embryos developed into fertile adults. Although these adults had brown eyes, 39 percent of them produced offspring with normal red eyes (Figure 18–25). Subsequent generations of offspring from these red-eyed flies also had red eyes, indicating that the gene had been incorporated into the *Drosophila* chromosomes in a stable fashion. Early indications are that transposons may prove to be the vectors of choice.

The hope is, of course, that it may someday be possible to correct human genetic defects. The immediate goal is the development of more reliable vectors to carry genes into somatic cells and, most important, increased knowledge of the regulatory factors that control their functions. Scientists in this field are cautious —but optimistic.

18–25 *The red-eyed fruit fly at the right is the offspring of the brown-eyed fly at the left.* Drosophila *transposons bearing a gene for red eyes were injected into the brown-eyed fly when it was an early embryo. Transposons with the gene for red eyes were incorporated into chromosomes of the cells that ultimately formed its gametes. The gene for red eyes was therefore passed on to its offspring.*

SUMMARY

The eukaryotic chromosome differs in many ways from the chromosome of prokaryotes. Its DNA may take the form of B-DNA (the right-handed helix described by Watson and Crick), A-DNA (a less tightly coiled right-handed helix), or Z-DNA (a left-handed helix). Eukaryotic DNA is always associated with proteins, which constitute more than half the weight of the chromosome. Most of these proteins are histones, which are relatively small, positively charged molecules. The DNA molecule wraps around cores made of eight molecules of histone to form nucleosomes, which are the basic packaging units of eukaryotic DNA.

Biologists are beginning to understand some aspects of the regulation of gene expression in eukaryotes. During embryonic development, different groups of genes are activated or inactivated in different types of cells. According to several lines of evidence, gene expression is correlated with the degree of condensation of the chromosome. Condensed chromatin may take the form of either euchromatin, which is loosely packed, or heterochromatin, which is tightly packed. Another factor thought to be involved in gene regulation is the methylation of cytosine nucleotides, which takes place after replication. A variety of specific regulatory proteins, still very poorly understood, are also thought to play key roles in the regulation of gene expression.

Eukaryotes have far more DNA than prokaryotes. The amount of DNA is constant in the cells of any given species but is not correlated with the size, complexity, or position on the evolutionary scale of the organism. Eukaryotic cells seem to have a great excess of DNA, much of which appears to be "meaningless."

In complex multicellular eukaryotes, the coding sequence of most structural genes is not continuous but contains introns, which are also known as intervening sequences. Although introns are transcribed into RNA in the nucleus, they are not present in the mRNA in the cytoplasm and thus are not translated into protein. The segments that are present in the cytoplasmic mRNA and are translated into protein are known as exons.

Hybridization and sequencing studies have revealed three classes of eukaryotic DNA. Multiple repeats of short nucleotide sequences, characteristically arranged in tandem, are known as simple-sequence DNA. Simple-sequence DNA is associated with the tightly coiled heterochromatin in the region of the centromere. Longer repeats, usually dispersed throughout the chromosomes, are known as intermediate-repeat DNA. Intermediate-repeat DNA includes multiple copies of the genes coding for the rRNAs, tRNAs, and histones. The third class, single-copy DNA, makes up about 70 percent of the chromosomal DNA in humans. Current data indicate that as little as 1 percent of the human genome may be translated into protein.

Some structural genes, such as those coding for the polypeptide chains of hemoglobin molecules, form gene families. The individual genes of the family differ slightly in their nucleotide sequences; as a consequence, the proteins for which they code differ slightly in structure and biological properties. Some members of gene families are not expressed, presumably because of deleterious mutations; these DNA sequences are known as pseudogenes. Gene families are believed to have their origins in gene duplications that occurred as a result of recombination "errors," followed by different mutations in different copies of the gene.

Transcription in eukaryotes differs from that in prokaryotes in a number of respects. Several different RNA polymerases are involved, as well as a multiplicity of regulatory proteins. Also, in eukaryotes, structural genes are not grouped in operons as they often are in prokaryotes; the transcription of each gene is regulated separately, and each gene produces an RNA transcript containing the encoded information for a single product. RNA transcripts are processed in the nucleus to produce the mature messenger RNA molecules that move through the nuclear pores into the cytoplasm. This processing includes addition of a methyl-guanine cap to the 5′ end of the molecule, addition of a poly-A tail to the 3′ end, and removal of the introns. Alternative splicing of identical RNA transcripts in different types of cells can produce different mRNA molecules and different polypeptides.

As is the case with prokaryotes, the eukaryotic genome contains a surprising array of movable genetic elements. Functional antibody genes are formed during the differentiation of lymphocytes by the rearrangement of gene sequences coding for different parts of the antibody molecule. Viruses, including both DNA viruses and RNA retroviruses, can become integrated into the eukaryotic chromosome; when integrated into the eukaryotic chromosome, they are known as proviruses. The incorporation of genetic information carried by RNA retroviruses depends on the transcription of the RNA into DNA, which is catalyzed by reverse transcriptase. Eukaryotic transposons resemble those of prokaryotes in that, by becoming inserted in the genome, they activate or inactivate genes, either by disrupting the coding sequences or by interfering with regulation. As in prokaryotes, the transposon may not actually move but instead may generate a copy that becomes integrated elsewhere in the genome. Many eukaryotic transposons differ from those of prokaryotes most notably in that the transposon is first transcribed to RNA and then back to DNA before insertion elsewhere in the chromosome. Current evidence suggests that the evolutionary origin of eukaryotic transposons is to be found in retroviruses.

Cancer, according to present evidence, is caused by alterations in the function of some normal cellular genes; such genes are known as oncogenes. These alterations in gene function may be caused by mutations, by changes in the regulation of gene expression, or both. Viruses can cause cancer by inserting an oncogene into a chromosome or by disrupting gene regulation. Elucidation of this role of viruses has served to unify conflicting hypotheses concerning the causes of cancer.

Limited success has been achieved in the transfer of genes to eukaryotic cells growing in test tubes, to fertilized eggs of the mouse, and to *Drosophila* embryos. In each of these cases, foreign genes have been incorporated and expressed in the new host. Such studies are leading to increased understanding of the regulatory factors governing the expression of eukaryotic genes.

QUESTIONS

1. Distinguish among the following: B-DNA/Z-DNA; nucleosome/nucleolus; euchromatin/heterochromatin; intron/exon; gene/pseudogene/oncogene.

2. In what ways are the chromosomes of prokaryotes and eukaryotes similar? In what ways are they different?

3. Describe the three classes of eukaryotic DNA. What are some of the functions that have been identified for each class?

4. What might be the advantage to an organism of having multiple copies of the genes for the rRNAs, tRNAs, and histones?

5. Although a human cell contains considerably fewer than 1 million structural genes, a human individual is, according to immunologists, capable of making at least 100 million different kinds of antibodies. How is this possible?

6. What is the function of the intron in the assembly of an mRNA molecule coding for a polypeptide chain of an antibody molecule?

7. Most bacteriophages lack introns, but introns have been found in the DNA viruses of eukaryotes. The RNA viruses of eukaryotes, however, lack introns. What do these findings suggest about the origin of viruses?

8. When DNA is heated, the hydrogen bonds between paired bases break and the complementary strands separate, denaturing the molecule. As shown in the electron micrograph below, when a DNA molecule containing the five histone genes was heated slightly, only the spacer sequences were denatured, revealing the location of the histone-coding regions. The genes coding for histones are rich in guanine and cytosine, while the spacers between them are rich in adenine and thymine. How does this explain the appearance of the DNA molecule? (*Hint:* Think about the chemistry of the DNA molecule and, if necessary, review Figure 14–11b on page 290.)

0.25 μm

Human Genetics: Past, Present, and Future

The principles of genetics are, of course, the same for humans as they are for members of any other diploid, eukaryotic species. In practice, however, there are some important differences. With the exception of those of us who belong to royal families (Figure 19–1), most people do not have information about their forebears that extends back over more than three generations. By contrast, any individual fruit fly in Morgan's laboratory had a pedigree that went back many generations. Breeding experiments, so readily performed with pea plants, are not possible with humans and, even if they were, the small numbers of offspring and the long generation time would make such investigations impractical. Consequently, in the past most of our knowledge about human genetics has come from the observation of abnormalities with a hereditary pattern—some trivial curiosities, such as the man with the urine black as ink (page 301), and some life-threatening, such as sickle cell anemia. Thus, until very recently, the flow of information has been from medical science to basic genetics. Now, however, the advances in molecular genetics made possible by recombinant DNA technology have reached the point of practical application in human genetics. They have revolutionized the understanding of many genetic defects while simultaneously providing new means for their diagnosis and new hopes for cure and prevention.

19–1 *Queen Victoria (seated center) and some of her immediate family. Seventeen of the people in this photograph, which was taken in 1894, are her direct descendants. These include Princess Irene of Prussia, standing to the right of Victoria and wearing a feather boa, and, to the left of Victoria, Alexandra (also wearing a boa), the future Tsarina of Russia. Nicholas II, to become the last Tsar of Russia, is standing beside Alexandra. Both Irene and Alexandra were carriers of a sex-linked recessive allele for hemophilia, a blood-clotting disorder.*

19–2 *The normal diploid chromosome number of a human being is 46, 22 pairs of autosomes and two sex chromosomes. The autosomes are grouped by size (A, B, C, etc.), and then the probable homologues are paired. A normal woman has two X chromosomes and a normal man, shown here, an X and a Y.*

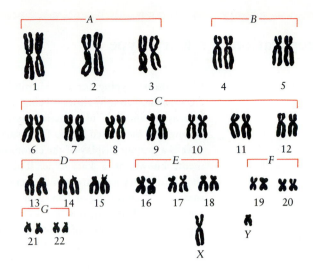

THE HUMAN KARYOTYPE

The number of chromosomes in the human species is 46, 44 autosomes and two sex chromosomes. Although cytologists began counting human chromosomes in the 1890s, it took a long time for them to arrive at the correct figure. The nucleus is small and the number of chromosomes large (as compared, for instance, to a fruit fly's four). Counts were made with tissues taken from corpses, often executed criminals; after death, the chromosomes tend to clump together, resulting in falsely low counts. In the 1920s, a cytologist was able to obtain fresh tissue, from the the testes of three patients at a state mental institution who were castrated for "excessive self-abuse." Working with this preparation, he reported finding a diploid number of 48. So seemingly authoritative was his statement and so technically difficult the problem that other cytologists also reported 48—for more than 30 years. It was not until techniques became available for growing cells in tissue culture and spreading them out for observation (see essay) that the correct number was revealed; this was in 1956, three years after Watson and Crick published their structure of DNA.

A graphic (or photographic) representation of the chromosomes present in the nucleus of a single somatic cell of a particular organism is known as a **karyotype.** From a karyotype, as shown in Figure 19–2, we can determine the number, size, and shape of the chromosomes and identify the homologous pairs. As you can see, however, some of the smaller chromosomes are quite similar in appearance. In these cases, staining to reveal banding patterns, as shown in Figure 19–3, makes it possible to distinguish similarly sized chromosomes and to identify homologues.

19–3 *A standard map of the banding patterns of chromosomes 8 through 11 in the human karyotype as determined both at the metaphase stage (black bands) and at the early prophase stage of mitosis (colored bands). The early prophase chromosomes are much longer and thinner than the metaphase chromosomes, and many more bands can therefore be detected. All of the bands shown here are those that stain with a specific reagent. Note how these chromosomes, which are similar in size and shape, can readily be distinguished by their banding patterns.*

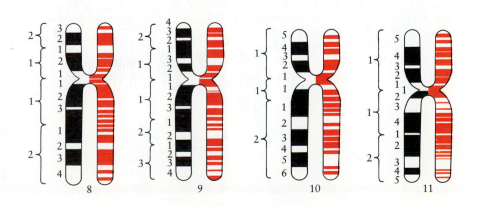

Preparation of a Karyotype

Chromosome typing for the identification of gross chromosomal abnormalities is being carried out at an increasing number of genetic counseling centers throughout the United States. The result of the procedure is a graphic display of the chromosome complement, known as a karyotype. The chromosomes shown in a karyotype are mitotic metaphase chromosomes, each consisting of two sister chromatids held together at their centromeres. To prepare a karyotype, cells in the process of dividing are interrupted at metaphase by the addition of colchicine, a drug that prevents the subsequent steps of mitosis from taking place by interfering with the spindle microtubules. After treating and staining, the chromosomes are photographed, enlarged, cut out, and arranged according to size. Chromosomes of the same size are paired according to centromere position, which results in different "arm" lengths. From the karyotype, certain abnormalities, such as an extra chromosome or piece of a chromosome, can be detected.

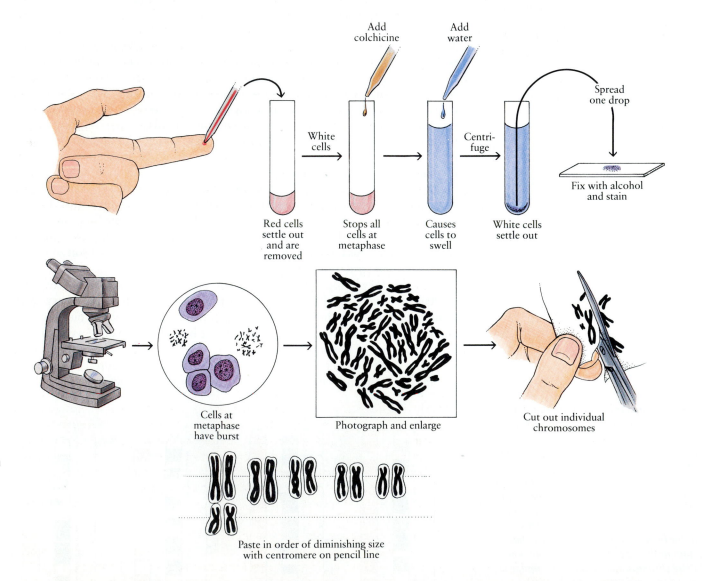

Add colchicine

Add water

Spread one drop

White cells

Centri-fuge

Fix with alcohol and stain

Red cells settle out and are removed

Stops all cells at metaphase

Causes cells to swell

White cells settle out

Cells at metaphase have burst

Photograph and enlarge

Cut out individual chromosomes

Paste in order of diminishing size with centromere on pencil line

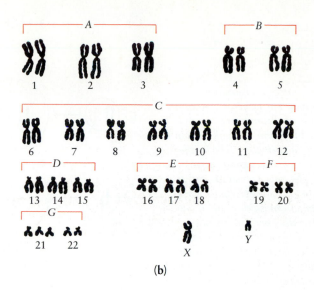

(b)

19–4 (a) *Although children with Down's syndrome share certain physical characteristics, there is a wide range of mental capacity among these individuals.* **(b)** *The karyotype of a male with Down's syndrome caused by nondisjunction. Note that there are three chromosomes 21.*

Chromosome Abnormalities

Certain genetic diseases are caused by abnormalities in the number or structure of chromosomes so severe that they can be detected in the karyotype. For example, from time to time, usually because of "mistakes" during meiosis or mitosis, homologous chromosomes or their chromatids may not separate. This phenomenon is known as **nondisjunction**. In meiosis, the results of nondisjunction are gametes with one or more chromosomes too many and other gametes with one or more chromosomes too few. A gamete with too few chromosomes (unless the missing chromosome is a sex chromosome) cannot produce a viable embryo. Sometimes, although rarely, a cell with too many chromosomes can produce a viable embryo; the result is an individual with one or more extra chromosomes in every cell of his or her body. In the vast majority of such cases, however, the fetus is spontaneously aborted early in pregnancy, an event that occurs in 15 to 20 percent of recognized pregnancies.

Individuals with additional autosomal chromosomes always have widespread abnormalities; with the exception of those with Down's syndrome, those that are not stillborn typically survive only a few months. Among the few who survive, most are mentally retarded and those who survive to maturity are usually sterile. They frequently have abnormalities of the heart and other organs as well.

Other abnormalities that may be visible in the karyotype are deletions and translocations. A deletion is simply the loss of a portion of a chromosome. A translocation (page 277) occurs when a deleted portion of one chromosome is transferred to and becomes part of another, nonhomologous chromosome.

Down's Syndrome

One of the most familiar conditions resulting from an abnormality in an autosomal chromosome is Down's syndrome, named after the physician who first described it. Because it usually involves more than one defect, it is referred to as a syndrome, a group of disorders that occur together. Down's syndrome includes, in most cases, a short, stocky body type with a thick neck; mental retardation, ranging from mild to severe in different individuals; a large tongue, resulting in speech defects; an increased susceptibility to infections; and, often, abnormalities of the heart and other organs. Individuals with Down's syndrome who survive into their thirties or forties also have a high probability of developing a form of senility similar to Alzheimer's disease (to be discussed in Chapter 43).

Down's syndrome arises when an individual has three, rather than two, copies of chromosome 21. In about 95 percent of the cases, the cause of the genetic abnormality is nondisjunction during formation of a parental gamete, resulting in 47 chromosomes, with an extra copy of chromosome 21 in the cells of the affected individual (Figure 19–4).

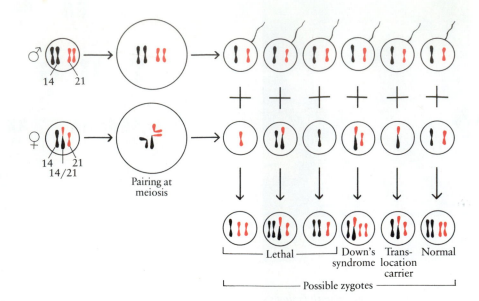

19–5 *Transmission of translocation Down's syndrome. The father, top row, has normal pairs of chromosomes 21 and 14, and each of his sperm cells will contain a normal 21 and a normal 14. The mother (shown here as the translocation carrier) has one normal 14, one normal 21, and a translocation 14/21. She herself appears normal, but her chromosomes cannot pair normally at meiosis. There are six possibilities for the offspring of these parents; the infant will (1) die before birth (three of the six possibilities), (2) have Down's syndrome, (3) be a translocation carrier like the mother, or (4) be normal. Tests for the chromosomal abnormality can be made in prospective parents and in the fetus before birth.*

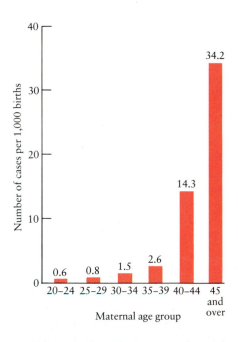

19–6 *The frequencies of births of infants with Down's syndrome in relation to the ages of the mothers. The number of cases shown for each age group represents the occurrence of Down's syndrome in every 1,000 live births by mothers in that group. As you can see, the risk of having a child with Down's syndrome increases rapidly after the mother's age exceeds 40. An increased risk is also thought to occur after the father's age exceeds 55.*

Down's syndrome may also result from a translocation in the chromosomes of one of the parents. The person with Down's syndrome caused by translocation usually has a third chromosome 21 (or, at least, most of it) attached to a larger chromosome, most often chromosome 14. Such an individual, although he or she has only 46 chromosomes, has the functional equivalent of a third chromosome 21.

When cases of Down's syndrome due to translocation are studied, it is usually found that one parent, although phenotypically normal, has only 45 separate chromosomes. One chromosome is usually composed of most of chromosomes 14 and 21 joined together. The possible genetic makeups of the offspring of this parent are diagrammed in Figure 19–5. Three out of the six possible combinations are lethal. One of the remaining three will produce Down's syndrome, one will be normal, and one will be an asymptomatic carrier of the 14/21 translocation. Thus, parents who have a child with Down's syndrome are advised to have their karyotypes prepared. If either parent has the translocation, they are warned that they are at a high risk of having another child with Down's syndrome and that half of their normal children will be carriers of the translocation.

It has been known for many years that Down's syndrome and a number of other defects involving nondisjunction are more likely to occur among infants born to older women (Figure 19–6); the reasons for this are not known. Recent studies, however, have also indicated that in about 25 percent of the cases of Down's syndrome due to nondisjunction, the extra chromosome comes from the father rather than the mother.

Abnormalities in the Sex Chromosomes

Nondisjunction may also produce individuals with unusual numbers of sex chromosomes. An *XY* combination in the twenty-third pair, as you know, produces maleness, but so do *XXY*, *XXXY*, and even *XXXXY*. These latter males, however, are usually sexually underdeveloped and sterile. *XXX* combinations sometimes produce normal females, but many of the *XXX* women and almost all *XO* women (women with only one X chromosome) are sterile.

Chromosome Deletions

Small chromosome deletions can also result in congenital defects or other illnesses. For example, a small deletion on the short arm of chromosome 11 is

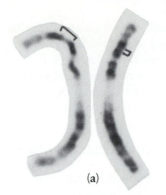

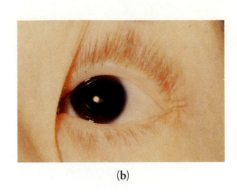

(a)   (b)

19-7 (a) *A chromosomal abnormality associated with cancer. The chromosomes shown here have been stained to reveal banding patterns. The chromosome on the left is normal. The one on the right has a deletion, shown by the smaller size of the bracket. Such deletions have been found in children with Wilms' tumor.* (b) *The left eye of a 15-year-old boy who has this chromosomal deletion and who developed Wilms' tumor in infancy. Note the absence of an iris. An older half-brother and a maternal aunt also had aniridia and developed Wilms' tumor at an early age. Another brother and the boy's mother are phenotypically normal. Analysis of the mother's chromosomes revealed that although she carries the deletion in chromosome 11, the missing segment is present in her cells in chromosome 2. Almost all other chromosomal abnormalities associated with cancer have occurred only in somatic cells and are not inherited.*

associated with Wilms' tumor, a cancer of the kidney found in infants and young children. Associated with this same deletion is a condition known as aniridia, the congenital absence of the iris of the eye (Figure 19–7). Not everyone with the deletion develops Wilms' tumor, but children with both aniridia and the chromosome deletion are at a high risk. By testing children with aniridia for the chromosome deletion, medical researchers can detect and treat Wilms' tumor at an early stage before it has caused any symptoms.

Prenatal Detection

A procedure known as **amniocentesis** makes possible the prenatal detection of Down's syndrome and a number of other genetic conditions in the fetus. A thin needle is inserted through the mother's abdominal wall and through the membranes that enclose the fetus, and a sample of the amniotic fluid surrounding the fetus is withdrawn (Figure 19–8). While the procedure is being performed, a sonogram of the fetus is displayed on a monitor, enabling the physician to identify its precise location. Although amniocentesis must be done with great care, it is simple, quick, and usually harmless. The amniotic fluid contains living cells sloughed off by the fetus. These cells, grown in tissue culture, can provide mitotic cells from which a karyotype can be made.

19-8 (a) *Amniocentesis. The position of the fetus is first determined by ultrasound. Then a needle is inserted into the amniotic cavity, and fluid containing fetal cells is withdrawn into a syringe. The cells are grown in tissue culture and then are analyzed for chromosomal abnormalities and other genetic defects. The procedure is usually not performed until the sixteenth week of pregnancy, to ensure both that there are enough fetal cells in the amniotic cavity to make detection possible and that there is sufficient amniotic fluid that removal of the small amount necessary for the test will not endanger the fetus.*

(b) *A sonogram of the uterus of a pregnant woman carrying a four-month-old fetus. The solid black regions enclosed by the muscular uterine walls are the fluid surrounding the fetus. The fetus is lying on its back, with its head at the left and its right arm in the foreground. Toward the right, you can see portions of the umbilical cord extending from the abdomen of the fetus up to the thickened tissues of the placenta.*

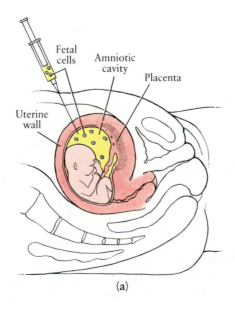

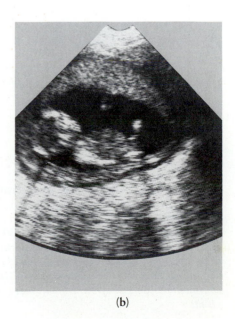

Fetal cells   Amniotic cavity   Placenta   Uterine wall

(a)   (b)

More recently, a technique has been developed for collecting cells from the chorion, one of the fetal membranes. The advantage of this method is that the test can be performed as early as the eighth week of pregnancy.

In most cases, at this stage of our knowledge, treatment of conditions detected by prenatal testing is not possible. Parents are faced with the difficult decision of whether or not to abort the affected fetus.

PKU, SICKLE CELL ANEMIA, AND OTHER RECESSIVES

Many inherited disorders are, like the white flowers in Mendel's pea plants, the result of the coming together of two recessive alleles. Individuals heterozygous for the gene are usually symptom-free; in the heterozygote enough of the particular protein is produced from the normal allele to make up for the recessive allele, which codes for either a defective, poorly functioning polypeptide or for none at all. Thus, the clinical symptoms show up only in the recessive homozygote.

Phenylketonuria

One of the best-studied examples of a genetic disorder inherited as a Mendelian recessive is phenylketonuria, or PKU, as it is commonly known. Individuals with PKU lack the enzyme that normally converts the amino acid phenylalanine to tyrosine (Figure 19–9). When this enzyme is missing or deficient, phenylalanine and its abnormal breakdown products accumulate in the bloodstream and urine. These breakdown products are harmful to the cells of the developing nervous system and can result in profound mental retardation.

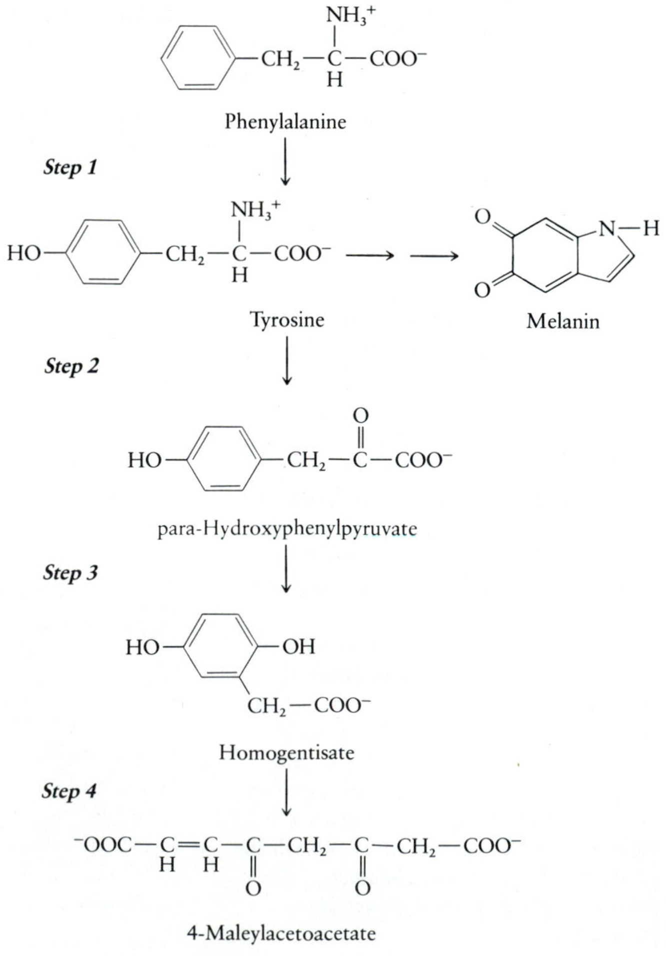

19–9 *Some steps in the pathway for the breakdown of the amino acids phenylalanine and tyrosine. If the enzyme that catalyzes Step 4, the conversion of homogentisate into 4-maleylacetoacetate, is missing, alkaptonuria results (page 301). If the enzyme that catalyzes Step 1, the conversion of phenylalanine to tyrosine, is defective, the result is an accumulation of phenylalanine and the disease known as phenylketonuria (PKU). If one of the enzymes that converts tyrosine to melanin is defective, albinism results.*

PKU is caused by a recessive allele in the homozygous state. About 1 in every 15,000 infants born in the United States is homozygous for this allele. Such infants usually appear healthy and normal at birth, but after the first few months the symptoms of the disease set in, and without treatment severe mental retardation usually results. Many never learn to walk or talk and are subject to periodic convulsions and seizures. Most afflicted individuals must be hospitalized for their entire lives, which, in untreated persons, is seldom more than 30 years.

It is not yet known how the high levels of phenylalanine and derivative compounds bring about the neurological symptoms. However, the knowledge we do have is enough to effectively treat infants with PKU and prevent the symptoms from appearing. Most states now require routine tests of all newborn babies in order to detect PKU homozygotes. Those identified at birth are put on a special diet containing low amounts of phenylalanine—enough to supply dietary needs but not enough to permit toxic accumulations. As a result, they are able to develop normally.

Albinism

Albinism—the lack of pigmentation in skin, hair, and eyes—is due to an inability to make the brown pigment melanin. Melanin is produced in pigment cells via an enzymatic pathway from the amino acid tyrosine; as shown in Figure 19–9, it is a product of the same pathway associated with the breakdown of phenylalanine. Most albinos lack one of the enzymes necessary to produce melanin. Other albinos have the enzyme, which is, however, unable to enter the pigment cells; as a consequence, the tyrosine within these cells is not acted upon by the enzyme and melanin is not produced. Both forms of albinism are inherited as autosomal recessives.

Tay-Sachs Disease

Tay-Sachs disease is an autosomal recessive condition resulting in degeneration of the nervous system. As with PKU, Tay-Sachs homozygotes appear normal at birth and through the early months. However, by about eight months, symptoms of severe listlessness become evident. Blindness usually occurs within the first year. Afflicted children rarely survive past their fifth year. The biochemical basis of this disease is now, at least in part, understood. Homozygous individuals lack an enzyme, N-acetyl-hexosaminidase, which breaks down a lipid known as GM_2 ganglioside. The enzyme is normally found in the lysosomes of brain cells and plays a crucial role in keeping GM_2 ganglioside from accumulating. In the child lacking the enzyme, the lysosomes of the brain cells fill with this lipid and swell, and the cells die (Figure 19–10). There is no therapy yet available for Tay-Sachs disease.

While Tay-Sachs disease is a rare disorder in the general population (1 in 300,000 births), until recently it has had a much higher incidence (1 in 3,600 births) among Jews of Eastern and Central European extraction (Ashkenazic Jews), who make up more than 90 percent of the American Jewish population. It is estimated that among this population approximately 1 in 28 individuals is a heterozygous carrier of the Tay-Sachs allele. The development of a blood test measuring levels of the enzyme that is deficient in Tay-Sachs disease has made it possible for prospective parents to determine if they are carriers (heterozygotes have half the normal levels of the enzyme). Since its development, this test has been so extensively utilized by prospective parents in the American Jewish population that the incidence of Tay-Sachs births has now dropped dramatically; in fact, in the United States today, the majority of the Tay-Sachs babies born are to non-Jewish parents.

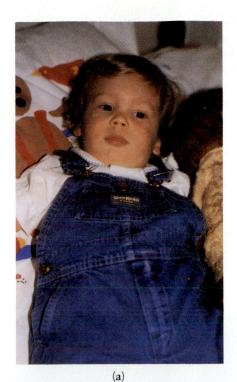

(a)

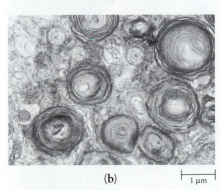

(b)   ⊢―――⊣ 1 μm

19–10 (a) *In this 17-month-old child with Tay-Sachs disease, deterioration of the brain, already begun, progressed rapidly. The child died before his sixth birthday.* (b) *Tay-Sachs disease is caused by the absence of an enzyme involved with lipid metabolism. Without the enzyme, harmful lipid deposits accumulate in the lysosomes of brain cells, as shown in this micrograph.*

Sickle Cell Anemia

The allele that, in the homozygous recessive condition, is responsible for sickle cell anemia apparently originated in Africa. For reasons that we shall discuss in Chapter 47, it has been maintained by natural selection at a very high frequency in the populations of certain regions of Africa. As a consequence, sickle cell anemia occurs at a high frequency among blacks. In the United States about 9 percent of blacks are heterozygous for the sickle cell allele, and about 0.2 percent are homozygous for the allele and therefore have the symptoms of sickle cell anemia.

Sickle cell anemia, you will recall, is due to a single amino acid substitution in the beta chains of the hemoglobin molecule (page 80). When the oxygen concentration is low, sickle cell hemoglobin becomes insoluble and forms bundles of stiff fibers. These fibers distort the shape of the red blood cells, making them more fragile; premature degradation of the red blood cells causes the anemia. Also, the loss of flexibility of the red blood cells (which are normally very flexible) makes it difficult for them to make their way through small blood vessels. Blocking of the blood vessels in the joints and in vital organs by these abnormal red cells is both painful and life-threatening.

Individuals heterozygous for the sickle cell allele are generally symptomless. Their hemoglobin, however, contains both normal and sickle cell beta chains. As with PKU, the "good" allele makes enough normal hemoglobin that the effects of the "bad" allele are not discernible. However, if blood samples are treated in ways that remove oxygen from all the hemoglobin molecules, some of the blood cells of a heterozygote will sickle (Figure 19–11). Thus it is possible to detect heterozygotes quite easily. If two heterozygotes have children, there is 1 chance in 4 that they will have a child with sickle cell anemia, and a fifty-fifty chance that they will have a child who is, like themselves, a heterozygote and so a carrier of the allele.

19–11 *Scanning electron micrograph of deoxygenated blood from an individual heterozygous for the sickle cell allele. As you can see, some—but not all—of the red blood cells have sickled.*

5μm

Discovering the cause of a disease, conventional wisdom dictates, should lead us directly to a cure. The cause of sickle cell anemia is known down to the last nucleotide, but no effective treatments are yet available.

Sickle cell hemoglobin is not the only known genetic alteration of the hemoglobin molecule. More than 100 hereditary variants have now been found, and about 20 of these cause disease.

19–12 *In the past, dwarfs were frequently attendants at the royal courts of Europe. This 1656 painting by Velasquez,* Las Meninas, *shows the Infanta attended by her maids of honor (las meninas), including, at the right, the dwarfs Maribarbola and Pertusato. Also in this most famous of Velasquez's works can be seen the artist himself at the left and the images of Philip IV and his queen caught in the mirror above the Infanta's head. Notice how the viewer is drawn into the scene; it is as if you had just opened the door and everyone looked up.*

DWARFS AND OTHER DOMINANTS

In terms of the numbers of individuals affected, serious medical problems caused by autosomal dominants are rare, simply because severely afflicted individuals are typically not able to reproduce. One of the more common disorders caused by a dominant allele is achondroplastic dwarfism (Figure 19–12). Although achondroplastic dwarfs are less likely to have children than are other individuals, there is apparently a high rate of mutation in the gene involved, causing the condition to reappear.

Perhaps the most familiar autosomal dominant is Huntington's disease. Huntington's is the disease that afflicted the folk singer and songwriter Woodie Guthrie. It is progressive, involving the destruction of brain cells; death usually occurs 10 to 20 years after the onset of symptoms. Huntington's disease is caused by a single dominant allele. As you can readily calculate, any child who has a parent with Huntington's disease has a 50 percent chance of inheriting the disease. Victims of Huntington's, however, usually have no symptoms of brain cell damage until they are past 30 years of age, and, until recently, there was no clinical way to tell who among those at risk would develop the disease and who would not. By the age at which symptoms first appear, individuals with the disease often have already had children who might then, as with the Guthrie family, spend years waiting to see whether they, too, would develop the disease. We shall have more to say about this later in the chapter.

SEX-LINKED TRAITS

Color Blindness

Can you distinguish the number in Figure 19–13? Eight percent of human males and 0.04 percent of human females cannot do so. The difference in these percentages is due to the fact that, in humans as in fruit flies (page 266), the Y chromosome carries much less genetic information than the X chromosome. Among the genes carried in humans on the X chromosome and not on the Y are genes affecting color discrimination.

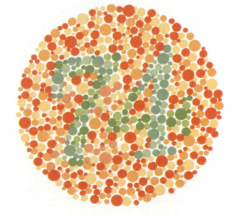

19–13 *A simple diagram used to detect red-green color blindness. Individuals with normal color vision can easily read the two-digit number embedded in the pattern of the dots. Individuals with red-green color blindness cannot distinguish the number.*

The ability to perceive color depends on three genes, coding for three different visual pigments, each responsive to light in a different region of the spectrum of visible light. One of these pigments is responsive to red wavelengths, another to green, and the third to blue. The gene coding for the pigment responsive to blue light is on an autosome, but the genes coding for the pigments responsive to red and green light are both on the X chromosome. In males, if the gene for green is defective, green cannot be distinguished from red, and, conversely, a defect in the gene for red results in red appearing as green. (The former is about three times more common than the latter.) In heterozygous females, the defective alleles are recessive to the normal alleles on the other X chromosome, and so vision is usually normal. However, as we noted in Chapter 13 (page 267), there are occasional cases in which a heterozygous female is color-blind in one eye but has normal color vision in the other eye. Complete red-green color blindness in females occurs only in those rare instances in which both X chromosomes carry the same defective allele.

If a woman carrying a defective allele on one X chromosome transmits that X chromosome to a daughter, the daughter will also have normal color vision if she receives an X chromosome with the normal allele from her father (that is, if he is not color-blind). If, however, the X chromosome with the defective allele is transmitted from mother to son, he will be color-blind since, lacking a second X chromosome, he has only the defective allele (Figure 19–14). As we noted in Chapter 13, traits such as eye color in *Drosophila* or color discrimination in humans, which are controlled by genes on the X chromosome, are said to be sex-linked.

Hemophilia

Another classic example of a sex-linked characteristic is the hemophilia that has afflicted some royal families of Europe since the nineteenth century. Hemophilia is a group of diseases in which the blood does not clot normally. Clotting occurs through a complex series of reactions in which each reaction depends on the presence of certain protein factors in the blood plasma. Failure to produce one essential plasma protein, known as Factor VIII, results in the most common form of hemophilia, hemophilia A, which is associated with a recessive allele of a gene carried on the X chromosome. In this type of hemophilia, even minor injuries carry the risk of the patient's bleeding to death. Persons with hemophilia A can be treated with Factor VIII extracted from normal human blood, but the cost is very high—an estimated $6,000 to $26,000 per year—and carries the risk of transmission of infectious diseases, including AIDS. Work is currently underway, using recombinant DNA techniques, to develop a genetically engineered Factor VIII that can be synthesized in bacteria, eliminating the risk of contamination with infectious agents.

19–14 *The pedigree of a family in which the mother has inherited one normal and one defective allele for red-green color discrimination. The normal allele is dominant, and she has normal color vision. However, half of her eggs (on the average) will carry the defective allele and half will carry the normal allele—and it is a matter of chance which kind is fertilized. Since her husband's Y chromosome, the one that determines a son rather than a daughter, carries no gene for color discrimination, the single allele the wife contributes (even though it is a recessive allele) will determine whether or not a son is color-blind. Therefore, half of her sons (on the average) will be color-blind. Assuming that her children marry individuals with X chromosomes with the normal alleles, the expected distribution of the trait among her grandchildren will be as shown in the F$_2$ generation. Note that all the daughters of a color-blind man will be carriers of the defective allele, and all his sons will be normal with respect to color discrimination, unless their mother is a carrier or is herself color-blind.*

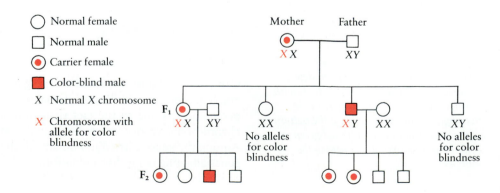

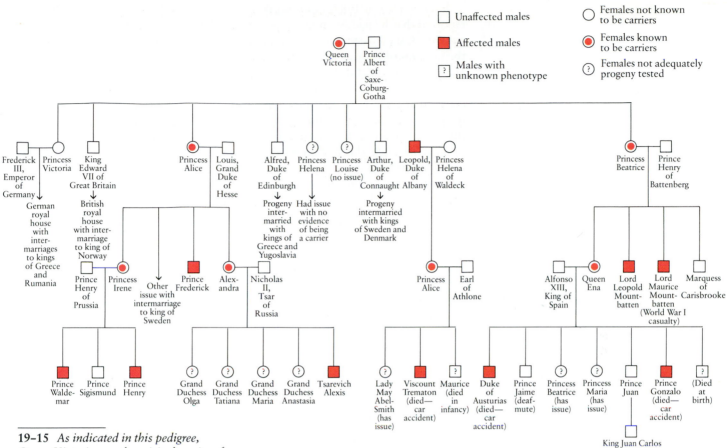

19-15 *As indicated in this pedigree, Queen Victoria was the original carrier of the allele for hemophilia that has afflicted male members of the royal families of Europe since the nineteenth century. The British royal family escaped the disease because King Edward VII, and consequently all his descendants, did not inherit the defective allele.*

Queen Victoria was probably the original carrier in her family (Figure 19–15). Prince Albert, Victoria's husband, could not have been the source because male-to-male inheritance of the disease, as with all X-linked characteristics, is impossible. Because none of Victoria's relatives other than her descendants were afflicted, we conclude that the mutation occurred on an X chromosome in one of her parents or in the cell line from which her own eggs were formed. One of her sons, Leopold, Duke of Albany, died of hemophilia at the age of 31. At least two of Victoria's daughters were carriers, since a number of *their* descendants were hemophiliacs. And so, through various intermarriages, the disease spread from throne to throne across Europe. Tsarevitch Alexis, the only son of Nicholas II and Alexandra, the last Tsar and Tsarina of Russia, inherited the allele for hemophilia from his mother. His parents' great concern for his health, which apparently distracted them from affairs of state, contributed to the turbulent events surrounding the Russian revolution.

Muscular Dystrophy

Muscular dystrophy is the name given to a group of diseases characterized by muscle wasting. The most common and severe type, Duchenne muscular dystrophy, affects cardiac muscle as well as skeletal muscle and is accompanied by mental retardation in about 30 percent of the cases. It is X-linked, occurring almost exclusively in males, with an incidence of about 1 in 3,500 newborn boys. The first symptoms usually develop in affected boys between the ages of 2 and 6, and most die by their early twenties.

In December of 1987, Louis Kunkel of the Harvard University Medical School reported that he had isolated the protein that is defective in muscular dystrophy patients. Called dystrophin, it accounts for only 0.002 percent of the protein in the muscles of normal individuals. It was totally absent in the two patients with Duchenne muscular dystrophy tested so far.

Although the role of this protein in normal muscle is not yet clear, some of the events that accompany its absence are. One of the most critical is a hardening of the muscles, a condition known as fibrosis. As a result of fibrosis, blood supply to the muscle cells is restricted, and they die. This poorly understood phenomenon is thought to cause the weakness and eventual death of patients with Duchenne muscular dystrophy.

Researchers are now in the process of cloning and sequencing the gene coding for dystrophin. This undertaking is made more difficult by the fact that it has proved to be the largest human gene ever known, with about 2 to 3 million base pairs, including some 60 exons and huge introns. Duchenne muscular dystrophy appears to be associated with deletions in this gene.

DIAGNOSIS OF GENETIC DISEASES: RFLPs

Sickle Cell Anemia

One of the first rewards of recombinant DNA technology, in terms of human genetics, has been the ability to diagnose many of the hereditary disorders we have just discussed. Sickle cell anemia, the first to be so diagnosed, is an instructive example. The DNA coding for beta globin, as we have seen, is one of the most extensively studied of human genes and was the first human disease-related gene to be cloned. In order to develop a diagnostic test for sickle cell anemia, radioactive copies of portions of the beta-globin nucleotide sequence were prepared as a probe. DNA from persons with normal hemoglobin and DNA from persons with sickle cell anemia were cleaved with the restriction enzyme *Hpa*I. The fragments produced were then exposed to the radioactive beta-globin probe. In persons with normal hemoglobin, it was found, the probe consistently hybridized with a fragment that was either 7,000 or 7,600 nucleotides long. By contrast, in 87 percent of persons with sickle cell anemia, the probe hybridized with a much longer fragment—some 13,000 nucleotides in length (Figure 19–16). The same result is seen with cells obtained by amniocentesis, thus providing the first prenatal screening test for one of the most common serious genetic disorders.

These markers are known as RFLPs, pronounced "rif-lips" and translated as "restriction-fragment-length polymorphisms." This means simply that inherited

19–16 *A test, using RFLPs, to detect the presence of the sickle cell allele. Treating human DNA with the restriction enzyme* Hpa*I produces three possible restriction fragments containing the gene for the beta chain of hemoglobin. In persons with the normal allele (gray) for beta globin, the fragments are either 7,000 or 7,600 nucleotides long. Among blacks in the United States with the sickle cell allele (color), the fragments are 13,000 nucleotides in length. A recognition sequence for the restriction enzyme* Hpa*I—present in the DNA of persons with the normal beta-globin allele—is missing in individuals carrying the sickle cell allele. The linkage between the sickle cell allele and the loss of the* Hpa*I site holds true only for populations in or originally from West Africa. Among blacks in or from East Africa, the sickle cell allele is associated with the 7,600-nucleotide fragment.*

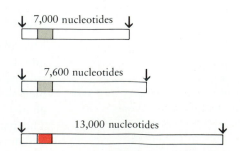

variations—mutations—in the nucleotide sequences of different individuals lead to differences in the lengths of the fragments produced by restriction enzymes. This occurs because the recognition sequence for the restriction enzyme has been eliminated or otherwise altered by the mutation. The principle involved here is gene linkage, the same principle that made possible the genetic mapping studies described in Chapter 13. Those studies in *Drosophila* confirmed that two genes that are close together on the same chromosome tend to stay together. The updated version is that two nucleotide sequences close together on the same DNA molecule tend to stay together, just as purple eyes and vestigial wings were inherited together in the fruit fly. In the *Drosophila* days, a marker gene was one that produced a detectable phenotypic change, such as a different wing shape or eye color. In the days of recombinant DNA technology, a marker may be "anonymous"—detectable only in the pattern produced by a restriction enzyme and a molecular probe. In the case of sickle cell RFLPs, the allele for sickle cell beta globin has become linked in some populations with another nucleotide sequence that alters the site for *Hpa*I recognition. The closer the allele and its marker are, the more accurate the diagnosis.

In the first family studied with this technique, both parents were known carriers of the sickle cell allele and had previously had a baby with sickle cell anemia. The woman was pregnant again, and the parents did not want to have another child with this extremely painful disease. DNA tests from each of the parents yielded both the short and the long fragments, as would be expected. Tests from their child with sickle cell anemia yielded only the long fragments. Tests of the fetal cells yielded both short and long fragments; the unborn child would be a carrier like the parents, but would not have the disease.

Huntington's Disease

James F. Gusella of the Massachusetts General Hospital, with a large team of coworkers, set out to find a diagnostic test for Huntington's disease based on the RFLP technique. Huntington's presented a new problem because neither the defective gene nor its normal allele had been identified—nor have they yet. What was available was a large library of cloned restriction fragments from human DNA. Gusella and coworkers assigned themselves the monumental task of working their way through this vast collection to find a Huntington's disease marker. They had astonishing good fortune. They found their marker within the first dozen restriction fragments tested, in a polymorphic restriction fragment produced by cleaving human DNA with the restriction enzyme *Hin*dIII. Four different patterns of fragments had been identified, known simply as A, B, C, and D, which were present in varying proportions in the normal population. Working with an American family in which some members had Huntington's disease, scientists screened the DNA from affected and normal individuals and found that, in all cases, the presence of Huntington's was associated with pattern A. This was a good start, but pattern A is the most common pattern, occurring in some 60 percent of the population, and the family was not a large one. More data were needed.

Gusella and his coworkers next turned their attention to Lake Maracaibo, Venezuela, where there is a family of more than 3,000 individuals, all apparently descended from one German sailor with Huntington's (Figure 19–17). For three years, the team in Venezuela, led by Nancy Wexler, interviewed family members to obtain the information needed to construct a pedigree, performed neurological examinations of family members, and collected skin and blood samples from 570 people. As the samples were collected, they were flown to Gusella's laboratory where a marker could be identified by correlation with the family pedigree and the results of the neurological examinations.

(a)

(b)

19–17 *The search for the Huntington's disease marker involved meticulous study of an isolated, inbred Venezuelan family. Pedigree data and neurological examinations were correlated with genetic studies of DNA. (a) Team leader Nancy Wexler (right) and other researchers recording pedigree information supplied by individuals of the afflicted family. (b) Two family members: a man in his forties with Huntington's disease and his daughter, who has a 50 percent chance of developing it.*

Witness for the Prosecution

An individual's DNA is as distinctive as a fingerprint and, in certain types of violent crime, more likely to be obtainable. The method for "DNA fingerprinting," which was devised by Alec Jeffreys of the University of Leicester in England, is basically simple. As we noted in Chapter 18, the eukaryotic genome contains many regions of simple-sequence DNA, identical short nucleotide sequences lined up in tandem and repeated thousands of times. Jeffreys found that the number of repeated units in such regions differs distinctively from individual to individual. (The only exceptions, as with fingerprints, are in the case of identical twins.) These regions can be excised from the total DNA by the use of appropriate restriction enzymes, placed on an electrophoretic gel, separated by length, denatured, and identified by a radioactive probe. When the process is completed, the end result, visible on x-ray film, looks like the bar code on a supermarket package.

Such a DNA bar code helped to convict Randall Jones, now on Death Row in Florida. Jones's car got stuck in the mud. In search of a tow, he found a young couple asleep in a pickup truck parked by a fishing ramp. He shot each of them in the head with a high-powered rifle, dragged their bodies into the woods, used the truck to pull out his car, and then went back and raped the woman. In such cases, standard blood or semen analysis can identify a suspect with a certainty of about 90 to 95 percent, leaving some room for argument. However, Jones's DNA pattern, which matched the sperm found in the victim's body, could occur in only one person out of 9.34 billion—about double the present population of the world. (Note that this test would also completely exonerate a defendant who was innocent.)

Often only very small samples of biological evidence are found at a crime scene. A gene amplification method known as PCR (polymerase chain reaction) has been developed that can take a minute fragment of DNA and, in a few hours, synthesize millions of copies. This method involves adding a short primer at each end of a selected DNA sequence, separating the two strands of the double helix by heating, and exposing them to a bacterial DNA polymerase that recognizes the primers. This enzyme is from a bacterium that thrives at high temperatures in hot springs and is not denatured by the temperatures used to denature the DNA. In the polymerase chain reaction, both strands of the DNA are copied simultaneously. If this is repeated for 20 cycles, the amount of DNA present is increased about one million times.

Gene amplification has made it possible to obtain DNA fingerprints from trace amounts of blood and semen and even from the root of a single hair. It has been used to speed up prenatal diagnosis of genetic disease and to detect latent virus infections. It also made possible the analysis of mitochondrial DNA from a woolly mammoth that died some 40,000 years ago (see Figure I–11, page 9).

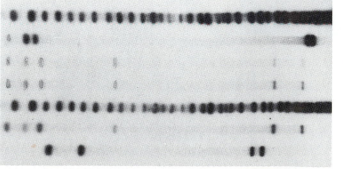

An x-ray film comparing the patterns produced by electrophoresis of restriction fragments of simple-sequence DNA from a rape victim, from the semen of the rapist, and from two suspects. The evidence revealed here by the use of a single radioactive probe strongly suggests that suspect 1 was, in fact, the rapist. In practice, three or four different probes are used, to provide conclusive identification.

Markers

Victim

Semen sample

Semen sample

Markers

Suspect 1

Suspect 2

This group, it was found, differed from the American family: in this family, the presence of Huntington's disease correlated with pattern C. Pattern C is much less common than pattern A, being present in only about 20 percent of the general population. Thus, the scientists were able to predict, in some cases, which individuals in the Venezuelan family would develop Huntington's later in life. For example, if a young person's mother did not have Huntington's and the father did, and the RFLP analysis of the mother did not reveal pattern C and the father's did, then the presence or absence of pattern C in the DNA of the person under study would predict with great certainty whether or not that person would be afflicted.

Actually, as it turns out, many members of the Venezuelan family do not want to know their future in terms of the disease. Similarly, Arlo Guthrie and his sisters have elected not to be tested. In the case of Huntington's, the most significant development may not be the possibility of earlier detection but rather the identification of the gene responsible. Work is now underway to locate the gene, with the hope of ultimately isolating and characterizing both the abnormal allele that causes Huntington's disease and its normal allele. It is hoped that this work will not only reveal the underlying abnormality that causes the brain cells to deteriorate some three to five decades after birth, but will also illuminate fundamental questions of the normal development and aging of the nervous system.

RFLPs have now been found for other hereditary diseases, including cystic fibrosis, the most common genetic disease in Caucasians.

DIAGNOSIS OF GENETIC DISEASES: RADIOACTIVE PROBES

As we noted earlier, the most common form of hemophilia is caused by defects in a plasma protein known as Factor VIII. It is possible to diagnose hemophilia in the fetus by testing for Factor VIII, but only after 20 weeks of gestation. A better test was sought, one that could be performed much earlier. The location of the gene coding for the factor was completely unknown, but, in this case, the protein had been isolated. From the known amino acid sequence of the protein, a short (36-nucleotide) fragment of RNA was synthesized and used as a radioactive probe to screen restriction fragments of the human genome. Starting with the few fragments to which the probe hybridized, it was possible to reconstruct the entire gene. It turned out to consist of 186,000 base pairs—26 exons separated by 25 introns. Once the gene was identified, it was found that mutations in the gene could be detected by RFLPs. Thus, it is now possible to identify those women who are carriers of the disease and also to detect hemophilia in fetuses at a very early stage of development.

Another example of the use of radioactive probes is a new and extremely precise test for sickle cell anemia. This test utilizes short synthetic radioactive probes that detect the single base-pair difference between the alleles for normal beta globin and sickle cell beta globin.

Until recently the only prenatal diagnosis for Duchenne muscular dystrophy was a test to determine the sex of the fetus. Parents who had a child with muscular dystrophy could elect to abort a male fetus. Now this form of muscular dystrophy is detected by a somewhat different use of a radioactive probe. A DNA probe has been synthesized that hybridizes with a portion of the nucleotide sequence that codes for dystrophin—the portion that is missing in children with the disease. Thus, the failure of the probe to find a complementary sequence among restriction fragments is diagnostic of the disease. Prospective parents, knowing whether their male fetus will be normal or will develop Duchenne muscular dystrophy, can therefore make an informed decision.

Some Ethical Dilemmas

Although the rapid advances in molecular genetics are contributing greatly to our understanding of human genetic diseases, the capacity to identify individuals either at risk of developing certain diseases or of transmitting them to their children is creating perplexing ethical dilemmas both for individuals and for our society as a whole. The four questions below cover situations that have either arisen already or are likely to arise by the year 2000. They were posed by Dr. Eric Lander, a human geneticist at the Whitehead Institute for Biomedical Research and Harvard University.

1. While the constitutional right to abortion is absolute, the choice is never easy. Imagine that you learned very early in a pregnancy that the child would certainly:

   (a) die within nine months from spinal muscular atrophy, a fatal genetic disorder;

   (b) suffer throughout life from cystic fibrosis, a painful chronic disease, and die at about age 20;

   (c) suffer from Huntington's disease at age 40 and die at about 50;

   (d) suffer from Alzheimer's disease at about 60;

   (e) be congenitally deaf;

   (f) be a dwarf, but otherwise healthy;

   (g) be predisposed to severe manic depression, which could be partially controlled by medication.

Would you choose to abort? (Assume that you are young enough that you may reasonably expect to have more children if you wish.) Regardless of your own choice, would you consider it *unethical* for another couple to do so? What principles underlie your choices?

2. Suppose you could learn with certainty whether you would suffer:

   (a) from Huntington's disease at about 40;

   (b) from Alzheimer's disease at about 60.

Would you want to know?

3. Should insurance companies have the right to:

   (a) charge higher premiums to individuals with higher risk of inherited disease?

   (b) know the results of testing for genetic predispositions?

   (c) refuse to cover a child whom prenatal tests show will suffer a severe genetic disease?

4. Suppose that 10 percent of the work force is particularly prone to cancer induced by an industrial chemical. Should an employer:

   (a) have the obligation to make the work place safe for this minority?

   (b) have the right to require pre-employment genetic screening?

   (c) have the right to refuse employment to such workers?

   (d) have the right to require such workers to pay for supplementary insurance?

Would your answer differ if the minority involved were 1 percent of the work force or 40 percent of the work force?

Although correction of the genetic defects is not possible at this time and treatment by gene replacement seems far distant, the development of radioactive probes for the detection of abnormal alleles has revolutionized medical genetic diagnosis. The use of radioactive probes is also revolutionizing forensic medicine, making it possible to conclusively identify the perpetrators of certain types of violent crime, while exonerating innocent suspects (see essay, page 396).

THE "BOOK OF MAN"

Plans are now underway for the largest and most extravagant enterprise ever undertaken in medical or biological research: the mapping and sequencing of the entire human genome. It has been compared in magnitude to the Manhattan Project, which brought forth the atomic bomb, and the space program, which culminated with a human footprint on the moon. Although initially a subject of some controversy and dissent, which was also true of the projects to which it is compared, it now has gained a momentum of its own and is clearly going forward.

There is action on two fronts. One multi-institutional enterprise is focused on mapping the human chromosomes. As in *Drosophila,* this involves finding marker genes—genes with different and detectable alleles—and documenting their rate of recombination in breeding populations. Ray White and his colleagues at the Howard Hughes Medical Center at the University of Utah, for example, have been tracing such markers through three generations of 60 Mormon families. (Mormons were chosen not only because they are in Utah, but also because their families are large and cohesive and, moreover, their genealogical records are meticulously maintained.) His group has mapped almost 500 genes. Another group, led by Helen Donis-Keller of Collaborative Research, Inc., a biotechnology company, has located several hundred markers in some 20 additional families. Her group has been conducting an intensive search for the gene responsible for cystic fibrosis and has identified some 60 markers on chromosome 7 alone, the chromosome to which this gene has been traced.

The second, and more controversial, effort involves the sequencing of the entire genome. When it was first proposed years ago, many molecular biologists were against it, mainly because of the great expense that would be involved— several billion dollars, it was estimated at the time. Much of the genome is "senseless," they pointed out, most funds available for research would have to be channeled into this one project, it would take decades to complete, and, perhaps most important of all, the creative talent of almost an entire generation of young scientists in the field would be wasted on this routine enterprise. However, the prevailing winds of opinion have now clearly shifted. The crucial change came about as a result of the development of methods for automatic sequencing of nucleotides, which is greatly decreasing the estimates of time and especially money required for the project. Also, as it has come closer to realization, there is mounting excitement about what reading this "book of man," as it has been called, might reveal. What is the mysterious role of the "senseless" DNA? How are genes regulated? (The answer to this question may hold the key to understanding and treating cancers and also to gene replacement therapy.) What further evolutionary relationships will be revealed among the human genes themselves? And among humans and other species? At this writing, the way in which this enterprise will be organized has not yet been decided, but a completion date has been set: the year 2000. That is less than a hundred years after the rediscovery of Mendel's work, which marks the beginning of genetics as a science, and less than 50 years after the announcement by Watson and Crick that opened up the field of molecular biology. And you will see it happen.

SUMMARY

The normal diploid number of human chromosomes is 46: 44 autosomes and two sex chromosomes, *XX* in females and *XY* in males. A karyotype is a graphic representation of a set of chromosomes. In preparing a karyotype, metaphase chromosomes are paired in homologues and the homologues are grouped by size. Nonhomologous chromosomes of similar size and shape can be distinguished by staining techniques that reveal chromosome banding.

Visible chromosomal abnormalities include extra chromosomes (usually a result of nondisjunction—the failure of two homologues to separate at the time of meiosis), translocations, and deletions. Down's syndrome is among the disorders associated with an extra chromosome; it may be caused either by nondisjunction or, less commonly, by translocation. Extra sex chromosomes can also result from nondisjunction and are usually associated with sterility. Aniridia and Wilms' tumor are defects associated with deletion of one small region of chromosome 11. A number of genetic defects can now be detected in the fetus by the use of amniocentesis, the collection of fetal cells from the amniotic fluid.

Many genetic diseases are the result of deficiencies or defects in enzymes or other critical proteins. These are caused, in turn, by mutations in the genes coding for the proteins. When the alleles resulting from the mutations are recessive, the diseases are apparent only in the homozygote. Such diseases include phenylketonuria, Tay-Sachs disease, sickle cell anemia, and a number of other disorders associated with variations in the hemoglobin molecules. Genetic conditions caused by autosomal dominants include a form of dwarfism and Huntington's disease.

Genetic defects expressed much more frequently in males than in females are caused by mutant alleles on the X chromosome. The mutant alleles responsible for sex-linked characteristics are usually recessive to the normal alleles and are thus not expressed in heterozygous females, who can, however, transmit them to their children. Human sex-linked characteristics include color blindness, hemophilia, and Duchenne muscular dystrophy.

Recombinant DNA technology is providing new means for early diagnosis of hereditary diseases. Among the most important tools for such diagnosis are RFLPs (restriction-fragment-length polymorphisms) and radioactive probes. RFLPs are the result of natural variations—mutations—that eliminate or alter the recognition sequence for a restriction enzyme. When such a mutation is associated with an allele causing a genetic disease, it can provide a diagnostic marker for that allele. Radioactive probes, which bind to either the normal or the mutant allele, can be used for detection and diagnosis when the nucleotide sequence of the allele is known or can be deduced.

A massive effort is now beginning to map and sequence the entire human genome. This project, which is expected to be completed by the year 2000, may answer many puzzling questions. If past experience is any guide, it is also likely to raise many new questions.

QUESTIONS

1. Nondisjunction can occur at the first meiotic division or the second. How do the effects differ? Include diagrams with your answers.

2. Describe the two types of chromosomal abnormalities that can cause Down's syndrome. With which type is it possible to identify unequivocally prospective parents who are at a higher-than-average risk of having a child with Down's syndrome? How?

3. If two healthy parents have a child with sickle cell anemia, what are their genotypes with respect to this allele? Having had one such child, what are their chances of having another child with the same disease?

4. What proportion of the children of the parents in Question 3 will be carriers of the sickle cell allele (that is, heterozygous)? What proportion of their children will not carry the allele? (Draw a Punnett square to diagram this problem.)

5. The first child of a normally pigmented man and woman is an albino. They plan to have three more children and want to know the probability that all will be normal. Calculate the probability that three normal children will be born.

6. In 1952 it was reported that two albinos, who had met at a school for the partially sighted, had married and had three children, all of whom had normal pigmentation. Assuming the children are not illegitimate, how do you explain that the children are normal?

7. In humans, either of two recessive alleles *(a or b)*, when homozygous, can cause congenital deafness. Hence, two people who are congenitally deaf could marry and have children, all of whom are normal. What would be the genotypes of the parents in such a situation?

8. A woman homozygous for the dominant alleles *A* and *B,* necessary for normal hearing, marries a man who is congenitally deaf. What are the possible genotypes of the man? What is the probability that this couple will have a deaf child?

9. A man with a particular hereditary disease marries a normal woman. They have six children, three girls and three boys. The girls all have the father's disease but the boys do not. What type of inheritance pattern is suggested? (Although this situation was not discussed in the text, you should be able to deduce the answer.)

10. Why is male-to-male inheritance of color blindness impossible? Under what conditions would color blindness be found in a woman? If she married a man who was not color-blind, what proportion of her sons would be color-blind? Of her daughters?

11. What is the probability that a woman with normal color vision whose father was color-blind but whose husband has normal color vision will have a color-blind son? A color-blind daughter?

12. How do we know that Prince Albert was not the source of the hemophilia in Queen Victoria's descendants?

13. A woman whose maternal grandfather was hemophiliac has parents who are clinically normal. She too seems normal, as does her husband. What are the chances that her first son will be normal? *(Hint:* Determine the genotype of the woman's mother and then the possible genotypes of the woman herself.)

SUGGESTIONS FOR FURTHER READING

Books

ALBERTS, BRUCE, DENNIS BRAY, JULIAN LEWIS, MARTIN RAFF, KEITH ROBERTS, and JAMES D. WATSON: *Molecular Biology of the Cell,* 2d ed., Garland Publishing, Inc., New York, 1988.

Chapters 5, 9, and 10 of this outstanding cell biology text provide an excellent discussion of contemporary molecular genetics.

BODMER, WALTER F., and L. L. CAVALLI-SFORZA: *Genetics, Evolution, and Man,* W. H. Freeman and Company, San Francisco, 1976.

Although now somewhat dated, this excellent undergraduate text, with an emphasis on human genetics, remains a valuable resource.

DARNELL, JAMES, HARVEY LODISH, and DAVID BALTIMORE: *Molecular Cell Biology,* Scientific American Books, New York, 1986.

A thorough review of gene structure and function. Chapter 7 discusses the techniques used in molecular biology.

GOODENOUGH, URSULA: *Genetics,* 3d ed., Saunders College/Holt, Rinehart and Winston, New York, 1984.

An up-to-date, general introductory text with emphasis on molecular genetics. Outstanding for its clarity of explanation.

HAWKINS, JOHN D.: *Gene Structure and Expression,* Cambridge University Press, New York, 1985.

A short synopsis on DNA and the organization of prokaryotic and eukaryotic genes.

JACOB, FRANÇOIS: *The Logic of Life: A History of Heredity,* Pantheon Books, New York, 1973.

Jacob's principal theme concerns the changes in the way people have looked at the nature of living beings. These changes, which are part of our total intellectual history, determine both the pace and direction of scientific investigation. The opening chapters are particularly brilliant.

JUDSON, HORACE F.: *The Eighth Day of Creation: Makers of the Revolution in Biology,* Simon and Schuster, New York, 1979.*

A comprehensive study of the human and scientific aspects of molecular biology from the 1930s through the mid-1970s. As events unfold, the story is told from each participant's point of view. We are treated to an enlightening and personal glimpse of the development of scientific thought.

LEHNINGER, ALBERT L.: *Principles of Biochemistry,* Worth Publishers, Inc., New York, 1982.

This introductory text is outstanding both for its clarity and for its consistent focus on the living cell. There are numerous medical and practical applications throughout.

LEWIN, BENJAMIN: *Genes III,* John Wiley & Sons, New York, 1987.

Revised frequently, Lewin's classic text provides an excellent update on the structure and function of genes. Chapters 29 and 30 present a concise discussion of movable genetic elements in both prokaryotes and eukaryotes.

NOSSAL, G. J. V.: *Reshaping Life,* Cambridge University Press, New York, 1985.

A short, excellent, easily read introduction to recombinant DNA techniques and the biotechnology industry.

OLBY, ROBERT: *The Path to the Double Helix,* University of Washington Press, Seattle, 1975.

An account, written by a professional historian of science, of twentieth-century genetics. Olby is interested not only in the scientific concepts and experiments but also in the various personalities involved and their effects on one another and on the course of scientific discovery.

OLIVER, STEPHEN G., and JOHN M. WARD: *A Dictionary of Genetic Engineering,* Cambridge University Press, Cambridge, 1985.

Short descriptions of the techniques and vocabulary used in modern molecular genetics are a help to any reader. This book is good to have as a reference when reading original articles in this field.

PETERS, JAMES A. (ed.): *Classic Papers in Genetics,* Prentice-Hall, Inc., Englewood Cliffs, N.J., 1959.*

Includes papers by most of the scientists responsible for the important developments in genetics: Mendel, Sutton, Morgan, Beadle and Tatum, Watson and Crick, and so on. You should find this book very interesting; the authors are surprisingly readable, and the papers give a feeling of immediacy that no modern account can achieve.

* Available in paperback.

RUSSELL, PETER J.: *Genetics*, Little, Brown, and Company, Boston, 1986.

> *This clearly written text covers the principles of genetics, using a problem-solving approach.*

SCIENTIFIC AMERICAN: *Molecules to Living Cells*, W. H. Freeman and Company, San Francisco, 1980.*

> *More than half of the articles in this collection from* Scientific American *are accounts of major breakthroughs in our understanding of the nucleic acids and their functions. Written for the general reader by the scientists who made the key discoveries, they are highly recommended.*

SCIENTIFIC AMERICAN: *Recombinant DNA*, W. H. Freeman and Company, San Francisco, 1978.*

> *The articles in this collection from* Scientific American *include discussions of the critical experiments that paved the way for modern molecular genetics.*

STRICKBERGER, MONROE W.: *Genetics*, 3d ed., The Macmillan Company, New York, 1986.

> *A cohesive account of the science, this book provides a broad coverage of classical genetics.*

STRYER, LUBERT: *Biochemistry*, 3d ed., W. H. Freeman and Company, New York, 1988.

> *An introductory text, with many examples of medical applications of biochemistry. Handsomely illustrated.*

WATSON, JAMES D.: *The Double Helix*, Atheneum Publishers, New York, 1968.*

> *"Making out" in molecular biology. A brash and lively book about how to become a Nobel laureate.*

WATSON, JAMES D., NANCY H. HOPKINS, JEFFREY W. ROBERTS, JOAN A. STEITZ, and ALAN M. WEINER: *Molecular Biology of the Gene*, 4th ed., vols. I and II, The Benjamin/Cummings Publishing Company, Menlo Park, Calif., 1987.

> Molecular Biology of the Gene *is a classic; now in its Fourth Edition, it continues to serve as a celebration of the extraordinary achievements of biology during the second half of the twentieth century. Well written and richly illustrated, the book covers an incredible array of topics, ranging from molecular biology of prokaryotes to that of eukaryotes and multicellular organisms.*

WATSON, JAMES D., JOHN TOOZE, and DAVID T. KURTZ: *Recombinant DNA: A Short Course*, W. H. Freeman and Company, New York, 1983.*

> *A short course in genetics, organized around the central theme of recombinant DNA. Clearly written and handsomely illustrated, it is accessible to anyone who enjoys reading* Scientific American.

Articles

ALBERTS, BRUCE M.: "The Function of the Hereditary Materials: Biological Catalysts Reflect the Cell's Evolutionary History," *American Zoologist*, vol. 26, pages 781–798, 1986.

ANDERSON, W. FRENCH: "Prospects for Human Gene Therapy," *Science*, vol. 226, pages 401–409, 1984.

ANDERSON, W. R., and E. G. DIACHMAKOS: "Genetic Engineering in Mammalian Cells," *Scientific American*, July 1981, pages 106–121.

ANGIER, NATALIE: "A Stupid Cell with all the Answers," *Discover*, November 1986, pages 71–83.

BISHOP, J. MICHAEL: "Oncogenes," *Scientific American*, March 1982, pages 80–92.

BROWN, DONALD D.: "Gene Expression in Eukaryotes," *Science*, vol. 211, pages 667–674, 1981.

CAMPBELL, A. M.: "How Viruses Insert Their DNA into the DNA of the Host Cell," *Scientific American*, December 1976, pages 102–113.

CECH, T. R.: "The Generality of Self-Splicing RNA: Relationship to Nuclear nRNA Splicing," *Cell*, January 1986, pages 207–210.

CHAMBON, PIERRE: "Split Genes," *Scientific American*, May 1981, pages 60–71.

CHILTON, MARY-DELL: "A Vector for Introducing New Genes into Plants," *Scientific American*, June 1983, pages 51–59.

COHEN, S. N., and J. A. SHAPIRO: "Transposable Genetic Elements," *Scientific American*, February 1980, pages 40–49.

DARNELL, JAMES E., JR.: "The Processing of RNA," *Scientific American*, October 1983, pages 90–100.

DARNELL, JAMES E., JR.: "RNA," *Scientific American*, October 1985, pages 68–87.

DAVIS, BERNARD D.: "Frontiers of the Biological Sciences," *Science*, vol. 209, pages 78–79, 1980.

DELISI, CHARLES: "The Human Genome Project," *American Scientist*, vol. 76, pages 488–493, 1988.

DE ROBERTIS, E. M., and J. B. GURDON: "Gene Transplantation and the Analysis of Development," *Scientific American*, December 1979, pages 74–82.

DICKERSON, RICHARD E.: "The DNA Helix and How It is Read," *Scientific American*, December 1983, pages 94–111.

EIGEN, MANFRED, WILLIAM GARDINER, PETER SCHUSTER and RUTHILD WINKLER-OSWATITSCH: "The Origin of Genetic Information," *Scientific American*, April 1981, pages 88–118.

FEDOROFF, NINA V.: "Transposable Genetic Elements in Maize," *Scientific American*, June 1984, pages 85–98.

FELSENFELD, GARY: "DNA," *Scientific American*, October 1985, pages 58–67.

FIDDES, J. C.: "The Nucleotide Sequence of a Viral DNA," *Scientific American*, December 1977, pages 54–67.

FISHER, ARTHUR: "Sinistral DNA," *Mosaic*, September 1983, pages 1–7.

GALLO, ROBERT C.: "The First Human Retrovirus," *Scientific American*, December 1986, pages 88–98.

---

* Available in paperback.

GILBERT, WALTER: "DNA Sequence and Gene Structure," *Science,* vol. 214, pages 1305–1312, 1981.

GILBERT, WALTER, and LYDIA VILLA-KAMAROFF: "Useful Proteins from Recombinant Bacteria," *Scientific American,* April 1980, pages 74–94.

GORDON, JON W., and FRANK H. RUDDLE: "Integration and Stable Germ Line Transmission of Genes Injected into Mouse Pronuclei," *Science,* vol. 214, pages 1344–1345, 1981.

GOULD, STEPHEN JAY: "Linnaean Limits," *Natural History,* August 1986, pages 16–23.

HALL, B. D.: "Mitochondria Spring Surprises," *Nature,* vol. 282, pages 129–130, 1979.

HOLZMAN, DAVID: "Ribosomal RNA: Where the Action Is," *Mosaic,* vol. 16, pages 32–39, 1985.

HOLZMAN, DAVID: "RNA: Messenger, Self-splicer, Catalyst . . .," *Mosaic,* vol. 15, pages 16–21, 1984.

KOLATA, GINA: "Two Disease-Causing Genes Found," *Science,* vol. 224, pages 669–670, 1986.

KORNBERG, ROGER D., and AARON KLUG: "The Nucleosome," *Scientific American,* February 1981, pages 52–64.

LAKE, JAMES A.: "The Ribosome," *Scientific American,* August 1981, pages 84–97.

LAWN, RICHARD M., and GORDON A. VEHAR: "The Molecular Genetics of Hemophilia," *Scientific American,* March 1986, pages 48–54.

LEWIN, ROGER: "Biggest Challenge Since the Double Helix," *Science,* vol. 212, pages 28–32, 1981.

LEWIN, ROGER: "Do Jumping Genes Make Evolutionary Leaps?" *Science,* vol. 213, pages 634–636, 1981.

LEWIN, ROGER: "On the Origin of Introns," *Science,* vol. 217, pages 921–922, 1982.

MARX, JEAN L.: "A Movable Feast in the Eukaryotic Genome," *Science,* vol. 211, pages 153–155, 1981.

MARX, JEAN L.: "Restriction Enzymes: Prenatal Diagnosis of Genetic Disease," *Science,* vol. 202, pages 1068–1069, 1978.

MILLER, JULIE ANN: "Building a Better Mouse," *BioScience,* February 1987, pages 103–106.

MILLER, JULIE ANN: "Switch-on Genes in Development," *Science News,* November 1985, page 205.

MOSES, PHYLLIS B., and NAM-HAI CHUA: "Light Switches for Plant Genes," *Scientific American,* April 1988, pages 88–93.

MURRAY, ANDREW W., and JACK W. SZOSTAK: "Artificial Chromosomes," *Scientific American,* November 1987, pages 62–68.

OLDROYD, DAVID: "Gregor Mendel: Founding-father of Modern Genetics," *Endeavour,* vol. 8, pages 29–31, 1984.

PATTERSON, DAVID: "The Causes of Down's Syndrome," *Scientific American,* April 1987, pages 52–60.

PINES, MAYA: "In the Shadow of Huntington's," *Science 84,* May 1984, pages 30–39.

PTASHNE, MARK, ALEXANDER D. JOHNSON, and CARL O. PALO: "A Genetic Switch in a Bacterial Virus," *Scientific American,* November 1982, pages 128–140.

RADMAN, MIROSLAV, and ROBERT WAGNER: "The High Fidelity of DNA Duplication," *Scientific American,* August 1988, pages 40–46.

RENSBERGER, BOYCE: "Cancer, the New Synthesis," *Science 84,* September 1984, pages 28–40.

RICH, A., and S. H. KIM: "The Three-Dimensional Structure of Transfer RNA," *Scientific American,* January 1978, pages 52–62.

SCHMID, CARL W., and WARREN R. JELINEK: "The *Alu* Family of Dispersed Repetition Sequences," *Science,* vol. 216, pages 1065–1070, 1982.

SMITH, MICHAEL: "The First Complete Nucleotide Sequencing of an Organism's DNA," *American Scientist,* vol. 67, pages 47–67, 1979.

STAHL, FRANKLIN W.: "Genetic Recombination," *Scientific American,* February 1987, pages 90–101.

STEITZ, JOAN A.: "Snurps," *Scientific American,* June 1988, pages 56–63.

TANGLEY, LAURA: "Gearing Up for Gene Therapy," *BioScience,* vol. 35, pages 8–10, 1985.

VARMUS, HAROLD: "Reverse Transcription," *Scientific American,* September 1987, pages 56–64.

WALTERS, LEROY: "The Ethics of Human Gene Therapy," *Nature,* vol. 320, pages 225–227, 1986.

WEINBERG, ROBERT A.: "Finding the Anti-Oncogene," *Scientific American,* September 1988, pages 44–51.

WEINBERG, ROBERT A.: "A Molecular Basis of Cancer," *Scientific American,* November 1983, pages 126–142.

WHITE, RAY, and JEAN-MARC LALOUEL: "Chromosome Mapping with DNA Markers," *Scientific American,* February 1988, pages 40–48.

---

* Available in paperback.

Appendixes

Metric Table

	QUANTITY	NUMERICAL VALUE	ENGLISH EQUIVALENT	CONVERTING ENGLISH TO METRIC
Length	kilometer (km)	1,000 (10^3) meters	1 km = 0.62 mile	1 mile = 1.609 km
	meter (m)	100 centimeters	1 m = 1.09 yards	1 yard = 0.914 m
			= 3.28 feet	1 foot = 0.305 m
	centimeter (cm)	0.01 (10^{-2}) meter	1 cm = 0.394 inch	1 foot = 30.5 cm
				1 inch = 2.54 cm
	millimeter (mm)	0.001 (10^{-3}) meter	1 mm = 0.039 inch	1 inch = 25.4 mm
	micrometer (μm)	0.000001 (10^{-6}) meter		
	nanometer (nm)	0.000000001 (10^{-9}) meter		
	angstrom (Å)	0.0000000001 (10^{-10}) meter		
Area	square kilometer (km²)	100 hectares	1 km² = 0.3861 square mile	1 square mile = 2.590 km²
	hectare (ha)	10,000 square meters	1 ha = 2.471 acres	1 acre = 0.4047 ha
	square meter (m²)	10,000 square centimeters	1 m² = 1.1960 square yards	1 square yard = 0.8361 m²
			= 10.764 square feet	1 square foot = 0.0929 m²
	square centimeter (cm²)	100 square millimeters	1 cm² = 0.155 square inch	1 square inch = 6.4516 cm²
Mass	metric ton (t)	1,000 kilograms = 1,000,000 grams	1 t = 1.103 tons	1 ton = 0.907 t
	kilogram (kg)	1,000 grams	1 kg = 2.205 pounds	1 pound = 0.4536 kg
	gram (g)	1,000 milligrams	1 g = 0.0353 ounce	1 ounce = 28.35 g
	milligram (mg)	0.001 gram		
	microgram (μg)	0.000001 gram		
Time	second (sec)	1,000 milliseconds		
	millisecond (msec)	0.001 second		
	microsecond (μsec)	0.000001 second		
Volume (solids)	1 cubic meter (m³)	1,000,000 cubic centimeters	1 m³ = 1.3080 cubic yards	1 cubic yard = 0.7646 m³
			= 35.315 cubic feet	1 cubic foot = 0.0283 m³
	1 cubic centimeter (cm³)	1,000 cubic millimeters	1 cm³ = 0.0610 cubic inch	1 cubic inch = 16.387 cm³
Volume (liquids)	kiloliter (kl)	1,000 liters	1 kl = 264.17 gallons	1 gal = 3.785 l
	liter (l)	1,000 milliliters	1 l = 1.06 quarts	1 qt = 0.94 l
				1 pt = 0.47 l
	milliliter (ml)	0.001 liter	1 ml = 0.034 fluid ounce	1 fluid ounce = 29.57 ml
	microliter (μl)	0.000001 liter		

Temperature Conversion Scale

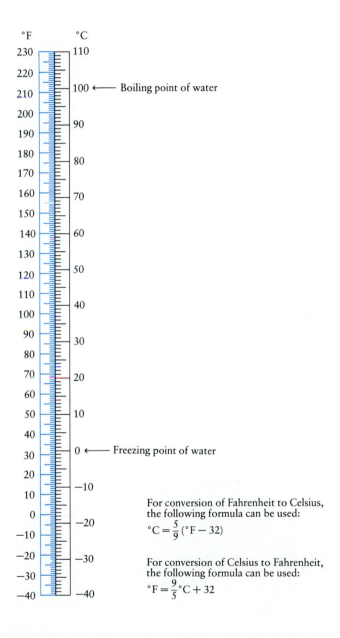

°F	°C	
230	110	
220		
210	100	← Boiling point of water
200	90	
190		
180	80	
170		
160	70	
150		
140	60	
130		
120	50	
110	40	
100		
90	30	
80		
70	20	
60		
50	10	
40		
30	0	← Freezing point of water
20		
10	−10	
0	−20	
−10		
−20	−30	
−30		
−40	−40	

For conversion of Fahrenheit to Celsius, the following formula can be used:

$$°C = \frac{5}{9}(°F - 32)$$

For conversion of Celsius to Fahrenheit, the following formula can be used:

$$°F = \frac{9}{5}°C + 32$$

Glossary

This list does not include units of measure or names of taxonomic groups, which can be found in Appendixes A and C, or terms that are used only once in the text and defined there.

abdomen: In vertebrates, the portion of the trunk containing visceral organs other than heart and lungs; in arthropods, the posterior portion of the body, made up of similar segments and containing the reproductive organs and part of the digestive tract.

abscisic acid (ABA) [L. *ab,* away, off + *scissio,* dividing]: A plant hormone with a variety of inhibitory effects; brings about dormancy in buds, maintains dormancy in many types of seeds, and effects stomatal closing; also known as the "stress hormone."

abscission [L. *ab,* away, off + *scissio,* dividing]: In plants, the dropping of leaves, flowers, fruits, or stems at the end of a growing season, as the result of formation of a two-layered zone of specialized cells (the abscission zone) and the action of a hormone (ethylene).

absorption [L. *absorbere,* to swallow down]: The movement of water and dissolved substances into a cell, tissue, or organism.

absorption spectrum: The characteristic pattern of the wavelengths (colors) of light that a particular pigment absorbs.

acetylcholine (asset-ill-**coal**-een): One of the principal chemicals (neurotransmitters) responsible for the transmission of nerve impulses across synapses.

acid [L. *acidus,* sour]: A substance that causes an increase in the number of hydrogen ions (H^+) in a solution and a decrease in the number of hydroxide ions (OH^-); having a pH of less than 7; the opposite of a base.

actin [Gk. *aktis,* a ray]: A protein, composed of globular subunits, that forms filaments that are among the principal components of the cytoskeleton. Also one of the two major proteins of muscle (the other is myosin); the principal constituent of the thin filaments.

action potential: A transient change in electric potential across a membrane; in nerve cells, results in conduction of a nerve impulse; in muscle cells, results in contraction.

action spectrum: The characteristic pattern of the wavelengths (colors) of light that elicit a particular reaction or response.

activation energy: The energy that must be possessed by atoms or molecules in order to react.

active site: The region of an enzyme surface that binds the substrate during the reaction catalyzed by the enzyme.

active transport: The energy-requiring transport of a solute across a cell membrane (or a membrane of an organelle) from a region of lower concentration to a region of higher concentration (that is, against a concentration gradient).

adaptation [L. *adaptare,* to fit]: (1) The evolution of features that make a group of organisms better suited to live and reproduce in their environment. (2) A peculiarity of structure, physiology, or behavior that aids the organism in its environment.

adaptive radiation: The evolution from a primitive and unspecialized ancestor of a number of divergent forms, each specialized to fill a distinct ecological niche; associated with the opening up of a new biological frontier.

adenosine diphosphate (ADP): A nucleotide consisting of adenine, ribose, and two phosphate groups; formed by the removal of one phosphate from an ATP molecule.

adenosine monophosphate (AMP): A nucleotide consisting of adenine, ribose, and one phosphate group; can be formed by the removal of two phosphates from an ATP molecule; in its cyclic form, functions as a "second messenger" for a number of vertebrate hormones and neurotransmitters.

adenosine triphosphate (ATP): The nucleotide that provides the energy currency for cell metabolism; composed of adenine, ribose, and three phosphate groups. On hydrolysis, ATP loses one phosphate group and one hydrogen ion to become adenosine diphosphate (ADP), releasing energy in the process. ATP is formed from ADP and inorganic phosphate in an enzymatic reaction that traps energy released by catabolism or energy captured in photosynthesis.

ADH: Abbreviation of antidiuretic hormone.

adhesion [L. *adhaerere,* to stick to]: The holding together of molecules of different substances.

ADP: Abbreviation of adenosine diphosphate.

adrenal gland [L. *ad,* near + *renes,* kidney]: A vertebrate endocrine gland. The cortex (outer surface) is the source of cortisol, aldosterone, and other steroid hormones; the medulla (inner core) secretes adrenaline and noradrenaline.

adrenaline: A hormone, produced by the medulla of the adrenal gland, that increases the concentration of sugar in the blood, raises blood pressure and heartbeat rate, and increases muscular power and resistance to fatigue; also a neurotransmitter across synaptic junctions. Also called epinephrine.

adventitious [L. *adventicius,* not properly belonging to]: Referring to a structure arising from an unusual place, such as roots growing from stems or leaves.

aerobic [Gk. *aēr,* air + *bios,* life]: Any biological process that can occur in the presence of molecular oxygen (O_2).

afferent [L. *ad,* near + *ferre,* to carry]: Bringing inward to a central part, applied to nerves and blood vessels.

agar: A gelatinous material prepared from certain red algae that is used to solidify nutrient media for growing microorganisms.

aldosterone [Gk. *aldainō,* to nourish + *stereō,* solid]: A hormone produced by the adrenal cortex that affects the concentration of ions in the blood; it stimulates the reabsorption of sodium and the excretion of potassium by the kidney.

alga, *pl.* **algae** (**al**-gah, **al**-jee): A unicellular or simple multicellular eukaryotic photosynthetic organism lacking multicellular sex organs.

alkaline: Pertaining to substances that increase the number of hydroxide ions (OH^-) in a solution; having a pH greater than 7; basic; opposite of acidic.

allantois [Gk. *allant,* sausage]: One of the four extraembryonic membranes that form during the development of reptiles, birds, and mammals.

allele frequency: The proportion of a particular allele in a population.

alleles (al-eels) [Gk. *allelon,* of one another]: Two or more different forms of a gene. Alleles occupy the same position (locus) on homologous chromosomes and are separated from each other at meiosis.

allopatric speciation [Gk. *allos,* other + *patra,* fatherland, country]: Speciation that occurs as the result of the geographic separation of a population of organisms.

allosteric interaction [Gk. *allos,* other + *stereō,* solid, shape]: An interaction involving an enzyme that has two binding sites, the active site and a site into which another molecule, an allosteric effector, fits; the binding of the effector changes the shape of the enzyme and activates or inactivates it. Allosteric interactions also play a role in transport processes involving integral membrane proteins.

alternation of generations: A sexual life cycle in which a haploid (*n*) phase alternates with a diploid (2*n*) phase. The gametophyte (*n*) produces gametes (*n*) by mitosis. The fusion of gametes yields zygotes (2*n*). Each zygote develops into a sporophyte (2*n*) that forms haploid spores (*n*) by meiosis. Each haploid spore forms a new gametophyte, completing the cycle.

altruism: Self-sacrifice for the benefit of others; any form of behavior that increases the fitness of the recipient while reducing the fitness of the altruistic individual.

alveolus, *pl.* **alveoli** [L. dim. of *alveus,* cavity, hollow]: One of the many small air sacs within the lungs in which the bronchioles terminate. The thin walls of the alveoli contain numerous capillaries and are the site of gas exchange between the air in the alveoli and the blood in the capillaries.

amino acids (am-ee-no) [Gk. *Ammon,* referring to the Egyptian sun god, near whose temple ammonium salts were first prepared from camel dung]: Organic molecules containing nitrogen in the form of —NH₂ and a carboxyl group, —COOH, bonded to the same carbon atom; the "building blocks" of protein molecules.

ammonification: The process by which decomposers break down proteins and amino acids, releasing the excess nitrogen in the form of ammonia (NH_3) or ammonium ion (NH_4^+).

amnion (am-neon) [Gk. dim. of *amnos,* lamb]: One of the four extraembryonic membranes that form during the development of reptiles, birds, and mammals; it encloses a fluid-filled space, the amniotic cavity, that surrounds the developing embryo.

amniote egg: An egg that is isolated and protected from the environment during the period of its development by a series of extraembryonic membranes and, often, a more or less impervious shell; the amniote eggs of birds and many reptiles are completely self-sufficient, requiring only oxygen from the outside.

amoeboid [Gk. *amoibē,* change]: Moving or feeding by means of pseudopodia (temporary cytoplasmic protrusions from the cell body).

AMP: Abbreviation of adenosine monophosphate.

anabolism [Gk. *ana,* up + -*bolism* (as in metabolism)]: Within a cell or organism, the sum of all biosynthetic reactions (that is, chemical reactions in which larger molecules are formed from smaller ones).

anaerobe [Gk. *an,* without + *aēr,* air + *bios,* life]: Cell that can live without free oxygen; obligate anaerobes cannot live in the presence of oxygen; facultative anaerobes can live with or without oxygen.

anaerobic [Gk. *an,* without + *aēr,* air + *bios,* life]: Applied to a process that can occur without oxygen, such as fermentation; also applied to organisms that can live without free oxygen.

analogous [Gk. *analogos,* proportionate]: Applied to structures similar in function but different in evolutionary origin, such as the wing of a bird and the wing of an insect.

anaphase (anna-phase) [Gk. *ana,* up + *phasis,* form]: In mitosis and meiosis II, the stage in which the chromatids of each chromosome separate and move to opposite poles; in meiosis I, the stage in which homologous chromosomes separate and move to opposite poles.

androgens [Gk. *andros,* man + *genos,* origin, descent]: Male sex hormones; any chemical with actions similar to those of testosterone.

angiosperms (an-jee-o-sperms) [Gk. *angeion,* vessel + *sperma,* seed]: The flowering plants. Literally, a seed borne in a vessel; thus, any plant whose seeds are borne within a matured ovary (fruit).

anisogamy [Gk. *aniso,* unequal + *gamos,* marriage]: Sexual reproduction in which one gamete is larger than the other; both gametes are motile.

annual plant [L. *annus,* year]: A plant that completes its life cycle (from seed germination to seed production) and dies within a single growing season.

antennae: Long, paired sensory appendages on the head of many arthropods.

anterior [L. *ante,* before, toward, in front of]: The front end of an organism.

anther [Gk. *anthos,* flower]: In flowering plants, the pollen-bearing portion of a stamen.

antheridium, *pl.* **antheridia:** In bryophytes and some vascular plants, the multicellular sperm-producing organ.

anthropoid [Gk. *anthropos,* man, human]: A higher primate; includes monkeys, apes, and humans.

antibiotic [Gk. *anti,* against + *bios,* life]: An organic compound, inhibitory or toxic to other species, that is formed and secreted by an organism.

antibody [Gk. *anti,* against]: A globular protein, synthesized by a B lymphocyte, that is complementary to a foreign substance (antigen) with which it combines specifically.

anticodon: In a tRNA molecule, the three-nucleotide sequence that base pairs with the mRNA codon for the amino acid carried by that particular tRNA; the anticodon is complementary to the mRNA codon.

antidiuretic hormone (ADH) [Gk. *anti,* against + *diurgos,* thoroughly wet + *hormaein,* to excite]: A peptide hormone synthesized in the hypothalamus that inhibits urine excretion by inducing the reabsorption of water from the nephrons of the kidneys; also called vasopressin.

antigen [Gk. *anti,* against + *genos,* origin, descent]: A foreign substance, usually a protein or polysaccharide, that, when bound to a complementary antibody displayed on the surface of a B lymphocyte or to a complementary T-cell receptor, stimulates an immune response.

aorta (a-ore-ta) [Gk. *aeirein,* to lift, heave]: The major artery in blood-circulating systems; the aorta sends blood to the other body tissues.

apical dominance [L. *apex,* top]: In plants, the hormone-mediated influence of a terminal bud in suppressing the growth of axillary buds.

apical meristem [L. *apex,* top + Gk. *meristos,* divided]: In vascular plants, the growing point at the tip of the root or stem.

arboreal [L. *arbor,* tree]: Tree-dwelling.

archegonium, *pl.* **archegonia** [Gk. *archegonos,* first of a race]: In bryophytes and some vascular plants, the multicellular egg-producing organ.

archenteron [Gk. *arch,* first, or main + *enteron,* gut]: The main cavity within the early embryo (gastrula) of many animals; lined with endoderm, it opens to the outside by means of the blastopore and ultimately becomes the digestive tract.

artery: A vessel carrying blood from the heart to the tissues; arteries are usually thick-walled, elastic, and muscular. A small artery is known as an arteriole.

artificial selection: The breeding of selected organisms for the purpose of producing descendants with desired characteristics.

ascus, *pl.* **asci** (as-kus, as-i) [Gk. *askos,* wineskin, bladder]: In the fungi of division Ascomycota, a specialized cell within which two haploid nuclei fuse to produce a zygote that immediately divides by meiosis; at maturity, an ascus contains ascospores.

asexual reproduction: Any reproductive process, such as budding or the division of a cell or body into two or more approximately equal parts, that does not involve the union of gametes.

atmospheric pressure [Gk. *atmos,* vapor + *sphaira,* globe]: The weight of the earth's atmosphere over a unit area of the earth's surface.

atom [Gk. *atomos,* indivisible]: The smallest particle into which a chemical element can be divided and still retain the properties characteristic of the element; consists of a central core, the nucleus, containing protons and neutrons, and electrons that move around the nucleus.

atomic number: The number of protons in the nucleus of an atom; equal to the number of electrons in the neutral atom.

atomic weight: The average weight of all the isotopes of an element relative to the weight of an atom of the most common isotope of carbon (^{12}C), which is by convention assigned the integral value of 12; approximately equal to the number of protons plus neutrons in the nucleus of an atom.

ATP: Abbreviation of adenosine triphosphate, the principal energy-carrying compound of the cell.

ATP synthetase: The enzyme complex in the inner membrane of the mitochondrion and the thylakoid membrane of the chloroplast through which protons flow down the gradient established in the first stage of chemiosmotic coupling; the site of formation of ATP from ADP and inorganic phosphate during oxidative phosphorylation and photophosphorylation.

atrioventricular node [L. *atrium*, yard, court, hall + *ventriculus*, the stomach + *nodus*, knot]: A group of slow-conducting fibers in the atrium of the vertebrate heart that are stimulated by impulses originating in the sinoatrial node (the pacemaker) and that conduct impulses to the bundle of His, a group of fibers that stimulate contraction of the ventricles.

atrium, *pl.* **atria** (a-tree-um) [L., yard, court, hall]: A thin-walled chamber of the heart that receives blood and passes it on to a thick, muscular ventricle.

autonomic [Gk. *autos*, self + *nomos*, usage, law]: Self-controlling, independent of outside influences.

autonomic nervous system [Gk. *autos*, self + *nomos*, usage, law]: In the peripheral nervous system of vertebrates, the neurons and ganglia that are not ordinarily under voluntary control; innervates the heart, glands, visceral organs, and smooth muscle; subdivided into the sympathetic and parasympathetic divisions.

autosome [Gk. *autos*, self + *soma*, body]: Any chromosome other than the sex chromosomes. Humans have 22 pairs of autosomes and one pair of sex chromosomes.

autotroph [Gk. *autos*, self + *trophos*, feeder]: An organism that is able to synthesize all needed organic molecules from simple inorganic substances (e.g., H_2O, CO_2, NH_3) and some energy source (e.g., sunlight); in contrast to heterotroph. Plants, algae, and some groups of prokaryotes are autotrophs.

auxin [Gk. *auxein*, to increase + *in*, of, or belonging to]: One of a group of plant hormones with a variety of growth-regulating effects, including promotion of cell elongation.

auxotroph [L. *auxillium*, help + Gk. *trophos*, feeder]: A mutant with a defect in the enzymatic pathway for the synthesis of a particular molecule, which must therefore be supplied for normal growth.

axil [Gk. *axilla*, armpit]: The upper angle between a twig or leaf and the stem from which it grows.

axillary [Gk. *axilla*, armpit]: In botany, term applied to buds or branches occurring in the axil of a leaf.

axis: An imaginary line passing through a body or organ around which parts are symmetrically aligned.

axon [Gk. *axon*, axle]: A long process of a neuron, or nerve cell, that is capable of rapidly conducting nerve impulses over great distances.

B lymphocyte: A type of white blood cell capable of becoming an antibody-secreting plasma cell; a B cell.

bacteriophage [L. *bacterium* + Gk. *phagein*, to eat]: A virus that parasitizes a bacterial cell.

bark: In plants, all tissues outside the vascular cambium in a woody stem.

basal body [Gk. *basis*, foundation]: A cytoplasmic organelle of animals and some protists, from which cilia or flagella arise; identical in structure to the centriole, which is involved in mitosis and meiosis in animals and some protists.

base: A substance that causes an increase in the number of hydroxide ions (OH^-) in a solution and a decrease in the number of hydrogen ions (H^+); having a pH of more than 7; the opposite of an acid. *See* Alkaline.

base-pairing principle: In the formation of nucleic acids, the requirement that adenine must always pair with thymine (or uracil) and guanine with cytosine.

basidium, *pl.* **basidia** (ba-sid-ium) [L., a little pedestal]: A specialized reproductive cell of the fungi of division Basidiomycota, often club-shaped, in which nuclear fusion and meiosis occur; homologous with the ascus.

behavior: All of the acts an organism performs, as in, for example, seeking a suitable habitat, obtaining food, avoiding predators, and seeking a mate and reproducing.

biennial [L. *biennium*, a space of two years; *bi*, twice + *annus*, year]: Occurring once in two years; a plant that requires two years to complete its reproductive cycle; vegetative growth occurs in the first year, sexual reproduction and death in the second.

bilateral symmetry [L. *bi*, twice, two + *lateris*, side; Gk. *summetros*, symmetry]: A body form in which the right and left halves of an organism are approximate mirror images of each other.

bile: A yellow secretion of the vertebrate liver, temporarily stored in the gallbladder and composed of organic salts that emulsify fats in the small intestine.

binary fission [L. *binarius*, consisting of two things or parts + *fissus*, split]: Asexual reproduction by division of the cell or body into two equal, or nearly equal, parts.

binomial system [L. *bi*, twice, two + Gk. *nomos*, usage, law]: A system of naming organisms in which the name consists of two parts, with the first designating genus and the second, species; originated by Linnaeus.

biogeochemical cycle [Gk. *bios*, life + *geō*, earth + *chēmeia*, alchemy; *kyklos*, circle, wheel]: The cyclic path of an inorganic substance, such as carbon or nitrogen, through an ecosystem. Its geological components are the atmosphere, the crust of the earth, and the oceans, lakes, and rivers; its biological components are producers, consumers, and detritivores, including decomposers.

biological clock [Gk. *bios*, life + *logos*, discourse]: Proposed internal factor(s) in organisms that governs functions that occur rhythmically in the absence of external stimuli.

biomass [Gk. *bios*, life]: Total weight of all organisms (or some group of organisms) living in a particular habitat or place.

biome: One of the major types of distinctive plant formations; for example, the grassland biome, the tropical rain forest biome, etc.

biosphere [Gk. *bios*, life + *sphaira*, globe]: The zones of air, land, and water at the surface of the earth occupied by living things.

biosynthesis [Gk. *bios*, life + *synthesis*, a putting together]: Formation by living organisms of organic compounds from elements or simple compounds.

blade: (1) The broad, expanded part of a leaf. (2) The broad, expanded photosynthetic part of the thallus of a multicellular alga or a simple plant.

blastocoel [Gk. *blastos*, sprout + *koilos*, a hollow]: The fluid-filled cavity in the interior of a blastula.

blastocyst [Gk. *blastos*, sprout + *kystis*, sac]: The blastula stage of a developing mammal; consists of an inner cell mass that will give rise to the embryo proper and a double layer of cells, the trophoblast, that is the precursor of the chorion.

blastodisc [Gk. *blastos*, sprout + *discos*, a round plate]: Disklike area on the surface of a large, yolky egg that undergoes cleavage and gives rise to the embryo.

blastomere [Gk. *blastos*, sprout + *meris*, part of, portion]: One of many cells produced by cleavage of the fertilized egg.

blastopore [Gk. *blastos*, sprout + *poros*, a way, means, path]: In the gastrula stage of an embryo, the opening that connects the archenteron with the outside; represents the future mouth in some animals (protostomes), the future anus in others (deuterostomes).

blastula [Gk. *blastos,* sprout]: An animal embryo after cleavage and before gastrulation; usually consists of a fluid-filled sphere, the walls of which are composed of a single layer of cells.

bond strength: The strength with which a chemical bond holds two atoms together; conventionally measured in terms of the amount of energy, in kilocalories per mole, required to break the bond.

botany [Gk. *botanikos,* of herbs]: The study of plants.

Bowman's capsule: In the vertebrate kidney, the bulbous unit of the nephron, which surrounds the glomerulus. In filtration, the initial process in urine formation, blood plasma is forced from the glomerular capillaries into Bowman's capsule.

brainstem: The most posterior portion of the vertebrate brain; includes medulla, pons, and midbrain.

bronchus, *pl.* **bronchi** (bronk-us, bronk-eye) [Gk. *bronchos,* windpipe]: One of a pair of respiratory tubes branching into either lung at the lower end of the trachea; it subdivides into progressively finer passageways, the bronchioles, culminating in the alveoli.

bud: (1) In plants, an embryonic shoot, including rudimentary leaves, often protected by special bud scales. (2) In animals, an asexually produced outgrowth that develops into a new individual.

buffer: A combination of H^+-donor and H^+-acceptor forms of a weak acid or a weak base; a buffer prevents appreciable changes of pH in solutions to which small amounts of acids or bases are added.

bulb: A modified bud with thickened leaves adapted for underground food storage.

bulk flow: The overall movement of a fluid induced by gravity, pressure, or an interplay of both.

bundle of His: In the vertebrate heart, a group of muscle fibers that carry impulses from the atrioventricular node to the walls of the ventricles; the only electrical bridge between the atria and the ventricles.

C_3 pathway: *See* Calvin cycle.

C_4 pathway: The set of reactions by which some plants initially fix carbon in the four-carbon compound oxaloacetic acid; the carbon dioxide is later released in the interior of the leaf and enters the Calvin cycle. Also known as the Hatch-Slack pathway.

callus [L. *callos,* hard skin]: In plants, undifferentiated tissue; a term used in tissue culture, grafting, and wound healing.

calorie [L. *calor,* heat]: The amount of energy in the form of heat required to raise the temperature of 1 gram of water 1°C; in making metabolic measurements the kilocalorie (Calorie) is generally used. A Calorie is the amount of heat required to raise the temperature of 1 kilogram of water 1°C.

Calvin cycle: The set of reactions in which carbon dioxide is reduced to carbohydrate during the second stage of photosynthesis.

calyx [Gk. *kalyx,* a husk, cup]: Collectively, the sepals of a flower.

CAM photosynthesis: *See* Crassulacean acid metabolism.

capillaries [L. *capillaris,* relating to hair]: Smallest thin-walled blood vessels through which exchanges between blood and the tissues occur; connect arteries with veins.

capillary action: The movement of water or any liquid along a surface; results from the combined effect of cohesion and adhesion.

capsid: The protein coat surrounding the nucleic acid core of a virus.

capsule (kap-sul) [L. *capsula,* a little chest]: (1) A slimy layer around the cells of certain bacteria. (2) The sporangium of a bryophyte.

carbohydrate [L. *carbo,* charcoal + *hydro,* water]: An organic compound consisting of a chain or ring of carbon atoms to which hydrogen and oxygen are attached in a ratio of approximately 2:1; carbohydrates include sugars, starch, glycogen, cellulose, etc.

carbon cycle: Worldwide circulation and reutilization of carbon atoms, chiefly due to metabolic processes of living organisms. Inorganic carbon, in the form of carbon dioxide, is incorporated into organic compounds by photosynthetic organisms; when the organic compounds are broken down

in respiration, carbon dioxide is released. Large quantities of carbon are "stored" in the seas and the atmosphere, as well as in fossil fuel deposits.

carbon fixation: The second stage of photosynthesis; energy stored in ATP and NADPH by the energy-capturing reactions of the first stage is used to reduce carbon from carbon dioxide to simple sugars.

cardiovascular system [Gk. *kardio,* heart + L. *vasculum,* a small vessel]: In animals, the heart and blood vessels.

carnivore [L. *caro, carnis,* flesh + *voro,* to devour]: Predator that obtains its nutrients and energy by eating meat.

carotenoids [L. *carota,* carrot]: A class of pigments that includes the carotenes (yellows, oranges, and reds) and the xanthophylls (yellow); accessory pigments in photosynthesis.

carpel [Gk. *karpos,* fruit]: A leaflike floral structure enclosing the ovule or ovules of angiosperms, typically divided into ovary, style, and stigma; a flower may have one or more carpels, either single or fused. A single carpel or a group of fused carpels is also known as a pistil.

carrying capacity: In ecology, the average number of individuals of a particular population that the environment can support under a particular set of conditions.

cartilage [L. *cartilago,* gristle]: A connective tissue in skeletons of vertebrates; forms much of the skeleton of adult lower vertebrates and immature higher vertebrates.

Casparian strip (after Robert Caspary, German botanist): In the roots of plants, a thickened, waxy strip that extends around and seals the walls of endodermal cells, restricting the diffusion of solutes across the endodermis into the vascular tissues of the root.

catabolism [Gk. *katabole,* throwing down]: Within a cell or organism, the sum of all chemical reactions in which large molecules are broken down into smaller parts.

catalyst [Gk. *katalysis,* dissolution]: A substance that lowers the activation energy of a chemical reaction by forming a temporary association with the reacting molecules; as a result, the rate of the reaction is accelerated. Enzymes are catalysts.

category [Gk. *katēgoria,* category]: In a hierarchical classification system, the level at which a particular group is ranked.

cell [L. *cella,* a chamber]: The structural unit of organisms, surrounded by a membrane and composed of cytoplasm and, in eukaryotes, one or more nuclei. In most plants, fungi, and bacteria, there is a cell wall outside the membrane.

cell cycle: A regular, timed sequence of the events of cell growth and division through which dividing cells pass.

cell membrane: The outer membrane of the cell; also called the plasma membrane.

cell plate: In the dividing cells of most plants (and in some algae), a flattened structure that forms at the equator of the mitotic spindle in early telophase; gives rise to the middle lamella.

cell theory: All living things are composed of cells; cells arise only from other cells. No exception has been found to these two principles since they were first proposed well over a century ago.

cellulose [L. *cellula,* a little cell]: The chief constituent of the cell wall in all plants and some protists; an insoluble complex carbohydrate formed of microfibrils of glucose molecules.

cell wall: A plastic or rigid structure, produced by the cell and located outside the cell membrane in most plants, algae, fungi, and prokaryotes; in plant cells, it consists mostly of cellulose.

central nervous system: In vertebrates, the brain and spinal cord; in invertebrates it usually consists of one or more cords of nervous tissue plus their associated ganglia.

centriole (sen-tree-ole) [Gk. *kentron,* center]: A cytoplasmic organelle identical in structure to a basal body; flagellated cells and all animal cells, including those without flagella, have centrioles at the spindle poles during division.

centromere (sen-tro-mere) [Gk. *kentron*, center + *meros*, a part]: Region of constriction of chromosome that holds sister chromatids together.

cerebellum [L. dim. of *cerebrum*, brain]: A subdivision of the vertebrate brain that lies above the brainstem and behind and below the cerebrum; functions in coordinating muscular activities and maintaining equilibrium.

cerebral cortex [L. *cerebrum*, brain]: A thin layer of neurons and glial cells forming the upper surface of the cerebrum, well developed only in mammals; the seat of conscious sensations and voluntary muscular activity.

cerebrum [L., brain]: The portion of the vertebrate brain occupying the upper part of the skull, consisting of two cerebral hemispheres united by the corpus callosum; coordinates most activities.

character displacement: A phenomenon in which species that live together in the same environment tend to diverge in those characteristics that overlap; exemplified by Darwin's finches.

chelicera, *pl.* **chelicerae** [Gk. *cheilos*, the edge, lips + *cheir*, arm]: First pair of appendages in horseshoe crabs, sea spiders, and arachnids; usually take the form of pincers or fangs.

chemical reaction: An interaction among atoms, ions, or molecules that results in the formation of new combinations of atoms, ions, or molecules; the making or breaking of chemical bonds.

chemiosmotic coupling: The mechanism by which ADP is phosphorylated to ATP in mitochondria and chloroplasts. The energy released as electrons pass down an electron transport chain is used to establish a proton gradient across an inner membrane of the organelle; when protons subsequently flow down this electrochemical gradient, the potential energy released is captured in the terminal phosphate bonds of ATP.

chemoreceptor: A sensory cell or organ that responds to the presence of a specific chemical stimulus; includes smell and taste receptors.

chemosynthetic: Applied to autotrophic bacteria that use the energy released by specific inorganic reactions to power their life processes, including the synthesis of organic molecules.

chemotactic [Gk. *cheimō*, storm + *taxis*, arrangement, order]: Of an organism, capable of responding to a chemical stimulus by moving toward or away from it.

chiasma, *pl.* **chiasmata** (kye-az-ma) [Gk., a cross]: Connection between paired homologous chromosomes at meiosis; the site at which crossing over and genetic recombination occur.

chitin (kye-tin) [Gk. *chitōn*, a tunic, undergarment]: A tough, resistant, nitrogen-containing polysaccharide present in the exoskeleton of arthropods, the epidermal cuticle or other surface structures of many other invertebrates, and in the cell walls of fungi.

chlorophyll [Gk. *chloros*, green + *phyllon*, leaf]: A class of green pigments that are the receptors of light energy in photosynthesis.

chloroplast [Gk. *chloros*, green + *plastos*, formed]: A membrane-bound, chlorophyll-containing organelle in eukaryotes (algae and plants) that is the site of photosynthesis.

chorion (core-ee-on) [Gk., skin, leather]: The outermost extraembryonic membrane of developing reptiles, birds, and mammals; in placental mammals it contributes to the structure of the placenta.

chromatid (crow-ma-tid) [Gk. *chrōma*, color]: Either of the two strands of a replicated chromosome, which are joined at the centromere.

chromatin [Gk. *chrōma*, color]: The deeply staining complex of DNA and histone proteins of which eukaryotic chromosomes are composed.

chromosome [Gk. *chrōma*, color + *soma*, body]: The structure that carries the genes. Eukaryotic chromosomes are visualized as threads or rods of chromatin, which appear in a contracted form during mitosis and meiosis, and are otherwise enclosed in a nucleus. Prokaryotic chromosomes consist of a closed circle of DNA with which a variety of proteins are associated. Viral chromosomes are linear or circular molecules of DNA or RNA.

chromosome map: A diagram of the linear order of the genes on a chromosome.

cilium, *pl.* **cilia** (silly-um) [L., eyelash]: A short, thin structure embedded in the surface of some eukaryotic cells, usually in large numbers and arranged in rows; has a highly characteristic internal structure of two inner microtubules surrounded by nine pairs of outer microtubules; involved in locomotion and the movement of substances across the cell surface.

circadian rhythms [L. *circa*, about + *dies*, day]: Regular rhythms of growth or activity that occur on an approximately 24-hour cycle.

cladogenesis [Gk. *clados*, branch + *genesis*, origin]: The splitting of an evolutionary lineage into two or more separate lineages; one of the principal patterns of evolutionary change; also known as splitting evolution.

class: A taxonomic grouping of related, similar orders; category above order and below phylum.

cleavage: The successive cell divisions of a fertilized egg of an animal to form a multicellular blastula.

cline [Gk. *klinein*, to lean]: A graded series of changes in some characteristic within a species, correlated with some gradual change in temperature, humidity, or other environmental factor over the geographic range of the species.

cloaca [L., sewer]: The exit chamber from the digestive system in reptiles and birds; also may serve as the exit for the reproductive and urinary systems.

clone [Gk. *klon*, twig]: A line of cells, all of which have arisen from the same single cell by repeated cell divisions; a population of individuals derived by asexual reproduction from a single ancestor.

cnidocyte (ni-do-site) [Gk. *knide*, nettle + *kytos*, vessel]: A stinging cell containing a nematocyst; characteristic of cnidarians.

coadaptive gene complex: A group of genes that collectively produce coordinated phenotypic characteristics; when linked together on one chromosome, known as a supergene.

cochlea [Gk. *kochlias*, snail]: Part of the inner ear of mammals; concerned with hearing.

codominance: In genetics, the phenomenon in which the effects of both alleles at a particular locus are apparent in the phenotype of the heterozygote.

codon (code-on): Basic unit ("letter") of the genetic code; three adjacent nucleotides in a molecule of DNA or mRNA that form the code for a specific amino acid or for polypeptide chain termination.

coelom (see-loam) [Gk. *koilos*, a hollow]: A body cavity formed between layers of mesoderm and in which the digestive tract and other internal organs are suspended.

coenocytic (see-no-sit-ik) [Gk. *koinos*, shared in common + *kytos*, a vessel]: An organism or part of an organism consisting of many nuclei within a common cytoplasm.

coenzyme [L. *co*, together + Gk. *en*, in + *zyme*, leaven]: A nonprotein organic molecule that plays an accessory role in enzyme-catalyzed processes, often by acting as a donor or acceptor of a substance involved in the reaction. NAD$^+$, FAD, and coenzyme A are common coenzymes.

coevolution [L. *co*, together + *e-*, out + *volvere*, to roll]: The simultaneous evolution of adaptations in two or more populations that interact so closely that each is a strong selective force on the other.

cofactor: A nonprotein component that plays an accessory role in enzyme-catalyzed processes; some cofactors are ions, and others are coenzymes.

cohesion [L. *cohaerere*, to stick together]: The attraction or holding together of molecules of the same substance.

cohesion-tension theory: A theory accounting for the upward movement of water in plants. According to this theory, transpiration of a water molecule results in a negative (below 1 atmosphere) pressure in the leaf cells, inducing the entrance from the vascular tissue of another water molecule, which, because of the cohesive property of water, pulls with it a chain of water molecules extending up from the cells of the root tip.

coleoptile (coal-ee-op-tile) [Gk. *koleon*, sheath + *ptilon*, feather]: The sheath enclosing the apical meristem and leaf primordia of a germinating monocot.

collagen [Gk. *kolla*, glue]: A fibrous protein in bones, tendons, and other connective tissues.

collenchyma [Gk. *kolla*, glue]: In plants, a type of supporting cell with an irregularly thickened primary cell wall; alive at maturity.

colony: A group of organisms of the same species living together in close association.

commensalism [L. *com*, together + *mensa*, table]: *See* Symbiosis.

community: All of the populations of organisms inhabiting a common environment and interacting with one another.

companion cell: In angiosperms, a specialized parenchyma cell associated with a sieve-tube member and arising from the same mother cell as the sieve-tube member.

competition: Interaction between members of the same population or of two or more populations using the same resource, often present in limited supply; interference competition involves fighting or other direct interactions, whereas exploitative competition involves the removal or preemption of a resource.

competitive exclusion: The hypothesis that two species with identical ecological requirements cannot coexist stably in the same locality and the species that is more efficient in utilizing the available resources will exclude the other; also known as Gause's principle, after the Russian biologist G. F. Gause.

complementary DNA (cDNA): DNA molecules synthesized by reverse transcriptase from an RNA template.

compound [L. *componere*, to put together]: A chemical substance composed of two or more kinds of atoms in definite ratios.

compound eye: In arthropods, a complex eye composed of many separate elements, each with light-sensitive cells and a lens that can form an image.

condensation [L. *co*, together + *densare*, to make dense]: A type of chemical reaction in which two molecules join to form one larger molecule, simultaneously splitting out a molecule of water. The biosynthetic reactions in which monomers (e.g., monosaccharides, amino acids) are joined to form polymers (e.g., polysaccharides, polypeptides) are condensation reactions.

conditioning: A form of learning in which one stimulus (the conditional stimulus) comes to be associated with and elicit the same response as another stimulus (the unconditional stimulus).

cone: (1) In plants, the reproductive structure of a conifer. (2) In vertebrates, a type of photoreceptor cell in the retina, concerned with the perception of color and with the most acute discrimination of detail.

conjugation [L. *conjugatio*, a joining, connection]: The sexual process in some unicellular organisms by which genetic material is transferred from one cell to another by cell-to-cell contact.

connective tissues: Supporting or packing tissues that lie between groups of nerves, glands, and muscle cells, and beneath epithelial cells, in which the cells are irregularly distributed through a relatively large amount of extracellular material; include bone, cartilage, blood, and lymph.

consumer, in ecological systems: A heterotroph that derives its energy from living or freshly killed organisms or parts thereof. Primary consumers are herbivores; higher-level consumers are carnivores.

continental drift: The gradual movement of the earth's continents that has occurred over hundreds of millions of years.

continuous variation: A gradation of small differences in a particular trait, such as height, within a population; occurs in traits that are controlled by a number of genes.

convergent evolution [L. *convergere*, to turn together; *evolutio*, to unfold]: The independent development of similarities between unrelated groups, such as porpoises and sharks, resulting from adaptation to similar environments.

cork [L. *cortex*, bark]: A secondary tissue that is a major constituent of bark in woody and some herbaceous plants; made up of flattened cells, dead at maturity; restricts gas and water exchange and protects the vascular tissues from injury.

cork cambium [L. *cortex*, bark + *cambium*, exchange]: The lateral meristem that produces cork.

corolla (ko-role-a) [L. dim. of *corona*, wreath, crown]: Petals, collectively; usually the conspicuously colored flower parts.

corpus callosum [L., callous body]: In the vertebrate brain, a tightly packed mass of myelinated nerve fibers connecting the two cerebral hemispheres.

corpus luteum [L., yellowish body]: An ovarian structure that secretes estrogens and progesterone, which maintain the uterus during pregnancy. It develops from the remaining cells of the ruptured follicle following ovulation.

cortex [L., bark]: (1) The outer, as opposed to the inner, part of an organ, as in the adrenal gland. (2) In a stem or root, the primary tissue bounded externally by the epidermis and internally by the central cylinder of vascular tissue.

cosmid: In recombinant DNA technology, a vector constructed of a DNA segment flanked by cohesive regions (COS regions) of the bacteriophage lambda.

cotyledon (cottle-ee-don) [Gk. *kotyledon*, a cup-shaped hollow]: A leaflike structure of the embryo of a seed plant; contains stored food used during germination.

countercurrent exchange: An anatomical device for manipulating gradients so as to maximize uptake (or minimize loss) of O_2, heat, etc.

coupled reactions: In cells, the linking of endergonic (energy-requiring) reactions to exergonic (energy-releasing) reactions that provide enough energy to drive the endergonic reactions forward.

covalent bond [L. *con*, together + *valere*, to be strong]: A chemical bond formed as a result of the sharing of one or more pairs of electrons.

Crassulacean acid metabolism: A process by which some species of plants in hot, dry climates take in carbon dioxide during the night, fixing it in organic acids; the carbon dioxide is released during the day and used immediately in the Calvin cycle.

cristae: The "shelves" formed by the intricate folding of the inner membrane of the mitochondrion.

cross-fertilization: Fusion of gametes formed by different individuals; as opposed to self-fertilization.

crossing over: During meiosis, the exchange of genetic material between paired chromatids of homologous chromosomes.

cuticle (ku-tik-l) [L. *cuticula*, dim. of *cutis*, the skin]: (1) In plants, a layer of waxy substance (cutin) on the outer surface of epidermal cell walls. (2) In animals, the noncellular, outermost layer of many invertebrates.

cyclic AMP: A form of adenosine monophosphate (AMP) in which the atoms of the phosphate group form a ring; functions in chemical communication in slime molds, in positive regulation of operons, and as a "second messenger" for a number of vertebrate hormones and neurotransmitters.

cytochromes [Gk. *kytos*, vessel + *chrōma*, color]: Heme-containing proteins that participate in electron transport chains; involved in cellular respiration and photosynthesis.

cytokinesis [Gk. *kytos*, vessel + *kinesis*, motion]: Division of the cytoplasm of a cell following nuclear division.

cytokinin [Gk. *kytos*, vessel + *kinesis*, motion]: One of a group of chemically related plant hormones that promote cell division, among other effects.

cytoplasm (sight-o-plazm) [Gk. *kytos*, vessel + *plasma*, anything molded]: The living matter within a cell, excluding the genetic material.

cytoskeleton: A network of filamentous protein structures within the cytoplasm that maintains the shape of the cell, anchors its organelles, and is involved in cell motility; includes microtubules, actin filaments, and intermediate filaments.

deciduous [L. *decidere*, to fall off]: Refers to plants that shed their leaves at a certain season.

decomposers: Specialized detritivores, usually bacteria or fungi, that consume such substances as cellulose and nitrogenous waste products. Their metabolic processes release inorganic nutrients, which are then available for reuse by plants and other organisms.

denaturation: The loss of the native configuration of a macromolecule resulting, for instance, from heat treatment, extreme pH changes, chemical treatment, or other denaturing agents. It is usually accompanied by a loss of biological activity.

dendrite [Gk. *dendron*, tree]: A process of a neuron, typically branched, that receives stimuli from other cells.

denitrification: The process by which certain bacteria living in poorly aerated soils break down nitrates, using the oxygen for their own respiration and releasing nitrogen back into the atmosphere.

density-dependent factors: Factors affecting the birth rate or mortality rate of a population, the effects of which vary with the density (number of individuals per unit area or volume) of the population; include resources for which members of the same or different populations compete, predation, and disease.

density-independent factors: Factors affecting the birth rate or mortality rate of a population, the effects of which are independent of the density of the population; often involve weather-related events.

deoxyribonucleic acid (DNA) (dee-ox-y-rye-bo-new-**clay**-ick): The carrier of genetic information in cells, composed of two complementary chains of nucleotides wound in a double helix; capable of self-replication as well as coding for RNA synthesis.

dermis [Gk. *derma*, skin]: The inner layer of the skin, beneath the epidermis.

desmosome [Gk. *desmos*, bond + *soma*, body]: A type of cell-cell junction that provides mechanical strength in animal tissues; consists of a plaque of dense fibrous material between adjacent cells, with clusters of filaments looping in and out from the cytoplasm of the two cells.

detritivores [L. *detritus*, worn down, worn away + *voro*, to devour]: Organisms that live on dead and discarded organic matter; include large scavengers, smaller animals such as earthworms and some insects, as well as decomposers (fungi and bacteria).

deuterostome [Gk. *deuteros*, second + *stoma*, mouth]: An animal in which the anus forms at or near the blastopore in the developing embryo and the mouth forms secondarily elsewhere; echinoderms and chordates are deuterostomes. Deuterostomes are also characterized by radial cleavage during the earliest stages of development and by enterocoelous formation of the coelom.

development: The progressive production of the phenotypic characteristics of a multicellular organism, beginning with the fertilization of an egg.

diaphragm [Gk. *diaphrassein*, to barricade]: In mammals, a sheetlike tissue (tendon and muscle) forming the partition between the abdominal and thoracic cavities; functions in breathing.

dicotyledon (dye-cottle-ee-don) [Gk. *di*, double, two + *kotyledon*, a cup-shaped hollow]: A member of the class of flowering plants having two seed leaves, or cotyledons, among other distinguishing features; often abbreviated as dicot.

diencephalon [Gk. *di*, two + *enkephalos*, brain]: One of the two principal subdivisions of the vertebrate forebrain; the posterior portion of the forebrain, it contains the thalamus and the hypothalamus.

differentiation: The developmental process by which a relatively unspecialized cell or tissue undergoes a progressive (usually irreversible) change to a more specialized cell or tissue.

diffusion [L. *diffundere*, to pour out]: The net movement of suspended or dissolved particles down a concentration gradient as a result of the random spontaneous movements of individual particles; the process tends to distribute the particles uniformly throughout a medium.

digestion [L. *digestio*, separating out, dividing]: The breakdown of complex, usually insoluble foods into molecules that can be absorbed into the body and used by the cells.

dikaryon (dye-**care**-ee-on) [Gk. *di*, two + *karyon*, kernel]: A cell or organism with paired but not fused nuclei derived from different parents; found among the fungi of divisions Ascomycota and Basidiomycota.

dioecious (dye-ee-shus) [Gk. *di*, two + *oikos*, house]: In angiosperms, having the male (staminate) and female (carpellate) flowers on different individuals of the same species.

diploid [Gk. *di*, double, two + *ploion*, vessel]: The condition in which each autosome is represented twice (2*n*); in contrast to haploid (*n*).

disaccharide [Gk. *di*, two + *sakcharon*, sugar]: A carbohydrate molecule composed of two monosaccharide monomers; examples are sucrose, maltose, and lactose.

diurnal [L. *diurnus*, of the day]: Applied to organisms that are active during the daylight hours.

division: A taxonomic grouping of related, similar classes; a high-level category below kingdom and above class. Division is generally used in the classification of prokaryotes, algae, fungi, and plants, whereas an equivalent category, phylum, is used in the classification of protozoa and animals.

DNA: Abbreviation of deoxyribonucleic acid.

dominant allele: An allele whose phenotypic effect is the same in both the heterozygous and homozygous conditions.

dormancy [L. *dormire*, to sleep]: A period during which growth ceases and metabolic activity is greatly reduced; dormancy is broken when certain requirements, for example, of temperature, moisture, or day length, are met.

dorsal [L. *dorsum*, the back]: Pertaining to or situated near the back; opposite of ventral.

dorsal lip: The tissue on the dorsal edge of the blastopore of the vertebrate embryo; the prospective chordamesoderm, it functions as an organizer, inducing undifferentiated cells to follow a specific course of development.

double fertilization: A phenomenon unique to the angiosperms, in which the egg and one sperm nucleus fuse (resulting in a 2*n* fertilized egg, the zygote) and simultaneously the second sperm nucleus fuses with the two polar nuclei (resulting in a 3*n* endosperm nucleus).

duodenum (duo-dee-num) [L. *duodeni*, twelve each—from its length, about 12 fingers' breadth]: The upper portion of the small intestine in vertebrates, where food is digested into molecules that can be absorbed by intestinal cells.

ecological niche: A description of the roles and associations of a particular species in the community of which it is a part; the way in which an organism interacts with all of the biotic and abiotic factors in its environment.

ecological pyramid: A graphic representation of the quantitative relationships of numbers of organisms, biomass, or energy flow between the trophic levels of an ecosystem. Because large amounts of energy and biomass are dissipated at every trophic level, these diagrams nearly always take the form of pyramids.

ecological succession: The gradual process by which the species composition of a community changes.

ecology [Gk. *oikos*, home + *logos*, a discourse]: The study of the interactions of organisms with their physical environment and with each other and of the results of such interactions.

ecosystem [Gk. *oikos*, home + *systema*, that which is put together]: The organisms in a community plus the associated abiotic factors with which they interact.

ecotype [Gk. *oikos*, home + L. *typus*, image]: A locally adapted variant of a species, differing genetically from other ecotypes of the same species.

ectoderm [Gk. *ecto*, outside + *derma*, skin]: One of the three embryonic tissue layers of animals; it gives rise to the outer covering of the body, the sensory receptors, and the nervous system.

ectotherm [Gk. *ecto*, outside + *therme*, heat]: An organism, such as a reptile, that maintains its body temperature by taking in heat from the environment or giving it off to the environment. *See also* Poikilotherm.

effector [L. *ex*, out of + *facere*, to make]: Cell, tissue, or organ (such as muscle or gland) capable of producing a response to a stimulus.

efferent [L. *ex*, out of + *ferre*, to bear]: Carrying away from a center, applied to nerves and blood vessels.

egg: A female gamete, which usually contains abundant cytoplasm and yolk; nonmotile and often larger than a male gamete.

electric potential: The difference in the amount of electric charge between a region of positive charge and a region of negative charge. The establishment of electric potentials across cell and organelle membranes makes possible a number of phenomena, including the chemiosmotic synthesis of ATP, the conduction of nerve impulses, and muscle contraction.

electron: A subatomic particle with a negative electric charge equal in magnitude to the positive charge of the proton but with a much smaller mass; normally found within orbitals surrounding the atom's positively charged nucleus.

electron acceptor: Substance that accepts or receives electrons in an oxidation-reduction reaction, becoming reduced in the process.

electron carrier: A specialized molecule, such as a cytochrome, that can lose and gain electrons reversibly, alternately becoming oxidized and reduced.

electron donor: Substance that donates or gives up electrons in an oxidation-reduction reaction, becoming oxidized in the process.

electron transport: The movement of electrons down a series of electron-carrier molecules that hold electrons at slightly different energy levels; as electrons move down the chain, the energy released is used to form ATP from ADP and phosphate. Electron transport plays an essential role in the final stage of cellular respiration and in the energy-capturing reactions of photosynthesis.

element: A substance composed only of atoms of the same atomic number and that cannot be decomposed by ordinary chemical means.

embryo [Gk. *en*, in + *bryein*, to swell]: The early developmental stage of an organism produced from a fertilized egg; a young organism before it emerges from the seed, egg, or body of its mother. In humans, refers to the first two months of intrauterine life. *See* Fetus.

embryo sac: The female gametophyte of a flowering plant, contained within an ovule; typically consists of seven cells with a total of eight haploid nuclei.

endergonic [Gk. *endon*, within + *ergon*, work]: Energy-requiring, as in a chemical reaction; applied to an "uphill" process.

endocrine gland [Gk. *endon*, within + *krinein*, to separate]: Ductless gland whose secretions (hormones) are released into the extracellular spaces, from which they diffuse into the circulatory system; in vertebrates, includes pituitary, sex glands, adrenal, thyroid, and others.

endocytosis [Gk. *endon*, within + *kytos*, vessel]: A cellular process in which material to be taken into the cell induces the membrane to form a vacuole enclosing the material; the vacuole is released into the cytoplasm. Includes phagocytosis (endocytosis of solid particles), pinocytosis (endocytosis of liquids), and receptor-mediated endocytosis.

endoderm [Gk. *endon*, within + *derma*, skin]: One of the three embryonic tissue layers of animals; it gives rise to the epithelium that lines certain internal structures, such as most of the digestive tract and its outgrowths, most of the respiratory tract, and the urinary bladder, liver, pancreas, and some endocrine glands.

endodermis [Gk. *endon*, within + *derma*, skin]: In plants, a layer of specialized cells, one cell thick, that lies between the cortex and the vascular tissues in young roots. The Casparian strip of the endodermis prevents diffusion of solutes across the root.

endometrium [Gk. *endon*, within + *metrios*, of the womb]: The glandular lining of the uterus in mammals; thickens in response to secretion of estrogens and progesterone; one of its two principal layers is sloughed off in menstruation.

endoplasmic reticulum [Gk. *endon*, within + *plasma*, from cytoplasm; L. *reticulum*, network]: An extensive system of membranes present in most eukaryotic cells, dividing the cytoplasm into compartments and channels; often coated with ribosomes.

endorphin: One of a group of small peptides with morphine-like properties; produced by the vertebrate brain.

endosperm [Gk. *endon*, within + *sperma*, seed]: In plants, a $3n$ tissue containing stored food that develops from the union of a sperm nucleus and the two nuclei of the central cell of the female gametophyte; found only in angiosperms.

endothelium [Gk. *endon*, + *thele*, nipple]: A type of epithelial tissue that forms the walls of the capillaries and the inner lining of arteries and veins.

endotherm [Gk. *endon*, within + *therme*, heat]: An organism, such as a bird or a mammal, that maintains its body temperature internally through metabolic processes. *See also* Homeotherm.

enterocoelous [Gk. *enteron*, gut + *koilos*, a hollow]: Formation of the coelom during embryonic development as cavities within mesoderm originating from outpocketings of the primitive gut; characteristic of deuterostomes.

entropy [Gk. *en*, in + *trope*, turning]: A measure of the randomness or disorder of a system.

enzyme [Gk. *en*, in + *zyme*, leaven]: A globular protein molecule that accelerates a specific chemical reaction.

epidermis [Gk. *epi*, on or over + *derma*, skin]: In plants and animals, the outermost layers of cells.

epinephrine: *See* Adrenaline.

episome: A plasmid that has become incorporated into a bacterial chromosome.

epistasis [Gk., a stopping]: Interaction between two nonallelic genes in which one of them interferes with or modifies the phenotypic expression of the other.

epithelial tissue [Gk. *epi*, on or over + *thele*, nipple]: In animals, a type of tissue that covers a body or structure or lines a cavity; epithelial cells form one or more regular layers with little intercellular material.

equilibrium [L. *aequus*, equal + *libra*, balance]: The state of a system in which no further net change is occurring; result of counterbalancing forward and backward processes.

erythrocyte (eh-rith-ro-site) [Gk. *erythros*, red + *kytos*, vessel]: Red blood cell, the carrier of hemoglobin.

estrogens [Gk. *oistros*, frenzy + *genos*, origin, descent]: Female sex hormones, which are the predominant secretions of the ovarian follicle during the preovulatory phase of the menstrual cycle; also produced by the corpus luteum and the placenta.

estrus [Gk. *oistros*, frenzy]: The mating period in female mammals, characterized by ovulation and intensified sexual activity.

ethology [Gk. *ethos*, habit, custom + *logos*, discourse]: The comparative study of patterns of animal behavior, with emphasis on their adaptive significance and evolutionary origin.

ethylene: A simple hydrocarbon ($H_2C = CH_2$) that functions as a plant hormone; plays a role in fruit ripening and leaf abscission.

eukaryote (you-car-ry-oat) [Gk. *eu*, good + *karyon*, nut, kernel]: A cell having a membrane-bound nucleus, membrane-bound organelles, and chromosomes in which DNA is combined with histone proteins; an organism composed of such cells.

eusocial [Gk. *eu*, good + L. *socius*, companion]: Applied to animal societies, such as those of certain insects, in which sterile individuals work on behalf of reproductive individuals.

evolution [L. *e-*, out + *volvere*, to roll]: Changes in the gene pool from one generation to the next as a consequence of processes such as mutation, natural selection, nonrandom mating, and genetic drift.

exergonic [Gk. *ex*, out of + *ergon*, work]: Energy-yielding, as in a chemical reaction; applied to a "downhill" process.

exocrine glands [Gk. *ex*, out of + *krinein*, to separate]: Glands, such as digestive glands and sweat glands, that secrete their products into ducts.

exocytosis [Gk. *ex*, out of + *kytos*, vessel]: A cellular process in which particulate matter or dissolved substances are enclosed in a vacuole and transported to the cell surface; there, the membrane of the vacuole fuses with the cell membrane, expelling the vacuole's contents to the outside.

exon: A segment of DNA that is transcribed into RNA and expressed; dictates the amino acid sequence of part of a polypeptide.

exoskeleton: The outer supporting covering of the body; common in arthropods.

exponential growth: In populations, the increasingly accelerated rate of growth due to the increasing number of individuals being added to the reproductive base. Exponential growth is very seldom approached or sustained in natural populations.

expressivity: In genetics, the degree to which a particular genotype is expressed in the phenotype of individuals with that genotype.

extinct [L. *exstinctus,* to be extinguished]: No longer existing.

extraembryonic membranes: In reptiles, birds, and mammals, membranes formed from embryonic tissues that lie outside the embryo proper, protecting it and aiding metabolism; include amnion, chorion, allantois, and yolk sac.

F_1 (first filial generation): The offspring resulting from the crossing of plants or animals of a parental generation.

F_2 (second filial generation): resulting from crossing members of the F_1 generation among themselves.

facilitated diffusion: The transport of substances across a cell or organelle membrane from a region of higher concentration to a region of lower concentration by protein molecules embedded in the membrane; driven by the concentration gradient.

Fallopian tube: *See* Oviduct.

family: A taxonomic grouping of related, similar genera; the category below order and above genus.

fatty acid: A molecule consisting of a —COOH group and a long hydrocarbon chain; fatty acids are components of fats, oils, phospholipids, glycolipids, and waxes.

feedback systems: Control mechanisms whereby an increase or decrease in the level of a particular factor inhibits or stimulates the production, utilization, or release of that factor; important in the regulation of enzyme and hormone levels, ion concentrations, temperature, and many other factors.

fermentation: The breakdown of organic compounds in the absence of oxygen; yields less energy than aerobic processes.

fertilization: The fusion of two haploid gamete nuclei to form a diploid zygote nucleus.

fetus [L., pregnant]: An unborn or unhatched vertebrate that has passed through the earliest developmental stages; a developing human from about the second month of gestation until birth.

fibril [L. *fibra,* fiber]: Any minute, threadlike structure within a cell.

fibrous protein: Insoluble structural protein in which the polypeptide chain is coiled along one dimension. Fibrous proteins constitute the main structural elements of many animal tissues.

filament [L. *filare,* to spin]: (1) A chain of cells. (2) In plants, the stalk of a stamen.

filtration: The first stage of kidney function; blood plasma is forced, under pressure, out of the glomerular capillaries into Bowman's capsule, through which it enters the renal tubule.

fission: *See* Binary fission.

fitness: The genetic contribution of an individual to succeeding generations relative to the contributions of other individuals in the population.

fixed action pattern: A behavior that appears substantially complete the first time the organism encounters the relevant stimulus; tends to be highly stereotyped, rigid, and predictable.

flagellum, *pl.* **flagella** (fla-jell-um) [L. *flagellum,* whip]: A long, threadlike organelle found in eukaryotes and used in locomotion and feeding; has an internal structure of nine pairs of microtubules encircling two central microtubules.

flower: The reproductive structure of angiosperms; a complete flower includes sepals, petals, stamens (male structures), and carpels (female structures).

food chain: A sequence of organisms related to one another as prey and predator.

food web: A set of interactions among organisms, including producers, consumers (herbivores and carnivores), and detritivores, through which energy and materials move within a community or ecosystem.

fossil [L. *fossilis,* dug up]: The remains of an organism, or direct evidence of its presence (such as tracks). May be an unaltered hard part (tooth or bone), a mold in a rock, petrification (wood or bone), unaltered or partially altered soft parts (a frozen mammoth).

founder effect: Type of genetic drift that occurs as the result of the founding of a population by a small number of individuals.

fovea [L., pit]: A small area in the center of the retina in which cones are concentrated; the area of sharpest vision.

free energy change: The total energy change that results from a chemical reaction or other process (such as evaporation); takes into account changes in both heat and entropy.

frequency-dependent selection: A type of natural selection that decreases the frequency of more common phenotypes in a population and increases the frequency of less common phenotypes.

fruit [L. *fructus,* fruit]: In angiosperms, a matured, ripened ovary or group of ovaries and associated structures; contains the seeds.

function [L. *fungor,* to busy oneself]: Characteristic role or action of a structure or process in the normal metabolism or behavior of an organism.

gametangium, *pl.* **gametangia** [Gk. *gamein,* to marry + L. *tangere,* to touch]: A unicellular or multicellular structure in which gametes are produced.

gamete (gam-meet) [Gk., wife]: A haploid reproductive cell whose nucleus fuses with that of another gamete of an opposite mating type or sex (fertilization); the resulting cell (zygote) may develop into a new diploid individual or, in some protists and fungi, may undergo meiosis to form haploid somatic cells.

gametophyte: In organisms that have alternation of haploid and diploid generations (all plants and some green algae), the haploid *(n)* gamete-producing generation.

ganglion, *pl.* **ganglia** (gang-lee-on) [Gk. *ganglion,* a swelling]: Aggregation of nerve cell bodies; in vertebrates, refers to an aggregation of nerve cell bodies located outside the central nervous system.

gap junction: A junction between adjacent animal cells that allows the passage of materials between the cells.

gastric [Gk. *gaster,* stomach]: Pertaining to the stomach.

gastrovascular cavity [Gk. *gaster,* stomach + L. *vasculum,* a small vessel]: A digestive cavity with only one opening, characteristic of the phyla Cnidaria (jellyfish, hydra, corals, etc.) and Ctenophora (comb jellies, sea walnuts); water circulating through the cavity supplies dissolved oxygen and carries away carbon dioxide and other waste products.

gastrula [Gk. *gaster,* stomach]: An animal embryo in the process of gastrulation; the stage of development during which the blastula, with its single layer of cells, turns into a three-layered embryo, made up of ectoderm, mesoderm, and endoderm, often enclosing an archenteron.

Gause's principle: *See* Competitive exclusion.

gene [Gk. *genos,* birth, race; L. *genus,* birth, race, origin]: A unit of heredity in the chromosome; a sequence of nucleotides in a DNA molecule that performs a specific function, such as coding for an RNA molecule or a polypeptide.

gene flow: The movement of alleles into or out of a population.

gene pool: All the alleles of all the genes of all the individuals in a population.

genetic code: The system of nucleotide triplets in DNA and RNA that carries genetic information; referred to as a code because it determines the amino acid sequence in the enzymes and other protein molecules synthesized by the organism.

genetic drift: Evolution (change in allele frequencies) owing to chance processes.

genetic isolation: The absence of genetic exchange between populations or species as a result of geographic separation or of premating or postmating mechanisms (behavioral, anatomical, or physiological) that prevent reproduction.

genome: The complete set of chromosomes, with their associated genes.

genomic DNA (gDNA): DNA fragments produced by the action of restriction enzymes on the DNA of a cell or organism.

genotype (jean-o-type): The genetic constitution of an individual cell or organism with reference to a single trait or a set of traits; the sum total of all the genes present in an individual.

genus, *pl.* **genera** (jean-us) [L. *genus,* race, origin]: A taxonomic grouping of closely related species.

geologic eras: See Table 24–1, pages 496–497.

germ cells [L. *germinare,* to bud]: Gametes or the cells that give rise directly to gametes

germination [L. *germinare,* to bud]: In plants, the resumption of growth or the development from seed or spore.

germ layer: A layer of distinctive cells in an embryo; an embryonic tissue layer. The majority of multicellular animals have three germ layers: ectoderm, mesoderm, and endoderm.

gibberellins (jibb-e-**rell**-ins) [Fr. *gibberella,* genus of fungi]: A group of chemically related plant growth hormones, whose most characteristic effect is stem elongation in dwarf plants and bolting.

gill: The respiratory organ of aquatic animals, usually a thin-walled projection from some part of the external body surface or, in vertebrates, from some part of the digestive tract.

gland [L. *glans, glandis,* acorn]: A structure composed of modified epithelial cells specialized to produce one or more secretions that are discharged to the outside of the gland.

globular protein [L. dim. of *globus,* a ball]: A polypeptide chain folded into a roughly spherical shape.

glomerulus (glom-**mare**-u-lus) [L. *glomus,* ball]: In the vertebrate kidney, a cluster of capillaries enclosed by Bowman's capsule; blood plasma minus large molecules filters through the walls of the glomerular capillaries into the renal tubule.

glucagon [Gk. *glukus,* sweet + *agō,* to lead toward]: Hormone produced in the pancreas that acts to raise the concentration of blood sugar.

glucose [Gk. *glukus,* sweet]: A six-carbon sugar ($C_6H_{12}O_6$); the most common monosaccharide in animals.

glycogen [Gk. *glukus,* sweet + *genos,* race or descent]: A complex carbohydrate (polysaccharide); one of the main stored food substances of most animals and fungi; it is converted into glucose by hydrolysis.

glycolipids [Gk. *glukus,* sweet + *lipos,* fat]: Organic molecules similar in structure to fats, but in which a short carbohydrate chain rather than a fatty acid is attached to the third carbon of the glycerol molecule; as a result, the molecule has a hydrophilic "head" and a hydrophobic "tail." Glycolipids are important constituents of cell and organelle membranes.

glycolysis (gly-**coll**-y-sis) [Gk. *glukus,* sweet + *lysis,* loosening]: The process by which a glucose molecule is changed anaerobically to two molecules of pyruvic acid with the liberation of a small amount of useful energy; catalyzed by cytoplasmic enzymes.

Golgi complex (**goal**-jee): An organelle present in many eukaryotic cells; consists of flat, membrane-bound sacs, tubules, and vesicles. It functions as a processing, packaging, and distribution center for substances that the cell manufactures.

gonad [Gk. *gone,* seed]: Gamete-producing organ of multicellular animals; ovary or testis.

granulocyte [L., *grānum,* grain or seed + Gk. *kytos,* vessel]: A type of phagocytic white blood cell involved in the inflammatory response; characterized by numerous lysosomes that give the cell a granular appearance under the light microscope. Granulocytes are classified on the basis of their staining properties as neutrophils, eosinophils, or basophils.

granum, *pl.* **grana** [L., grain or seed]: In chloroplasts, stacked membrane-bound disks (thylakoids) that contain chlorophylls and carotenoids and are the sites of the light-trapping reactions of photosynthesis.

gravitropism [L. *gravis,* heavy + Gk. *trope,* turning]: The direction of growth or movement in which the force of gravity is the determining factor; also called geotropism.

gross productivity: A measure of the rate at which energy is assimilated by the organisms in a trophic level, a community, or an ecosystem.

ground tissues: In leaves and young roots and stems, all tissues other than the epidermis and the vascular tissues.

guard cells: Specialized epidermal cells surrounding a pore, or stoma, in a leaf or green stem; changes in turgor of a pair of guard cells cause opening and closing of the pore.

gymnosperm [Gk. *gymnos,* naked + *sperma,* seed]: A seed plant in which the seeds are not enclosed in an ovary; the conifers are the most familiar group.

habitat [L. *habitare,* to live in]: The place in which individuals of a particular species can usually be found.

habituation [L. *habitus,* condition]: A response to a repeated stimulus in which the stimulus comes to be ignored and a previous behavior pattern is restored; one of the simplest forms of learning.

half-life: The average time required for the disappearance or decay of one-half of any amount of a given substance.

haploid [Gk. *haploos,* single + *ploion,* vessel]: Having only one set of chromosomes *(n),* in contrast to diploid (2*n*); characteristic of eukaryotic gametes, of gametophytes in plants, and of some protists and fungi.

Hardy-Weinberg equilibrium: The steady-state relationship between relative frequencies of two or more alleles in an idealized population; both the allele frequencies and the genotype frequencies will remain constant from generation to generation in a population breeding at random in the absence of evolutionary forces.

haustorium, *pl.* **haustoria** [L. *haustus,* from *haurire,* to drink, draw]: A projection from a parasitic oomycete, fungus, or plant that functions as a penetrating and absorbing structure.

heat of vaporization: The amount of heat required to change a given amount of a liquid into a gas; 540 calories are required to change 1 gram of liquid water into vapor.

heme [Gk. *haima,* blood]: The iron-containing group of heme proteins such as hemoglobin and the cytochromes.

hemocoel [Gk. *haima,* blood + *koilos,* a hollow]: A blood-filled space within the tissues; characteristic of animals with an incomplete circulatory system, such as mollusks and arthropods.

hemoglobin [Gk. *haima,* blood + L. *globus,* a ball]: The iron-containing protein in vertebrate blood that carries oxygen.

hemophilia [Gk. *haima,* blood + *philios,* friendly]: A group of hereditary diseases characterized by failure of the blood to clot and consequent excessive bleeding from even minor wounds.

hepatic [Gk. *hēpatikos,* liver]: Pertaining to the liver.

herbaceous (her-**bay**-shus) [L. *herba,* grass]: In plants, nonwoody.

herbivore [L. *herba,* grass + *vorare,* to devour]: A consumer that eats plants or other photosynthetic organisms to obtain its food and energy.

heredity [L. *herres, heredis,* heir]: The transmission of characteristics from parent to offspring.

hermaphrodite [Gk. *Hermes* and *Aphrodite*]: An organism possessing both male and female reproductive organs; hermaphrodites may or may not be self-fertilizing.

heterosis [Gk. *heteros,* other, different]: Hybrid vigor; the overall superiority of the hybrid over either parent.

heterotroph [Gk. *heteros,* other, different + *trophos,* feeder]: An organism that must feed on organic materials formed by other organisms in order to obtain energy and small building-block molecules; in contrast to autotroph. Animals, fungi, and many unicellular organisms are heterotrophs.

heterozygote [Gk. *heteros*, other + *zugōtos*, a pair]: A diploid organism that carries two different alleles at one or more genetic loci.

heterozygote superiority: The greater fitness of an organism heterozygous at a given genetic locus as compared with either homozygote.

hibernation [L. *hiberna*, winter]: A period of dormancy and inactivity, varying in length, depending on the species, and occurring in dry or cold seasons. During hibernation, metabolic processes are greatly slowed and, even in mammals, body temperature may drop to just above freezing.

histones: A group of five relatively small, basic polypeptide molecules found bound to the DNA of eukaryotic cells.

homeostasis (home-e-o-**stay**-sis) [Gk. *homos*, same or similar + *stasis*, standing]: Maintenance of a relatively stable internal physiological environment or internal equilibrium in an organism.

homeotherm [Gk. *homos*, same or similar + *therme*, heat]: An organism, such as a bird or mammal, capable of maintaining a stable body temperature independent of the environment.

hominid [L. *homo*, man]: Humans and closely related primates; includes modern and fossil forms, such as the australopithecines, but not the apes.

hominoid [L. *homo*, man]: Hominids and the apes.

homologues [Gk. *homologia*, agreement]: Chromosomes that carry corresponding genes and associate in pairs in the first stage of meiosis; each member of the pair is derived from a different parent.

homology [Gk. *homologia*, agreement]: Similarity in structure and/or position, assumed to result from a common ancestry, regardless of function, such as the wing of a bird and the foreleg of a mammal.

homozygote [Gk. *homos*, same or similar + *zugōtos*, a pair]: A diploid organism that carries identical alleles at one or more genetic loci.

hormone [Gk. *hormaein*, to excite]: An organic molecule secreted, usually in minute amounts, in one part of an organism that regulates the function of another tissue or organ.

host: (1) An organism on or in which a parasite lives. (2) A recipient of grafted tissue.

hybrid [L. *hybrida*, the offspring of a tame sow and a wild boar]: (1) Offspring of two parents that differ in one or more inheritable characteristics. (2) Offspring of two different varieties or of two different species.

hydrocarbon [L. *hydro*, water + *carbo*, charcoal]: An organic compound consisting of only carbon and hydrogen.

hydrogen bond: A weak molecular bond linking a hydrogen atom that is covalently bonded to another atom (usually oxygen, nitrogen, or fluorine) to another oxygen, nitrogen, or fluorine atom of the same or another molecule.

hydrolysis [L. *hydro*, water + Gk. *lysis*, loosening]: Splitting of one molecule into two by addition of H^+ and OH^- ions from water.

hydrophilic [L. *hydro*, water + Gk. *philios*, friendly]: Having an affinity for water; applied to polar molecules or polar regions of large molecules.

hydrophobic [L. *hydro*, water + Gk. *phobos*, fearing]: Having no affinity for water; applied to nonpolar molecules or nonpolar regions of molecules.

hypertonic [Gk. *hyper*, above + *tonos*, tension]: Of two solutions of different concentration, the solution that contains the higher concentration of solute particles; water moves across a selectively permeable membrane into a hypertonic solution.

hypha [Gk. *hyphe*, web]: A single tubular filament of a fungus or an oomycete; the hyphae together make up the mycelium, the matlike "body" of a fungus.

hypothalamus [Gk. *hypo*, under + *thalamos*, inner room]: The region of the vertebrate brain just below the thalamus; responsible for the integration of many basic behavioral patterns that involve correlation of neural and endocrine functions.

hypothesis [Gk. *hypo*, under + *tithenai*, to put]: A temporary working explanation or supposition based on accumulated facts and suggesting some general principle or relation of cause and effect; a postulated solution to a scientific problem that must be tested and if not validated, discarded.

hypotonic [Gk. *hypo*, under + *tonos*, tension]: Of two solutions of different concentration, the solution that contains the lower concentration of solute particles; water moves across a selectively permeable membrane from a hypotonic solution.

imbibition [L. *imbibere*, to drink in]: The capillary movement of water into germinating seeds and into substances such as wood and gelatin, which swell as a result.

immune response: A highly specific defensive reaction of the body to invasion by a foreign substance or organism; consists of a primary response in which the invader is recognized as foreign, or "not-self," and eliminated and a secondary response to subsequent attacks by the same invader. Mediated by two types of lymphocytes: B lymphocytes, which mature in the bone marrow and are responsible for antibody production, and T lymphocytes, which mature in the thymus and are responsible for cell-mediated immunity.

immunoglobulins: Complex, highly specific globular proteins synthesized by B lymphocytes; include both circulating antibodies and antibodies displayed on the surface of B lymphocytes prior to activation.

imprinting: A rapid and extremely narrow form of learning, common in birds and important in species recognition, that occurs during a very short critical period in the early life of an animal; depends on exposure to particular characteristics of the parent or parents.

inbreeding: The mating of individuals that are closely related genetically.

inclusive fitness: The relative number of an individual's alleles that are passed on from generation to generation, either as a result of his or her own reproductive success, or that of related individuals.

incomplete dominance: In genetics, the phenomenon in which the effects of both alleles at a particular locus are apparent in the phenotype of the heterozygote.

independent assortment: *See* Mendel's second law.

induction [L. *inducere*, to induce]: (1) In genetics, the phenomenon in which the presence of a substrate initiates transcription and translation of the genes coding for the enzymes required for its metabolism. (2) In embryonic development, the process in which one tissue or body part causes the differentiation of another tissue or body part.

inflammatory response: A nonspecific defensive reaction of the body to invasion by a foreign substance or organism; involves phagocytosis by white blood cells and is often accompanied by accumulation of pus and an increase in the local temperature.

innate releasing mechanism: In ethology, an area within an animal's brain that is hypothesized to respond to a specific stimulus, setting in motion, or "releasing," the sequence of movements that constitute a fixed action pattern.

insertion sequences: Relatively short sequences of DNA that can produce copies of themselves that become incorporated at other sites in the same chromosome or in other chromosomes; also known as simple transposons.

insulin: A peptide hormone, produced by the vertebrate pancreas, that acts to lower the concentration of glucose in the blood.

interferon: A protein made by virus-infected cells that inhibits viral multiplication.

intermediate filaments: Fibrous protein filaments that form part of the cytoskeleton; found in greatest density in cells subject to mechanical stress.

interneuron: Neuron that transmits signals from one neuron to another within a local region of the central nervous system; may receive signals from and transmit signals to many different neurons.

interphase: The portion of the cell cycle that occurs before mitosis or meiosis can take place; includes the G_1, S, and G_2 phases.

intron: A segment of DNA that is transcribed into RNA but is removed enzymatically from the RNA molecule before the mRNA enters the cytoplasm and is translated; also known as an intervening sequence.

invagination [L. *in*, in + *vagina*, sheath]: The local infolding of a layer of tissue, especially in animal embryos, so as to form a depression or pocket opening to the outside.

inversion: A chromosomal aberration in which a double break occurs and a segment is turned 180° before it is reincorporated into the chromosome.

ion (eye-on): Any atom or small molecule containing an unequal number of electrons and protons and therefore carrying a net positive or net negative charge.

ionic bond: A chemical bond formed as a result of the mutual attraction of ions of opposite charge.

isogamy [Gk. *isos*, equal + *gamos*, marriage]: Sexual reproduction in which both gametes are motile and are structurally alike.

isolating mechanisms: Mechanisms that prevent genetic exchange between individuals of different populations or species; they prevent mating or successful reproduction even when mating occurs; may be behavioral, anatomical, or physiological.

isotonic [Gk. *isos*, equal + *tonos*, tension]: Having the same concentration of solutes as another solution. If two isotonic solutions are separated by a selectively permeable membrane, there will be no net flow of water across the membrane.

isotope [Gk. *isos*, equal + *topos*, place]: Atom of an element that differs from other atoms of the same element in the number of neutrons in the atomic nucleus; isotopes thus differ in atomic weight. Some isotopes are unstable and emit radiation.

karyotype [Gk. *kara*, the head + *typos*, stamp or print]: The general appearance of the chromosomes of an organism with regard to number, size, and shape.

keratin [Gk. *karas*, horn]: One of a group of tough, fibrous proteins formed by certain epidermal tissues and especially abundant in skin, claws, hair, feathers, and hooves.

kidney: In vertebrates, the organ that regulates the balance of water and solutes in the blood and the excretion of nitrogenous wastes in the form of urine.

kinetic energy [Gk. *kinetikos*, putting in motion]: Energy of motion.

kinetochore [Gk. *kinetikos*, putting in motion + *choros*, chorus]: Disk-shaped protein structure within the centromere to which spindle fibers are attached during mitosis or meiosis.

kingdom: A taxonomic grouping of related, similar phyla or divisions; the highest-level category in biological classification.

kin selection: The differential reproduction of lineages of related individuals—that is, different groups of related individuals of a species reproduce at different rates; leads to an increase in the frequency of alleles shared by members of the groups with the greatest reproductive success.

Krebs cycle: Stage of cellular respiration in which acetyl groups are broken down into carbon dioxide; molecules reduced in the process can be used in ATP formation.

lagging strand: In DNA replication, the 3′ to 5′ strand of the DNA double helix, synthesized as a series of Okazaki fragments in the 5′ to 3′ direction; these segments are subsequently linked to one another in condensation reactions catalyzed by the enzyme DNA ligase.

lamella (lah-**mell**-ah) [L. dim. of *lamina*, plate or leaf]: Layer, thin sheet.

larva [L., ghost]: An immature animal that is anatomically very different from the adult; examples are caterpillars and tadpoles.

lateral meristem [L. *latus, lateris,* side + Gk. *meristos,* divided]: In vascular plants, one of the two rings of tissue (vascular cambium and cork cambium) that produce new cells for secondary growth.

leaching: The dissolving of minerals and other elements in soil or rocks by the downward movement of water.

leading strand: In DNA replication, the 5′ to 3′ strand of the DNA double helix, which is synthesized continuously.

learning: The process that leads to modification in individual behavior as the result of experience.

leucoplast [Gk. *leukos*, white + *plastes*, molder]: In plant cells, a colorless organelle that serves as a starch repository; usually found in cells not exposed to light, such as those in roots and the internal tissues of stems.

leukocyte [Gk. *leukos*, white + *kytos*, vessel]: White blood cell; principal types include granulocytes, monocytes and macrophages, and lymphocytes.

lichen: Organism composed of a fungus and a green alga or a cyanobacterium that are symbiotically associated.

life cycle: The entire span of existence of any organism from time of zygote formation (or asexual reproduction) until it itself reproduces.

limbic system [L. *limbus*, border]: Neuron network forming a loop around the inside of the brain and connecting the hypothalamus to the cerebral cortex; thought to be circuit by which drives and emotions are translated into complex actions and to play a role in the consolidation of memory.

linkage: The tendency for certain alleles to be inherited together because they are located on the same chromosome.

linkage group: A pair of homologous chromosomes.

lipid [Gk. *lipos*, fat]: One of a large variety of organic substances that are insoluble in polar solvents, such as water, but that dissolve readily in nonpolar organic solvents; includes fats, oils, waxes, steroids, phospholipids, glycolipids, and carotenes.

locus, *pl.* **loci** [L., place]: In genetics, the position of a gene in a chromosome. For any given locus, there may be a number of possible alleles.

logistic growth: A pattern of population growth in which growth is rapid when the population is small, gradually slows as the population approaches the carrying capacity of its environment, and then oscillates as the population stabilizes at or near its maximum size; it is one of the simplest growth patterns observed for populations in nature.

loop of Henle (after F. G. J. Henle, German pathologist): A hairpin-shaped portion of the renal tubule of mammals in which a hypertonic urine is formed by processes of diffusion and active transport.

lumen [L., light]: The cavity of a tubular structure, such as endoplasmic reticulum or a blood vessel.

lymph [L. *lympha*, water]: Colorless fluid derived from blood by filtration through capillary walls in the tissues; carried in special lymph ducts.

lymphatic system: The system through which lymph circulates; consists of lymph capillaries, which begin blindly in the tissues, and a network of progressively larger vessels that empty into the vena cava; also includes the lymph nodes, spleen, thymus, and tonsils.

lymph node [L. *lympha*, water + *nodus*, knot]: A mass of spongy tissues, separated into compartments; located throughout the lymphatic system, lymph nodes remove dead cells, debris, and foreign particles from the circulation; also are sites at which foreign antigens are displayed to immunologically active cells.

lymphocyte [L. *lympha*, water + Gk. *kytos*, vessel]: A type of white blood cell involved in the immune response; B lymphocytes differentiate into antibody-producing plasma cells, whereas cytotoxic T lymphocytes lyse diseased eukaryotic cells; other T lymphocytes interact with both cytotoxic T lymphocytes and with B lymphocytes.

lysis [Gk., a loosening]: Disintegration of a cell by rupture of its cell membrane.

lysogenic bacteria (lye-so-**jenn**-ick) [Gk. *lysis*, a loosening + *genos*, race or descent]: Bacteria carrying a bacteriophage integrated into the bacterial chromosome. The virus may subsequently set up an active cycle of infection, causing lysis of the bacterial cells.

lysosome [Gk. *lysis*, loosening + *soma*, body]: A membrane-bound organelle in which hydrolytic enzymes are segregated.

macromolecule [Gk. *makros*, large + L. dim. of *moles*, mass]: An extremely large molecule; refers specifically to proteins, nucleic acids, polysaccharides, and complexes of these.

macrophage [Gk. *makros*, large + *phagein*, to eat]: A type of phagocytic white blood cell important in both the inflammatory and immune responses.

major histocompatibility complex (MHC): In mammals, a group of at least 20 different genes, each with multiple alleles, coding for the protein components of the antigens that are displayed on nucleated cells and that serve to identify "self."

mandibles [L. *mandibula*, jaw]: In crustaceans, insects, and myriapods, the appendages immediately posterior to the antennae; used to seize, hold, bite, or chew food.

mantle: In mollusks, the outermost layer of the body wall or a soft extension of it; usually secretes a shell.

marine [L. *marini(us)*, from *mare*, the sea]: Living in salt water.

marsupial [Gk. *marsypos*, pouch, little bag]: A mammal in which the female has a ventral pouch or folds surrounding the nipples; the premature young leave the uterus and crawl into the pouch, where each one attaches itself by the mouth to a nipple until development is completed.

matrix: The dense solution in the interior of the mitochondrion, surrounding the cristae; contains enzymes, phosphates, coenzymes, and other molecules involved in cellular respiration.

mechanoreceptor: A sensory cell or organ that receives mechanical stimuli such as those involved in touch, pressure, hearing, and balance.

medulla (med-**dull**-a) [L., the innermost part]: (1) The inner, as opposed to the outer, part of an organ, as in the adrenal gland. (2) The most posterior region of the vertebrate brain; connects with the spinal cord.

medusa: The free-swimming, bell- or umbrella-shaped stage in the life cycle of many cnidarians; a jellyfish.

megaspore [Gk. *megas*, great, large + *spora*, a sowing]: In plants, a haploid (n) spore that develops into a female gametophyte.

meiosis (my-o-sis) [Gk. *meioun*, to make smaller]: The two successive nuclear divisions in which a single diploid (2n) cell forms four haploid (n) nuclei, and segregation, crossing over, and reassortment of the alleles occur; gametes or spores may be produced as a result of meiosis.

Mendel's first law: The factors for a pair of alternative characters are separate, and only one may be carried in a particular gamete (genetic segregation). In modern form: Alleles segregate in meiosis.

Mendel's second law: The inheritance of a pair of factors for one trait is independent of the simultaneous inheritance of factors for other traits, such factors "assorting independently" as though there were no other factors present (later modified by the discovery of linkage). Modern form: The alleles of unlinked genes assort independently.

menstrual cycle [L. *mensis*, month]: In humans and certain other primates, the cyclic, hormone-regulated changes in the condition of the uterine lining; marked by the periodic discharge of blood and disintegrated uterine lining through the vagina. Mammals with a menstrual cycle lack a well-defined period of estrus.

meristem [Gk. *merizein*, to divide]: The undifferentiated plant tissue, including a mass of rapidly dividing cells, from which new tissues arise.

mesenteries [Gk. *mesos*, middle + *enteron*, gut]: Double layers of mesoderm that suspend the digestive tract and other internal organs within the coelom.

mesoderm [Gk. *mesos*, middle + *derma*, skin]: In animals, the middle layer of the three embryonic tissue layers. In vertebrates, includes the chordamesoderm, which gives rise to the notochord and skeletal muscle, and the lateral plate mesoderm, which gives rise to the circulatory system, most of the excretory and reproductive systems, the lining of the coelom, and the outer covering of the internal organs.

mesophyll [Gk. *mesos*, middle + *phyllon*, leaf]: The internal tissue of a leaf, sandwiched between two layers of epidermal cells; consists of palisade parenchyma and spongy parenchyma cells.

messenger RNA (mRNA): A class of RNA molecules, each of which is complementary to one strand of DNA and which serves to carry the genetic information from the chromosome to the ribosomes, where it is translated into protein.

metabolism [Gk. *metabole*, change]: The sum of all chemical reactions occurring within a cell or organism.

metamere [Gk. *meta*, middle + *meros*, part]: One of a linear series of similar body segments.

metamorphosis [Gk. *metamorphoun*, to transform]: Abrupt transition from larval to adult form, such as the transition from tadpole to adult frog.

metaphase [Gk. *meta*, middle + *phasis*, form]: The stage of mitosis or meiosis during which the chromosomes lie in the equatorial plane of the spindle.

microbe [Gk. *mikros*, small + *bios*, life]: A microscopic organism.

micronutrient [Gk. *mikros*, small + L. *nutrire*, to nourish]: An inorganic nutrient required in only minute amounts for plant growth, such as iron, chlorine, copper, manganese, zinc, molybdenum, and boron.

microspore [Gk. *mikros*, small + *spora*, a sowing]: In plants, a haploid (n) spore that develops into a male gametophyte; in seed plants, it becomes a pollen grain.

microtubule [Gk. *mikros*, small + L. dim. of *tubus*, tube]: An extremely small hollow tube composed of two types of globular protein subunits. Among their many functions, microtubules make up the internal structure of cilia and flagella.

middle lamella: In plants, distinct layer between adjacent cell walls, rich in pectins and other polysaccharides; derived from the cell plate.

mimicry [Gk. *mimos*, mime]: The superficial resemblance in form, color, or behavior of certain organisms (mimics) to other more powerful or more protected ones (models), resulting in protection, concealment, or some other advantage for the mimic.

mineral: A naturally occurring element or inorganic compound.

mitochondrion, *pl.* **mitochondria** [Gk. *mitos*, thread + *chondros*, cartilage or grain]: An organelle, bound by a double membrane, in which the reactions of the Krebs cycle, terminal electron transport, and oxidative phosphorylation take place, resulting in the formation of CO_2, H_2O, and ATP from acetyl CoA and ADP. Mitochondria are the organelles in which most of the ATP of the eukaryotic cell is produced.

mitosis [Gk. *mitos*, thread]: Nuclear division characterized by chromosome replication and formation of two identical daughter nuclei.

mole [L. *moles*, mass]: The amount of an element equivalent to its atomic weight expressed in grams, or the amount of a substance equivalent to its molecular weight expressed in grams.

molecular weight: The sum of the atomic weights of the constituent atoms in a molecule.

molecule [L. dim. of *moles*, mass]: A particle consisting of two or more atoms held together by chemical bonds; the smallest unit of a compound that displays the properties of the compound.

molting: Shedding of all or part of an organism's outer covering; in arthropods, periodic shedding of the exoskeleton to permit an increase in size.

monocotyledon [Gk. *monos*, single + *kotyledon*, a cup-shaped hollow]: A member of the class of flowering plants having one seed leaf, or cotyledon, among other distinguishing features; often abbreviated as monocot.

monocyte [Gk. *monos*, single + *kytos*, vessel]: A type of circulating white blood cell that, in the presence of infectious organisms or other foreign invaders, becomes transformed into a macrophage.

monoecious (mo-nee-shus) [Gk. *monos*, single + *oikos*, house]: In angiosperms, having the male and female structures (the stamens and the carpels, respectively) on the same individual but on different flowers.

monomer [Gk. *monos*, single + *meros*, part]: A simple, relatively small molecule that can be linked to others to form a polymer.

monosaccharide [Gk. *monos*, single + *sakcharon*, sugar]: A simple sugar, such as glucose, fructose, ribose.

monotreme [Gk. *monos*, single + *trēma*, hole]: A nonplacental mammal, such as the duckbilled platypus, in which the female lays shelled eggs and nurses the young.

morphogenesis [Gk. *morphe*, form + *genesis*, origin]: The development of size, form, and other structural features of organisms.

morphological [Gk. *morphe*, form + *logos*, discourse]: Pertaining to form and structure, at any level of organization.

motor neuron: Neuron that conducts nerve impulses from the central nervous system to an effector, which is typically a muscle or a gland.

muscle fiber: Muscle cell; a long, cylindrical, multinucleated cell containing numerous myofibrils, which is capable of contraction when stimulated.

mutagen [L. *mutare*, to change + *genus*, source or origin]: A chemical or physical agent that increases the mutation rate.

mutant [L. *mutare*, to change]: An organism carrying a gene that has undergone a mutation.

mutation [L. *mutare*, to change]: The change of a gene from one allelic form to another; an inheritable change in the DNA sequence of a chromosome.

mutualism [L. *mutuus*, lent, borrowed]: *See* Symbiosis.

mycelium [Gk. *mykes*, fungus]: The mass of hyphae forming the body of a fungus.

mycorrhizae [Gk. *mykes*, fungus + *rhiza*, root]: Symbiotic associations between particular species of fungi and the roots of vascular plants.

myelin sheath [Gk. *myelinos*, full of marrow]: A lipid-rich layer surrounding the long axons of neurons in the vertebrate nervous system; in the peripheral nervous system, made up of the membranes of Schwann cells.

myofibril [Gk. *mys*, muscle + L. *fibra*, fiber]: Contractile element of a muscle fiber, made up of thick and thin filaments arranged in sarcomeres.

myoglobin [Gk. *mys*, muscle + L. *globus*, a ball]: An oxygen-binding, heme-containing globular protein found in muscles.

myosin [Gk. *mys*, muscle]: One of the principal proteins in muscle; makes up the thick filaments.

NAD: Abbreviation of nicotinamide adenine dinucleotide, a coenzyme that functions as an electron acceptor.

natural selection: A process of interaction between organisms and their environment that results in a differential rate of reproduction of different phenotypes in the population; can result in changes in the relative frequencies of alleles and genotypes in the population—that is, in evolution.

nectar [Gk. *nektar*, the drink of the gods]: A sugary fluid that attracts insects to plants.

negative feedback: A control mechanism whereby an increase in some substance inhibits the process leading to the increase; also known as feedback inhibition.

nematocyst [Gk. *nema, nematos*, thread + *kyst*, bladder]: A threadlike stinger, containing a poisonous or paralyzing substance, found in the cnidocyte of cnidarians.

nephridium, *pl.* **nephridia** [Gk. *nephros*, kidney]: A tubular excretory structure found in many invertebrates.

nephron [Gk. *nephros*, kidney]: The functional unit of the kidney in reptiles, birds, and mammals; a human kidney contains about 1 million nephrons.

nerve: A group or bundle of nerve fibers with accompanying connective tissue, located in the peripheral nervous system. A bundle of nerve fibers within the central nervous system is known as a tract.

nerve fiber: A filamentous process extending from the cell body of a neuron and conducting the nerve impulse; an axon.

nerve impulse: A rapid, transient, self-propagating change in electric potential across the membrane of an axon.

nervous system: All the nerve cells of an animal; the receptor-conductor-effector system; in humans, the nervous system consists of the central nervous system (brain and spinal cord) and the peripheral nervous system.

net productivity: In a trophic level, a community, or an ecosystem, the amount of energy (in calories) stored in chemical compounds or the increase in biomass (in grams or metric tons) in a particular period of time; it is the difference between gross productivity and the energy used by the organisms in respiration.

neural groove: Dorsal, longitudinal groove that forms in a vertebrate embryo; bordered by two neural folds; preceded by the neural-plate stage and followed by the neural-tube stage.

neural plate: Thickened strip of ectoderm in early vertebrate embryos that forms along the dorsal side of the body and gives rise to the central nervous system.

neural tube: Primitive, hollow, dorsal nervous system of the early vertebrate embryo; formed by fusion of neural folds around the neural groove.

neuromodulator: A chemical agent that is released by a neuron and diffuses through a local region of the central nervous system, acting on neurons within that region; generally has the effect of modulating the response to neurotransmitters.

neuron [Gk., nerve]: Nerve cell, including cell body, dendrites, and axon.

neurosecretory cell: A neuron that releases one or more hormones into the circulatory system.

neurotransmitter: A chemical agent that is released by a neuron at a synapse, diffuses across the synaptic cleft, and acts upon a postsynaptic neuron or muscle or gland cell and alters its electrical state or activity.

neutron (new-tron): An uncharged particle with a mass slightly greater than that of a proton. Found in the atomic nucleus of all elements except hydrogen, in which the nucleus consists of a single proton.

niche: *See* Ecological niche.

nitrification: The oxidation of ammonia or ammonium to nitrites and nitrates, as by nitrifying bacteria.

nitrogen cycle: Worldwide circulation and reutilization of nitrogen atoms, chiefly due to metabolic processes of living organisms; plants take up inorganic nitrogen and convert it into organic compounds (chiefly proteins), which are assimilated into the bodies of one or more animals; bacterial and fungal action on nitrogenous waste products and dead organisms return nitrogen atoms to the inorganic state.

nitrogen fixation: Incorporation of atmospheric nitrogen into inorganic nitrogen compounds available to plants, a process that can be carried out only by some soil bacteria, many free-living and symbiotic cyanobacteria, and certain symbiotic bacteria in association with legumes.

nitrogenous base: A nitrogen-containing molecule having basic properties (tendency to acquire an H^+ ion); a purine or pyrimidine.

nocturnal [L. *nocturnus*, of night]: Applied to organisms that are active during the hours of darkness.

node [L. *nodus*, knot]: In plants, a joint of a stem; the place where branches and leaves are joined to the stem.

nondisjunction [L. *non*, not + *disjungere*, to separate]: The failure of chromatids to separate during meiosis, resulting in one or more extra chromosomes in some gametes and correspondingly fewer in others.

noradrenaline: A hormone, produced by the medulla of the adrenal gland, that increases the concentration of sugar in the blood, raises blood pressure and heartbeat rate, and increases muscular power and resistance to fatigue; also one of the principal neurotransmitters; also called norepinephrine.

norepinephrine: *See* noradrenaline.

notochord [Gk. *noto*, back + L. *chorda*, cord]: A dorsal rodlike structure that runs the length of the body and serves as the internal skeleton in the embryos of all chordates; in most adult chordates the notochord is replaced by a vertebral column that forms around (but not from) the notochord.

nuclear envelope [L. *nucleus*, a kernel]: The double membrane surrounding the nucleus within a eukaryotic cell.

nucleic acid: A macromolecule consisting of nucleotides; the principal types are deoxyribonucleic acid (DNA) and ribonucleic acid (RNA).

nucleoid: In prokaryotic cells, the region of the cell in which the chromosome is localized.

nucleolus (new-klee-o-lus) [L., a small kernel]: A small, dense region visible in the nucleus of nondividing eukaryotic cells; consists of rRNA molecules, ribosomal proteins, and loops of chromatin from which the rRNA molecules are transcribed.

nucleosome [L. *nucleus*, a kernel + Gk. *soma*, body]: A complex of DNA and histone proteins that forms the fundamental packaging unit of eukaryotic DNA; its structure resembles a bead on a string.

nucleotide [L. *nucleus*, a kernel]: A molecule composed of a phosphate group, a five-carbon sugar (either ribose or deoxyribose), and a purine or pyrimidine base; nucleotides are the building blocks of nucleic acids.

nucleus [L., a kernel]: (1) The central core of an atom, containing protons and neutrons, around which electrons move. (2) The membrane-bound structure characteristic of eukaryotic cells that contains the genetic information in the form of DNA organized into chromosomes. (3) A group of nerve cell bodies within the central nervous system.

ocellus, *pl.* **ocelli** [L. dim. of *oculus*, eye]: A simple light receptor common among invertebrates.

Okazaki fragments (after R. Okazaki, Japanese geneticist): In DNA replication, the discontinuous segments in which the 3′ to 5′ strand (the lagging strand) of the DNA double helix is synthesized; typically 1,000 to 2,000 nucleotides long in prokaryotes, and 100 to 200 nucleotides long in eukaryotes.

olfactory [L. *olfacere*, to smell]: Pertaining to smell.

oligosaccharins [Gk. *oligo*, few + *sakcharon*, sugar]: Short carbohydrate chains released from plant cell walls in response to a variety of stimuli, including injury; hypothesized to play a role in regulation of plant growth and development.

ommatidium, *pl.* **ommatidia** [Gk. *ommos*, eye]: The single visual unit in the compound eye of arthropods; contains light-sensitive cells and a lens able to form an image.

omnivore [L. *omnis*, all + *vorare*, to devour]: An organism that "eats everything"; for example, an animal that eats both plants and meat.

oncogene [Gk. *onkos*, tumor + *genos*, birth, race]: One of a group of eukaryotic genes that closely resemble normal genes of the cells in which they are found and that are thought to play a role in the development of cancer; their gene products appear to be regulatory proteins, involved in the control of either cell growth or cell division.

oocyte (o-uh-sight) [Gk. *oion*, egg + *kytos*, vessel]: A cell that gives rise by meiosis to an ovum.

oogamy (oh-og-amy) [Gk. *oion*, egg + *gamos*, marriage]: Sexual reproduction in which one of the gametes, usually the larger, is not motile.

operator: A segment of DNA that interacts with a repressor protein to regulate the transcription of the structural genes of an operon.

operon [L. *opus*, *operis*, work]: In the bacterial chromosome, a segment of DNA consisting of a promoter, an operator, and a group of adjacent structural genes; the structural genes, which code for products related to a particular biochemical pathway, are transcribed onto a single mRNA molecule, and their transcription is regulated by a single repressor protein.

opportunistic species: Species characterized by high reproduction rates, rapid development, early reproduction, small body size, and uncertain adult survival.

orbital [L. *orbis*, circle, disk]: In the current model of atomic structure, the volume of space surrounding the atomic nucleus in which an electron will be found 90 percent of the time.

order: A taxonomic grouping of related, similar families; the category below class and above family.

organ [Gk. *organon*, tool]: A body part composed of several tissues grouped together in a structural and functional unit.

organelle [Gk. *organon*, instrument, tool]: A formed body in the cytoplasm of a cell.

organic [Gk. *organon*, instrument, tool]: Pertaining to (1) organisms or living things generally, or (2) compounds formed by living organisms, or (3) the chemistry of compounds containing carbon.

organism [Gk. *organon*, instrument, tool]: Any living creature, either unicellular or multicellular.

organizer [Gk. *organon*, instrument, tool]: In vertebrates, the part of an embryo capable of inducing undifferentiated cells to follow a specific course of development; in particular, the dorsal lip of the blastopore.

osmosis [Gk. *osmos*, impulse, thrust]: The diffusion of water across a selectively permeable membrane (a membrane that permits the free passage of water but prevents or retards the passage of a solute). In the absence of other factors that affect the water potential, the net movement of water is from the side containing a lower concentration of solute to the side containing a higher concentration.

osmotic potential [Gk. *osmos*, impulse, thrust]: The tendency of water to move across a selectively permeable membrane into a solution; it is determined by measuring the pressure required to stop the osmotic movement of water into the solution; the higher the solute concentration, the greater the osmotic potential of the solution.

ovary [L. *ovum*, egg]: (1) In animals, the egg-producing organ. (2) In flowering plants, the enlarged basal portion of a carpel or a fused carpel, containing the ovule or ovules; the ovary matures to become the fruit.

oviduct [L. *ovum*, egg + *ductus*, duct]: The tube serving to transport the eggs to the outside or to the uterus; also called uterine tube or Fallopian tube (in humans).

ovulation: In animals, release of an egg or eggs from the ovary.

ovule [L. dim. of *ovum*, egg]: In seed plants, a structure composed of a protective outer coat, a tissue specialized for food storage, and a female gametophyte with an egg cell; becomes a seed after fertilization.

ovum, *pl.* **ova** [L., egg]: The egg cell; female gamete.

oxidation: Gain of oxygen, loss of hydrogen, or loss of an electron by an atom, ion, or molecule. Oxidation and reduction take place simultaneously, with the electron lost by one reactant being transferred to another reactant.

oxidative phosphorylation: The process by which the energy released as electrons pass down the mitochondrial electron transport chain in the final stage of cellular respiration is used to phosphorylate (add a phosphate group to) ADP molecules, thereby yielding ATP molecules.

pacemaker: *See* Sinoatrial node.

paleontology [Gk. *palaios*, old + *onta*, things that exist + *logos*, discourse]: The study of the life of past geologic times, principally by means of fossils.

palisade cells [L. *palus*, stake + *cella*, a chamber]: In plant leaves, the columnar, chloroplast-containing parenchyma cells of the mesophyll.

pancreas (pang-kree-us) [Gk. *pan*, all + *kreas*, meat, flesh]: In vertebrates, a small, complex gland located between the stomach and the duodenum, which produces digestive enzymes and the hormones insulin and glucagon.

parasite [Gk. *para*, beside, akin to + *sitos*, food]: An organism that lives on or in an organism of a different species and derives nutrients from it.

parasitism: *See* Symbiosis.

parasympathetic division [Gk. *para*, beside, akin to]: A subdivision of the autonomic nervous system of vertebrates, with centers located in the brain and in the most anterior and most posterior parts of the spinal cord; stimulates digestion; generally inhibits other functions and restores the body to normal following emergencies.

parenchyma (pah-renk-ee-ma) [Gk. *para*, beside, akin to + *en*, in + *chein*, to pour]: A plant tissue composed of living, thin-walled, randomly arranged cells with large vacuoles; usually photosynthetic or storage tissue.

parthenogenesis [Gk. *parthenon*, virgin + *genesis*, birth]: The development of an organism from an unfertilized egg.

pellicle [L. dim. of *pellis*, skin]: A flexible series of protein strips inside the cell membrane of many protists.

penetrance: In genetics, the proportion of individuals with a particular genotype that show the phenotype ascribed to that genotype.

peptide bond [Gk. *pepto*, to soften, digest]: The type of bond formed when two amino acids are joined end to end; the acidic group (— COOH) of one amino acid is linked covalently to the basic group (—NH$_2$) of the next, and a molecule of water (H$_2$O) is removed.

perennial [L. *per*, through + *annus*, year]: A plant that persists in whole or in part from year to year and usually produces reproductive structures in more than one year.

pericycle [Gk. *peri*, around + *kyklos*, circle]: One or more layers of cells completely surrounding the vascular tissues of the root; branch roots arise from the pericycle.

peripheral nervous system [Gk. *peripherein*, to carry around]: All of the neurons and axons outside the central nervous system, including both motor neurons and sensory neurons; consists of the somatic nervous system and the autonomic nervous system.

peristalsis [Gk. *peristellein*, to wrap around]: Successive waves of muscular contraction in the walls of a tubular structure, such as the digestive tract or an oviduct; moves the contents, such as food or an egg cell, through the tube.

peritoneum [Gk. *peritonos*, stretched over]: A membrane that lines the body cavity and forms the external covering of the visceral organs.

permeable [L. *permeare*, to pass through]: Penetrable by molecules, ions, or atoms; usually applied to membranes that let given solutes pass through.

peroxisome: A membrane-bound organelle in which enzymes catalyzing peroxide-forming and peroxide-destroying reactions are segregated; in plant cells, the site of photorespiration.

petiole (pet-ee-ole) [Fr., from L. *petiolus*, dim. of *pes, pedis*, a foot]: The stalk of a leaf, connecting the blade of the leaf with the branch or stem.

pH: A symbol denoting the concentration of hydrogen ions in a solution; pH values range from 0 to 14; the lower the value, the more acidic a solution, that is, the more hydrogen ions it contains; pH 7 is neutral, less than 7 is acidic, more than 7 is alkaline.

phagocytosis [Gk. *phagein*, to eat + *kytos*, vessel]: Cell "eating." *See* Endocytosis.

phenotype [Gk. *phainein*, to show + *typos*, stamp, print]: Observable characteristics of an organism, resulting from interactions between the genotype and the environment.

pheromone (fair-o-moan) [Gk. *phero*, to bear, carry]: Substance secreted by an animal that influences the behavior or development of other animals of the same species, such as the sex attractants of moths, the queen substance of honey bees.

phloem (flow-em) [Gk. *phloos*, bark]: Vascular tissue of higher plants; conducts sugars and other organic molecules from the leaves to other parts of the plant; in angiosperms, composed of sieve-tube members, companion cells, other parenchyma cells, and fibers.

phospholipids: Organic molecules similar in structure to fats, but in which a phosphate group rather than a fatty acid is attached to the third carbon of the glycerol molecule; as a result, the molecule has a hydrophilic "head" and a hydrophobic "tail." Phospholipids form the basic structure of cell and organelle membranes.

phosphorylation: Addition of a phosphate group or groups to a molecule.

photon [Gk. *photos*, light]: The elementary particle of light and other electromagnetic radiations.

photoperiodism [Gk. *photos*, light]: The response to relative day and night length, a mechanism by which organisms measure seasonal change.

photophosphorylation [Gk. *photos*, light + *phosphoros*, bringing light]: The process by which the energy released as electrons pass down the electron transport chain between photosystems II and I during photosynthesis is used to phosphorylate ADP to ATP.

photoreceptor [Gk. *photos*, light]: A cell or organ capable of detecting light.

photorespiration [Gk. *photos*, light + L. *respirare*, to breathe]: The oxidation of carbohydrates in the presence of light and oxygen; occurs when the carbon dioxide concentration in the leaf is low in relation to the oxygen concentration.

photosynthesis [Gk. *photos*, light + *syn*, together + *tithenai*, to place]: The conversion of light energy to chemical energy; the synthesis of organic compounds from carbon dioxide and water in the presence of chlorophyll, using light energy.

phototropism [Gk. *photos*, light + *trope*, turning]: Movement in which the direction of the light is the determining factor, such as the growth of a plant toward a light source; a curving response to light.

phyletic change [Gk. *phylon*, race, tribe]: The changes taking place in a single lineage of organisms over a long period of time; one of the principal patterns of evolutionary change.

phylogeny [Gk. *phylon*, race, tribe]: Evolutionary history of a taxonomic group. Phylogenies are often depicted as "evolutionary trees."

phylum, *pl.* **phyla** [Gk. *phylon*, race, tribe]: A taxonomic grouping of related, similar classes; a high-level category below kingdom and above class. Phylum is generally used in the classification of protozoa and animals, whereas an equivalent category, division, is used in the classification of prokaryotes, algae, fungi, and plants.

physiology [Gk. *physis*, nature + *logos*, a discourse]: The study of function in cells, organs, or entire organisms; the processes of life.

phytochrome [Gk. *phyton*, plant + *chrōma*, color]: A plant pigment that is a photoreceptor for red or far-red light and is involved with a number of developmental processes, such as flowering, dormancy, leaf formation, and seed germination.

phytoplankton [Gk. *phyton*, plant + *planktos*, wandering]: Aquatic, free-floating, microscopic, photosynthetic organisms.

pigment [L. *pigmentum*, paint]: A colored substance that absorbs light over a narrow band of wavelengths.

pinocytosis [Gk. *pinein*, to drink + *kytos*, vessel]: Cell "drinking." *See* Endocytosis.

pituitary [L. *pituita*, phlegm]: Endocrine gland in vertebrates; the anterior lobe is the source of tropic hormones, growth hormone, and prolactin and is regulated by secretions of the hypothalamus; the posterior lobe stores and releases oxytocin and ADH produced by the hypothalamus.

placenta [Gk. *plax*, a flat object]: A tissue formed as the result of interactions between the inner lining of the mammalian uterus and the extraembryonic chorion; serves as the connection through which exchanges of nutrients and wastes occur between the blood of the mother and that of the embryo.

plankton [Gk. *planktos*, wandering]: Small (mostly microscopic) aquatic and marine organisms found in the upper levels of the water, where light is abundant; includes both photosynthetic (phytoplankton) and heterotrophic (zooplankton) forms.

planula [L. dim. of *planus*, a wanderer]: The ciliated, free-swimming type of larva formed by many cnidarians.

plasma [Gk., form or mold]: The clear, colorless fluid component of vertebrate blood, containing dissolved ions, molecules, and plasma proteins; blood minus the blood cells.

plasma cell: An antibody-producing cell resulting from the differentiation and proliferation of a B lymphocyte that has interacted with an antigen complementary to the antibodies displayed on its surface; a mature plasma cell can produce from 3,000 to 30,000 antibody molecules per second.

plasma membrane: The membrane surrounding the cytoplasm of a cell; the cell membrane.

plasmid: In prokaryotes, an extrachromosomal, independently replicating, small, circular DNA molecule.

plasmodesma, *pl.* **plasmodesmata** [Gk. *plassein*, to mold + *desmos*, band, bond]: In plants, a minute, cytoplasmic thread that extends through pores in cell walls and connects the cytoplasm of adjacent cells.

plastid [Gk. *plastos*, formed or molded]: A cytoplasmic, often pigmented, organelle in plant cells; includes leucoplasts, chromoplasts, and chloroplasts.

platelet (plate-let) [Gk. *platus*, flat]: In mammals, a round or biconcave disk suspended in the blood and involved in the formation of blood clots.

pleiotropy (plee-o-trope-ee) [Gk. *pleios*, more + *trope*, a turning]: The capacity of a gene to affect a number of different phenotypic characteristics.

poikilotherm [Gk. *poikilos*, changeable + *therme*, heat]: An organism with a body temperature that varies with that of the environment.

polar [L. *polus*, end of axis]: Having parts or areas with opposed or contrasting properties, such as positive and negative charges, head and tail.

polar body: Minute, nonfunctioning cell produced during those meiotic divisions that lead to egg cells; contains a nucleus but very little cytoplasm.

polar covalent bond: A covalent bond in which the electrons are shared unequally between the two atoms; the resulting polar molecule has regions of slightly negative and slightly positive charge.

pollen [L., fine dust]: In seed plants, spores consisting of an immature male gametophyte and a protective outer covering.

pollination [L. *pollen*, fine dust]: The transfer of pollen from the anther to a receptive surface of a flower.

polygenic inheritance [Gk. *polus*, many + *genos*, race, descent]: The determination of a given characteristic, such as weight or height, by the interaction of many genes.

polymer [Gk. *polus*, many + *meris*, part or portion]: A large molecule composed of many similar or identical molecular subunits.

polymorphism [Gk. *polus*, many + *morphe*, form]: The presence in a single population of two or more phenotypically distinct forms of a trait.

polyp [Gk. *polus*, many + *pous*, foot]: The sessile stage in the life cycle of cnidarians.

polypeptide [Gk. *polus*, many + *pepto*, to soften, digest]: A molecule consisting of a long chain of amino acids linked together by peptide bonds.

polyploid [Gk. *polus*, many + *ploion*, vessel]: Cell with more than two complete sets of chromosomes per nucleus.

polyribosome: Two or more ribosomes together with a molecule of mRNA that they are simultaneously translating; a polysome.

polysaccharide [Gk. *polus*, many + *sakcharon*, sugar]: A carbohydrate polymer composed of monosaccharide monomers in long chains; includes starch, cellulose.

polysome: *See* Polyribosome.

population: Any group of individuals of one species that occupy a given area at the same time; in genetic terms, an interbreeding group of organisms.

population bottleneck: Type of genetic drift that occurs as the result of a population being drastically reduced in numbers by an event having little to do with the usual forces of natural selection.

portal system [L. *porta*, gate]: In the circulatory system, a circuit in which blood flows through two distinct capillary beds, connected by either veins or arteries, before entering the veins that return it to the heart.

posterior: Of or pertaining to the rear, or tail, end.

potential energy: Energy in a potentially usable form that is not, for the moment, being used; often called "energy of position."

predator [L. *praedari*, to prey upon; from *prehendere*, to grasp, seize]: An organism that eats other living organisms.

pressure-flow hypothesis: A hypothesis accounting for sap flow through the phloem system. According to this hypothesis, the solution containing nutrient sugars moves through the sieve tubes by bulk flow, moving into and out of the sieve tubes by active transport and diffusion.

prey [L. *prehendere*, to grasp, seize]: An organism eaten by another organism.

primary growth: In plants, growth originating in the apical meristem of the shoots and roots, as contrasted with secondary growth; results in an increase in length.

primary structure of a protein: The amino acid sequence of a protein.

primate: A member of the order of mammals that includes anthropoids and prosimians.

primitive [L. *primus*, first]: Not specialized; at an early stage of evolution or development.

primitive streak [L. *primus*, first]: The thickened, dorsal, longitudinal strip of ectoderm and mesoderm in early avian, reptilian, and mammalian embryos; equivalent to the blastopore in other forms.

procambium [L. *pro*, before + *cambium*, exchange]: In plants, a primary meristematic tissue; gives rise to vascular tissues of the primary plant body and to the vascular cambium.

producer, in ecological systems: An autotrophic organism, usually a photosynthesizer, that contributes to the net primary productivity of a community.

progesterone [L. *progerere*, to carry forth or out + *steiras*, barren]: In mammals, a steroid hormone produced by the corpus luteum that prepares the uterus for implantation of the embryo; also produced by the placenta during pregnancy.

prokaryote [L. *pro*, before + Gk. *karyon*, nut, kernel]: A cell lacking a membrane-bound nucleus and membrane-bound organelles; a bacterium or a cyanobacterium.

promoter: Specific segment of DNA to which RNA polymerase attaches to initiate transcription of mRNA from an operon.

prophage: A bacterial virus (bacteriophage) integrated into a host chromosome.

prophase [Gk. *pro*, before + *phasis*, form]: An early stage in nuclear division, characterized by the condensing of the chromosomes and their movement toward the equator of the spindle. Homologous chromosomes pair up during meiotic prophase.

proprioceptor [L. *proprius*, one's own]: Receptor that senses movements, position of the body, or muscle strength.

prosimian [L. *pro*, before + *simia*, ape]: A lower primate; includes lemurs, lorises, tarsiers, and bush babies, as well as many fossil forms.

prostaglandins [Gk. *prostas*, a porch or vestibule + L. *glans*, acorn]: A group of fatty acids that function as chemical messengers; synthesized in most, possibly all, cells of the body; thought to play key roles in fertilization and in triggering the onset of both menstruation and labor.

prostate gland [Gk. *prostas*, a porch or vestibule + L. *glans*, acorn]: A mass of muscle and glandular tissue surrounding the base of the urethra in male mammals; the vasa deferentia merge with ducts from the seminal vesicles, enter the prostate gland, and there merge with the urethra. The prostate gland secretes an alkaline fluid that has a stimulating effect on the sperm as they are released.

protein [Gk. *proteios*, primary]: A complex organic compound composed of one or more polypeptide chains, each made up of many (about 100 or more) amino acids linked together by peptide bonds.

proton: A subatomic particle with a single positive charge equal in magnitude to the charge of an electron and with a mass slightly less than that of a neutron; a component of every atomic nucleus.

protoplasm [Gk. *protos*, first + *plasma*, anything molded]: Living matter.

protostome [Gk. *protos*, first + *stoma*, mouth]: An animal in which the mouth forms at or near the blastopore in the developing embryo; mollusks, annelids, and arthropods are protostomes. Protostomes are also characterized by spiral cleavage during the earliest stages of development and by schizocoelous formation of the coelom.

provirus: A virus of a eukaryote that has become integrated into a host chromosome.

pseudocoelom [Gk. *pseudes*, false + *koilos*, a hollow]: A body cavity consisting of a fluid-filled space between the endoderm and the mesoderm; characteristic of the nematodes.

pseudopodium [Gk. *pseudes*, false + *pous*, pod-, foot]: A temporary cytoplasmic protrusion from an amoeboid cell, which functions in locomotion or in feeding by phagocytosis.

pulmonary [L. *pulmonis*, lung]: Pertaining to the lungs.

pulmonary artery [L. *pulmonis*, lung]: In birds and mammals, an artery that carries deoxygenated blood from the right ventricle of the heart to the lungs, where it is oxygenated.

pulmonary vein [L. *pulmonis*, lung]: In birds and mammals, a vein that carries oxygenated blood from the lungs to the left atrium of the heart, from which blood is pumped into the left ventricle and from there to the body tissues.

punctuated equilibrium: A model of the mechanism of evolutionary change that proposes that long periods of no change ("stasis") are punctuated by periods of rapid speciation, with natural selection acting on species as well as on individuals.

Punnett square: The checkerboard diagram used for analysis of allele segregation.

pupa [L., girl, doll]: A developmental stage of some insects, in which the organism is nonfeeding, immotile, and sometimes encapsulated or in a cocoon; the pupal stage occurs between the larval and adult phases.

purine [Gk. *purinos*, fiery, sparkling]: A nitrogenous base, such as adenine or guanine, with a characteristic two-ring structure; one of the components of nucleic acids.

pyramid, ecological: *See* Ecological pyramid.

pyramid of energy: A diagram of the energy flow between the trophic levels of an ecosystem; plants or other autotrophs (at the base of the pyramid) represent the greatest amount of energy, herbivores next, then primary carnivores, secondary carnivores, etc.

pyrimidine: A nitrogenous base, such as cytosine, thymine, or uracil, with a characteristic single-ring structure; one of the components of nucleic acids.

quaternary structure of a protein: The overall structure of a globular protein molecule that consists of two or more polypeptide chains.

queen: In social insects (ants, termites, and some species of bees and wasps), the fertile, or fully developed, female whose function is to lay eggs.

radial symmetry [L. *radius*, a spoke of a wheel + Gk. *summetros*, symmetry]: The regular arrangement of parts around a central axis such that any plane passing through the central axis divides the organism into halves that are approximate mirror images; seen in cnidarians, ctenophorans, and adult echinoderms.

radiation [L. *radius*, a spoke of a wheel, hence, a ray]: Energy emitted in the form of waves or particles.

radioactive isotope: An isotope with an unstable nucleus that stabilizes itself by emitting radiation.

receptor: A protein or glycoprotein molecule with a specific three-dimensional structure, to which a substance (for example, a hormone, a neurotransmitter, or an antigen) with a complementary structure can bind; typically displayed on the surface of a membrane. Binding of a complementary molecule to a receptor may trigger a transport process or a change in processes occurring within the cell.

recessive allele [L. *recedere*, to recede]: An allele whose phenotypic effect is masked in the heterozygote by that of another, dominant allele.

reciprocal altruism: Performance of an altruistic act with the expectation that the favor will be returned.

recognition sequence: A specific sequence of nucleotides at which a restriction enzyme cleaves a DNA molecule.

recombinant DNA: DNA formed either naturally or in the laboratory by the joining of segments of DNA from different sources.

recombination: The formation of new gene combinations; in eukaryotes, may be accomplished by new associations of chromosomes produced during sexual reproduction or crossing over; in prokaryotes, may be accomplished through transformation, conjugation, or transduction.

reduction [L. *reducere*, to lead back]: Loss of oxygen, gain of hydrogen, or gain of an electron by an atom, ion, or molecule; oxidation and reduction take place simultaneously, with the electron lost by one reactant being transferred to another.

reflex [L. *reflectere*, to bend back]: Unit of action of the nervous system involving a sensory neuron, often one or more interneurons, and one or more motor neurons.

relay neuron: Neuron that transmits signals between different regions of the central nervous system.

releaser: In ethology, a stimulus that functions as a communication signal between members of the same species and that sets in motion, or

"releases," the sequence of movements that constitute a fixed action pattern.

renal [L. *renes*, kidneys]: Pertaining to the kidney.

replication fork: In DNA synthesis, the Y-shaped structure formed at the point where the two strands of the original molecule are being separated and the complementary strands are being synthesized.

repressor [L. *reprimere*, to press back, keep back]: In genetics, a protein that binds to the operator, preventing RNA polymerase from attaching to the promoter and transcribing the structural genes of the operon; coded by a segment of DNA known as the regulator.

resolving power [L. *resolvere*, to loosen, unbind]: The ability of a lens to distinguish two lines as separate.

respiration [L. *respirare*, to breathe]: (1) In aerobic organisms, the intake of oxygen and the liberation of carbon dioxide. (2) In cells, the oxygen-requiring stage in the breakdown and release of energy from fuel molecules.

resting potential: The difference in electric potential (about 70 millivolts) across the membrane of an axon at rest.

restriction enzymes: Enzymes that cleave the DNA double helix at specific nucleotide sequences.

reticular activating system [L. *reticulum*, a network]: A brain circuit involved with alertness and direction of attention to selected events; includes the reticular formation, a core of tissue that runs centrally through the brainstem, and neurons in the thalamus.

reticulum [L., network]: A fine network (e.g., endoplasmic reticulum).

retina [L. dim. of *rete*, net]: The light-sensitive layer of the vertebrate eye; contains several layers of neurons and photoreceptor cells (rods and cones); receives the image formed by the lens and transmits it to the brain via the optic nerve.

retrovirus [L., turning back]: An RNA virus that codes for an enzyme, reverse transcriptase, that transcribes the RNA into DNA.

reverse transcriptase: An enzyme that transcribes RNA into DNA; found only in association with retroviruses.

rhizoid [Gk. *rhiza*, root]: Rootlike anchoring structure in fungi and non-vascular plants.

rhizome [Gk. *rhizoma*, mass of roots]: In vascular plants, a horizontal stem growing along or below the surface of the soil; may be enlarged for storage or may function in vegetative reproduction.

ribonucleic acid (RNA) (rye-bo-new-clay-ick): A class of nucleic acids characterized by the presence of the sugar ribose and the pyrimidine uracil; includes mRNA, tRNA, and rRNA. RNA is the genetic material of many viruses.

ribosomal RNA (rRNA): A class of RNA molecules found, along with characteristic proteins, in ribosomes; transcribed from DNA of the chromatin loops that form the nucleolus.

ribosome: A small organelle composed of protein and ribonucleic acid; the site of translation in protein synthesis; in eukaryotic cells, often bound to the endoplasmic reticulum. Many ribosomes attached to a single strand of mRNA are called a polyribosome, or polysome.

RNA: Abbreviation of ribonucleic acid.

rod: Photoreceptor cell found in the vertebrate retina; sensitive to very dim light, responsible for "night vision."

root: The descending axis of a plant, normally below ground and serving both to anchor the plant and to take up and conduct water and minerals.

root hair: An extremely fine cytoplasmic extension of an epidermal cell of a young root; root hairs greatly increase the surface area for the uptake of water and minerals.

saprobe [Gk. *sapros*, rotten, putrid + *bios*, life]: An organism that feeds on nonliving organic matter.

sarcolemma [Gk. *sarx*, the flesh + *lemma*, husk]: The specialized cell membrane surrounding a muscle cell (muscle fiber); capable of propagating action potentials.

sarcomere [Gk. *sarx*, the flesh + *meris*, part of, portion]: Functional and structural unit of contraction in striated muscle.

sarcoplasmic reticulum [Gk. *sarx*, the flesh + *plasma*, from cytoplasm + L. *reticulum*, network]: The specialized endoplasmic reticulum that encases each myofibril of a muscle cell.

schizocoelous [Gk. *schizo*, to split + *koilos*, a hollow]: Formation of the coelom during embryonic development by a splitting of the mesoderm; characteristic of protostomes.

sclerenchyma [Gk. *skleros*, hard]: In plants, a type of supporting cell with thick, often lignified, secondary walls; may be alive or dead at maturity; includes fibers and sclereids.

secondary sex characteristics: Characteristics of animals that distinguish between the two sexes but that do not produce or convey gametes; includes facial hair of the human male and enlarged hips and breasts of the female.

secondary structure of a protein: The simple structure (often a helix, a sheet, or a cable) resulting from the spontaneous folding of a polypeptide chain as it is formed; maintained by hydrogen bonds and other weak forces.

secretion [L. *secermere*, to sever, separate]: (1) Product of any cell, gland, or tissue that is released through the cell membrane and that performs its function outside the cell that produced it. (2) The stage of kidney function in which, through active transport processes, molecules remaining in the blood plasma are selectively removed from the peritubular capillaries and pumped into the filtrate in the renal tubule.

seed: A complex structure formed by the maturation of the ovule of seed plants following fertilization; upon germination, a seed develops into a new sporophyte; generally consists of seed coat, embryo, and a food reserve.

segregation: *See* Mendel's first law.

selectively permeable [L. *seligere*, to gather apart + *permeare*, to go through]: Applied to membranes that permit passage of water and some solutes but block passage of most solutes; semipermeable.

self-fertilization: The union of egg and sperm produced by a single hermaphroditic organism.

self-pollination: The transfer of pollen from anther to stigma in the same flower or to another flower of the same plant, leading to self-fertilization.

semen [L., seed]: Product of the male reproductive system; includes sperm and the sperm-carrying fluids.

seminal vesicles [L. *semen*, seed + *vesicula*, a little bladder]: In male mammals, small vesicles, the ducts of which merge with the vasa deferentia as they enter the prostate gland; they produce an alkaline, fructose-containing fluid that suspends and nourishes the sperm cells.

sensory neuron: A neuron that conducts impulses from a sensory receptor to the central nervous system or central ganglion.

sensory receptor: A cell, tissue, or organ that detects internal or external stimuli.

septum [L., fence]: A partition, or cross wall, that divides a structure, such as a fungal hypha, into compartments.

sessile [L. *sedere*, to sit]: Attached; not free to move about.

sex chromosomes: Chromosomes that are different in the two sexes and that are involved in sex determination.

sex-linked trait: An inherited trait, such as color discrimination, determined by a gene located on a sex chromosome and that therefore shows a different pattern of inheritance in males and females.

sexual reproduction: Reproduction involving meiosis and fertilization.

sexual selection: A type of natural selection that acts on characteristics of direct consequence in obtaining a mate and successfully reproducing; thought to be the chief cause of sexual dimorphism, the striking phenotypic differences between the males and females of many species.

shoot: The aboveground portions, such as the stem and leaves, of a vascular plant.

sieve cell: A long, slender cell of the phloem of gymnosperms; involved in transport of sugars synthesized in the leaves to other parts of the plant.

sieve tube: A series of sugar-conducting cells (sieve-tube members) found in the phloem of angiosperms.

sinoatrial node [L. *sinus*, fold, hollow + *atrium*, yard, court, hall + *nodus*, knot]: Area of the vertebrate heart that initiates the heartbeat; located where the superior vena cava enters the right atrium; the pacemaker.

smooth muscle: Nonstriated muscle; lines the walls of internal organs and arteries and is under involuntary control.

social dominance: A hierarchical pattern of social organization involving domination of some members of a group by other members in a relatively orderly and long-lasting pattern.

society [L. *socius*, companion]: An organization of individuals of the same species in which there are divisions of resources, divisions of labor, and mutual dependence; a society is held together by stimuli exchanged among members of the group.

sociobiology: The study of the biological basis of social behavior.

solution: A homogeneous mixture of the molecules of two or more substances; the substance present in the greatest amount (usually a liquid) is called the solvent, and the substances present in lesser amounts are called solutes.

somatic cells [Gk. *soma*, body]: The differentiated cells composing body tissues of multicellular plants and animals; all body cells except those giving rise to gametes.

somatic nervous system [Gk. *soma*, body]: In vertebrates, the motor and sensory neurons of the peripheral nervous system that control skeletal muscle; the "voluntary" system, as contrasted with the "involuntary," or autonomic, nervous system.

somite: One of the blocks, or segments, of tissue into which the chorda-mesoderm is divided during differentiation of the vertebrate embryo.

specialized: (1) Of cells, having particular functions in a multicellular organism. (2) Of organisms, having special adaptations to a particular habitat or mode of life.

speciation: The process by which new species are formed.

species, *pl.* **species** [L., kind, sort]: A group of organisms that actually (or potentially) interbreed in nature and are reproductively isolated from all other such groups; a taxonomic grouping of anatomically similar individuals (the category below genus).

species-specific: Characteristic of (and limited to) a particular species.

specific: Unique; for example, the proteins in a given organism, the enzyme catalyzing a given reaction, or the antibody to a given antigen.

specific heat: The amount of heat (in calories) required to raise the temperature of 1 gram of a substance 1°C. The specific heat of water is 1 calorie per gram.

sperm [Gk. *sperma*, seed]: A mature male sex cell, or gamete, usually motile and smaller than the female gamete.

spermatid [Gk. *sperma*, seed]: Each of four haploid (*n*) cells resulting from the meiotic divisions of a spermatocyte; each spermatid becomes differentiated into a sperm cell.

spermatocytes [Gk. *sperma*, seed + *kytos*, vessel]: The diploid (2*n*) cells formed by the enlargement of the spermatogonia; they give rise by meiotic division to the spermatids.

spermatogonia [Gk. *sperma*, seed + *gonos*, a child, the young]: The unspecialized diploid (2*n*) cells on the walls of the testes that, by meiotic division, become spermatocytes, then spermatids, then sperm cells.

sphincter [Gk. *sphinktēr*, a band]: A circular muscle surrounding the opening of a tubular structure or the juncture of different regions of a tubular structure (e.g., the pyloric sphincter, at the juncture of the stomach and the small intestine); contraction of the sphincter closes the passageway, and relaxation opens it.

spinal cord: Part of the vertebrate central nervous system; consists of a thick, dorsal, longitudinal bundle of nerve fibers extending posteriorly from the brain.

spindle: In dividing cells, the structure formed of microtubules that extends from pole to pole; the spindle fibers appear to maneuver the chromosomes into position during metaphase and to pull the newly separated chromosomes toward the poles during anaphase.

spiracle [L. *spirare*, to breathe]: One of the external openings of the respiratory system in terrestrial arthropods.

splitting evolution: *See* Cladogenesis.

sporangiophore (spo-ran-ji-o-for) [Gk. *spora*, seed + *phore*, from *phorein*, to bear]: A specialized hypha or a branch bearing one or more sporangia.

sporangium, *pl.* **sporangia** [Gk. *spora*, seed]: A unicellular or multicellular structure in which spores are produced.

spore [Gk. *spora*, seed]: An asexual reproductive or resting cell capable of developing into a new organism without fusion with another cell; in contrast to a gamete.

sporophyll [Gk. *spora*, seed + *phyllon*, the leaves]: Spore-bearing leaf. The carpels and stamens of flowers are modified sporophylls.

sporophyte [Gk. *spora*, seed + *phytos*, growing]: In organisms that have alternation of haploid and diploid generations (all plants and some green algae), the diploid (2*n*) spore-producing generation.

stamen [L., a thread]: The male structure of a flower, which produces microspores or pollen; usually consists of a stalk, the filament, bearing a pollen-producing anther at its tip.

starch [M.E. *sterchen*, to stiffen]: A class of complex, insoluble carbohydrates, the chief food-storage substances of plants; composed of 1,000 or more glucose units and readily broken down enzymatically into these units.

statocyst [Gk. *statos*, standing + *kystis*, sac]: An organ of balance, consisting of a vesicle containing granules of sand (statoliths) or some other material that stimulates sensory cells when the organism moves.

stem: The aboveground part of the axis of vascular plants, as well as anatomically similar portions below ground (such as rhizomes).

stem cells: The common, self-regenerating cells in the marrow of long bones that give rise, by differentiation and division, to red blood cells and all of the different types of white blood cells.

stereoscopic vision [Gk. *stereōs*, solid + *optikos*, pertaining to the eye]: Ability to perceive a single, three-dimensional image from the simultaneous but separate images delivered to the brain by each eye.

steroid: One of a group of lipids having four linked carbon rings and, often, a hydrocarbon tail; cholesterol, sex hormones, and the hormones of the adrenal cortex are steroids.

stigma [Gk. *stigme*, a prick mark, puncture]: In plants, the region of a carpel serving as a receptive surface for pollen grains, which germinate on it.

stimulus [L., goad, incentive]: Any internal or external change or signal that influences the activity of an organism or of part of an organism.

stoma, *pl.* **stomata** [Gk., mouth]: A minute opening in the epidermis of leaves and stems, bordered by guard cells, through which gases pass.

strategy [Gk. *strategein*, to maneuver]: A group of related traits, evolved under the influence of natural selection, that solve particular problems encountered by living organisms; often includes anatomical, physiological, and behavioral characteristics.

striated muscle [L., from *striare*, to groove]: Skeletal voluntary muscle and cardiac muscle. The name derives from the striped appearance, which reflects the arrangement of contractile elements.

stroma [Gk., a bed, from *stronnymi*, to spread out]: A dense solution that fills the interior of the chloroplast and surrounds the thylakoids.

structural gene: Any gene that codes for a protein; in distinction to regulatory genes.

style [L. *stilus*, stake, stalk]: In angiosperms, the stalk of a carpel, down which the pollen tube grows.

substrate [L. *substratus*, strewn under]: (1) The foundation to which an organism is attached. (2) A substance on which an enzyme acts.

succession: *See* Ecological succession.

sucrose: Cane sugar; a common disaccharide found in many plants; a molecule of glucose linked to a molecule of fructose.

sugar: Any monosaccharide or disaccharide.

supergene: *See* Coadaptive gene complex.

surface tension: A tautness of the surface of a liquid, caused by the cohesion of the molecules of liquid. Water has an extremely high surface tension.

symbiosis [Gk. *syn*, together with + *bioonai*, to live]: An intimate and protracted association between two or more organisms of different species. Includes mutualism, in which the association is beneficial to both; commensalism, in which one benefits and the other is neither harmed nor benefited; and parasitism, in which one benefits and the other is harmed.

sympathetic division: A subdivision of the autonomic nervous system, with centers in the midportion of the spinal cord; slows digestion; generally excites other functions.

sympatric speciation [Gk. *syn*, together with + *patra*, fatherland, country]: Speciation that occurs without geographic isolation of a population of organisms; usually occurs as the result of hybridization accompanied by polyploidy; may occur in some cases as a result of disruptive selection.

synapse [Gk. *synapsis*, a union]: A specialized junction between two neurons where the activity in one influences the activity in another; may be chemical or electrical, excitatory or inhibitory.

syngamy (sin-gamy) [Gk. *syn*, with + *gamos*, a marriage]: The union of gametes in sexual reproduction; fertilization.

synthesis [Gk. *syntheke*, a putting together]: The formation of a more complex substance from simpler ones.

synthetic theory: The currently prevailing theory of the mechanism of evolutionary change; combines the Darwinian two-step model of variation and selection with the principles of Mendelian genetics.

systematics [Gk. *systema*, that which is put together]: Scientific study of the kinds and diversity of organisms and of the relationships among them.

T lymphocyte: A type of white blood cell arising from precursors in the thymus gland and, upon maturation, involved in cell-mediated immunity and interactions with B lymphocytes; a T cell.

tagmosis [Gk. *tagma*, arrangement, order + *-osis*, process]: The formation of groups of segments (metameres) into body regions (tagmata) with functional differences.

taxon, *pl.* **taxa** [Gk. *taxis*, arrange, put in order]: A particular group, ranked at a particular categorical level, in a hierarchical classification scheme; for example, *Drosophila* is a taxon at the categorical level of genus.

taxonomy [Gk. *taxis*, arrange, put in order + *nomos*, law]: The study of the classification of organisms; the ordering of organisms into a hierarchy that reflects their essential similarities and differences.

telencephalon [Gk. *tēl*, far off + *enkephalos*, brain]: One of the two principal subdivisions of the vertebrate forebrain; the anterior portion of the forebrain, it contains the cerebrum and the olfactory bulbs.

telophase [Gk. *telos*, end + *phasis*, form]: The last stage in mitosis and meiosis, during which the chromosomes become reorganized into two new nuclei.

temperate bacteriophage: A bacterial virus that may become incorporated into the host-cell chromosome.

template: A pattern or mold guiding the formation of a negative or complement.

tentacles [L. *tentare*, to touch]: Long, flexible protrusions located about the mouth of many invertebrates; usually prehensile or tactile.

territory: An area or space occupied and defended by an individual or a group; trespassers are attacked (and usually defeated); may be the site of breeding, nesting, food gathering, or any combination thereof.

tertiary structure of a protein: A complex structure, usually globular, resulting from further folding of the secondary structure of a protein;

forms spontaneously due to attractions and repulsions among amino acids with different charges on their R groups.

testcross: A mating between a phenotypically dominant individual and a homozygous recessive "tester" to determine the genetic constitution of the dominant phenotype, that is, whether it is homozygous or heterozygous for the relevant gene.

testis, *pl.* **testes** [L., witness]: The sperm-producing organ; also the source of the male sex hormone testosterone.

testosterone [Gk. *testis,* testicle + *steiras,* barren]: A steroid hormone secreted by the testes in higher vertebrates and stimulating the development and maintenance of male sex characteristics and the production of sperm; the principal androgen.

tetrad [Gk. *tetras,* four]: In genetics, a pair of homologous chromosomes that have replicated and come together in prophase I of meiosis; consists of four chromatids.

thalamus [Gk. *thalamos,* chamber]: A part of the vertebrate forebrain just posterior to and tucked below the cerebrum; the main relay center between the brainstem and the higher brain centers.

thallus [Gk. *thallos,* a young twig]: A simple plant or algal body without true roots, leaves, or stems.

theory [Gk. *theorein,* to look at]: A generalization based on many observations and experiments; a verified hypothesis.

thermodynamics [Gk. *therme,* heat + *dynamis,* power]: The study of transformations of energy. The first law of thermodynamics states that, in all processes, the total energy of a system plus its surroundings remains constant. The second law states that all natural processes tend to proceed in such a direction that the disorder or randomness of the system increases.

thorax [Gk., breastplate]: (1) In vertebrates, that portion of the trunk containing the heart and lungs. (2) In crustaceans and insects, the fused, leg-bearing segments between head and abdomen.

thylakoid [Gk. *thylakos,* a small bag]: A flattened sac, or vesicle, that forms part of the internal membrane structure of the chloroplast; the site of the light-trapping reactions of photosynthesis and of photophosphorylation; stacks of thylakoids collectively form the grana.

thyroid [Gk. *thyra,* a door]: An endocrine gland of vertebrates, located in the neck; source of an iodine-containing hormone (thyroxine) that increases the metabolic rate and affects growth.

tight junction: A junction between adjacent animal cells that prevents materials from leaking through the tissue; for example, intestinal epithelial cells are surrounded by tight junctions.

tissue [L. *texere,* to weave]: A group of similar cells organized into a structural and functional unit.

tonoplast [Gk. *tonos,* stretching, tension + *plastos,* formed, molded]: In plant cells, the membrane surrounding the vacuole.

trachea, *pl.* **tracheae (trake-ee-a)** [Gk. *tracheia,* rough]: An air-conducting tube. (1) In insects and some other terrestrial arthropods, a system of chitin-lined air ducts. (2) In terrestrial vertebrates, the windpipe.

tracheid (tray-key-idd) [Gk. *tracheia,* rough]: In vascular plants, an elongated, thick-walled conducting and supporting cell of xylem, characterized by tapering ends and pitted walls without true perforations.

tract: A group or bundle of nerve fibers with accompanying connective tissue, located within the central nervous system.

transcription [L. *trans,* across + *scribere,* to write]: The enzymatic process by which the genetic information contained in one strand of DNA is used to specify a complementary sequence of bases in an RNA molecule.

transduction [L. *trans,* across + *ducere,* to lead]: (1) The transfer of genetic material (DNA) from one cell to another by a virus. (2) The conversion of one form of energy into another form of energy; for example, the conversion of the energy of a chemical stimulus into the energy of an action potential.

transfer RNA (tRNA) [L. *trans,* across + *ferre,* to bear or carry]: A class of small RNAs (about 80 nucleotides each) with two functional sites; one recognizes a specific activated amino acid; the other carries the nucleotide triplet (anticodon) for that amino acid. Each type of tRNA accepts a specific activated amino acid and transfers it to a growing polypeptide chain as specified by the nucleotide sequence of the mRNA being translated.

transformation [L. *trans,* across + *formare,* to shape]: A genetic change produced by the incorporation into a cell of DNA from the external medium.

translation [L. *trans,* across + *latus,* that which is carried]: The process by which the genetic information present in a strand of mRNA directs the sequence of amino acids during protein synthesis.

translocation [L. *trans,* across + *locare,* to put or place]: (1) In plants, the transport of the products of photosynthesis from a leaf to another part of the plant. (2) In genetics, the breaking off of a piece of chromosome with its reattachment to a nonhomologous chromosome.

transpiration [L. *trans,* across + *spirare,* to breathe]: In plants, the loss of water vapor from the stomata.

transposon [L. *transponere,* to change the position of]: A DNA sequence carrying one or more genes that is capable of moving from one location in the chromosomes to another. Simple transposons, also known as insertion sequences, carry only the genes essential for transposition; complex transposons carry genes that code for additional proteins.

tritium: A radioactive isotope (^{3}H) of hydrogen with a half-life of 12.5 years.

trophic level [Gk. *trophos,* feeder]: The position of a species in the food web or chain, that is, its feeding level; a step in the movement of biomass or energy through an ecosystem.

trophoblast [Gk. *trophos,* feeder + *blastos,* sprout]: In the early mammalian embryo (the blastocyst), a double layer of cells that surrounds the inner cell mass and subsequently gives rise to the chorion.

tropic [Gk. *trope,* a turning]: Pertaining to behavior or action brought about by specific stimuli, for example, phototropic ("light-oriented") motion, gonadotropic ("stimulating the gonads") hormone.

tuber [L. *tuber,* bump, swelling]: A much-enlarged, short, fleshy underground stem, such as that of the potato.

turgor [L. *turgere,* to swell]: The pressure exerted on the inside of a plant cell wall by the fluid contents of the cell; the interior of the cell is hypertonic in relation to the fluids surrounding it and so gains water by osmosis.

urea [Gk. *ouron,* urine]: An organic compound formed in the vertebrate liver; principal form of disposal of nitrogenous wastes by mammals.

ureter [Gk. from *ourein,* to urinate]: The tube carrying urine from the kidney to the cloaca (in reptiles and birds) or to the bladder (in amphibians and mammals).

urethra [Gk. from *ourein,* to urinate]: The tube carrying urine from the bladder to the exterior of mammals.

uric acid [Gk. *ouron,* urine]: An insoluble nitrogenous waste product that is the principal excretory product in birds, reptiles, and insects.

urine [Gk. *ouron,* urine]: The liquid waste filtered from the blood by the kidney and stored in the bladder pending elimination through the urethra.

uterine tube: *See* Oviduct.

uterus [L., womb]: The muscular, expanded portion of the female reproductive tract modified for the storage of eggs or for housing and nourishing the developing embryo.

vacuole [L. *vacuus,* empty]: A membrane-bound, fluid-filled sac within the cytoplasm of a cell.

vagus nerve [L. *vagus,* wandering]: A nerve arising from the medulla of the vertebrate brain that innnervates the heart and visceral organs; carries parasympathetic fibers.

vaporization [L. *vapor,* steam]: The change from a liquid to a gas; evaporation.

vascular [L. *vasculum,* a small vessel]: Containing or concerning vessels that conduct fluid.

vascular bundle: In plants, a group of longitudinal supporting and conducting tissues (xylem and phloem).

vascular cambium [L. *vasculum,* a small vessel + *cambium,* exchange]: In plants, a cylindrical sheath of meristematic cells that divide mitotically, producing secondary phloem to one side and secondary xylem to the other, but always with a cambial cell remaining.

vas deferens, *pl.* **vasa deferentia** (vass **deff**-er-ens) [L. *vas,* a vessel + *deferre,* to carry down]: In mammals, the tube carrying sperm from a testis to the urethra.

vector [L., carrier]: In recombinant DNA, a small, self-replicating DNA molecule, or a portion thereof, into which a DNA segment can be spliced and introduced into a cell; generally a plasmid, a bacteriophage, or a cosmid.

vein [L. *vena,* a blood vessel]: (1) In plants, a vascular bundle forming part of the framework of the conducting and supporting tissue of a leaf. (2) In animals, a blood vessel carrying blood from the tissues to the heart. A small vein is known as a venule.

vena cava (vee-na **cah**-va) [L., blood vessel + hollow]: A large vein that brings blood from the tissues to the right atrium of the four-chambered mammalian heart. The superior vena cava collects blood from the forelimbs, head, and anterior or upper trunk; the inferior vena cava collects blood from the posterior body region.

ventral [L. *venter,* belly]: Pertaining to the undersurface of an animal that holds its body in a horizontal position; to the front surface of an animal that holds its body erect.

ventricle [L. *ventriculus,* the stomach]: A muscular chamber of the heart that receives blood from an atrium and pumps blood out of the heart, either to the lungs or to the body tissues.

vertebral column [L. *vertebra,* joint]: The backbone; in nearly all vertebrates, it forms the supporting axis of the body and protects the spinal cord.

vesicle [L. *vesicula,* a little bladder]: A small, intracellular membrane-bound sac.

vessel [L. *vas,* a vessel]: A tubelike element of the xylem of angiosperms; composed of dead cells (vessel members) arranged end to end. Its function is to conduct water and minerals from the soil.

viable [L. *vita,* life]: Able to live.

villus, *pl.* **villi** [L., a tuft of hair]: In vertebrates, one of the minute, fingerlike projections lining the small intestine that serve to increase the absorptive surface area of the intestine.

virus [L., slimy, liquid, poison]: A submicroscopic, noncellular particle composed of a nucleic acid core and a protein coat; parasitic; reproduces only within a host cell.

viscera [L., internal organs]: The collective term for the internal organs of an animal.

vitamin [L. *vita,* life]: Any of a number of unrelated organic substances that cannot be synthesized by a particular organism and are essential in minute quantities for normal growth and function.

water cycle: Worldwide circulation of water molecules, powered by the sun. Water evaporates from oceans, lakes, rivers, and, in smaller amounts, soil surfaces and bodies of organisms; water returns to the earth in the form of rain and snow. Of the water falling on land, some flows into rivers that pour water back into the oceans and some percolates down through the soil until it reaches a zone where all pores and cracks in the rock are filled with water (groundwater); the deep groundwater eventually reaches the oceans, completing the cycle.

water potential: The potential energy of water molecules; regardless of the reason (e.g., gravity, pressure, concentration of solute particles) for the water potential, water moves from a region where water potential is greater to a region where water potential is lower.

wild type: In genetics, the phenotype that is characteristic of the vast majority of individuals of a species in a natural environment.

worker: A member of the nonreproductive laboring caste of social insects.

xanthophyll [Gk. *xanthos,* yellow + *phyllon,* leaf]: In algae and plants, one of a group of yellow pigments; a member of the carotenoid group.

xylem [Gk. *xylon,* wood]: A complex vascular tissue through which most of the water and minerals are conducted from the roots to other parts of the plant; consists of tracheids or vessel members, parenchyma cells, and fibers; constitutes the wood of trees and shrubs.

yolk: The stored food in egg cells that nourishes the embryo.

yolk sac: In developing reptiles and birds, the extraembryonic membrane that surrounds and encloses the yolk; performs a nutritive function. In mammals, the extraembryonic membrane in which the germ cells are set aside very early in development.

zoology [Gk. *zoe,* life + *logos,* a discourse]: The study of animals.

zooplankton [Gk. *zoe,* life + *plankton,* wanderer]: A collective term for the nonphotosynthetic organisms present in plankton.

zygote (zi-goat) [Gk. *zygon,* yolk, pair]: The diploid (2*n*) cell resulting from the fusion of male and female gametes (fertilization); a zygote may either develop into a diploid individual by mitotic divisions or may undergo meiosis to form haploid (*n*) individuals that divide mitotically to form a population of cells.

Illustration Acknowledgments

Illustration Acknowledgments

Page ix © Rita Summers/Colorado Nature Photographic Studio; **Page xi** *(top)* © Stephen Dalton/NHPA; *(bottom, left to right)* © Jen & Des Bartlett/Bruce Coleman; © Michael Medford/Wheeler Pictures; Carolina Biological Supply Company; **Page xii** *(left to right)* © Larry West; © Grant Heilman Photography; © Paul W. Johnson/Biological Photo Service; **Page xiii** *(left to right)* Ripon Microslides; Nelson Max & Richard Dickerson; © Tony Mendoza/The Picture Cube; **Page xv** *(left to right)* © Brian Parker/Tom Stack & Associates; © Kim Taylor/Bruce Coleman Ltd.; © Larry West; **Page xvii** © E. R. Degginger/Earth Scenes; © James L. Castner; © Larry West; **Page xviii** *(left to right)* Lennart Nilsson, THE BODY VICTORIOUS. New York: Delacorte Press. Boehringer Ingelheim International GmbH; © E. R. Degginger/Animals Animals; Antone G. Jacobson; **Page xx** *(left to right)* © Frans Lanting; © Jen & Des Bartlett/Bruce Coleman; © Jim Brandenburg; **Page xxii** *(left to right)* M. A. Chappell/Animals Animals; © James L. Castner; © Wolfgang Kaehler

I-1 © Francisco Erize/Bruce Coleman; **Page 1** © Raymond A. Mendez/Animals Animals; **I-2** The Royal College of Surgeons of England; **I-3** © Chip & Rosa Maria Peterson; **I-4** (a), (b) © J. Fennell/Bruce Coleman; (c) © W. H. Hodge/Peter Arnold; **I-5** American Museum of Natural History; **I-6** © Breck P. Kent/Animals Animals; **I-7** (a) Christopher Ralling; (b) Medical Illustration Unit, The Royal College of Surgeons of England; **I-9** © Frans Lanting/Bruce Coleman Ltd.; **I-10** (a) The Granger Collection; (b) Ann Ronan Picture Library; **Page 8** The Royal College of Surgeons of England; **I-11** (a) Field Museum, Photo Researchers; (b) © S. Robinson/NHPA; **I-12** (a) Rare Books Division, New York Public Library; (b) John Mais; **I-13** The Bettmann Archive; **I-14** © Laura Riley/Bruce Coleman; **I-15** (a) © Eric V. Gravé; (b) © J. Robert Waaland/Biological Photo Service; **I-16** © Larry West; **I-17** © Bruce Coleman; **I-18** © E. S. Ross; **I-19** Terry Erwin & Linda Sims, Smithsonian Institution; **I-20** (a) © Clem Haagner/Bruce Coleman; (b) © Raymond A. Mendez/Animals Animals

Page 20 National Optical Astronomy Observatories; **Page 21 & 1-1** © Sea Studio, Inc./Peter Arnold; **Page 23** © Jen & Des Bartlett/Bruce Coleman; **1-2** © John Cancalosi; **Page 26** (a) © John D. Cunningham/Visuals Unlimited; (b) © M. Walker/NHPA; (c) © Mary M. Thacher/Photo Researchers; **Page 27** (d) © Mitch Reardon/Photo Researchers; (e) © Jeff Foott; (f) © Dwayne M. Reed; (g) © Stephen Dalton/Photo Researchers; **1-5** © Bruce Coleman; **1-7** © Charles M. Falco/Science Source, Photo Researchers; **1-8** © Jen & Des Bartlett/Bruce Coleman; **1-13** © Runk & Schoenberger/Grant Heilman Photography; **1-14** H. Berg & G. Forté; **1-15** (a) © Frieder Sauer/Bruce Coleman; (b) George I. Schwartz; (c) © Biology Media/Photo Researchers; (d) © Manfred Kage/Peter Arnold

Page 40 © Michael Medford/Wheeler Pictures; **2-1** © E. R. Degginger/Bruce Coleman; **2-4** Fritz Polking; **2-5** © Runk & Shoenberger/Grant Heilman Photography; **2-7** (b) © Brian Milne/Earth Scenes; **Page 51** (b) © John D. Cunningham/Visuals Unlimited; (c) © Michael Medford/Wheeler Pictures; **2-11** Jeanne M. Riddle

Page 55 © Herbert B. Parsons/Photo NATS; **3-1** © E. R. Degginger/Earth Scenes; **3-3** After DuPraw, E. J. (1968). *Cell and molecular biology.* New York: Academic Press, Inc.; **Page 58** John M. Sieburth; **3-7** © Charles & Elizabeth Schwartz/Animals Animals; **3-9** (a), (b), (c) After Lehninger, A. L. (1975). *Biochemistry,* 2d ed. New York: Worth Publishers, Inc.; (d)

L. M. Biedler; (e) J. C. Warren; **3-10** (c) R. D. Preston; **3-11** (b) © Herbert B. Parsons/Photo NATS; **3-13** © Caroline Kroeger/Animals Animals; **3-16** B. E. Juniper; **Page 71** (a), (b) © Sloop-Ober/Visuals Unlimited; **3-20** Sequence information from Lehninger, A. L. (1982). *Principles of biochemistry* (p. 135). New York: Worth Publishers, Inc.; **3-21** (b) After Wilson, E. O., *et al.* (1977). *Life, cells, organisms, populations.* Sunderland, MA: Sinauer Associates, Inc.; **3-22** (b) Computer graphics modeling and photography by Arthur J. Olson, Ph.D., Research Institute of Scripps Clinic, La Jolla, CA 92037, © 1988; **3-23** (a) After Alberts, B., Bray, D., Lewis, J., Raff, M., Roberts, K. & Watson, J. D. (1983). *Molecular biology of the cell.* New York: Garland Publishing Company; (b) Daniel Friend; **3-24** (a) © Anthony Bannister/NHPA; (b) © Robert L. Dunne/Bruce Coleman; **3-25** (a) After Karp, G. (1979). *Cell biology.* New York: McGraw-Hill Book Company; (b) © Manfred Kage/Peter Arnold; **3-27** Adapted from Dickerson, R. E. & Geis, I. (1969). *The structure and action of proteins.* Menlo Park, CA: W. A. Benjamin, Inc. Copyright 1969 by Dickerson & Geis; **3-28** (a), (b) Margaret Clark; **Page 83** After Lehninger, A. L. (1982). *op. cit.*

Page 84 © David M. Phillips; **4-1** Brent McCown; **4-2** Big Bear Solar Observatory; **Page 86** Pasteur Institute & The Rockefeller University Press; **4-4** © S. Johannson & Frank Lane/Bruce Coleman; **4-5** Sidney W. Fox; **4-6** © S. M. Awramik/Biological Photo Service; **4-8** (b) A. Ryter; **4-9** (b) Lang, N. J. (1965). *Journal of Phycology,* 1, 127–134; **4-10** (b) George Palade; **4-12** (b) Michael A. Walsh; **4-13** (b) Keith Porter; **4-14** After Alberts, *et al.* (1983). *op. cit.;* **4-15** (a), (b), (c) David M. Phillips; **4-17** After Alberts, *et al.* (1983). *op. cit.;* **4-18** *Ibid;* **4-19** (a)–(d) Keith Roberts & James Barnett

Page 102 Osborn, M. (October 1985). The molecules of life. *Scientific American;* **5-2** (a) © Eric V. Gravé/Photo Researchers; (b) © M. Schliwa/Visuals Unlimited; **5-3** © J. Robert Waaland/Biological Photo Service; **5-4** J. D. Robertson; **5-6** Adapted in part from *Scientific American* (February 1984), p. 81, and in part from Darnell, J., Lodisch, H. & Baltimore, D. (1986). *Molecular cell biology.* New York: W. H. Freeman and Company; **5-7** (a, *photo*) Myron C. Ledbetter; (b) After Albershamm, P. (April 1975). *Scientific American;* **5-8** (a) Daniel Friend; (b) Nigel Unwin; (c) Barbara J. Stevens & Hewson Swift; **5-9** Ursula Goodenough; **5-10** (a) © Doug Wechsler; (b) Mia Tegner & David Epel; **5-13** Osborn, M. (October 1985). The molecules of life. *Scientific American;* **5-14** (a), (b), (c) *Ibid;* **Page 114** Adapted from Darnell, *et al.* (1986). *op. cit.;* **5-17** Peter Webster; **5-18** (a), (b) Don Fawcett; **5-19** Don Fawcett/Photo Researchers; **5-20** (b) Flickinger, C. J. (1975). *Journal of Cell Biology,* 49, 221.; **5-22** Birgit Satir; **5-23** (a) Don Fawcett; (b) G. Decker; **5-24** (b) Keith Porter; **5-25** (a) Roland R. Dute; (b) Myron C. Ledbetter; **5-26** David Stetler; **5-28** © Manfred Kage/Peter Arnold; **5-29** Don Fawcett; **5-30** Gregory Antipa; **5-31** (b) Peter Satir; **5-32** © David M. Phillips/Visuals Unlimited

Page 127 © M. I. Walker/NHPA; **6-2** Keith R. Porter; **6-3** © Frederick J. Dodd/Peter Arnold; **6-5** (a) G. M. Hughes; (b) After Schmidt-Nielsen, K. (1979). *Animal physiology,* 2d ed. Cambridge: Cambridge University Press; **6-6** (a) © M. I. Walker/NHPA; (b), (c) © Thomas Eisner; **6-7** After Lehninger, A. L. (1975). *Biochemistry,* 2d ed. New York: Worth Publishers, Inc.; **6-9** (b) Daniel Branton; **6-10** Adapted from Raven, P. E., Eichhorn, S. E. & Evert, R. F. (1986). *Biology of plants,* 4th ed. (Fig. 4-9). New York: Worth Publishers, Inc.; **6-11** Adapted from Alberts, *et al.*

(1983). *Molecular biology of the cell* (p. 289). New York: Garland Publishing Company; **6-12** Keith R. Porter/Photo Researchers; **6-14** Adapted from Raven, *et. al.* (1986). *op. cit.*; **6-15** Adapted from *Scientific American* (May 1984), p. 54; **6-16** (a)-(d) Gregory Antipa; **6-17** (a)-(d) Perry, M. M. & Gilbert, A. B. (1979). *Journal of Cell Science*, 39, 257–272; **6-18** (a) Ray F. Evert; (b) Peter K. Hepler; **Page 141** (a), (b), (c) K. T. Raper; (d)-(g) © David Scharf/Peter Arnold; **6-19** (a) N. Bernard Gilula; (b) Adapted in part from Darnell *et al.* (1986). *Molecular cell biology* (Fig. 14-63). New York: W. H. Freeman and Company; in part from *Scientific American* (May 1978), p. 150; and in part from *Scientific American* (October 1985), p. 106.

Page 144 Carolina Biological Supply Company; **7-1** John Mais; **7-3** L. P. Wisniewski & K. Hirschhorn; **7-5** © David M. Phillips/Visuals Unlimited; **7-6** Carolina Biological Supply Company; **7-7** After Alberts, *et al.* (1983). *Molecular biology of the cell* (Fig. 11-11, p. 619). New York: Garland Publishing Company; **7-8** (a) Adapted from Alberts, *et al., op. cit.* (Fig. 11-47, p. 652); (b) M. J. Schibler; **7-9** (a) Andrew S. Bajer; (b), (c) Adapted from Alberts, *et al., op. cit.* (Fig. 11-48, p. 652); **7-10, 7-11,** and **7-12** Andrew S. Bajer; **7-13** (a)-(d) Carolina Biological Supply Company; **7-14** (a), (b) Beams, H. W. & Kessel, R. G. (1976). *American Scientist*, 64, 279; **7-15** James Cronshaw

Page 158 © Grant Heilman Photography; **Page 159 & 8-1** © Bruce Coleman; **Page 161** © Larry West; **8-2** © Zig Leszczynski/Animals Animals; **8-3** ©Alain Eurard/Photo Researchers; **Page 163** Lotte Jacobi; **8-5** (a)-(d) Miami Seaquarium; **8-7** Jeremy Pickett-Heaps; **8-8** After Lehninger, A. L. (1975). *Biochemistry*, 2d ed., New York: Worth Publishers, Inc.; **8-10** (a) *Ibid*; (b) William Goddard III; **8-16** Adapted from Lehninger, A. L. (1982). *Principles of biochemistry* (pp. 212 & 219). New York: Worth Publishers, Inc.; **8-18** V. Lennard; **8-19** After Watson, J. D., *et al.* (1970). *Molecular biology of the gene*, 2d ed., Menlo Park, CA: The Benjamin/Cummings Publishing Company; **8-22** After Lehninger (1975). *op. cit.*; **Page 179** (a) © Robert Pearcy/Animals Animals; (b) © Bill Curtsinger/Rapho, Photo Researchers; **8-24** (a) © Larry West; (b) © M. P. Price/Bruce Coleman; **Page 185** © R. D. Estes

Page 186 © Grant Heilman Photography; **9-1** (a) © Bob Evans/Peter Arnold; (b) Gary Robinson; (c) Bray, R. (1978). *Science*, 200, 333–334, © 1978 by AAAS; **9-5** (b) © Grant Heilman Photography; **9-7** Adapted from Darnell, *et al.* (1986). *Molecular cell biology*. New York: W. H. Freeman and Company; **Page 193** R. H. Kirschner; **9-8** (b) Lester J. Reed from Boyer, P. D., ed. (1970). *The enzymes, Vol. 1*. New York: Academic Press, Inc., **9-9** Adapted from Vander, A. J., Sherman, J. H. & Luciano, D. (1969). *Human physiology*. New York: McGraw-Hill Book Company; **9-10** After Lehninger, A. L. (1975). *Biochemistry*, 2d ed. New York: Worth Publishers, Inc.; **9-13** After Takano, T., Kallai, O. B., Swanson, R. and Dickerson, R. E. (1973). *Journal of biological chemistry*, 248, 5244; **9-14** Adapted from Lehninger, A. L. (1982). *Principles of biochemistry*. New York: Worth Publishers, Inc.; **9-15** After Alberts, B., *et al.* (1983). *Molecular biology of the cell*. New York: Garland Publishing Company; **9-16** (a) After Lehninger, A. L. (1982). *op. cit.*; (b) John N. Telford; **9-18** After Lehninger, A. L. (1975). *op. cit.*; **Page 202** (a), (b) Lieber, C. S. (March 1976). The metabolism of alcohol. *Scientific American*; **9-21** After Lehninger, A. L. (1975). *op. cit.*

Page 206 © Paul W. Johnson & J. McN. Sieburth/Biological Photo Service; **10-1** © J. Metzner/Peter Arnold; **10-4** (a), (b) Pearse, V. & Buchsbaum, R. (1987). *Living invertebrates*. Pacific Grove, CA: The Boxwood Press; photo by Karl J. Marschall; **10-8** Micrograph by Oxford Scientific Films/Bruce Coleman; **10-10** (a) A. D. Greenwood; (b) L. K. Shumway; **10-13** After Lehninger, A. L. (1975). *Biochemistry*, 2d ed. New York: Worth Publishers, Inc.; **Page 217** © Paul W. Johnson & J. McN. Sieburth/Biological Photo Service; **Page 221** Stoeckenius, W. (June 1976). The purple membrane of salt-loving bacteria. *Scientific American*; **10-17** Pallard, S. G. & Kozlowski, T. T. (1980). *New Phytologist*, 85, 363-368; **10-21** Ray F. Evert

Page 232 Edward Hicks, *Noah's Ark* (detail), The Philadelphia Museum of Art: Bequest of Lisa Norris Elkins; **Page 233 & 11-1** Computer Graphics Laboratory, University of California, San Francisco. © Regents of the University of California; **Page 235** Courtesy LKB Productions, Sweden; **11-2** (a), (b), (c) The Granger Collection; **11-3** The Bettmann

Archive; **11-4** The Granger Collection; **11-6** Adapted from von Frisch, K. (1964). *Biology.* translated by Jane Oppenheimer. New York: Harper & Row Publishers, Inc.; **11-12** Bill Ratcliffe; **11-13** (a) Dr. V. Orel, The Moravian Museum; (b) Courtesy LKB Productions, Sweden

Page 249 Ripon Microslides; **12-1** (a) © John Bova/Photo Researchers; (b) Arnold Sparrow, Brookhaven National Laboratory; **12-9** (a), (b) William Marks; **12-10** (a), (b) Mary E. Clutter; **12-12** After DuPraw, E. J. (1968). *Cell and molecular biology*. New York: Academic Press, Inc.; **Page 259** After Moore, J. L. (1963). *Heredity and development*. New York: Oxford University Press; **12-14** B. John; **Page 262** Ripon Microslides

Page 263 Ralph G. Somes, University of Connecticut; **13-1** Columbiana Collection, Rare Book and Manuscript Library, Columbia University; **13-4** © Jeremy Burgess/Science Photo Library, Photo Researchers; **Page 267** © George F. Godfrey/Animals Animals; **13-7** © John Chiasson/Gamma-Liaison; **13-9** (a) © Richard Kolar/Animals Animals; (b) © Grant Heilman Photography; (c), (d) © Jane Burton/Bruce Coleman; **13-10** (a)-(d) Ralph G. Somes, University of Connecticut **13-11** After Ayala, F. J. & Kiger, J. A., (1980). *Modern genetics*. Menlo Park, CA: The Benjamin/Cummings Publishing Company; **13-13** (a) After E. D. Merrell, 1964; (b) *Journal of Heredity*, 15 (1914); **13-14** Photograph by F. B. Hutt (1930). *Journal of Genetics*, 22, 126; **13-15** B. John; **13-20** Beth Myers, courtesy of William Marks

Page 281 Nelson Max & Richard Dickerson; **14-1** Kleinschmidt, A. K., Land, D., Jacherts, D. & Zahn, R. K. (1962). *Biochemica Biophysica Acta*, 61, 857–864; **14-2** (a), (b) © Bruce Iverson; **14-3** After Koob, D. D. & Bogs, W. E. (1972). *The nature of life*. Reading, MA: Addison-Wesley Publishing Co.; **14-5** From Cairns, J., Stent, G. S., & Watson, J. D., eds. (1966). *Phage and the origins of molecular biology*. Cold Spring Harbor, NY: Cold Spring Harbor Laboratory of Quantitative Biology; **14-6** M. Wurtz, Biozentrum, University of Basel/Science Photo Library, Photo Researchers; **14-8** ©Lee D. Simon/Science Photo Library, Photo Researchers; **14-9** (a) From Watson, J. D. (1968). *The double helix*. New York: Atheneum Publishers; (b) Vittorio Luzzati; **14-10** (b) Nelson Max & Richard Dickerson; **Page 291** Watson (1968). *op. cit.*; **14-13** After Lehninger, A. L. (1975). *Biochemistry*, 2d ed. New York: Worth Publishers, Inc.; **14-14** *Ibid*; **14-16** (a-d) Bernhard Hirt; **14-17** Blumenthal, A. B., Kreigstein, H. J. & Hogness, D. S. (1973). *Cold Spring Harbor Symposium on Quantitative Biology*, 38, 205; **Page 300** C. M. Plork, Museum of Comparative Zoology, Harvard University

Page 301 © K. G. Murti/Visuals Unlimited; **15-1** © Gary R. Robinson/Visuals Unlimited; **15-6** (b) © K. G. Murti/Visuals Unlimited; **15-14** Hans Ris; **15-15** (b) Miller, O. L., Hamkalo, B. A. & Thomas, C. A. (1970). *Science*, 69, 392–395. © 1970 by AAAS

Page 319 Paris Match; **16-1** Jack Griffith; **16-5** Adapted from Darnell, J., Lodish, H. & Baltimore, G. (1986). *Molecular cell biology* (p. 278). New York: Scientific American Books; **16-6** Paris Match; **16-11** (a) Palchaudhuri, S., Bell, E. & Salton, M. R. J. (1975). *Infection and Immunity*, 11, 1141; (b), (c), (d) T. Kakefuda; **16-12** Judith Carnahan & Charles Brinton, Jr.; **16-13** After Ayala, F. J. & Kiger, J. A., (1980). *Modern genetics*. Menlo Park, CA: Benjamin/Cummings Publishing Co., Inc.; **16-14** *Ibid*; **16-15** Cohen, S. N. (December 27, 1969). *Nature*; **16-16** (a)-(d) B. Menge, J. V. D. Brock, H. Wunderli, K. Lickfield, M. Wurtz & E. Kellenberger; **16-20** Based on Alberts, *et al.* (1983). *Molecular biology of the cell*. New York: Garland Publishing Company; **16-22** Watson, *et al.,* (1987). *Molecular biology of the gene* (p. 335). Menlo Park, CA: Benjamin/Cummings Publishing Co., Inc.; **16-23** Stanley Falkow

Page 340 Eli Lilly and Co.; **17-1** John C. Fiddes & Howard M. Goodman; **17-2** Adapted from Alberts, *et al.* (1983). *Molecular biology of the cell*. New York: Garland Publishing Company; **17-3** Adapted from Lehninger, A. L. (1982). *Principles of biochemistry* (p. 863, Fig. 28-4). New York: Worth Publishers, Inc.; **17-5** Adapted from Watson, J. D., Tooze, J. & Durtz, D. T. (1983). *Recombinant DNA* (p. 64, Fig. 5-5). New York: W. H. Freeman and Company; **17-6** (a) Stanley N. Cohen; **17-7** (a), (b), (c) Huntington Potter & David Dressler, *Life Magazine* © 1980 Time Inc.; **17-8** Adapted from Fristrom, J. W. & Spieth, P. T. (1980). *Principles of genetics* (p. 346, Fig. 12-20). New York & Concord: Chiron Press; **17-9** Adapted from Alberts, *et al.* (1983). *Molecular biology of the cell* (Fig.

4-56). New York: Garland Publishing Company; **17-10** (a) Jack Griffith; (b), (c) Adapted from Darnell, *et al.* (1986). *Molecular cell biology* (p. 247, Fig. 17-22). New York: W. H. Freeman and Company; (d) Daniel Nathans; **17-11** After Lehninger, A. L. (1982). *op. cit.;* **17-12** After *City of Hope Quarterly, 7,* 2. (Winter, 1978); **17-13** Eli Lilly and Co.; **17-14** Eugene W. Nester; **17-15** Keith Wood, University of California, San Diego

Page 355 Robert Noonan; **18-1** E. J. DuPraw; **18-2** Rich, A., *et al.* (1981). *Science, 211,* 171–176. © 1981 by AAAS; **18-3** (a) Victoria Foe; **18-4** (a) Barbara Hamkalo; (b), (c) Adapted from Alberts, *et al.* (1983). *Molecular biology of the cell* (Fig. 8-5). New York: Garland Publishing Company; **18-5** Victoria Foe; **18-6** After Alberts, *et al.* (1983). *op. cit.;* **18-8** James German; **18-9** George T. Rudkin; **18-10** (*photo*) James German; (*art*) After Chambon, P. (May 1981). Split genes. *Scientific American,* 60–71; **18-11** After Lewin, R. (1981). *Science, 212,* 28–32; **18-12** H. C. MacGregor; **18-13** (a), (b) Ullrich Scheer, W. W. Franke & M. F. Trendelenberg; **Page 366** Don Fawcett; **18-14** After Rahbar, S. (Winter 1982). Abnormalities of human hemoglobin. *City of Hope Quarterly, 11,* 2.; **18-15** After Goodenough, U. (1983). *Genetics,* 3rd ed. New York: Saunders College/Holt, Rinehart & Winston; **Page 370** Grabowski, P. J. & Cech, T. R. (1981). *Cell, 23,* 467–476; **18-18** After Lehninger, A. L. (1982). *Principles of biochemistry.* New York: Worth Publishers, Inc.; **18-19** *Ibid;* **18-20** UPI/Bettmann Newsphotos; **18-21** Adapted from Darnell, *et al.* (1986). *Molecular cell biology.* New York: W. H. Freeman and Company; **Page 375** Dr. Mary Clutter, Cold Spring Harbor Laboratory Research Archives; **18-22** (a), (b) R. D. Goldman; **18-23** (a) after Jon W. Gordon & Frank Ruddle; (b) Jon W. Gordon & Frank Ruddle; **18-24** R. L. Brinster; **18-25** Robert Noonan; **Page 381** R. Portman & M. L. Birnsteil

Page 382 © Tony Mendoza/The Picture Cube; **19-1** Gernsheim Collection, Humanities Research Center, University of Texas at Austin;

19-3 After Yunis, J. J. (1976). *Science, 191,* 1268-1270. © 1976 by AAAS; **19-4** (a) © Tony Mendoza/The Picture Cube; **19-7** (a), (b) Jorge J. Yunis; **19-8** (b) Courtesy Nell Ubbelohde & William Stryk; **19-9** After Stryer, L. (1988). *Biochemistry,* 3rd ed. (p. 512, Fig. 21-21). New York: W. H. Freeman and Company; **19-10** (a) Courtesy Jan Chalker; (b) John S. O'Brien; **19-11** Margaret Clark; **19-12** Scala/Art Resource; **19-13** Richmond Products, Boca Raton; **19-15** After Lerner, I. M. (1968). *Heredity, evolution, and society.* New York: W. H. Freeman and Company; **19-17** (a), (b) Steve Uzzell III; **Page 396** Lifecodes Corporation

Page 404 Donald R. Perry; **Page 405 & 20-1** © Michael Fogden/Animals Animals; **Page 407** © Brian Parker/Tom Stack & Associates; **20-2** (a) © Larry West; (b) © L. Campbell/NHPA; (c) John Shaw/NHPA; **20-3** (a) © John H. Gerard/DPI; (b) © Hans Reinhard/Bruce Coleman; **20-4** The Bettmann Archive; **20-6** (a) © Leonard Lee Rue III/Photo Researchers; (b) © Brian Parker/Tom Stack & Associates; (c) © Kevin Schafer/Tom Stack & Associates; (d) © Dale & Marian Zimmerman/Animals Animals; (e) © Tom McHugh/Photo Researchers; **20-8** R. M. Kristensen; **20 9 & 20-10** After Dobzhansky, T. (1977). *Evolution.* New York: W. H. Freeman and Company; **20-11** (a) After Lehninger, A. L. (1982). *Principles of biochemistry.* New York: Worth Publishers, Inc.; (b) Data provided by Lai-Su L. Yeh, Protein Identification Resource, National Biomedical Research Foundation, Georgetown University Medical Center; **20-12** Adapted from Sibley & Ahlquist (February 1986). *Scientific American,* 85; **Page 422** (a) © Ralph A. Reinhold/Animals Animals; (b) © Zig Leszczynski/Animals Animals; (c) Adapted from O'Brien, S. J., *et. al.* (September 12, 1985). *Nature, 317,* 141; **20-13** Adapted from Sibley & Ahlquist (February 1986). *Scientific American,* 92; **20-14** (a) © Manfred Kage/Peter Arnold; (b) © Eric V. Gravé; (c) © Robert P. Carr/Bruce Coleman; (d) © Marion Patterson/Black Star; (e) © Dwight R. Kuhn

Index

Index